Nelson Yuan-sheng Kiang

OXFORD MEDICAL PUBLICATIONS

Ventilatory and Phonatory Control Systems

An International Symposium

PARTICIPANTS IN THE SYMPOSIUM

Front row (left to right): J. Widdicombe, Y. Murakami, J. Malcolm, J. A. M. Martin, J. A. Kirchner, J. K. F. Anthony, A. Mårtensson, W. J. Gould, F. Brodnitz, Beverly Bishop, A. A. Viljanen, Ruth E. M. Bowden, S. Godfrey.

At rear (left to right): F. Winckel, B. D. Wyke, E. J. M. Campbell, S. Freedman, G. Cavagna, D. F. Proctor, L. Edström, C. Lindquist, J. Mead, A. B. Otis, C. von Euler, M. I. Cohen, P. Ladefoged, F. Plum, D. H. Klatt, B. Emery, F. Moran, R. R. Sonnenschein, K. N. Stevens.

Absent: A. J. Fourcin, D. B. Fry, A. I. Pack, J. G. Scadding, D. Whitteridge.

The photograph was taken in the Vestibule of the Royal College of Surgeons of England, in front of the statue of John Hunter (1728–1793) that was sculpted by H.

Ventilatory and Phonatory Control Systems

An International Symposium

EDITED BY

BARRY WYKE, M.B., B.S., M.D.,

Neurological Laboratory,
Royal College of Surgeons of England

LONDON

Oxford University Press

NEW YORK TORONTO

1 9 7 4

Oxford University Press, Ely House, London W.1
GLASGOW NEW YORK TORONTO MELBOURNE WELLINGTON
CAPE TOWN IBADAN NAIROBI DAR ES SALAAM LUSAKA ADDIS ABABA
DELHI BOMBAY CALCUTTA MADRAS KARACHI LAHORE DACCA
KUALA LUMPUR SINGAPORE HONG KONG TOKYO

ISBN 0 19 264158 1

Text set in 11/12 pt. Photon Times Roman, printed by photolithography, and bound in Great Britain at The Pitman Press, Bath

Contents

Preface

This volume is a record of the proceedings of an International Symposium held in London at the Royal College of Surgeons of England from 24th to 28th January, 1972, to discuss the current status of and possible future research developments in the related fields of ventilatory and phonatory control. For this purpose, an interdisciplinary assembly of distinguished research workers in respiratory physiology and medicine, in neurology and neurophysiology, in linguistics and phonetics, in laryngology, in anatomy and in bio-engineering was convened from a number of different countries to present up-to-date accounts of their own researches and to make rigorous critical assessments of the contributions of their colleagues.

The need for such an assembly at this time arose from realization that although each of these disparate disciplines had made significant contributions to the understanding of the complex problems of the control of the neuromuscular systems involved in breathing, speaking and singing in the last decade, no attempt had been made to integrate this wealth of data into a coherent whole since the Conference held at the New York Academy of Sciences in November, 1966. Furthermore, the Symposium was intended to reinforce the wisdom of the statement made at the New York Conference by the late Professor Wallace O. Fenn (whose lamented death just prior to the London Symposium precluded his intended participation therein) to the effect that 'vocalization has evolved as an ingenious exploitation of the breathing mechanism for the purpose of meaningful communication', and to emphasize the importance of considering together the mechanisms that control breathing and speaking.

Generous financial support for the Symposium, and for the publication of this volume, was provided by the U.S.V. Pharmaceutical Corporation of Tuckahoe, New York, U.S.A. through the good offices of Dr. W. J. Gould of New York City, for which all the participants are most grateful. Thanks are also due to the President and Council of the Royal College of Surgeons of England for permission to hold the meeting in the College, and for the excellent facilities provided; and the Editor is deeply grateful to his secretary (Miss G. Manwaring), to Mrs. G. Forsythe (Conference Secretary), to Mr. J. Ramcharan (Head Technician) and to Mr. J. Higgins (Conference Steward) for their devoted assistance. He is also indebted to Dr. J. C. Gregory (Oxford University Press) for his helpful co-operation in preparing this volume for publication, to his fellow participants for the willing promptness with which they delivered their contributions for publication, and to all the authors and publishers who have given permission for the reproduction of illustrations from previously published work (individual acknowledgement of which is contained in the legends to the relevant Figures).

Barry Wyke

London, 1972

Participants

Names marked (*) indicate those who presented the Communications.

ANTHONY, J. K. F., C.Eng., M.I.E.E.	*Lecturer in Linguistics, University of Edinburgh, Scotland.*
*BISHOP, BEVERLY, Ph.D.	*Associate Professor of Physiology, State University of New York, Buffalo, New York, U.S.A.*
*BOWDEN, RUTH E. M., D.Sc., M.R.C.S., L. R. C. P.	*Professor of Anatomy, Royal Free Hospital School of Medicine, University of London, England.*
BRODNITZ, F., M.D.	*Attending Physician, Department of Otolaryngology, Mt. Sinai Hospital, New York, U.S.A.*
*CAMPBELL, E. J. M., B.Sc., Ph.D., M.D., F.R.C.P.	*Professor of Medicine, McMaster University, Hamilton, Ontario, Canada.*
*CAVAGNA, G. A., M.D.	*Assistant Professor of Biophysics, Laboratory of Physiology, University of Milan, Italy.*
*COHEN, M. I., Ph.D.	*Associate Professor of Physiology, Albert Einstein College of Medicine, Yeshira University, New York, U.S.A.*
EDSTRÖM, L., M.D.	*Assistant, Neurological Department, Karolinska Sjukhuset, Stockholm, Sweden.*
EMERY, B., M.Sc.	*Research Scientist, Centre for Respiratory Investigation, Glasgow Royal Infirmary, Scotland.*
*VON EULER, C., M.D.	*Associate Professor and Head of Division of Brain Research, Nobel Institute for Neurophysiology, Karolinska Institutet, Stockholm, Sweden.*
*FOURCIN, A. J., Ph.D.	*Reader in Experimental Phonetics, Department of Phonetics and Linguistics, University College, London, England.*

*FREEDMAN, S., B.Sc., M.B., Ph.D., M.R.C.P.	*Assistant Lecturer in Medicine, Royal Postgraduate Medical School, London, England.*
FRY, D. B., Ph. D.	*Professor of Phonetics, University College, London, England.*
*GODFREY, S., M.D., Ph.D., M.R.C.P.	*Senior Lecturer and Consultant Paediatrician, Institute of Child Health, The Hammersmith Hospital, London, England.*
*GOULD, W. J., M.D., F.A.C.S.	*Clinical Associate Professor, New York University, and Director of the Department of Otolaryngology, Lenox Hill Hospital, New York, U.S.A.*
*KIRCHNER, J. A., M.D., M.S., F.A.C.S.	*Professor of Otolaryngology, Department of Surgery, Yale University School of Medicine, New Haven, Connecticut, U.S.A.*
*KLATT, D. H., Ph.D.	*Research Associate, Department of Electronics and Electrical Engineering, and Member of Speech Communication Research Group, Massachusetts Institute of Technology, Cambridge, Massachusetts, U.S.A.*
*LADEFOGED, P., Ph.D.	*Professor of Linguistics, University of California at Los Angeles, Los Angeles, California, U.S.A.*
*LINDQUIST, C., M.D.	*Research Associate, Department of Physiology, Karolinska Institutet, Stockholm, Sweden.*
*MALCOLM, J. E., *O.B.E.*, B.Sc., M.B., F.R.C.S.	*Cade Professor of Surgery, Royal Air Force Hospital, Ely, Cambridge, England.*
MÅRTENSSON, A., M.D.	*Lecturer, Neurological Department, Karolinska Sjukhuset, Stockholm, Sweden.*
MARTIN, J. A. M., M.B., B.S., F.R.C.S.	*Director, Nuffield Hearing and Speech Centre (Royal National Throat, Nose and Ear Hospital), London, England.*
*MEAD, J., M.D.	*Professor of Physiology, Harvard University School of Public Health, Boston, Massachusetts, U.S.A.*

MORAN, F., M.Sc., M.B., F.R.C.P. (Ed.) — *Honorary Lecturer in Medicine, University of Glasgow, and Physician-in-Charge, Centre for Respiratory Investigation, Glasgow Royal Infirmary, Scotland.*

MURAKAMI, Y., M.D. — *Head of Department of Otolaryngology, National Tochigi Hospital, Utsonomiya, Japan.*

*OTIS, A. B., Ph.D. — *Professor and Chairman, Department of Physiology, University of Florida College of Medicine, Gainesville, Florida, U.S.A.*

*PACK, A. I., M.B., B.Ch. — *Anderson Research Fellow, Centre for Respiratory Investigation, Glasgow Royal Infirmary, Scotland.*

PARISER, G., M.D. — *Representing U.S.V. Pharmaceutical Corporation (European Division).*

*PLUM, F., M.D. — *Professor of Neurology, Cornell University Medical College, and Neurologist-in-Chief, New York Hospital, New York, U.S.A.*

*PROCTOR, D. F., M.D., F.A.C.S. — *Professor of Environmental Medicine, Johns Hopkins University School of Hygiene and Public Health, Baltimore, Maryland, U.S.A.*

SCADDING, J. G., M.D., F.R.C.P. — *Professor of Medicine, Institute of Diseases of the Chest (Brompton Hospital), University of London, and Vice-President of the Royal College of Physicians of London.*

SONNENSCHEIN, R. R., M.D. — *Professor of Physiology, University of California at Los Angeles, Los Angeles, California, U.S.A., and Liaison Scientist, U.S. Office of Naval Research, London, England.*

STEVENS, K. N., Ph.D. — *Professor of Electrical Engineering and Head of Speech Communication Research Group, Massachusetts Institute of Technology, Cambridge, Massachusetts, U.S.A.*

*VILJANEN, A. A., M.D. — *Instructor in Pulmonary Diseases, Helsinki University Central Hospital, Helsinki, Finland.*

WHITTERIDGE, D., F.R.S., D.M., F.R.C.P. — *Waynflete Professor of Physiology, University of Oxford, and Fellow of Magdalen College, Oxford, England.*

*WIDDICOMBE, J. G., M.A., D.Phil., D.M. — *University Lecturer in Physiology, University of Oxford, and Fellow of New College, Oxford, England.*

*WINCKEL, F., Dr-Ing. — *Professor and Director, Institute of Communications Science, Technical University of Berlin, Germany.*

*WYKE, B. D., M.B., B.S., M.D. — *Senior Lecturer in Applied Physiology, Institute of Basic Medical Sciences, University of London, and Head of the Neurological Laboratory, Royal College of Surgeons of England, London, England.*

Opening Address

Dr. Wyke: Ladies and Gentlemen, our Proceedings this morning commence with the Official Opening of this Symposium, and for this purpose I would like to present to you Sir Thomas Holmes Sellors, the President of the Royal College of Surgeons of England.

Sir Thomas Holmes Sellors: Ladies and Gentlemen. It is my most pleasant duty to welcome you formally to our College and to wish you all success in your meeting, which has such an interesting programme. We are happy that you should come to this College, which aims to continue the scientific approach to surgery established by John Hunter who, as you may well appreciate, is our 'Patron Saint'. Naturally the name of John Hunter is familiar to those of you who come from across the Atlantic, because many of Hunter's pupils are regarded as the founders of surgery in North America.

Now as regards your Congress—you will appreciate how much we are indebted to Dr. Gould of Lenox Hill Hospital, New York City. The idea of such a meeting was originally his, and Dr. Wyke has acted as co-ordinator dealing with the local arrangements and organization within the College. This is of course an international and multi-disciplined Symposium, for I understand that among you there are people who describe themselves as neurologists, physiologists, engineers and scientists. Each one of you is a specialist—but I hope that you are not confined to your ivory towers, and will show in such a meeting as this how communications can develop between the different disciplines in medicine.

Medicine is, I am afraid, tending to divide up into smaller and smaller compartments as a result of the vast increase of scientific knowledge. In other words, the specialists are tending to know more and more about less and less. When I was in North America quite recently, many of the leaders of surgery expressed their anxiety that surgeons were tending to become 'one operation' men, and in this country we are beginning to see this tendency already developing. Undoubtedly experience and continued practice does tend towards perfection, but I think there comes a point where anyone working in too narrow a field becomes impervious to outside influences and increases the tendency to isolation.

During this week I see an opportunity of bringing together people who have been working in different aspects of the control of ventilation and phonation. I myself, being a thoracic surgeon, have been interested in respiration and in its control, and I realize that this meeting gives people like myself an opportunity of acquiring great knowledge from the experts. You really are supporting a hobby horse of mine, which is to try and bring together the various fragments of medicine and to allow freer communication. Tradition has somewhat cast us into a certain mould, and in some branches of medicine the boundaries have got to be broken down. Where, I would ask, is the border line between medicine and surgery for a thoracic surgeon, for a cardiac surgeon, for a neurosurgeon and for a gastro-enterologist? I think that your Symposium is going to be of immense value in bridging similar types of gaps in your own fields of expertise.

I have been asked to open this Congress, which I now do, and I am certain that you will have a most profitable and successful meeting under the leadership of Drs. Gould and Wyke.

Symbols, Abbreviations and Units

V	= gas volume
$\dot{V}$	= gas volume per unit time
V_T	= tidal volume
V_I	= inspired gas volume
V_E	= expired gas volume
V_A	= alveolar gas volume
T_I	= duration of inspiration
T_E	= duration of expiration
f	= respiratory frequency (breaths per minute)
V_A	= alveolar ventilation per minute
VC	= vital capacity
TLC	= total lung capacity
RV	= residual lung volume
FRC	= functional residual capacity
P	= gas pressure
$\bar{\mathrm{P}}$	= mean gas pressure
$\mathrm{PO_2}$	= partial pressure of oxygen
$\mathrm{PCO_2}$	= partial pressure of carbon dioxide
$\mathrm{P}_A\mathrm{O}_2$	= alveolar oxygen pressure
$\mathrm{P}_A\mathrm{CO}_2$	= alveolar carbon dioxide pressure
V_{O_2}	= oxygen consumption per minute
Q	= volume of blood
Q	= volume flow of blood per unit time
l	= litre
ml	= millilitre
kg	= kilogramme

mg	= milligramme
μg	= microgramme
m	= metre
cm	= centimetre
mm	= millimetre
μ	= micron
cm H_2O	= centimetres of water pressure
mm Hg	= millimetres of mercury pressure
min	= minute
sec	= second
msec	= millisecond
Hz	= cycles per second
V	= volt
mV	= millivolt
μV	= microvolt
DC	= direct current
AC	= alternating current
db	= decibel
EEG	= electroencephalogram
EMG	= electromyogram
CRO	= cathode ray oscilloscope

NOTE: The symbolic conventions adopted for categorization of respiratory function follow those proposed by the Federation of American Societies for Experimental Biology. The phonetic symbols used are those of the International Phonetic Alphabet (Revised).

SECTION I

Basic Principles of Ventilatory and Phonatory Behaviour

CHAIRMAN

Dr. W. J. Gould

New York, U.S.A.

1 Muscular Activity in Normal and Abnormal Ventilation

E. J. M. CAMPBELL

Department of Medicine, McMaster University, Hamilton, Ontario, Canada

The motive force for all the acts with which this Symposium is concerned is ultimately provided by the muscles of breathing. A knowledge of the action, behaviour and function of these muscles is necessary before tackling more subtle matters and I conceive my prime duty in this talk to provide this firm foundation. However, I shall occasionally stray beyond this brief when I feel some comment or question may be germane.

ACTIVITY OF THE MAIN MUSCLE GROUPS

THE DIAPHRAGM

Electromyographic studies of normal subjects show the following behaviour in both vertebral and costal portions of the diaphragm.

During quiet breathing [Fig. 1.1], activity increases progressively throughout inspiration and decreases at the beginning of expiration to cease about half way through the phase. During increased ventilation, activity during inspiration is increased and the muscle relaxes early in the expiratory phase; but at high levels of ventilation, contraction may resume before the end of expiration.

The diaphragm is active during expulsive efforts, when it acts synergistically with the abdominal muscles to raise abdominal pressure in such events as defaecation, vomiting and parturition. The muscle is also active, although to a lesser extent, in coughing and sneezing.

During phonation the activity of the diaphragm diminishes and rapidly ceases during the first two or three seconds of an utterance, after a maximal inspiration.

Respiratory function

The diaphragm is the principal muscle of inspiration, but is not essential for breathing in that paralysis causes little loss of ventilatory capacity. A quantitative assessment of its importance is difficult because the mechanical techniques available register movement rather than force.

Most of the detailed examination of the evidence and review of the literature is contained in the appropriate parts of the monograph by Campbell *et alii* (1970). Citations in this review are restricted either to specific points or work which has appeared in the last two years. Anatomical details can be found in any standard text.

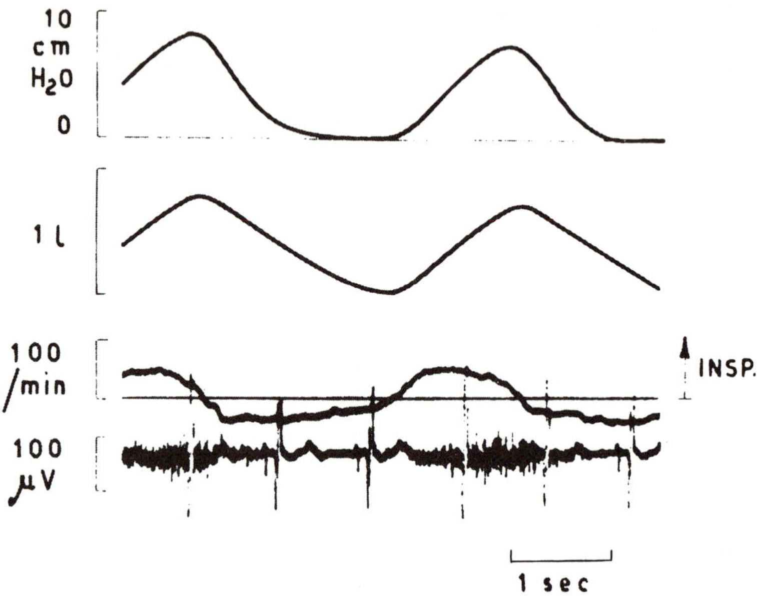

Fig. 1.1. Activity of the diaphragm during quiet breathing in a normal subject. Records from top to bottom: transdiaphragmatic pressure, tidal volume, airflow, EMG of the vertebral part of the diaphragm recorded from within the oesophageal lead. (From Agostoni, 1964.)

Provided the behaviour of the abdominal muscles is either constant or known, it is fair to interpret changes in abdominal pressure and movements of the abdomen or the diaphragm domes in terms of diaphragm activity, but it remains difficult to assess the contribution of the diaphragm to movements of the rib cage. In brief, it appears that the diaphragm probably contributes at least two-thirds of the tidal volume during both quiet and considerably increased ventilation. The proportion is probably less in the young than in the elderly, whose rib cages are less mobile (Rizzato and Marazzini, 1970).

THE INTERCOSTAL MUSCLES

In normal subjects, inspiratory activity during quiet breathing is limited to the parasternal intercostals (the interchondral portions of the internal intercostals) and to the external intercostal layer posteriorly in the upper spaces [Fig. 1.2]. During expiration, there is activity in the interosseous portion of the internal intercostals in the lower spaces.

As breathing is increased, either voluntarily or by chemical stimulation, there is a spread of the inspiratory activity throughout the external muscle layer (i.e. the external intercostals and the interchondral portion of the internal intercostals). During forced expiration, coughing and other expulsive events, expiratory activity occurs throughout the interosseous internal intercostals.

During phonation at high lung volumes, there is contraction of the external and interchondral muscles. At low lung volumes, there is contraction of the expiratory (interosseous internal) intercostals [see below and Fig. 1.5]

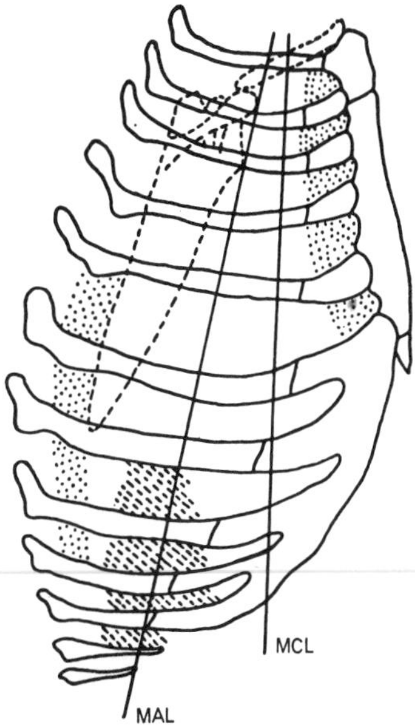

Fig. 1.2. Diagrammatic projection of the right side of the chest to show where intercostal activity is found in quiet breathing. MAL and MCL, mid-axillary line and mid-clavicular line. The dots indicate inspiratory and the dashes expiratory activity. (Modified from Taylor, 1960.)

Respiratory function

The respiratory function of the intercostal muscles is difficult to assess. In man they are clearly of much less importance than the diaphragm as over-all contributors to the force or work of breathing. It seems that they may be more important during phonation and this may be reflected in their greater proprioceptive control. They may also be important locally in the chest wall when respiratory forces interact with gravitational and postural forces. Recent studies of mechanical loading (Newsom Davis and Sears, 1970) show that the proprioceptive control of the intercostals is more complex than was suggested a decade ago.

MUSCLES OF THE ABDOMINAL WALL

Although anatomical examination and observation of a Turkish belly dancer clearly indicate that there must be subtle differences in action between the several muscles, for the purposes of this review they may be regarded as a group with an expiratory function by virtue of their action in raising abdominal pressure and depressing the rib cage.

In view of the forthcoming contribution by Dr. Beverly Bishop [see p. 12] I am simply going to summarize their behaviour in man [Fig. 1.3]. Under nearly all circumstances, there is some postural activity which wanes and waxes with inspiration and expiration. As breathing increases, inspiratory activity disappears; but activity late in expiration remains, and at very high levels of ventilation becomes very marked. The behaviour of the abdominal pressure reflects this pattern. These muscles are also active during all expiratory or expulsive efforts; and in most subjects they also appear to contract at the end of a maximum voluntary inspiration.

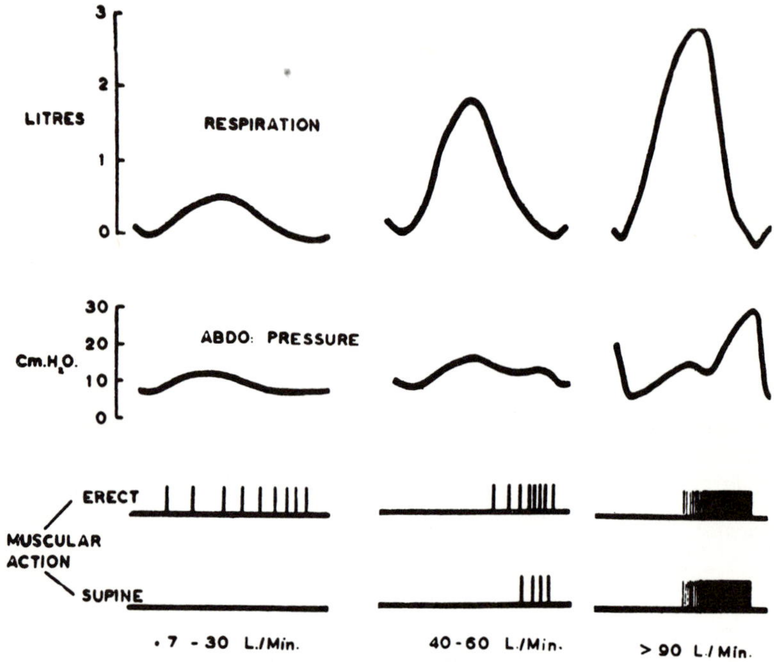

Fig. 1.3. Schematic pattern of the activity of the muscles of the abdominal wall, of the abdominal pressure and of the tidal volume at various levels of ventilation. (From Campbell, 1955.)

THE ACCESSORY MUSCLES

The only groups of accessory muscles worthy of consideration are the scaleni and the sternomastoids.

The Scaleni

In most postures in most subjects the scaleni have a continuous activity which waxes and wanes with inspiration and expiration. As ventilation is increased this pattern becomes intensified, with cessation of activity during expiration. This description might lead the scaleni to be classed with the 'ordinary' rather than with the 'accessory' muscles; but in many subjects, attention to the posture of the shoulders and neck will abolish this activity and the subjects are then capable of achieving considerable increases in ventilation without activity in the scaleni. The scaleni also contract during voluntary expiration efforts.

I am unaware of any detailed study of their behaviour during phonation, but my own observations suggest that they behave under most circumstances like the inspiratory intercostals, although they may contract like the expiratory intercostals during efforts requiring great expiratory force.

The Sternomastoids

The sternomastoids are usually inactive during quiet and increased breathing up to high ventilation. They contract towards the end of voluntary maximum inspiration at lung volumes corresponding to about 70–80 per cent of the vital capacity. During graded inspiratory pressures their activity is greater than one would expect on purely mechanical grounds.

THE OVERALL PATTERN OF MUSCULAR ACTIVITY AND ITS SIGNIFICANCE FOR THE ORGANIZATION OF RESPIRATORY CONTROL

Without wishing to trespass on the territory of neurological organization, I must emphasize that breathing at rest and up to considerable levels of ventilation sufficient to cover most of the activities of daily living is accomplished by the interaction between inspiratory muscles and passive forces [Fig. 1.4]. Two questions that are obviously germane to a Symposium such as this are: to what extent is there selective voluntary control of individual muscles or muscle groups? and, given the same mechanical conditions of pressure, volume and flow, is the pattern the same during such voluntary activities as phonation, as it is during ordinary breathing?

The answer to the first question is that the evidence is slight, and on the whole suggests that there is little selective voluntary control of the main muscle groups. Certainly, the great attention devoted to the diaphragm by singing teachers is in no way reflected in the contractile behaviour of this muscle. The answer to the second question is that there is some slight evidence that the intercostals and upper chest accessory muscles may be more active during phonation or during the voluntary maintenance of a fixed pressure than during involuntary breathing. Although this evidence is not very good, it does accord with the anatomical and neurophysiological evidence of greater proprioceptive mechanisms in these muscles than in the diaphragm.

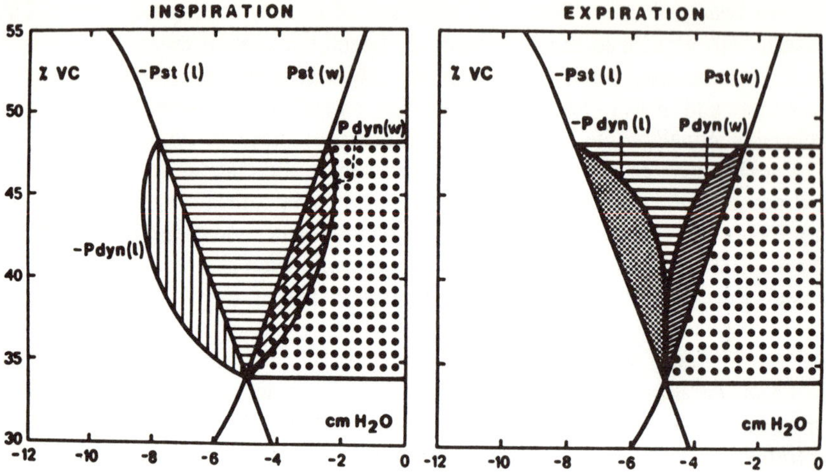

Fig. 1.4. Graphical analysis of the work done during a quiet breathing cycle through the mouth. *Inspiration*: The vertically-hatched area represents the work done by the inspiratory muscles to overcome the flow-resistance of the lung (gas and tissue)—this energy is lost as heat. The horizontally-hatched area represents the work done by the inspiratory muscles to overcome the elastic resistance of the lung, this energy being stored as elastic energy in the lung. The obliquely-hatched area represents the work done by the inspiratory muscles to overcome the flow-resistance of the chest wall—this energy is lost as heat. The coarsely-stippled area represents the elastic energy transferred from the chest wall to the lung. *Expiration*: The finely-stippled area represents the work done by the elastic energy stored in the lung to overcome the flow-resistance of the lung (gas and tissue)—this energy is lost as heat. The heavy horizontally-hatched area is the work done by the elastic energy stored in the lung to overcome the persistent activity of the inspiratory muscles—this energy is lost as heat. The obliquely-hatched area represents the work done by the elastic energy stored in the lung to overcome the flow-resistance of the chest wall—this energy is lost as heat. The coarsely-stippled area represents the elastic energy transferred from the lung to the chest wall. (From Campbell *et alii,* 1970.)

It must be emphasized that some or all of the respiratory muscles are used for four functions: breathing, posture, expulsion (coughing, defaecation, giving birth, etc.) and phonation. Usually they are being used for more than one of these purposes at any one time, and often there must be conflicting demands. On what information are such acts integrated and such demands reconciled? At what levels in the nervous system? I know of little solid information with which to answer these questions, and hope this Symposium will provide some.

RESPIRATORY BEHAVIOUR IN ABNORMAL (DISEASED) STATES

It is difficult to extract general principles of value to this Symposium from the scattered and questionable information currently available.

DISEASES OF THE LUNG AND CHEST WALL

In patients with airways obstruction, the characteristic pattern is of increase in the activity of the upper chest accessory muscles (scaleni and sternomastoids) and probably of the diaphragm. The intercostals and abdominal muscles seem to be less active.

I know of no studies of respiratory muscle behaviour in non-obstructive conditions affecting the chest and lung symmetrically, but from the few (unpublished) observations I have made 20 years ago, I doubt that it differs significantly in pattern from that seen in normal subjects. In localized or unilateral conditions it is a well known clinical fact that the chest moves less on the affected side; but, as far as I am aware, it is not known whether this simply reflects the mechanical consequences of disease or a change in the local behaviour of the muscles. This is a challenging question, because of its implication for local (ultimately spinal) control mediated perhaps by vagal afferents.

An interesting observation, because of its reminder to be careful in arguing from mechanical behaviour to muscular behaviour, is the example of ankylosing spondylitis in which the mobility of the rib cage is lost because of disease of the vertebrocostal joints. Josenhans and Wang (1971) have shown that in this condition all of the respiratory movement may be contributed by the diaphragm, but it remains unknown whether or not the intercostal muscles are inactive or just ineffective—although ineffective in moving the ribs, they may be effective in sustaining transthoracic pressure.

NEUROLOGICAL DISEASE

Peripheral or spinal disorders effecting particular nerves or muscle groups cause diminution or absence of action in the affected muscles with consequent increase in the activity of others. The most dramatic example of this is the maintenance of ventilation by the sternomastoids in lesions of the spinal cord above the phrenic outflow. Lesions of the cervical cord below the phrenic outflow provide a useful reminder of the interaction of the various muscle groups. In such tetraplegic patients the diaphragm causes a paradoxical inward movement of the costal margin during inspiration. This paradoxical movement is probably due either to failure of the paralysed intercostals to stabilize the rib cage or to failure of the abdominal muscles to sustain abdominal pressure (Moulton and Silver, 1970).

Disease of the brain stem affects the rhythm of breathing, and may cause such conditions as hiccup or diaphragmatic flutter. I suspect that central disorganization of breathing may affect the pattern of activity in the muscles, particularly during such acts as phonation, but I am unaware of any specific studies.

In hemiplegia there is diminution of the movement of the affected side, which is more apparent in the upper than the lower chest and in voluntary than in involuntary acts. These observations suggest that the descending pathways of voluntary respiratory acts may be different from those of normal breathing; but this point is less defensible for the diaphragm.

PHONATION

Figure 1.5 is offered as a bridge to other contributors to this Symposium, because it illustrates the interaction between inspiratory and expiratory muscular and passive elastic forces in phonation. Phonation beginning after a

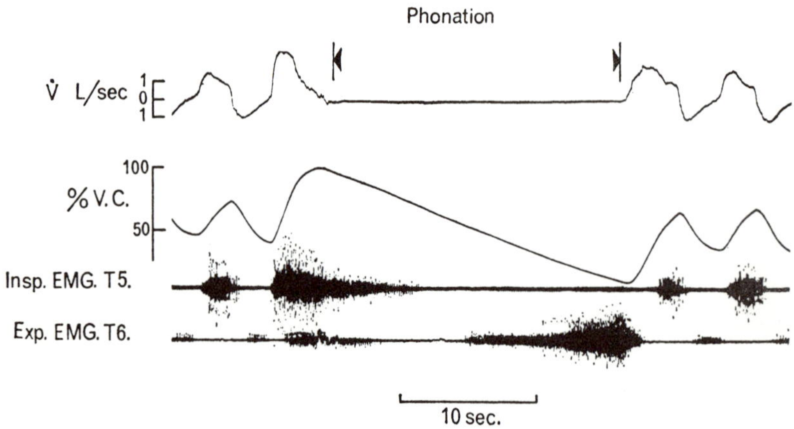

Fig. 1.5. Intercostal EMG activity during phonation. (From Sears and Newsom Davis, 1968.)

deep inspiration is produced by passive recoil, checked by a contraction of the inspiratory intercostals: the diaphragm is inactive. As the volume decreases below the resting (normal end-expiratory) chest volume, the motive force becomes supplied by expiratory muscles.

REFERENCES

Agostoni, E. (1964) Action of the respiratory muscles, in *Handbook of Physiology,* Section 3, Respiration, Vol. 1, pp. 377–86, ed. Fenn, W. O., and Rahn, H., Washington, D.C., American Physiological Society.

Campbell, E. J. M. (1955) The functions of the abdominal muscles in relation to the intra-abdominal pressure and the respiration, *Arch. Middx Hosp.,* **5,** 87–94.

Campbell, E. J. M., Agostoni, E., and Newsom Davis, J. (1970) *The Respiratory Muscles: Mechanics and Neural Control,* 2nd ed., London.

Josenhans, W. T., and Wang, S. C. (1971) Axial mass displacement by ventilation in man, *Proc. XXV Int. Congress (I.U.P.S.),* abstr. 836.

Moulton, A., and Silver, J. R. (1970) Chest movements in patients with traumatic injuries of the cervical cord, *Clin. Sci.,* **39,** 407–22.

Newsom Davis, J., and Sears, T. A. (1970) The proprioceptive reflex control of the intercostal muscles during their voluntary activation, *J. Physiol. (Lond.)*, **209,** 711–38.

Rizzato, G., and Marazzini, L. (1970) Thoracoabdominal mechanics in elderly men, *J. appl. Physiol.,* **28,** 457–64.

Sears, T. A., and Newsom Davis, J. (1968) The control of respiratory muscles during voluntary breathing, *Ann. N.Y. Acad. Sci.,* **155,** 183–90.

Taylor, A. (1960) The contribution of the intercostal muscles to the effort of respiration in man, *J. Physiol. (Lond.),* **151,** 390–402.

2 Abdominal Muscle Activity During Respiration

BEVERLY BISHOP

Department of Physiology, School of Medicine, State University of New York at Buffalo, Buffalo, N.Y., U.S.A.

The abdominal muscles participate in many motor activities including posture, expulsive functions, expiration and circulation (Chennells and Floyd, 1952; Bjurstedt, 1953; Campbell and Moran, 1958; Youmans *et alii,* 1963; Campbell *et alii,* 1970). Thus, it is not surprising that they have a relatively complex control system. It is the purpose of this paper to identify, analyse and assess some of the many afferent and central components known to play a rôle in the motor control of the abdominal wall.

POSTURAL MECHANISMS

SEGMENTAL INPUTS INFLUENCING ABDOMINAL MOTONEURONE EXCITABILITY

Neural mechanisms for co-ordinating the postural and expulsive functions of the abdominal wall appear to be located at the spinal level. *Bilateral rhizotomy* of the lower thoracic and upper lumbar dorsal roots abolishes all abdominal muscle activity in the anaesthetized cat, suggesting that the dorsal roots normally carry essential facilitatory signals to the abdominal motoneurone pool (Bishop, 1964).

In upright man one of the major functions of the abdominal wall is that of supporting the viscera against gravity. Like the antigravity limb muscles, the abdominal muscles are well endowed with muscle spindles (Coërs, 1959), whose presence has been verified histologically and electrophysiologically. *A tonic stretch reflex* is easily elicited from the abdominal muscles of a supine anaesthetized cat by placing a tongue depressor over the linea alba and adding weights to stretch the abdominal wall. As the muscle is lengthened by increasing the number of weights, additional abdominal motor units are progressively recruited into the tonic, asynchronous muscle response. A *phasic stretch reflex* is evoked by delivering a tap to the tongue depressor over the linea alba. As the strength of the tap is increased the amplitude of the synchronous, biphasic muscle action potential increases. A phasic stretch reflex may also be elicited electrically because of the convenient anatomical arrangement of the terminal

branches of the abdominal muscle nerves. Some of the terminal branches contain only afferent axons arising from muscle receptors. A shock to these axons while the muscle is tonically active produces not only a phasic reflex twitch but also a subsequent silent period lasting 40 msec or more, as shown in Fig. 2.1A. In this figure 15 sweeps of the CRO beam are superimposed and stored. Each sweep is triggered by the stimulator at randomly spaced intervals. Regardless of the ongoing tonic activity, the phasic reflex twitch (marked by an arrow) is always followed by a silent period of at least 40 msec. Although direct evidence is lacking, this silent period is most likely due to neural mechanisms similar to those governing limb motoneurones—including disfacilitation from decreased discharge of muscle spindles, autogenous inhibition from the discharge of Golgi tendon organs and recurrent inhibition via Renshaw cells (Granit *et alii,* 1966).

Spinal transection between T.1 and T.5 does not abolish either the tonic or phasic stretch reflexes, showing that the necessary neural mechanisms reside at the spinal level (Bishop, 1968a). However, spinal transection is followed by an increase in the thresholds and a reduction in the amplitudes of these reflexes, suggesting that neurones higher in the neural axis provide additional sources of facilitation to the abdominal motoneurone pool. This supraspinal control of abdominal motoneurones will be considered subsequently.

The skin reflex of the abdominal muscles, well-known to the neurologist as a clinical test for pyramidal tract lesions, can be elicited mechanically by rapidly stroking the abdomen with a blunt point or by electrically stimulating either the skin or the cutaneous afferent fibres (Kugelberg and Hagbarth, 1958). In many ways the abdominal skin reflex resembles the flexion reflex of the limbs. Contraction of the abdominal wall and simultaneous reciprocal inhibition of the antagonistic dorsal muscles serve to flex the trunk and retract the abdomen away from the noxious stimulus. Whether the reflex is elicited mechanically or electrically, the muscle response is polyphasic and quite variable, as exemplified by the first response in Fig. 2.1B. The polyphasic properties of the skin reflex probably reflect temporal and spatial dispersion of incoming signals over afferent fibres whose calibre spectrum is broad (Kugelberg and Hagbarth, 1958). An optimal skin reflex is recorded from the ipsilateral muscle underlying the stimulus. However, as the strength of the stimulus is increased more distant motor units participate. In Fig. 2.1B there are two abdominal muscle responses to a single shock delivered to a cutaneous branch of an abdominal nerve at the beginning of the trace. After a 10 msec latency there is a polyphasic burst (the abdominal skin reflex being marked by an arrow), followed by a short silent period which is terminated by a second, more synchronized burst of activity from the muscle. This second response is likely to be a stretch reflex occurring in response to the muscle's sudden relaxation following the first contraction. Whereas the second response is seen only following strong stimuli, the silent period following the response to cutaneous nerve stimulation is a standard aspect of the abdominal skin reflex (Kugelberg and Hagbarth, 1958). In man, sustained voluntary efforts show an interruption in their electrical activity following the skin reflex—suggesting that an active inhibition accounts for the silent period.

There are, of course, many other segmental inputs playing upon the abdominal motoneurone pool. For example, tonic activity in the abdominal muscle can be silenced by squeezing the ipsilateral hamstring muscles (Chennells and

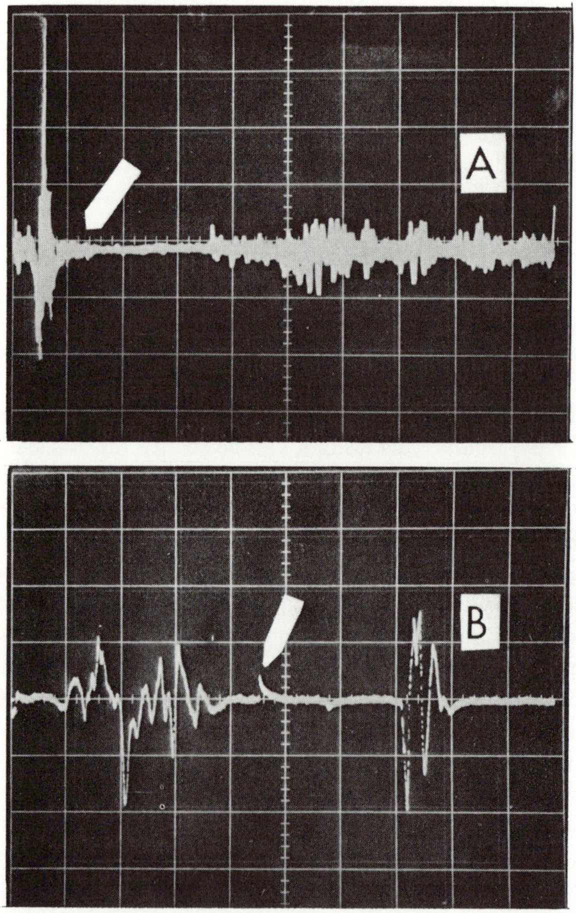

Fig. 2.1. Silent periods (arrows) occur in the EMG of the external oblique abdominal muscle following an electrically elicited monosynaptic stretch reflex [A] or a polysynaptic abdominal skin reflex [B]. In [A] randomly spaced responses are superimposed and stored on 15 consecutive sweeps of the CRO beam triggered at the same moment as a 0.4 volt, 1 msec pulse was delivered to a terminal branch of an abdominal muscle nerve. Sweep time, 20 msec/division and vertical sensitivity, 200 μv/division. In [B] a 0.6 volt, 1 msec pulse delivered to a cutaneous branch of an ipsilateral abdominal nerve evokes the early polyphasic skin response followed by a 25 msec pause and a second, more synchronized stretch reflex. Sweep time, 10 msec/division and vertical sensitivity, 100 μv/division.

Floyd, 1952). Another source of inhibition can be demonstrated by rotating the head to the ipsilateral side (unpublished observations). In addition, Hagbarth and Kugelberg (1958) have defined visceral inputs which have potent effects upon abdominal motoneurone excitability.

Thus, the segmental and intersegmental neural inputs playing upon the abdominal motoneurone pool are analogous to those controlling the limb muscles. The abdominal neurones, like other final common pathways, are the site of considerable neural integration by which the postural activities of the abdominal muscles are appropriately co-ordinated. However, the mechanisms for initiating and integrating the expiratory activities of the abdominal muscles are found in more cephalic neuronal networks concerned with co-ordination of respiratory and laryngeal systems.

RESPIRATORY MECHANISMS

SUPRASPINAL CONTROL OF ABDOMINAL MOTONEURONE EXCITABILITY

The neural mechanisms controlling limb and abdominal motoneurones lose their similarity above the spinal level. Unlike limb motoneurones, the abdominal motoneurones depend upon the medullary expiratory neurones as a major source of facilitation. Even though the abdominal muscles do not participate in quiet breathing or speaking, the phasic stretch reflex is significantly augmented during early expiration, at a time when the extrafusal muscle fibres are passively shortening (Bishop, 1968).

This cyclic facilitation may be readily demonstrated in the anaesthetized cat by delivering an electric shock of constant intensity to an afferent branch of the muscle nerve randomly throughout the respiratory cycle. As shown in Fig. 2.2, the amplitude of the electrically-elicited phasic stretch reflex is significantly greater during early expiration than during any other part of the respiratory cycle. This recruitment of motor units to the monosynaptic reflex during expiration shows that some motoneurones in the subliminal fringe during inspiration become part of the discharge zone during expiration. This facilitation of the abdominal motoneurone pool during early expiration occurs whether the spontaneous frequency of breathing is fast or slow. Since the facilitation is always locked to expiration in the intact animal, the medullary respiratory neurones seem a likely source. In fact, separation of the abdominal motoneurone pool from brain stem neurones either functionally (by artificial overventilation) or anatomically (by low cervical spinal transection) abolishes the expiratory facilitation. During this separation, a cyclic increase in the monosynaptic reflex now occurs during inspiration when lung volume is increasing and the abdominal wall is being passively stretched. This inspiratory facilitation is, of course, characteristic of the classical segmental stretch reflex. In the intact preparation during inspiration the segmental stretch reflex is suppressed and any facilitation of the abdominal motoneurones occurs during expiration.

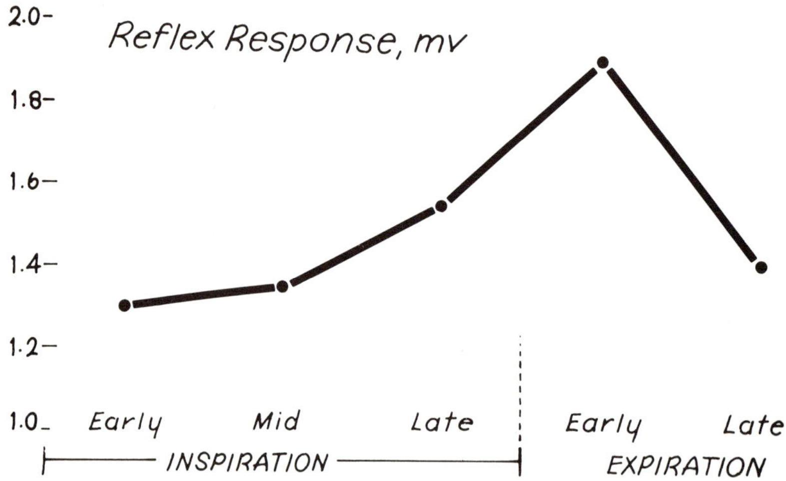

Fig. 2.2. The monosynaptic test of abdominal motoneurone excitability at different times in the respiratory cycle (abscissa). The amplitude of the abdominal muscle action potential (ordinate) produced by a constant intensity shock to a sensory nerve is greatest in early expiration.

This phasing of abdominal motoneurone excitability with respiration qualitatively parallels the respiratory control of the intercostal muscles (Critchlow and von Euler, 1963; von Euler and Critchlow, 1963; Sears, 1964a, 1964b). However, unlike the respiratory control of the intercostals, the expiratory facilitation of the abdominal motoneurones is usually subliminal during quiet breathing and can be unmasked only by intracellular recording or by a technique such as the monosynaptic test. Nonetheless, this cyclic respiratory control of the abdominal motoneurones becomes dominant whenever expiration is compromised (Bishop, 1967).

Active expiration

It is common knowledge that the abdominal muscles participate in expiration whenever their contribution to the expiratory effort is appropriate or required (Campbell, 1952; Campbell and Green, 1953a, 1953b; Campbell, 1957). The functional consequences of the active contraction of abdominal muscles during expiration is an increase in intra-abdominal pressure which, in turn, is transmitted across the relaxed diaphragm to cause a rise in the intrapulmonary and intrapleural pressures (Bishop, 1963). Hence, the lung is emptied faster and more completely than when expiration occurs passively as the result of the elastic recoil of chest and lungs (Rahn *et alii,* 1946).

AFFERENT SYSTEMS CONTROLLING ABDOMINAL EXPIRATORY ACTIVITY

To identify and study the components of the system controlling abdominal expiratory activity, it is necessary to have methods for initiating active expiration. One means of evoking expiratory activity in the anaesthetized animal is to elevate the pressure opposing expiration (Bishop, 1963, 1964, 1967). Whether the pressure opposes expiration only during the expiratory phase of the respiratory cycle (as in expiratory threshold loading) or opposes it throughout the respiratory cycle (as during continuous positive pressure breathing), the steady-state expiratory activity of the abdominal muscles increases [as shown in Fig. 2.3] with increases in the pressure opposing expiration (Bishop, 1957). The results of the following experiments disclose some of the neural mechanisms involved in the initiation and maintenance of abdominal expiratory activity.

Laparotomy, from the sternum to the pelvis, interferes with the mechanical usefulness of the expiratory contractions of the abdominal muscle, but does not abolish the active expiration of pressure breathing (Bishop, 1964).

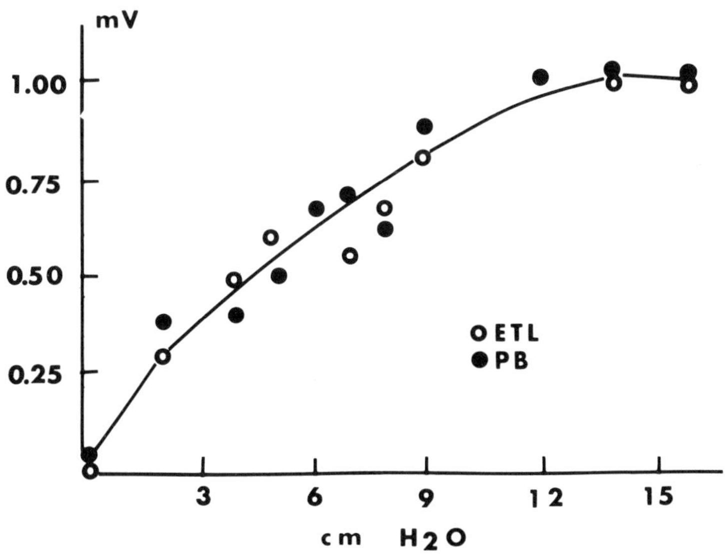

Fig. 2.3. Steady-state expiratory activity of the abdominal muscle in response to expiratory threshold loads (open symbols) and to continuous positive pressure breathing (filled symbols). The ordinate is the mean peak amplitude of the integrated EMG of the external oblique abdominal muscle. The abscissa is the pressure opposing expiration.

Abdominal vagotomy just below the diaphragm does not interrupt abdominal expiratory activity; but *bilateral vagotomy* in the neck completely abolishes all abdominal expiratory activity, indicating that the critical vagal receptors must lie within the thorax.

Thoracic receptors essential to the expiratory reflex

Paintal has reported that there are 30,000 fibres in the cervical vagus, 24,000 of which are afferent—arising from receptors in cardiovascular and other visceral structures, as well as in pulmonary structures (Paintal, 1963).

The rhythmic increase in abdominal pressure occurring during active expiration provides a pumping action aiding venous return (Bjurstedt, 1953). Thus, it might be predicted that thoracic receptors whose signals evoke abdominal expiratory activity lie in cardiovascular as well as in respiratory structures (Youmans *et alii,* 1963). To assess the influence of cardiovascular receptors in controlling abdominal motoneurone excitability, systemic blood pressure in cats was reduced by different experimental procedures to simulate the hypotension of pressure breathing in the absence of a concomitant pulmonary inflation. Hypotension, whether induced by acute hemorrhage, cardiac tamponade, occlusion of the superior vena cava or by drugs, always increases ventilation and dramatically increases diaphragm activity but never excites abdominal expiratory activity (Bishop, 1968). The marked increase in diaphragm activity during hypotension presumably is in response to hypoxia. Since there is no abdominal expiratory response to hypotension or hypoxia, pulmonary events rather than cardiovascular events or chemostimulation probably trigger the active expiration.

This hypothesis has recently been tested in our laboratory by comparing the reflex responses of the diaphragm and abdominal muscle in cats following pulmonary denervation with those in animals with cardiac denervation. Pulmonary denervation totally abolishes, whereas cardiac denervation does not interfere with abdominal expiratory activity (unpublished results, J. Russell's Ph.D. Dissertation). These observations suggest that pulmonary vagal afferent fibres are not only necessary but are also sufficient for mediating the abdominal expiratory reflex.

Effects of chemostimulation on abdominal expiratory activity.

As previously mentioned, hypoxia caused by hypotension rarely evokes abdominal expiratory activity in spite of marked increases in diaphragm activity and in the frequency and depth of breathing. There are other observations indicating that abdominal expiratory activity is quite independent of blood gas tensions. As shown by the example in Fig. 2.4, during a prolonged exposure to 15 cm H_2O pressure breathing, the end-tidal CO_2 gradually rises in the anaesthetized cat as a result of a reduction in tidal volume and a fall in minute ventilation (Bishop and Bachofen, 1972). In spite of this progressive hypercapnia, the level of the abdominal expiratory activity remains constant throughout the period of pressure breathing. In contrast, the diaphragm—which is initially depressed by the elevated airways pressure (the Hering-Breuer inflation reflex)—displays a progressive increase in activity until a new steady state is achieved. Other investigators have provided additional evidence indicating that the inspiratory neurones controlling the phrenic motoneurones are far more responsive to chemostimulation than are the expiratory neurones which drive the abdominal expiratory reflex (Flenley *et alii,* 1971; Milic-Emili and Tyler, 1963; Woldring, 1965).

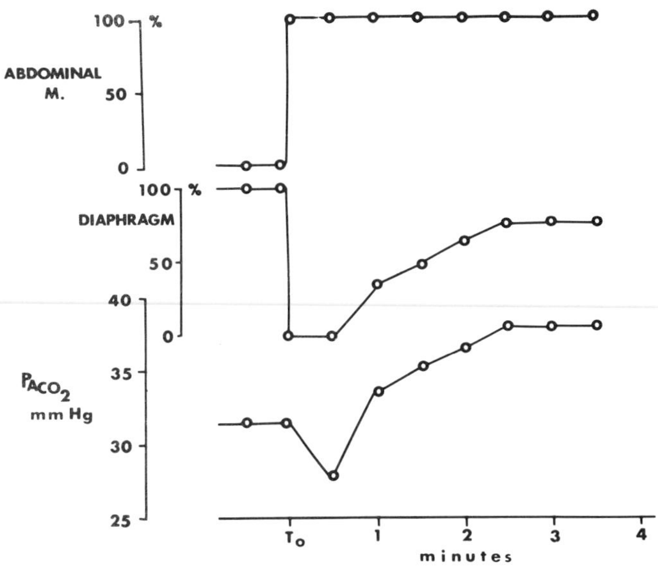

Fig. 2.4. Respiratory responses of an anaesthetized cat to 15 cm H_2O continuous positive breathing |PB|. End-tidal CO_2 tensions (bottom curve) were obtained from continuous recordings of instantaneous mass spectrometer analysis of airway gas sampled at the tracheal cannula. The fall in CO_2 with the onset of pressure breathing $|T_0|$ is due to the dilution of alveolar gas caused by lung inflation. The subsequent rise in CO_2 reflects the fall in minute ventilation. Diaphragm activity (middle curve) is the mean peak amplitude of the integrated EMG expressed as a percentage of control (i.e. just prior to T_0). From total inhibition at T_0, gradual return of diaphragm activity mirrors the rise in P_ACO_2. In contrast, abdominal expiratory activity (upper curve) appears on the first expiration of PB and remains constant thereafter. Abdominal activity is the mean peak amplitude of the integrated EMG expressed as a percentage of maximum.

In recent experiments on cats, the proprio- and chemoreceptor control of the diaphragm and of abdominal expiratory activity have been compared (Bishop and Bachofen, 1971). When the inspired gas is suddenly changed from air to 5.25 per cent CO_2 in air or to 12.4 per cent O_2 in N_2 breathing becomes faster and deeper, causing a marked increase in minute ventilation. This increased ventilation is achieved without participation of the abdominal muscles. Thus, chemostimulation alone is an inadequate stimulus for exciting active expiration.

Effects of chemostimulation on the abdominal expiratory reflex

Whether chemostimulation causes subliminal alterations in the excitability of the expiratory neurones driving the abdominal expiratory reflex was tested by measuring the effects of chemostimulation on the abdominal expiratory reflex to pressure breathing.

As previously indicated, continuous positive pressure breathing with air evokes the vagal-mediated abdominal expiratory reflex triggered by the pulmonary proprioceptors. To test the effect of chemostimulation on the abdominal expiratory reflex, the reservoir providing the inspired gas at any desired positive pressure was filled with either the hypercapnic (5.25 per cent CO_2) or the hypoxic (12.4 per cent O_2) gas mixtures.

Figure 2.5 shows that the abdominal expiratory activity in either anaesthetized cat or conscious man is adjusted in accordance with the elevated airways pressure and is essentially independent of the CO_2 or O_2 tensions of the inspired gas. Our interpretation of these results is that the expiratory neurones, in marked contrast to inspiratory neurones, receive little afferent information concerning the status of the blood gases. Instead, the major source

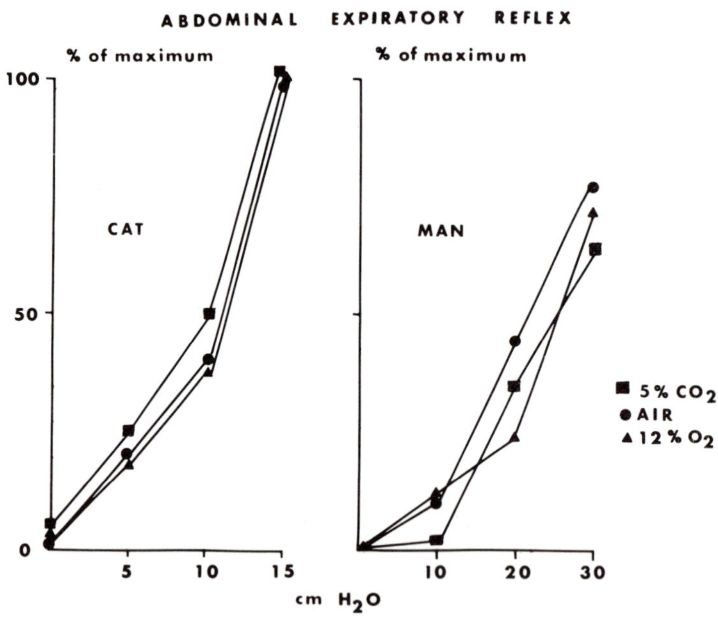

Fig. 2.5. Effect of inspired gases on the abdominal expiratory reflex response (ordinate) to three levels of continuous positive pressure breathing (abscissa) in cat (left) and man (right).

of information controlling expiratory discharges is derived from vagal impulses initiated by pulmonary receptors. Since the vagal signals report on lung volume and changes in the lung volume (Davis *et alii,* 1956), it appears that the function of abdominal expiratory activity is one of regulating lung volume rather than blood gases.

Negative feed-back control of lung volume

The vagal-mediated abdominal expiratory reflex provides a powerful negative feed-back circuit for controlling lung volume. Fig. 2.6 shows a simple scheme depicting the control circuit. Any condition causing an increase in lung volume increases the discharge frequency of lung proprioceptors. Their signals, propagated by vagal afferents, facilitate expiratory neurones. When temporal

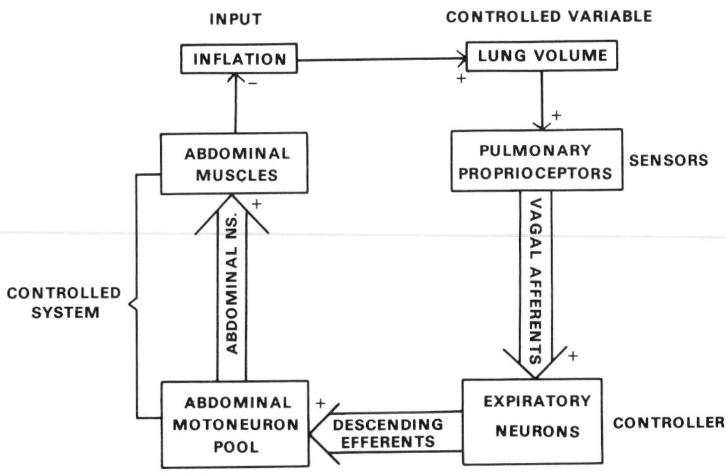

Fig. 2.6. A negative feed-back circuit for controlling lung volume. One of the major respiratory functions of the abdominal muscles is that of keeping lung volume optimal for efficient breathing.

and spatial summation are sufficient, some expiratory neurones discharge to activate abdominal expiratory activity via spinal motoneurones. Contraction of abdominal muscles squeezes the incompressible abdominal contents, forcing the relaxed diaphragm upward and thus reducing lung volume to one smaller than is achieved in the absence of abdominal contraction. Many advantages accrue from a controlled lung volume. The efficiency of respiration is greatest when tidal volumes are close to resting, at which point chest and lung compliance are optimal (Zechman *et alii,* 1958; Marshall, 1962). The vagal feed-back circuit also plays a major rôle in establishing the breathing pattern not only in terms of rate and depth, but in terms of the relative contributions of the various respiratory muscles. This partitioning of labour among the respiratory muscles influences: (1) air flow and therefore the distribution of tidal volume; and (2) cavity pressures and therefore the distribution of perfusion (Bishop and Bachofen, 1971). Hence, the vagal feed-back mechanism is an important factor determining $\dot{V}_1/\dot{Q}$—that is, the efficiency of gas exchange.

SUMMARY AND CONCLUSIONS

Neural mechanisms controlling and integrating the multiple activities of the abdominal muscles have been reviewed in this paper. In most respects, the spinal mechanisms controlling the tonic and phasic postural reflexes and the abdominal skin reflex are analogous to those controlling limb reflexes [Fig. 2.1]. However, the brain stem control of abdominal expiratory activities more closely resembles that of the intercostals, in that abdominal motoneurones are continuously subjected to cyclic excitability changes imposed by descending impulses from respiratory neurones [Fig. 2.2]. Even so, the respiratory control of the intercostals and diaphragm differs from that of the abdominal muscles in two important ways. First, expiratory facilitation of abdominal motoneurones is subliminal during quiet breathing. For expiratory facilitation to be sufficient to cause active expiration the dorsal roots, the abdominal motoneurones and spinal cord rostral to them, the brain stem and the pulmonary vagal afferents must be intact. Then, and only then, will a pressure opposing expiration initiate abdominal expiratory activity [Fig. 2.3]. Second, abdominal expiratory activity is independent of alveolar gas tensions [Figs. 2.4 and 2.5]. Whereas chemostimulation exerts powerful effects on intercostal and diaphragm activities, chemostimulation *per se* is unimportant in controlling the abdominal expiratory activity.

These observations suggest that the primary ventilatory function of active expiration is one of keeping lung volume, and hence the work of breathing, small. This negative feedback control of lung volume is imposed almost exclusively by the vagal pulmonary proprioceptors [Fig. 2.6].

As shown by Olafsson and Hyatt (1969), normal man usually does not waste expiratory effort; he ceases increasing expiratory effort when such effort no longer increases expiratory flow. Presumably the afferent and central systems discussed in this paper and others, yet to be defined, are responsible for this 'wisdom'.

REFERENCES

Bishop, B. (1963) Abdominal muscle and diaphragm activities and cavity pressures in pressure breathing, *J. appl. Physiol.,* **18,** 37–42.

Bishop, B. (1964) Reflex control of abdominal muscles during positive pressure breathing, *J. appl. Physiol.,* **19,** 224–32.

Bishop, B. (1967) Diaphragm and abdominal muscle responses to elevated airway pressures in the cat, *J. appl. Physiol.,* **22,** 959–65.

Bishop, B. (1968a) Neural regulation of abdominal muscle contractions, *Ann. N.Y. Acad. Sci.,* **155,** 191–200.

Bishop, B. (1968b) Diaphragm and abdominal muscle activity during induced hypotension, *J. appl. Physiol.,* **25,** 73–79.

Bishop, B., and Bachofen, H. (1971) Comparative influence of proprioceptors and chemoreceptors in the control of respiratory muscles, *XXV Internatl. Congr. Physiol. Sci. Satellite Symposium,* 'Neural Control of Breathing,' Warsaw. [In press.]

Bishop, B., and Bachofen, H. (1972) Vagal control of ventilation and respiratory muscles during elevated pressures in cat, *J. appl. Physiol.* **32,** 103–12.

Bjurstedt, H. (1953) Influence of the abdominal muscle tone on the circulatory response to positive pressure breathing in anesthetized dogs, *Acta physiol. scand.,* **29,** 145–62.

Campbell, E. J. M. (1952) An electromyographic study of the role of the abdominal muscles in breathing, *J. Physiol. (Lond.),* **117,** 222–33.

Campbell, E. J. M. (1957) The effects of increased resistance to expiration on the respiratory behaviour of the abdominal muscles and intra-abdominal pressure, *J. Physiol. (Lond.),* **136,** 556–62.

Campbell, E. J. M., Agostoni, E., and Newsom Davis, J. (1970) *The Respiratory Muscles: Mechanics and Neural Control,* 2nd. ed., Philadelphia.

Campbell, E. J. M., and Green, J. H. (1953a) The expiratory function of the abdominal muscles in man, *J. Physiol. (Lond.),* **120,** 409–18.

Campbell, E. J. M., and Green, J. H. (1953b) The variations in intra-abdominal pressure and the activity of the abdominal muscles during breathing: A study in man, *J. Physiol. (Lond.),* **122,** 282–90.

Campbell, E. J. M., and Moran, E. J. (1958) *The Respiratory Muscles and the Mechanics of Breathing,* pp. 28–39, Chicago.

Chennells, M., and Floyd, W. F. (1952) Reflex activity in abdominal and limb muscles. *J. Physiol. (Lond.),* **118,** 196–206.

Coërs, C. (1959) Structural organization of the motor nerve endings in mammalian muscle spindles and other striated muscle fibres, *Amer. J. phys. Med.,* **38,** 166.

Critchlow, V., and von Euler, C. (1963) Intercostal muscle spindle activity and its motor control, *J. Physiol. (Lond.),* **168,** 820–47.

Davis, H. L., Fowler, W. S., and Lambert, E. H. (1956) Effect of volume and rate of inflation and deflation on transpulmonary pressure and response of pulmonary stretch receptors, *Amer. J. Physiol.,* **187,** 558–66.

Eccles, R. M., Sears, T. A., and Shealy, C. N. (1962) Intracellular recording from respiratory motoneurones of the thoracic spinal cord of the cat, *Nature (Lond.),* **193,** 844–6.

von Euler, C., and Critchlow, V. (1963) Intercostal muscle spindle activity and its motor control, *J. Physiol. (Lond.),* **168,** 820–47.

Flenley, D. C., Pengelly, L. D., and Milic-Emili, J. (1971) Immediate effects of positive-pressure breathing on the ventilatory response to CO_2, *J. appl. Physiol.,* **30,** 7–12.

Granit, R., Kellerth, J.-O., and Szumski, A. J. (1966) Intracellular autogenetic effects of muscular contraction on extensor motoneurones: The silent period, *J. Physiol. (Lond.),* **182,** 484–503.

Hagbarth, K. E., and Kugelberg, E. (1958) Plasticity of the human abdominal skin reflex, *Brain,* **81,** 305–18.

Kugelberg, E., and Hagbarth, K. E. (1958) Spinal mechanism of the abdominal and erector spinal skin reflexes, *Brain,* **81,** 290–304.

Marshall, R. (1962) Relationships between stimulus and work of breathing at different lung volumes, *J. appl. Physiol.,* **17,** 917–21.

Milic-Emili, J., and Tyler, J. M. (1963) Relation between work output of respiratory muscles and end-tidal CO_2 tension, *J. appl. Physiol.,* **18,** 497–504.

Olafsson, S., and Hyatt, R. E. (1969) Ventilatory mechanics and expiratory flow limitation during exercise in normal subjects, *J. clin. Invest.,* **48,** 564–73.

Paintal, A. S. (1963) Vagal afferent fibres, *Ergebn. Physiol.,* **52,** 74–156.

Rahn, H., Otis, A. B., Chadwick, L. E., and Fenn, W. O. (1946) The pressure-volume diagram of the thorax and lung, *Amer. J. Physiol.,* **146,** 161–78.

Sears, T. A. (1964a) Investigations on respiratory motoneurones of the thoracic spinal cord, *Progr. Brain Res.* **12,** 259–73.

Sears, T. A. (1964b) The slow potentials of thoracic respiratory motoneurones and their relation to breathing, *J. Physiol. (Lond.),* **175,** 404–24.

Woldring, S. (1965) Interrelation between lung volume, arterial CO_2 tension, and respiratory activity, *J. appl. Physiol.,* **20,** 647–52.

Youmans, W. B., Murphy, Q. R., Turner, J. K., Davis, L. D., Briggs, D. I., and Hoye, A. S. (1963) Activity of abdominal muscles elicited from the circulatory system, *Amer. J. phys. Med.,* **42,** 1–70.

Zechman, F. W., Jr., Salzano, J., and Hall, F. G. (1958) Effect of cooling the cervical vagi on work of breathing, *J. appl. Physiol.,* **12,** 301–4.

DISCUSSION ON COMMUNICATIONS [1] AND [2]

Opened by J. Mead. Dr. Moran Campbell referred to the action of the diaphragm, and said I might be saying something about it. I think the greatest emphasis in this Symposium is going to be on phonation; and in phonation the diaphragm is for the most part (as shown originally by Draper, Ladefoged and Whitteridge (1959)[1]) quiet. So I thought I would be largely quiet about the diaphragm, and I intend to be—except on a couple of counts.

First, I wish to say some things that really relate in part to Dr. Beverly Bishop's thoughts about the abdominal muscles. Dr. Michael Goldman, in our Laboratory, has developed a thesis about the diaphragm's action which really puts in numbers notions that we have been teaching medical students for a long

[1] Draper, M. H., Ladefoged, P., and Whitteridge, D. (1959) Respiratory muscles in speech, *J. Speech Res.,* **2,** 16–27.

time; namely, that the diaphragm not only pushes the abdominal contents down but also pulls up on the rib cage and tends to expand it. I can't describe his theory now, or the experimental evidence leading up to it, but I will just mention a couple of its consequences. If tension in the diaphragm is inspiratory to the rib cage, then either active tension in the diaphragm or passive distension of the diaphragm by contraction of the abdominal muscles can be inspiratory to the rib cage and to the lungs.

Our own focus on the action of the abdominal muscles, with that in mind, has not been on their expiratory influence—which we feel is certainly marked in most of the acts we will talk about this week, but *not* in breathing. We think the principal respiratory action of the abdominal muscles may in fact be to tune the diaphragm for its optimum function—that is, to push it up to an optimal length for its inspiratory contraction. Personally, I also find your notion that lung volume may be defended by the abdominal muscles fits beautifully with your lovely exposition of the reflexes you have studied. But I don't think it fits with physiology in conscious man. I don't defend my lung volume when I get up in the morning—I allow it to get quite low during the night, and I don't keep it low by pushing in my abdominal muscles; and I also know that I do not defend lung volume when I am exposed to positive pressures—perhaps because I am a highly-trained and opinionated respiratory physiologist. But I think that if you take the average man and expose him to modest positive pressure, most subjects will shift their lung volume up without a murmur.

There are two other comments I want to make. One has to do with a statement that Dr. Moran Campbell made; namely, that there is little evidence of the ability voluntarily to separate the action of the diaphragm and the intercostal or other muscles. Perhaps Dr. Whitteridge would comment on that at some point; but in our experiments it didn't seem to take much training at all to be able to turn on one or the other muscle group at will, over a modest range of lung volumes.

Finally, I want to pick up Dr. Moran Campbell on a point I thought he was going to make, but didn't—but since I have a few slides about it, I will pick him up on the point anyway. He has been quoted as saying '. . . the subtlety of control required of the intercostal muscles during speech makes demands of the same order as those that are made in the small muscles of the hands in moving the fingers'. I would like to ask him: subtlety required for what purpose? For my part I am struck by the fine detail seen in the intercostal electromyogram on the one hand, and on the other by the relatively coarse nature of their combined mechanical output—namely, the fluctuations in pleural pressure that they produce—in which much of the detail disappears. I repeat, then: what is this apparently finely-controlled activity for?

For my own edification, I passed an oesophageal balloon in myself into a fairly quiet spot in terms of the cardiac pulsations and looked at this indirect measure of pleural pressure during various manoeuvres. First I pounded on my chest, and satisfied myself that the recording system was fast enough to pick up rapid events [Fig. D.1, upper tracing]: these transients lasted only about 60 msec, and were much shorter than anything I saw later. Incidentally, I noticed that the fundamental frequency of my phonation at about 140 Hz was countable on the tracing—which is further evidence that the system responds briskly.

I then did various things. Being a wind musician, I tried rapid single-tonguing up to 7 Hz and then double-tonguing up to 11.5–12 Hz [Fig. D.1, middle and lower tracings]. I found that the pleural pressure showed simultaneous fluctuations, but only small ones (not more than 10 per cent of the mean pressure); and I assumed that these reflected the activity of the muscles articulating the tonguing and not the synchronized activity of the muscles driving the thorax. This was borne out by my next observations. I said: '*Pah-Pah-Pah . . .*', slowly at first, with conscious emphasis on each '*Pah*', and then at increasing rates up to the highest I could attain [Fig. D.2]. The sound intensity remained about the same throughout, but the pleural pressure fluctuations—which were large, and synchronous with the *Pah's* at first—reduced in amplitude, and somewhere in the region of 5 to 6 *Pah's* per second essentially disappeared. I have assumed that the *Pah's* thereafter were produced with steady action of the muscles of the thorax, modulated by the articulating muscles of the head and neck. The highest frequency corresponded fairly closely to my single-tonguing maximum of about 6.5 to 7 per second.

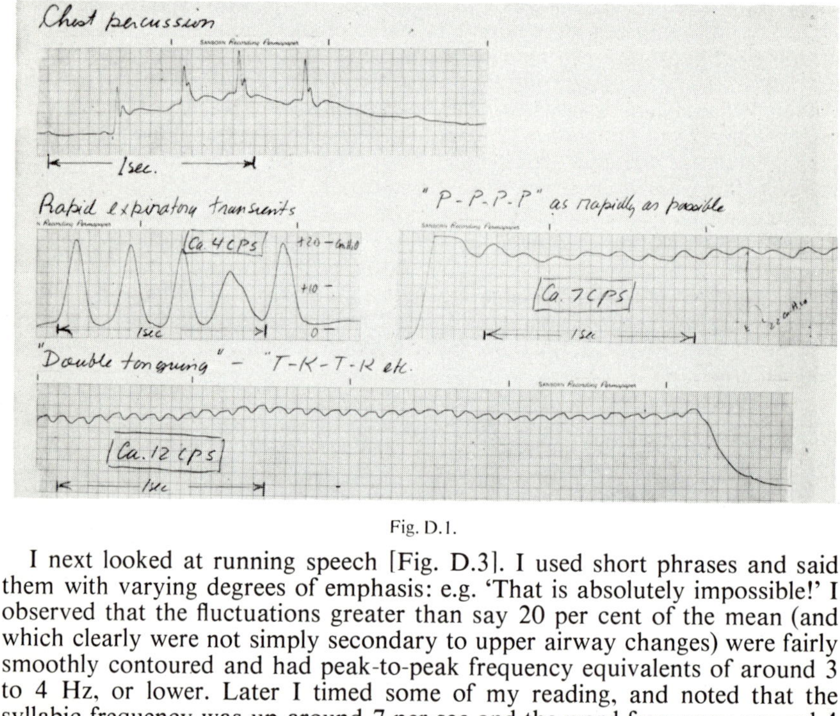

Fig. D.1.

I next looked at running speech [Fig. D.3]. I used short phrases and said them with varying degrees of emphasis: e.g. 'That is absolutely impossible!' I observed that the fluctuations greater than say 20 per cent of the mean (and which clearly were not simply secondary to upper airway changes) were fairly smoothly contoured and had peak-to-peak frequency equivalents of around 3 to 4 Hz, or lower. Later I timed some of my reading, and noted that the syllabic frequency was up around 7 per sec and the word frequency around 4 per sec [Fig. D.4]. I was stimulated to do this by a conversation I had had with Dr. Arthur Otis, who told me that he had been examining the syllabic frequency relative to maximum voluntary alternating movements: he is going to tell you about this later on (see Communication 20). I was merely interested to see whether or not the muscles of the thorax appeared to participate in the syllabic detail, and I have concluded that in myself, at least, they do not. I appear to modulate my thoracic muscles only with emphatic speech, and I don't seem to do this at a rate substantially higher than my 'word rate'. Now it doesn't seem

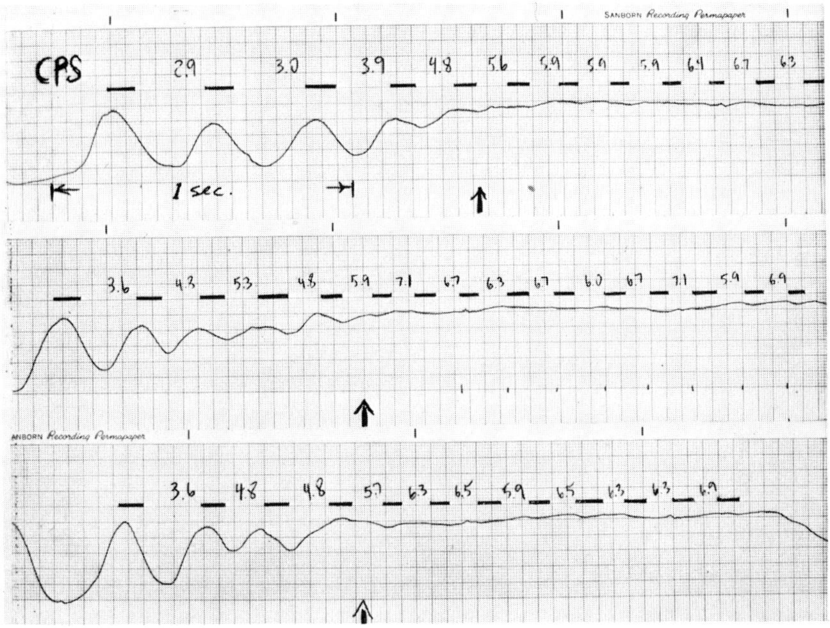
CPS
2.9
3.0
3.9
4.8
5.6
1 sec.
3.6
4.3
5.3
4.8
5.9
7.1
6.7
6.3
3.6
4.8
4.8
5.7
6.3
6.5

Fig. D.2.

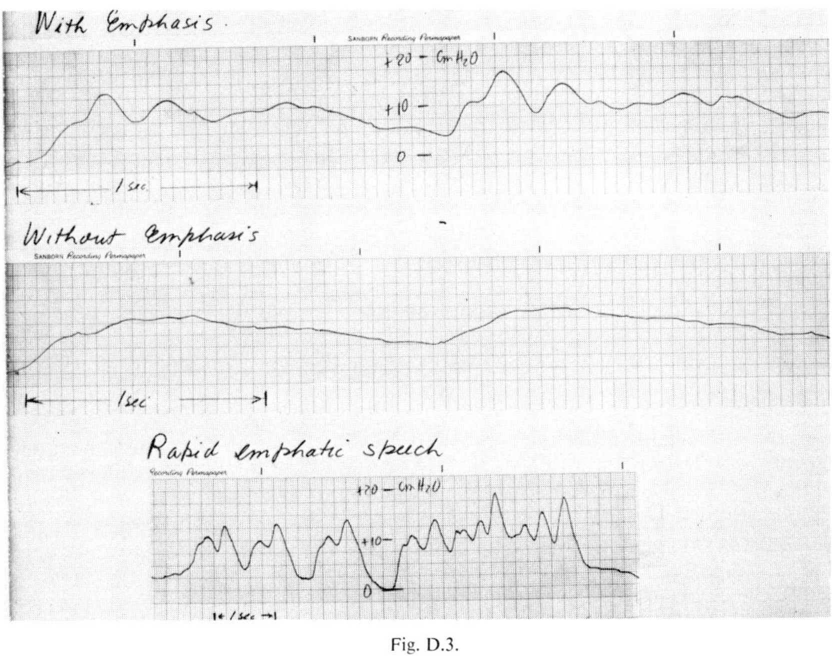
With Emphasis.
+20 – Cm H_2O
+10 –
0 –
1 sec.
Without Emphasis
1 sec
Rapid emphatic speech
+20 – Cm H_2O
+10–
0
1 sec

Fig. D.3.

to me that this ability—to squeeze somewhat harder at fairly frequent intervals—is comparable to the fine control of the muscles of the hand. But I would like to hear more of what Dr. Campbell has in mind regarding the subtlety of intercostal muscle control.

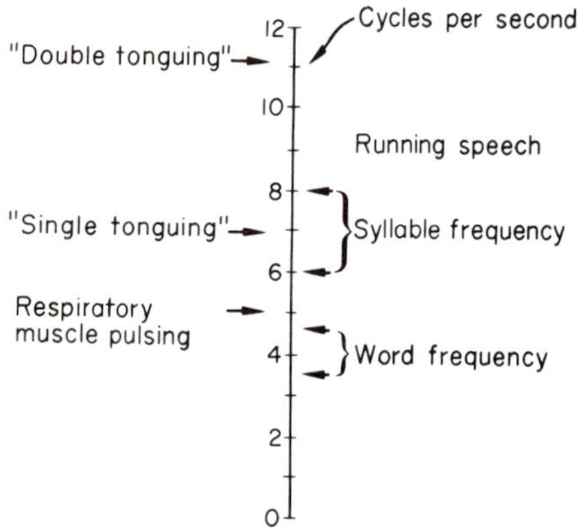

Fig. D.4.

Beverly Bishop. Beverly Bishop could not but rise to some of the comments made by Dr. Mead about the abdominal muscles—it is not the first time that I have had to defend their rôle. Youmans *et alii*[1] have been looking at the abdominal muscles in dogs for a number of years, and have taken quite a different viewpoint from mine. Youmans says that the abdominal muscles are *always* active, and that they are simply inhibited during inspiration. I take the opposite view (that which I have discussed this morning), having shown that they are activated during expiration and are usually quiet during inspiration. Even though the end results of these things may be similar, certainly the neural control depends on one's point of view—and I do want to get at the neural control.

Now I think it most interesting that Dr. Mead has suggested another function for the abdominal muscles (and I am always looking for functions for them); namely, that what they really do is 'tune the diaphragm to its optimal length'. I think that is just another way of saying that the abdominal muscles are participating in such a way as to make breathing more efficient: by keeping lung volume close to resting volume on the pressure/volume curve, the abdominal muscles contribute to the efficiency of respiration. Now I have another argument to this end from Dr. Mead's observations.

I think I must also say a word or two about his comments on my concept of the abdominal muscles playing a rôle in maintaining lung volume, for again I think we are taking different viewpoints of the same phenomenon. The abdominal muscles have nothing to do with regulation of lung volume at low lung volumes. If one's lung volume has decreased overnight, then when one gets up in the morning I am sure the abdominal muscles do not play much of a rôle in restoring the normal FRC—but what I am trying to stress is the possible 'safety' function of the abdominal muscles. For instance, if one takes a maximal in-

[1] Youmans, W. B., Murphy, Q. R., Turner, J. K., Davis, L. D., Briggs, D. I., and Hoyl, A. S. (1963) The abdominal compression reaction, *Amer. J. phys. Med.*, **42,** 1–70.

spiration these muscles become active at about 5 per cent of vital capacity—which may suggest that they are preventing over-inflation. I also know that if the expiratory reflex be abolished by pulmonary denervation, an anaesthetized cat cannot survive pressure breathing—so the expiratory activity which is initiated by positive pressure breathing is important to the cat.

Dr. Mead also mentioned that the abdominal muscles don't seem to play much of a rôle in man until you expose man to very high pressures. In a small series of six human subjects the abdominal muscles did contract during pressure breathing—although their pattern of activity was quite different from the patterns seen in anaesthetized cats. I am not trying to say these two responses are equivalent, as we don't yet know all the factors contributing to the differences between the responses of anaesthetized cats and conscious man; but I hope that future experiments will demonstrate the neural mechanisms accounting for these differences. Nevertheless, I do want to make the point that man does have this expiratory reflex which aids expiration, in that expirations proceed more rapidly and more deeply when the abdominal muscles are involved—in this sense, abdominal muscle activity tends to bring lung volume back to a reasonable level. I agree with Dr. Moran Campbell's stress on how unimportant the abdominal muscles are in man until ventilation reaches high levels: Dr. Campbell has done much more on man than I have.

Freedman. I would like to comment on the question of changes in lung volume at positive pressures. I wonder if Dr. Bishop could tell us about the time relations involved. I think it is reasonably well established (in naïve, conscious human subjects) that if one applies reasonably modest pressures (I have used something up to 15 cm H_2O) one gets an immediate increase in end-expiratory position—and end-expiratory position, as Flenley *et alii* (1971)[1] have shown, falls exactly on the relaxation pressure curve of the thorax at that point. In fact, it is a way of measuring total compliance. But as one keeps the pressure up, then after a period of 5, 10 or 15 minutes the end-expiratory position gradually returns towards its previous level: and therefore I wonder if Dr. Bishop could tell us at what time the measurements on which she based her curves were made, as I think this might be relevant to the obvious discrepancy.

The other comment I would like to make is that patients with emphysema over a long period of time increase their lung volume considerably. We have recently studied some patients with asthma who could change their lung volume by as much as a litre over a period of half an hour; and it would seem to me that if there is a control mechanism that is regulating lung volume in these people, then it is not a very active one.

Beverly Bishop. In regard to the first question, our experiments have all been acute as the maximum time for which we could study this response in conscious man was up to 4 minutes, because we were altering the inspired gas mixtures that were presented in a much more complicated experiment than I indicated. We do not have lung volume measures, as yet, on these subjects. I don't know what the adequate stimulus is for initiating the expiratory activity of the abdominal muscles: it is still an unanswered question. I can simply tell

---

[1] Flenley, D. C., Pengelly, L. D., and Milic-Emili, J. (1971) Immediate effect of positive pressure breathing on the ventilatory response to CO_2, *J. appl. Physiol.*, **30**, 7–11.

you that the pulmonary vagal afferents are absolutely essential to the response in anaesthetized animals.

In regard to your second question, I don't have any information on disease states. I think your comment about a gradual increase in lung volume over a period of time is probably accurate; but I think, perhaps, one thing we have not paid attention to is that a disease process itself may completely alter the control mechanisms. We should focus more attention on how receptor sensitivities, or the time course of reflex responses, may be altered as secondary events in disease processes.

von Euler. I should like to ask Dr. Bishop if she can tell us something about which receptors might be responsible for the effects she has described. You said that it might be possible to differentiate, on the afferent side, between inspiratory inhibition and the provocation of expiration. I should like to know whether your theory implies different types of receptor, or whether you can concede that the differentiation could be effected at central relay stations. Consider, for instance, the Hering-Breuer inflation reflex with which you have compared your observations—this reflex has two aspects, one being the termination of inspiration at the moment when the volume threshold is reached and the other is the prolongation of expiration. The termination of inspiration (in response to an inflation imposed during the inspiratory phase) seems to be something of an 'all-or-nothing' affair with a sharp threshold—no significant inspiratory inhibition occurs before complete arrest. The prolongation of the expiratory phase, on the other hand, depends in a graded manner (over a broad range) on the degree of inflation that remains in the lungs during the expiratory phase, especially in its middle portion.

Beverly Bishop. This is a very difficult question. Dr. von Euler is asking something I have been trying to solve for 10 years and I don't think I've solved it yet. We have shown within the past year that pulmonary vagal afferents are essential to the reflex, but that does not indicate the nature of the receptors. Clearly they have something to do with lung volume—but that is not the full story. You will recall that I showed that expiratory threshold loading and positive pressure breathing evoked similar levels of activity from the abdominal muscles during expiration, and yet these two experimental conditions are associated with very different lung volumes. So lung volume *per se* is not the answer; but unless you have information coming from the pulmonary receptors you don't get the expiratory activity, so they must play some role. This question came up for discussion in the recent Warsaw Symposium,[1] where it was suggested that a second category of pulmonary receptors, now being referred to as 'irritant' receptors to distinguish them from the classical pulmonary stretch receptors, might well be the critical receptors—I would like to have Dr. Widdicombe comment on this later [see Communication 9—*Ed.*]. I believe these receptors have finer diameter afferents than the pulmonary stretch receptors, and in some ways less is known about them in terms of the adequate stimuli which excite them.

I would also like to make a comment about the central mechanisms. I think the idea that Dr. von Euler threw out, that one could have the same afferent informa-

[1] *Neural Control of Breathing,* Satellite Symposium of XXV Intern. Congr. Physiol. Sciences, Warsaw, Poland, 2–6th August, 1971.

tion going in but that it may be handled by the 'central computer' in a different way and at different sites so that inspiration and expiration are dealt with separately at the centres, may well turn out to be the answer. I may have been looking in the wrong place for *the* receptors that excite expiratory activity in the abdominal muscles. I have a very simplistic view of the respiratory 'centre', and I came to this meeting to obtain more detailed knowledge about it. From my own observations, I think of the inspiratory neurones (those that drive the diaphragm) as sitting on a great highway scanning all kinds of incoming afferent information going by—and whether this information be from limb muscles that have become active during exercise or from chemoreceptors responding to changes in blood gases, they alter the activity of the diaphragm. In other words, the inspiratory neurones are extremely responsive to anything involving changes in these multiple input systems. On the other hand, the expiratory neurones that drive the abdominal muscles must be way off on a side road, a long way from this main highway—they seem to monitor only those signals that are pertinent to pressures to be overcome during expiration or to changes in lung volume or tidal volume, and they are unresponsive to the state of the blood gases. They're only listening in on certain input lines and not on others, and so may be activated only by a few of the many signals that are arriving at the main 'centre'.

Ruth Bowden. I wonder whether Dr. Campbell would consider that the differential action of the anterior and posterior parts of the intercostal muscles might be related to the function of preventing the intercostal spaces being sucked in. When one examines a patient with intercostal paralysis, one sees that with inspiration the spaces are sucked in; whereas normally the internal intercostal muscle behind the anterior intercostal membrane contracts to preserve the rigidity of the chest wall.

Another thing is—has anybody investigated the importance of the parietal pleura and peritoneum as possible sites of receptor mechanisms for reflex activation of the abdominal muscles during respiratory action? We are all familiar with the localized rigidity of the abdominal muscles that occurs when the related part of the peritoneum is involved in visceral infection.

The action of the diaphragm is of considerable interest in relation to the movments of the ribs, and I understood from some other work that the last rib, for example, does not move at all in normal respiration. Finally, in the study of the abdominal musculature has anybody considered the action of the quadratus lumborum in the total pattern? This is a large muscle that might be significant.

Campbell. In answer to the first question—do the intercostal muscles resist forces tending to blow the intercostal spaces out or suck them in—I think this is very likely, and such an action of the intercostal muscles needs more specific study. In the anterior chest wall this function seems to be performed by the interchondral portions of the internal intercostals rather than the external, so I don't think there is any conflict here—I think they are performing different functions in different parts of the chest.

With regard to the pleura being a source of relevant afferent activity I know of no information, but the idea is very tempting. With regard to the quadratus lumborum, this has been regularly suggested as an important muscle, being given different significance and different actions by different authorities. I tried to get at it electromyographically myself 20 years ago, but was never sure that I was recording from it: with surface electrodes one could never be certain, and with needle electrodes it was difficult to be sure that one was really in the muscle. The last point concerns the diaphragm and rib movements: the diaphragm raises the ribs because the costal fibres are vertical, lifting the ribs upwards and outwards by exerting a vertical force on the costal margin.

Widdicombe. In answer to Dr. Bishop about the possibility of two or more reflexes in the lungs that might be responsible for the effects she described, I want to say something about these tomorrow [see Communication 9], and will only take a minute now. If you inflate the lungs as you have been doing, we know that this stimulates at least two types of afferent endorgan—the pulmonary stretch receptors and the so-called 'irritant receptors'. In my present view, the positive contractions of the abdominal muscles that you have been getting are due to the irritant receptors—because these have a fairly high threshold and your inflations were sometimes fairly large; and, as far as we know, the irritant receptors may cause contractions of inspiratory and expiratory muscles (though the patterns of activity may differ in different muscles). On the other hand, most of the evidence about pulmonary stretch receptors suggests that with small lung inflations, which we know stimulate them, there is very little contraction of the abdominal muscles. This is certainly true of work with cats, and I think it fits with Dr. Moran Campbell's work on man. So possibly there *are* two mechanisms involved.

Whitteridge. There are fibres coming from the pleura and from the posterior mediastinum. They are all rapidly-adapting, as recorded by Holmes and Torrance (1955).[1] We always thought that they were responsible for the reflex finger vasoconstriction that you get with a deep breath, but they don't seem to be of much use in relation to Dr. Bowden's inquiry. Another thing is, I do agree very much with Dr. Mead about separating respiratory activity in the diaphragm and intercostals: we thought you could separate them up to about a litre, but you certainly can't beyond that.

---

[1] Holmes, O., and Torrance, R. W. (1955) Afferent fibres of the inferior cardiac nerve, *J. Physiol. (Lond.)*, **130**, 45P.

3 Speech as Breathing

J. MEAD

Department of Physiology, School of Public Health, Harvard University, Boston, Massachusetts, U.S.A.

AND

J. C. BUNN

Department of Ophthalmology, School of Medicine, University of Florida, Gainesville, Florida, U.S.A.

The title of our Symposium, Ventilatory and Phonatory Control Systems, is symbolized by the lady pumping the organ [Fig. 3.1]. In the case of *vox humana* she not only has to keep the organ going; she has to ventilate the hall as well, at the same time and with the same pump. We wondered how well she could do both jobs at once (Bunn and Mead, 1971). To that end we used an old technique for measuring respired volumes, apparently first used (Menzies, 1790) in Edinburgh nearly 200 years ago [Fig. 3.2] and again, in that same city, more recently by Draper and his associates in their classic studies of muscle activity during speech (Draper *et alii,* 1960). This method has the advantage of leaving the vocal instrument unemcumbered. Our only contribution has been an improved neck seal consisting of a rubber balloon-collar filled with plastic beads, which hardens on evacuation and forms a rigid but comfortable seal. Dr. Proctor will show you a picture of this [see Fig. 4.3, p. 42 in Communication 4].

To this apparatus we added a transparent dome through which we could pass gas mixtures containing CO_2 and thereby stimulate breathing. We had our subjects read aloud while exposed to air and then, unbeknownst to them, to 3–3.5 per cent CO_2 in air. This concentration causes approximately a doubling of minute ventilation in spontaneously breathing subjects. We wanted to see, first of all, how well air-breathing subjects ventilated their lungs while reading aloud, and second, we wanted to drive the respiratory system and compare the ventilatory response while reading aloud with that during spontaneous breathing.

All seven subjects ventilated slightly more while reading aloud in air than when breathing spontaneously. As a result, the partial pressure of CO_2 in end-tidal samples, which can be taken as a measure of the effectiveness of ventilation, averaged 37.5 mm Hg during reading as compared to 39 mm Hg during spontaneous breathing.

Fig. 3.1. Young lady pumping the air supply for a pipe organ.

Thus, in the circumstance which is probably the commonest of all; namely, talking at rest, the lady at the pump does both jobs very well—shading matters a bit on the side of too much ventilation. (To me this is vaguely reminiscent of the tendency for Covent Garden to begin the evening on the cool side. In both instances the notion that an effective control system is at work is illusory.) The limited capacity for increased ventilation during speech is shown by the CO_2 results. The average increase in ventilation per mm Hg rise in end-tidal PCO_2 was 3.6 litres per minute during spontaneous breathing and 1 litre per minute during reading aloud. This nearly four-fold reduction in responsiveness to chemical stimulation makes it clear that the happy marriage of speech and gas exchange at rest must run into difficulty when respiration is stimulated, as during exercise.

Fig. 3.2. Menzies' (1790) body chamber.

But these measurements tell us little that we did not know already. It is common experience that the unit of speech tends toward the paragraph at rest, the sentence during exercise, and the gasped phrase during heavy exertion. For unlike the nearly 15-fold range of ventilation from rest to maximal exercise, the flow range during phonation is limited. The voice easily moves from a near whisper to a full shout at much the same flow rate: indeed, the range of expiratory flow over which satisfactory phonation can be accomplished is far less than 15-fold.

The breathing pattern of five of our seven subjects probably reflects this limited capacity for increased flow during phonation. These subjects introduced non-phonated expirations, usually at the end of spoken phrases [Fig. 3.3]. Incidentally, this is the pattern of breathing shown by Bouhuys (1964) to be common for oboe players, who have a similar problem in more severe form—the flow of gas during sound production with the oboe is apparently particularly low.

Although ventilation during speech is clearly inadequate for gas exchange during heavy exertion, it was just about right at rest. Could this reflect relatively close control of ventilation during speech, but over a limited range? We played a trick that answers this question. The material the subjects read aloud had been specially prepared for us by Dr. Dennis Klatt to contain either an excess or a paucity of consonants requiring large volume increments. Thus 'Professor

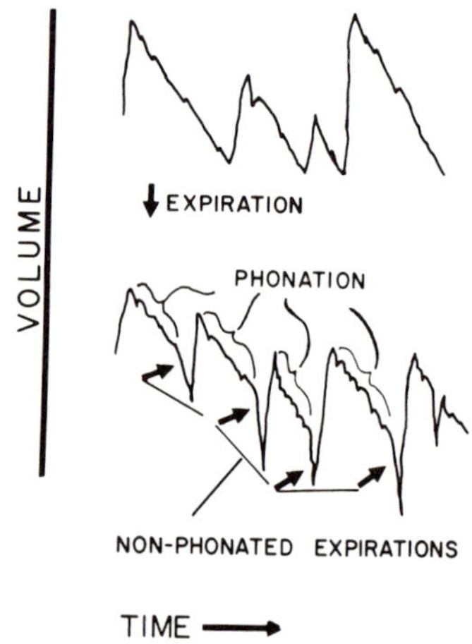

Fig. 3.3. Respiratory patterns during speech. The upper tracing is for a subject reading in air; the lower trace for a subject reading while exposed to 3.5 per cent CO_2. Note the non-phonated expirations in the lower trace.

Harris thinks his hypothesis is false; some tests which clarify his thinking should be suggested to him' requires more gas flow than 'one calendar year in New York may lead anyone away from the moral values governing many a year of honourable living'. Fig. 3.4 shows minute ventilation relative to that during spontaneous breathing measured during 5 minutes of reading of neutral material (designated [N]), the doctored material [L] corresponding to a preponderance of low volume consonants, and [H] to a preponderance of high volume consonants. Clearly the consonant content had a lasting effect on ventilation in individual instances. Four of the seven subjects increased their ventilation substantially on the 'high volume' material and maintained this increase throughout the reading period. This suggests that ventilation during speech is not under immediate control. How it came to pass that ordinary speech should be associated with ventilation which is suitable to the modest levels of activity commonly accompanying prolonged speech is a control problem of a different order.

Incidentally, the trick we played had an unexpected result which gives a convincing answer to another, older question: Does man exhibit true post-hyperventilation apnoea? The trouble has been than preconceptions probably influence results, prompting the suggestion that perhaps only respiratory physiologists have apnoea after hyperventilation. But our subjects had no idea that they were hyperventilating at all while reading. The question is convincingly answered. One of our subjects, the one with the largest increase in ventilation during speech, regularly had periods of apnoea lasting as long as 18 seconds immediately after reading, and two other subjects had briefer periods of apnoea immediately after reading the 'high volume' material.

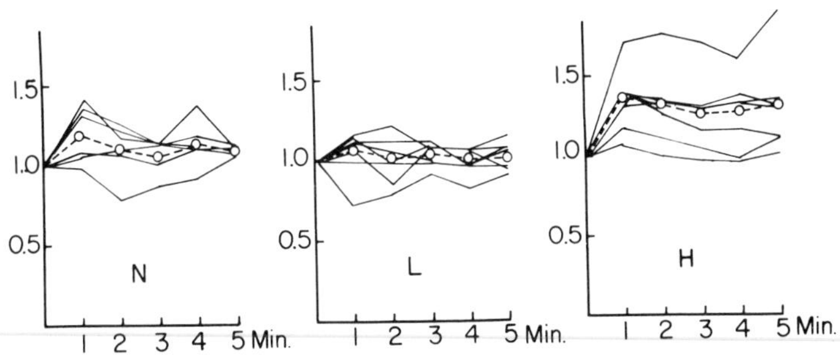

Fig. 3.4. Individual (———) and average (– – -O- – –) ventilations relative to spontaneous breathing during 5 minutes of reading aloud from texts containing normal [N], low [L] and high [H] volume consonants. (From Bunn and Mead, 1971.)

We conclude from our studies that however complicated ventilatory and phonatory control may be separately, their interaction is fairly simple; this because, although they use the same system, their requirements of this system are very different. In speech the details of the respiratory pattern mean everything, and the total volume displaced means relatively little. In breathing, the details of the respiratory pattern are completely unimportant so long as there is adequate compensation for dead space. The mutual demands of speech and breathing are nil until the maximum practical ventilation for phonation falls below the minimum required for respiratory comfort. Clearly one sure way to put an end to a conversation is to invite the offender for a brisk walk. Shall we go!

REFERENCES

Bouhuys, A. (1964) Lung volumes and breathing patterns in wind instrument players, *J. appl. Physiol.*, **19,** 967–75.

Bouhuys, A., Proctor, D. F., and Mead, J. (1966) Kinetic aspects of singing, *J. appl. Physiol.,* **21,** 483–96.

Bunn, J. C., and Mead, J. (1971) Control of ventilation during speech, *J. appl. Physiol.*, **31,** 870–2.

Draper, M. H., Ladefoged, P., and Whitteridge, D. (1960) Expiratory pressures and air flow during speech, *Brit. med. J.*, **1,** 1837–43.

Menzies, R. (1790) *Tentamen Physiologicum Inaugurale De Respiratione,* Edinburgh.

4 Breathing Mechanics During Phonation and Singing

DONALD F. PROCTOR

Department of Environmental Medicine,
The Johns Hopkins University
School of Hygiene and Public Health,
Baltimore, Maryland, U.S.A.

Since speech and singing are such ubiquitous phenomena in man we sometimes assume they require no more special training than walking. Yet we all know that whilst some speak in a clear and pleasant manner and some sing so well as to hold audiences spellbound, the speech of others is inaudible, unintelligible, and unpleasant to hear. If we recall the fact that the larynx developed phylogenetically (Negus, 1962) as a valve to prevent soiling the lungs during swallowing, and bear in mind the complex neuromuscular co-ordination involved in using this valve for phonation, it is not so surprising that, without special training, many never master the talent of effective speech and song.

The purpose of this paper is to examine the modifications in breathing mechanics required for speech and, especially, those employed by the well-trained singer.

It is immediately apparent that gross alterations in the pattern of quiet breathing are necessary for phonation. During speech the greater portion of air flow passes through the mouth rather than the nose: inspirations are brief and rapid: expirations are prolonged and slow; and airway pressures are more negative in inspiration and more positive in expiration. These variances from ordinary breathing are more pronounced in singing than in speaking.

SPEECH

Conversational speech is produced at or near the lung volume reached after a quiet expiration (FRC). Depending upon the phrasing, expirations may be only moderately below FRC; but, occasionally, they extend to 15–20 per cent of the vital capacity (Yanagilara *et alii*, 1966; Cavagna and Margaria, 1968; Klatt *et alii*, 1968). Hoshiko (1965) reported that most conversational phrases were initiated near to FRC and seldom used more than 20 per cent of the vital capacity. Flows vary between approximately 60–300 ml/sec, with an occasional brief burst for some consonants to as much as 1–2 l/sec (Subtelny and Worth, 1966), and subglottic pressures vary between 2–12 cm H_2O. Inspirations between phrases are rapid, and occur through parted lips but with a narrow oropharyngeal airway [Figs. 4.1 and 4.8].

In more formal speech, as in reading aloud (Bouhuys, Proctor and Mead, 1966) or public speaking, phrases may be initiated at larger lung volumes (60–90 per cent of vital capacity): most phrases are produced at volumes above FRC, but occasionally longer phrases involve volumes as low as 15 per cent of vital capacity [Fig. 4.2]. For loud speech and for some speech sounds somewhat higher subglottic pressures (10–20 cm H_2O) are required (Subtelny and Worth, 1966; Rubin *et alii*, 1967; Perkins and Koike, 1969).

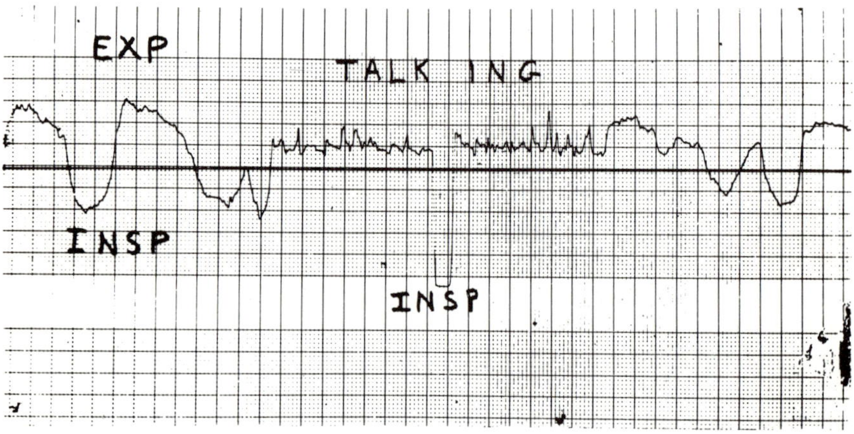

Fig 4.1. Pneumotachogram (airflow vertical axis, time horizontal) showing flows with quiet breathing to left, followed by conversational talking. Note that during talking most flow is only a third of the peak quiet breathing expiratory flow, and that the inspiration is brief and rapid. The signal goes off scale, but the flow is at least three times that with quiet inspiration.

The muscular activity accompanying speech is not so complex as with singing; but for clear enunciation and proper adjustment of sound intensity, fine control is necessary. Campbell (1968) has said that '. . . the subtlety of control required of the intercostal muscles during speech makes demands of the same order as those that are made in the small muscles of the hand'. We will return to this later.

Another difference between speech and singing is in the use of pitch. During speech there is a continual fluctuation in pitch, but with little variation from that basic to the individual's voice. During singing, pitch is held quite constant for more or less prolonged intervals and must be variable over approximately two octaves in the professional singer. A frequency of 128 Hz produces the low C of a bass voice whilst the coloratura soprano reaches a frequency of 1024 Hz for her high C.

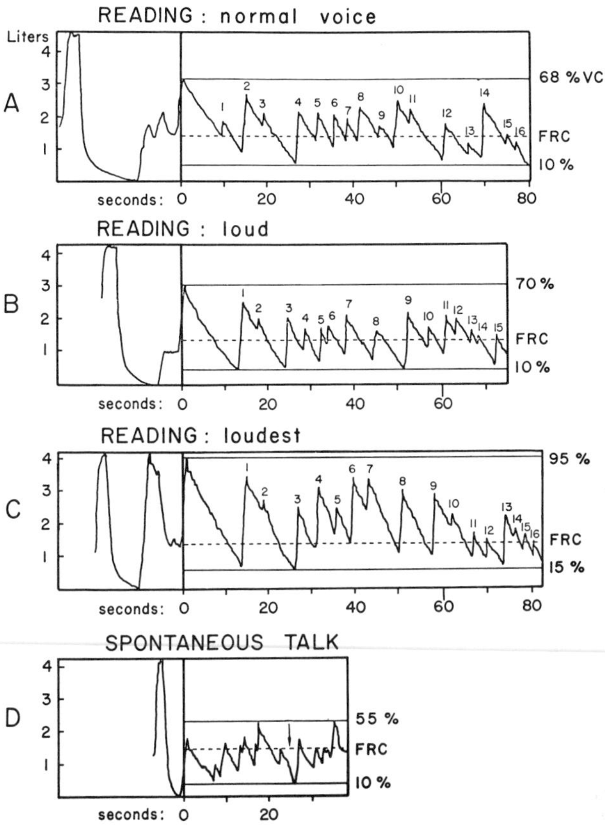

Fig. 4.2. Volume changes during quiet and loud reading and spontaneous talk. The vital capacity is shown to the left and the dashed line indicates the level of FRC. Note the larger volumes for loudest reading, comparable to public speaking, as well as the greater use of the nearly full lung. (From Bouhuys *et alii,* 1966.)

SINGING

Some years ago Drs. Jere Mead, Arend Bouhuys and myself conducted extensive studies of breathing mechanics during singing (and, to a lesser degree, during speech), and I will describe some of our findings (Bouhuys *et alii,* 1966; Bouhuys *et alii*, 1968; Mead *et alii*, 1968; Proctor, 1968).

For these experiments the subject swallowed two balloons for recording of oesophageal and gastric pressures and was seated in a body plethysmograph with a seal around the neck [Fig. 4.3].

With this technique, volume and air flow were recorded through the plethysmograph instead of a face mask; and as subglottic pressures were readily calculated by relating oesophageal pressure to known lung volumes, no transglottic transtracheal instrumentation was required (van den Berg, 1956). The subject's position was comfortable, the neck collar produced no constriction, and the nasal catheters leading to the balloons were well tolerated for periods of 2 hours or more. The record of pressure above and below the diaphragm enabled us to observe the activity of this important muscle.

Fig. 4.3. Subject seated in body box with catheters coming from nose attached to oesophageal and gastric balloons. (From Bouhuys *et alii,* 1966.)

AIRFLOW AND LUNG VOLUME

Expiratory flows during singing vary between just under 100 to over 500 ml/sec [Fig. 4.4]. This means that for a singer with a vital capacity of 5 l beginning a phrase at full lung volume, sound could continue for 10–50 sec if he were able to sing over the entire range of his vital capacity. Most song phrases are not over 10 sec in duration, and most singing is accomplished with flows of 100–200 ml/sec [Fig. 4.5]. Therefore, ordinarily the singer requires only 1–2 l of air to sing most phrases. But because he must habitually be prepared for longer phrases or those requiring higher flows, he learns to initiate song very near to full lung volume and to inspire as quickly and deeply as time permits between phrases. He also finds from experience that he can control

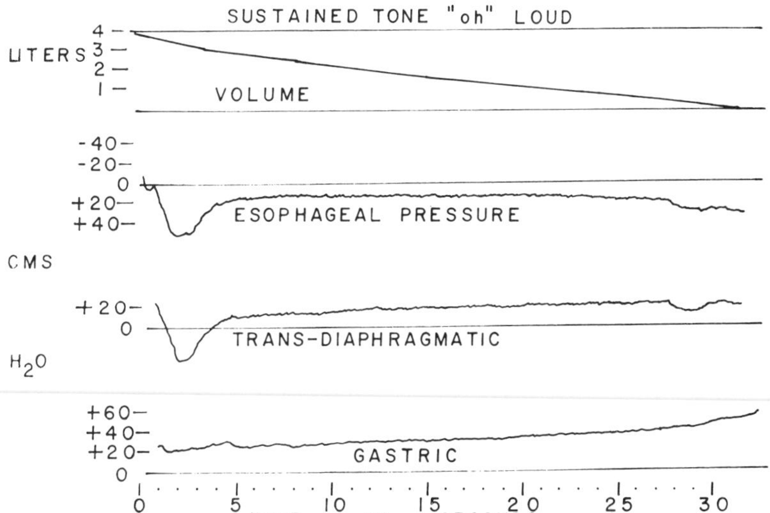

Fig. 4.4. Tracing of record taken during singing of sustained tone. The average flow was 125 ml/sec. The initial rise in oesophageal pressure is due to an oesophageal contraction.

sound production more easily at high lung volumes: for whereas the respiratory muscles are capable of producing pressures of 100 or more cm H_2O at full lung volume, at 10 per cent of the vital capacity they can produce only about 40 cm H_2O with full effort. Parenthetically we may say that the singer learns to accomplish these quick, deep inspirations without noticeable effort, as otherwise the audience would notice and resent violent use of all the accessory muscles of breathing.

As a result of training, the singer inspires at rates of 2.5–4 l/sec, is able to sing at any lung volume from the full lung down to about 5 per cent of vital capacity and can, when required, sustain a single tone for nearly a minute (Isshiki *et alii*, 1967) [Fig. 4.6]. The maximum amount of time during singing is spent at about 50–60 per cent of vital capacity, only 2–3 per cent at about 10 per cent or below, and 15–20 per cent above 75 per cent of vital capacity [Fig. 4.7].

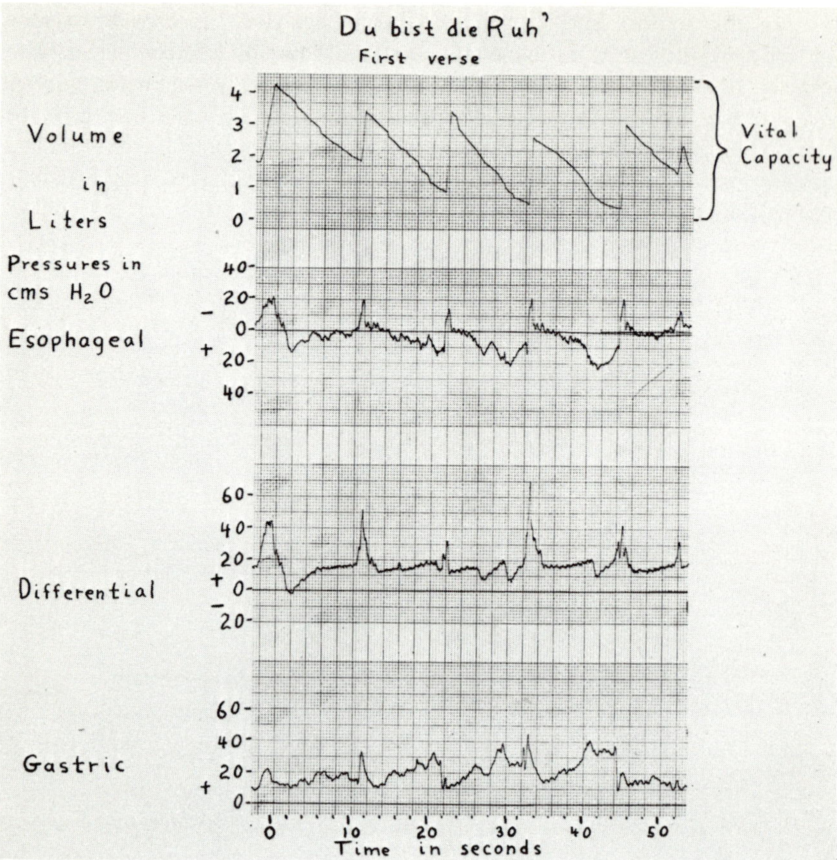

Fig. 4.5. Record during singing of first verse of Schubert's *Du bist die Ruh.* Note that nearly all of the vital capacity is used, and that the five phrases are sung in 52 sec. The third phrase requires nearly 3 l.

In contrast to speech, during singing the oropharyngeal airway is wide open during inspiration to reduce the resistance to rapid inspiratory flow [Fig. 4.8].

The flow of air required for singing is dependent upon the pitch and intensity of the sound, the enunciation of words, and the ability of the singer to avoid wasted airflow [Figs. 4.9 and 4.10]. In general, very low pitched sounds (and, to a lesser degree, those of very high pitch) require higher flows than tones in the singer's middle register. Loud tones (and, to a lesser degree, very soft sounds) require higher flows than those of moderate intensity. A sustained tone on a single vowel sound requires less flow than a phrase involving pitch changes or enunciation of words; but the trained singer learns to enunciate even windy consonants—such as in the German 'beschworst'— and to do rapid scales or other pitch changes with minimal airflow. Loss of air before sound is begun at the initiation of a phrase is a particular hazard which the singer learns to overcome, for he must initiate sound with the first escape of air from the lungs (Subtelny and Worth, 1966).

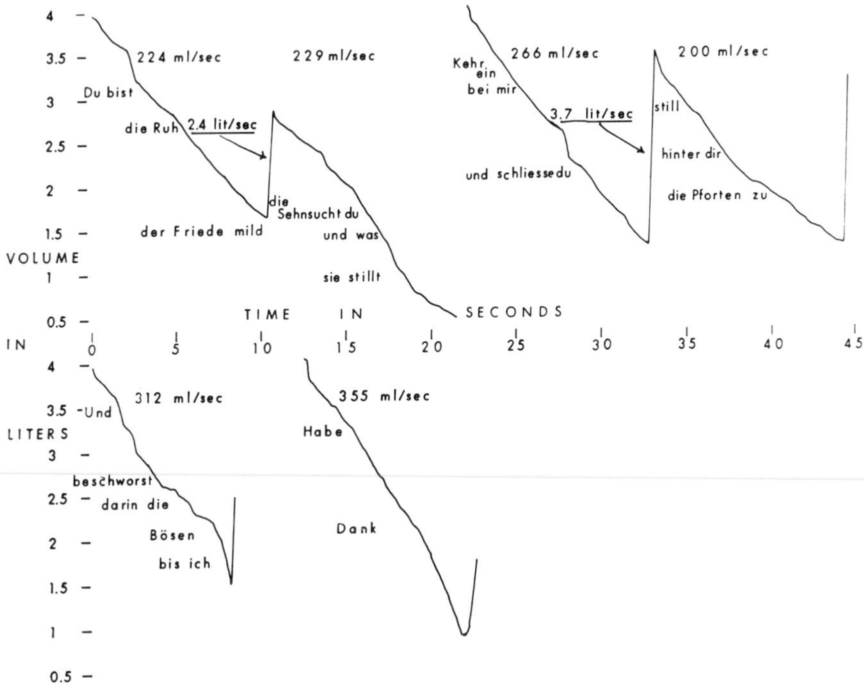

Fig. 4.6. Tracing of volume changes during singing of *Du bist die Ruh* and of two phrases from R. Strauss's *Zueignung*. Peak inspiratory flows are 3.7 l/sec, and flows during singing average from 200 ml/sec during a soft phrase in the first song to 355 ml/sec during a *crescendo* in the second song.

The flows employed for quick inspirations during singing are not near to those of which we are capable. Only rarely are they as high as 5 l/sec [Fig. 4.11]. Indeed, resort to very high inspiratory flows would dry the oral and laryngeal mucosa, with deleterious effects upon the vocal mechanism.

It is obvious that the ease with which singing is accomplished, and the duration of the phrases the singer can produce, depend upon minimizing air flow during phonation. For this reason exercises in which the singer learns to produce the longest possible sustained tone or the longest phrasing of scale singing are of value in learning how to avoid useless escape of air during phonation [Fig. 4.12].

Minute volumes of pulmonary ventilation during singing generally exceed those of quiet breathing, ranging over brief periods to as much as 20 l/min. Singers do not suffer from symptoms of hyperventilation; but how this adjustment occurs is not understood. Bouhuys (1964) has found no gross changes in blood pH or PCO_2 in wind instrument players during sessions of one half-hour or more.

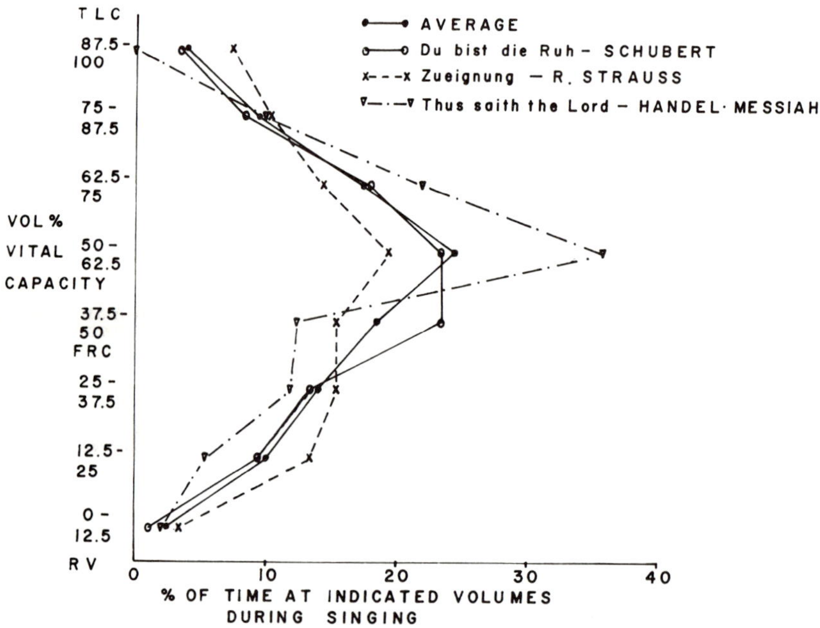

Fig. 4.7. Calculations made from records showing per cent of time spent at various lung volumes during singing songs of different character. *Thus Saith the Lord* is a recitative characterized by maximum use of mid-lung volumes, whilst the more dramatic *Zueignung* requires more even distribution of time through all lung volumes.

DICTION

Proper diction is of importance in both speech and singing if words are to be understood, to avoid needless expenditure of air, and to achieve the most effective resonance. Many public speakers attempt to reach their audiences with a loud voice when a quiet voice clearly enunciated would be most effective. Indeed, it is of interest that the diction required in some languages seems to result in a naturally beautiful singing voice without training—the Welsh are famous for this capacity, as are some Italians.

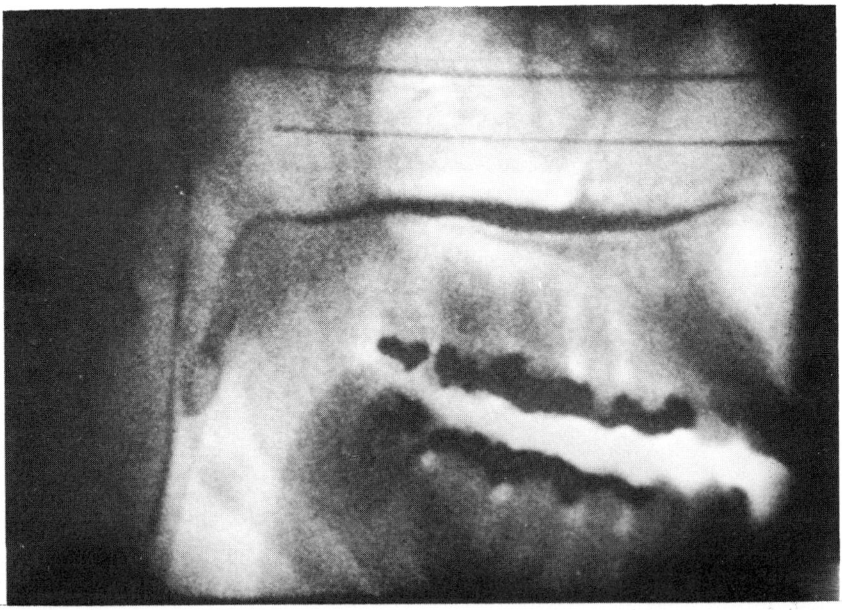

Fig. 4.8. One frame from a cinefluorograph taken during singing showing the wide oropharyngeal airway used for inspiration between phrases. Radio-opaque material marks the floor of the nose, palate and posterior pharynx.

AIRWAY PRESSURE

Alveolar pressure during ordinary breathing ranges from 0.5 to 2 cm H_2O, with greater changes for rapid breathing. The singer goes from a relatively high negative pressure for his quick inspirations to 5–20 cm H_2O positive pressure during singing (van den Berg, 1956; Rubin *et alii*, 1967; Mead *et alii*, 1968; Perkins and Koike, 1969). Rarely, the singer generates pressures as high as 40–70 cm H_2O for very loud tones [Fig. 4.13]. Again we should note that even these pressures are well below those of which we are capable (100 cm H_2O or more at full lung volume).

In general, we can say that sound intensity is varied through changes in subglottic pressure. Thus, for singing phrases requiring variations in loudness, the ability to produce fine adjustments of subglottic pressure is necessary. The sound-carrying quality of the voice is not so simply determined. Many untrained speakers are heard with difficulty at some distance even when they make maximum effort, whilst the trained singer is able to 'project' his voice in such a manner that even his softest tone is heard at the back of a large auditorium. This ability is related to the manner in which he uses his resonance cavities—and the speaker can be similarly trained.

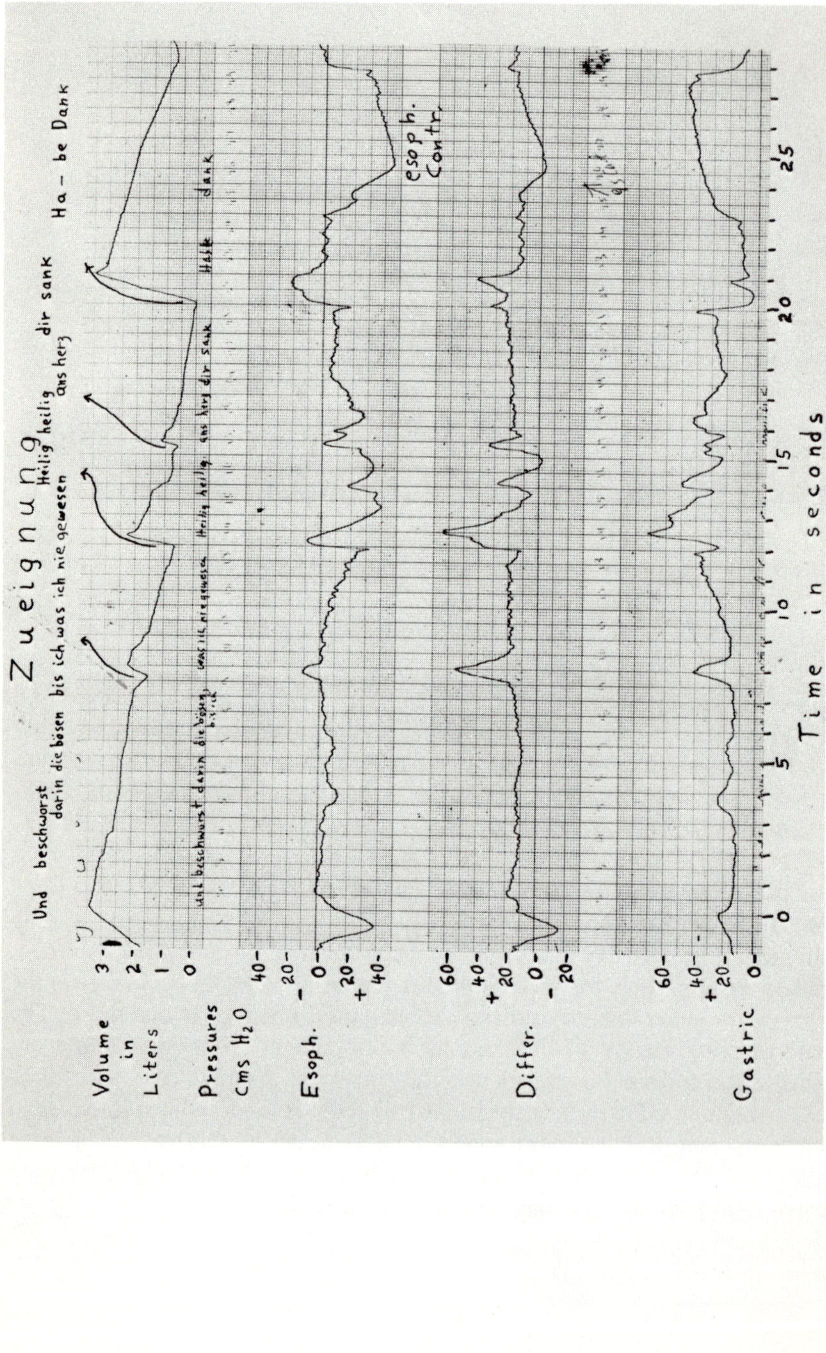

Fig. 4.9. Record during the singing of R. Strauss's *Zueignung*. Note the rapid flow with sounds such as '*ich*' at the end of the first phrase and near the beginning of the second, and the words '*Heilig*' and '*habe*'. (From Proctor, 1968.)

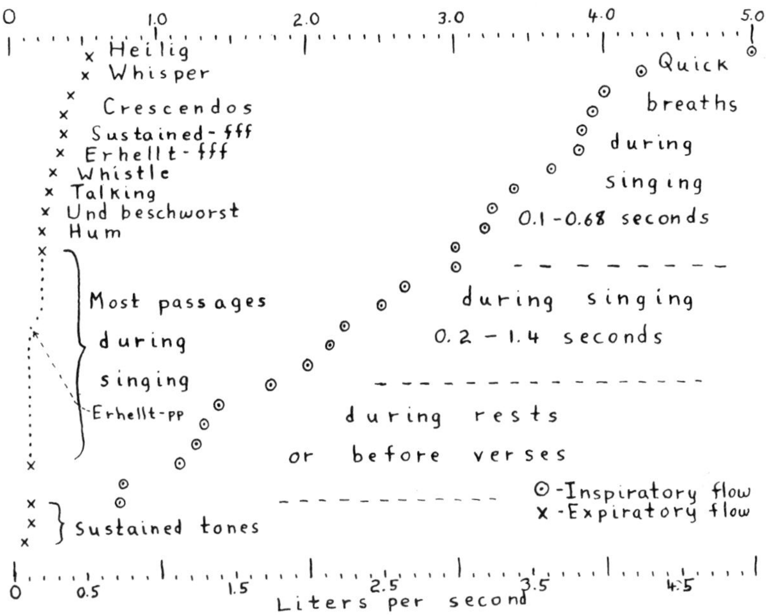

Fig. 4.10. Flows during singing, talking (ranging from less than 100 to just over 500 ml/sec) and during the inspirations between phrases—with low flows when time permits and very high flows when the song phrasing permitted only a short time for their accomplishment. (From Proctor, 1968.)

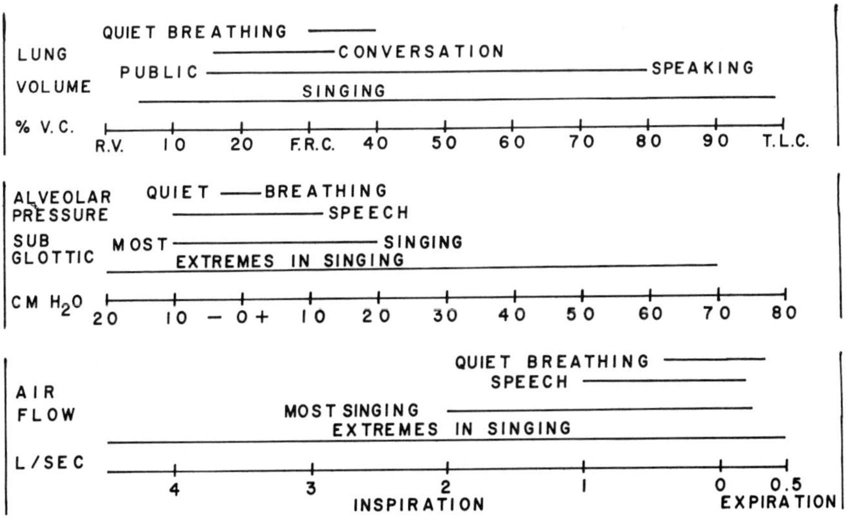

Fig. 4.11. Lung volume, airway pressure and airflow changes with quiet breathing, speech and singing. During speech and singing alveolar pressure and subglottic pressure are nearly identical because of the low velocity of airflow.

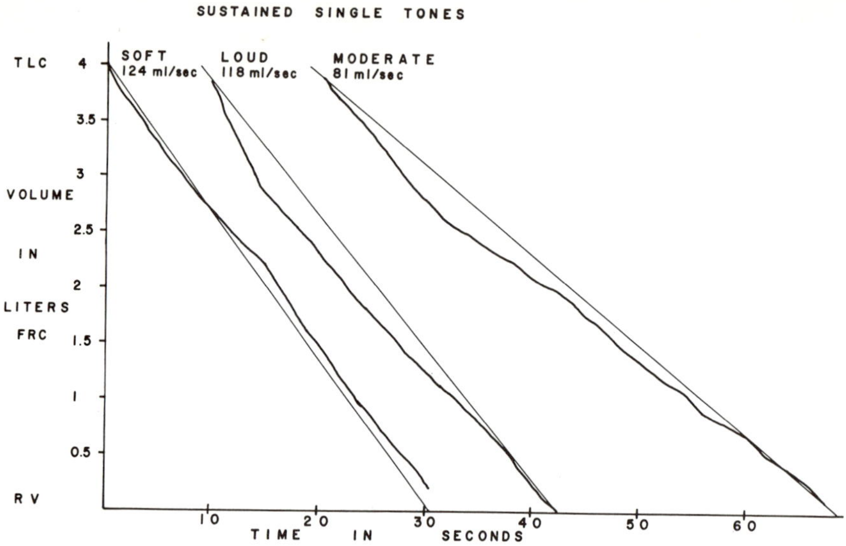

Fig. 4.12. Tracing of volume records during singing of three sustained tones at different intensities. Note that the minimum flow (81 ml/sec) is with the 'moderate' voice, and that flow tends to increase after initiation of tone. The straight lines mark what would have been a completely steady flow for each tone, very nearly achieved in the 'soft' tone.

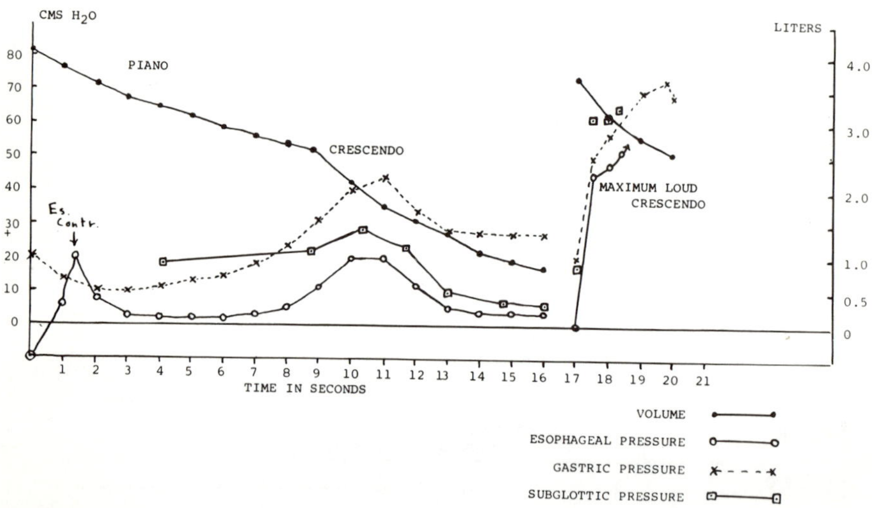

Fig. 4.13. Tracing of record of volume, oesophageal and gastric pressure, and chart of subglottic pressure calculated from oesophageal pressure at known lung volumes. Note that one *crescendo* is accomplished by going only from 20 to 30 cm H_2O, whilst the second very loud *crescendo* required about 65 cm H_2O. (From Proctor, 1968.)

Lindahl (1970) has described a patient who had cut his throat just above the larynx in such a manner that the vocal folds were clearly visible and projected outward rather than into the upper respiratory tract. When this patient talked, although the vocal folds could be seen to move normally, only a 'thin, reedy squeak' could be heard.

From this and other evidence it is clear that the vibrating glottis has little to do with the resonance of the voice. The shape of the oropharyngeal cavity and the patency of the nasal passage and the paranasal sinuses are important influences on voice quality.

MUSCULAR AND ELASTIC FORCES

For an understanding of the use of the respiratory muscles in phonation we must look at the elastic forces involved in breathing and their changes with lung volume [Fig. 4.14]. There are two main elastic forces—the lung's tendency to seek a volume somewhat less than residual volume, and the chest's tendency to seek a volume somewhat greater than FRC. The chest's elastic inspiratory force exactly equals the expiratory force of lung elasticity at FRC.

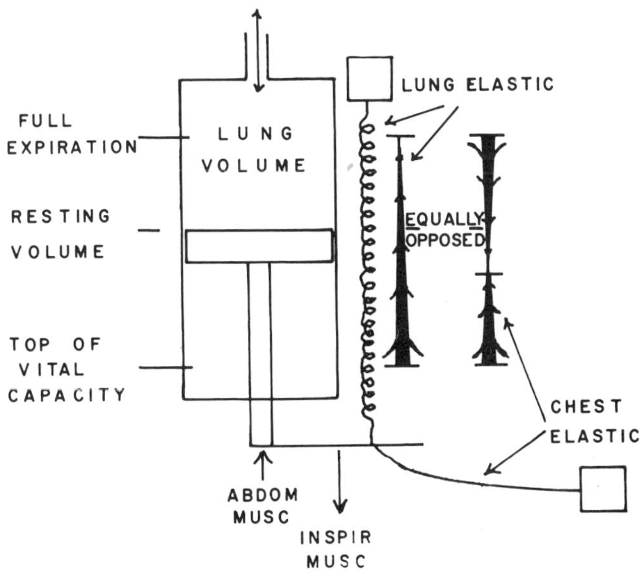

Fig. 4.14. Diagram of the elastic forces of breathing where chest elasticity is represented by an elastic strip attached to a fixed point (which might be considered the spinal column) and lung elasticity is represented by a coil spring (which is attached to what might be the lung hilum). (From Proctor, 1968.)

When air is expired below FRC, lung elasticity grows weaker and chest elasticity grows stronger. On breathing in from residual volume the lung's elastic tendency to collapse steadily increases, while the chest's inspiratory force decreases. At about 60 per cent of vital capacity the chest is at its resting volume; and, as lung volume is further increased toward the full vital capacity, both elastic forces grow stronger in an expiratory direction.

Thus, at full lung volume [Fig. 4.15] the combined elastic forces in the absence of any muscular effort result in a positive pressure of about 30 cm H_2O against the closed glottis. It is most important to realize that, since most

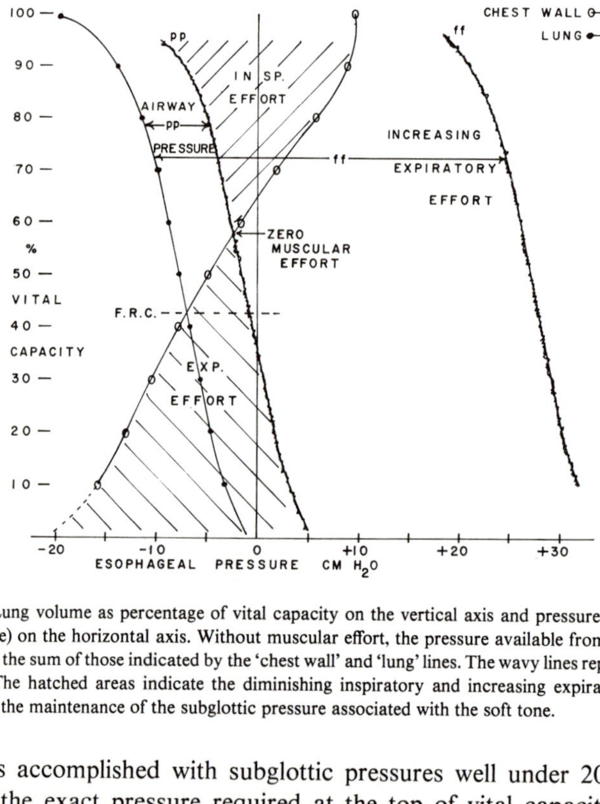

Fig. 4.15. Lung volume as percentage of vital capacity on the vertical axis and pressure (as seen from the pleural space) on the horizontal axis. Without muscular effort, the pressure available from the elastic forces alone equals the sum of those indicated by the 'chest wall' and 'lung' lines. The wavy lines represent a soft and a loud tone. The hatched areas indicate the diminishing inspiratory and increasing expiratory muscle effort required for the maintenance of the subglottic pressure associated with the soft tone.

singing is accomplished with subglottic pressures well under 20 cm H_2O, to produce the exact pressure required at the top of vital capacity inspiratory effort must counteract the elastic forces. If a higher subglottic pressure (30–50 cm H_2O) is required for loud tones, expiratory muscle effort will be necessary in just the proper amount for the pressure desired.

Let us consider the production of a tone requiring 8 cm H_2O subglottic pressure sustained from full lung volume to near residual volume [Fig. 4.15]. After filling the lungs, the glottis is closed and then quickly relaxed just sufficiently to maintain the cord tension and length for the desired pitch. At that moment inspiratory effort must be applied in exactly the right amount to result in a combined force on the airways producing a subglottic pressure of 8 cm H_2O. Immediately, as air escapes at let us say 200 ml/sec, this inspiratory effort must be neatly and steadily reduced to balance the loss of elastic recoil. At one point no respiratory muscle effort is required and immediately thereafter a finely adjusted and steadily increasing expiratory effort is needed, at first to supplement the elastic forces and finally to counteract their inspiratory direction.

Mead *et alii* (1968) have suggested an interesting analogy between the singer and a violinist. An artist with the violin can produce a beautifully steady tone while moving his bow across the string from one end to the other. Imagine his difficulties if springs were attached to either end of the bow in such a manner that they balanced one another at mid position: he would then have to learn to pass from pulling at one end of his stroke to pushing at the other.

Since the diaphragm is the strongest inspiratory muscle, one might expect it to be employed at the volume where inspiratory effort is required to reduce the effect of elastic recoil. Actually, our studies show that, during most speech or singing, the diaphragm is completely relaxed. Only at the beginning of sound production initiated at full lung volume is there occasional evidence of minimal diaphragmatic activity [Fig. 4.16]. The inspiratory force needed at high lung volumes is provided by the intercostals, which are probably capable of much finer control than is the diaphragm.

The ability of the singer to sustain a steady subglottic pressure or vary it at will (upon which sound intensity depends) is a result of fine control of inspiratory intercostal effort at high lung volumes, together with the use of the abdominal muscles acting across a relaxed diaphragm to supplement and then counteract elastic forces at lower lung volumes. The idea, deeply seated in the minds of many voice teachers, that the diaphragm is a key muscle during singing is wrong.

Some of this is reflected in pneumograph tracings made during singing [Fig. 4.17]. A more detailed and accurate record of the involved events requires study of the transdiaphragmatic pressure, or the use of other techniques (Hoshiko and Berger, 1965; Konno and Mead, 1967).

Recent work in measuring the lung volumes of professional singers, voice students and subjects without voice training has indicated a lower RV/TLC ratio in the former, professional singer group (Gould and Okamura, 1971). This change has been attributed to an increase in vital capacity at the expense of residual volume, although previous work had uncovered no significant difference in lung volumes between singers and non-singers (Heller *et alii*, 1960). If the former finding should be confirmed, it must mean that the exercise associated with long term voice training results in a strengthening of expiratory muscles or an increased flexibility of the thoracic cage, or that individuals with an innate unusually large vital capacity in relation to their total lung capacity tend to become professional singers.

The relationship between the exercise associated with singing and health and disease in medical history is the subject of two interesting papers by Finney (1968, 1971).

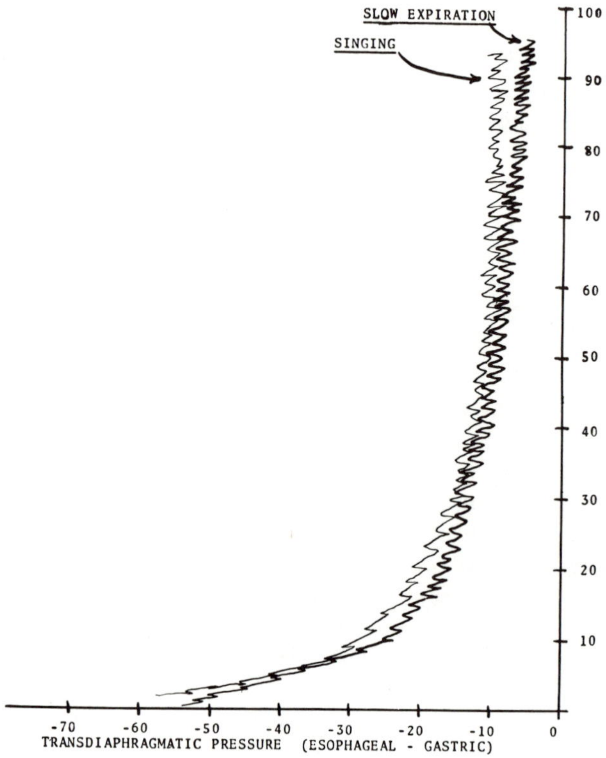

Fig. 4.16. Transdiaphragmatic pressure during an expiratory vital capacity manoeuvre with the glottis open (when the diaphragm is fully relaxed), and during singing of a sustained tone.

Leaving aside the activity of the intrinsic laryngeal musculature, one other point should be made (Hirano *et alii*, 1967). Sonninen (1956) has shown that '. . . one of the important functions of the extralaryngeal musculature is to assist the cricothyroid muscle in the production of high tones by lengthening and thinning the vocal cords'. In this manner the strap muscles of the neck serve as accessory respiratory muscles in singing.

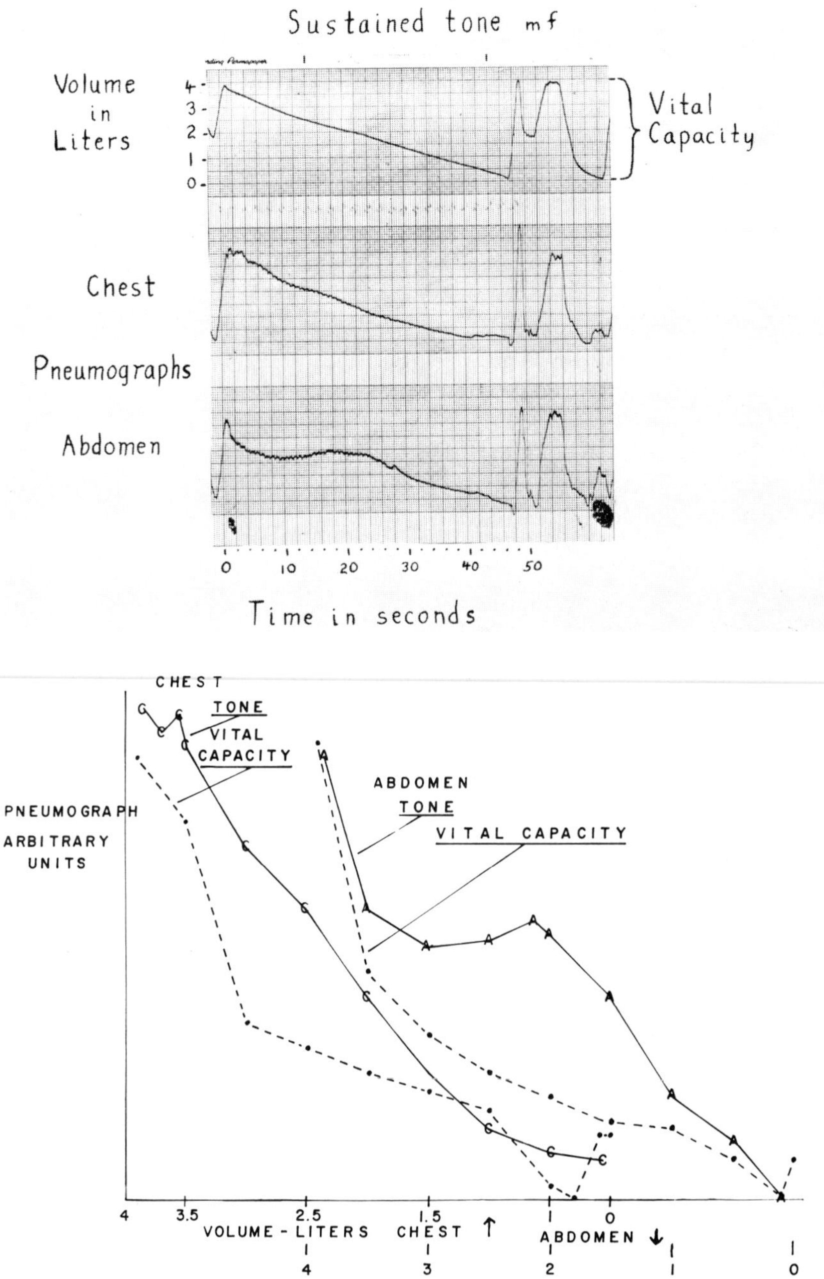

Fig. 4.17. Record of chest and abdominal pneumographs (above) and chart comparing relative motion (below) during singing of sustained tone and performance of vital capacity. It is obvious that the shape of the chest and abdomen are grossly different during singing, in comparison to breathing from the top of lung capacity to residual volume.

CONCLUSIONS

1. Both speech and singing require alterations in ordinary breathing mechanics, but to different degrees.
2. Human phonation is accomplished with airway pressures and airflows far less than those of which normal man is capable.
3. In conversational speech only about 20 per cent of the vital capacity is employed, but almost all of the vital capacity is used for singing.
4. The fine control of airway pressure and air flow required in either speech or singing is accomplished through a balanced effort of the inspiratory intercostal muscles and the abdominal muscles, with the diaphragm relaxed.
5. In singing, economic control of expiratory airflow is an essential concern.

Finally, it seems reasonable to say that whereas these considerations of breathing mechanics are of great importance to the singing teacher and his students, some of those who indulge in public speaking might also benefit from a more complete understanding of the complex events involved in voice production.

Editor's Note: This Communication included the showing of a cinéradiographic film of Dr. Proctor's upper airways during his singing of different types of song.

REFERENCES

Berg, J. van den (1956) Direct and indirect determination of the mean subglottic pressure, *Folia phoniat. (Basel),* **8,** 1–24.

Bouhuys, A. (1964) Lung volumes and breathing patterns in wind-instrument playing, *J. appl. Physiol.,* **19,** 967–75.

Bouhuys, A., Proctor, D. F., and Mead, J. (1966) Kinetic aspects of singing, *J. appl. Physiol.,* **21,** 483–96.

Bouhuys, A., Mead, J., Proctor, D. F., and Stevens, K. N. (1968) Pressure-flow events during singing, *Ann. N.Y. Acad. Sci.,* **155,** 165–76.

Campbell, E. J. M. (1968) The respiratory muscles, *Ann. N.Y. Acad. Sci.,* **155,** 135–40.

Cavagna, G. A., and Margaria, R. (1968) Airflow rates and efficiency changes during phonation, *Ann. N.Y. Acad. Sci.,* **155,** 152–64.

Finney, G. (1968) Vocal exercise in the sixteenth century related to theories of physiology and disease, *Bull. Hist. Med.,* **42,** 422–49.

Finney, G. (1971) Fear of exercising the lungs related to iatro-mechanics (1675–1750), *Bull Hist. Med.,* **45,** 341–66.

Gould, W. J., and Okamura, H. (1971) Personal communication.

Heller, S. S., Hicks, W. R., and Root, W. S. (1960) Lung volumes of singers, *J. appl. Physiol.,* **15,** 40–42.

Hirano, M., Koike, Y., and von Leden, H. (1967) The sternohyoid muscle during phonation, *Acta otolaryng.,* **64,** 500–7.

Hoshiko, M. S. (1965) Lung volume for initiation of phonation, *J. appl. Physiol.,* **20,** 480–2.

Hoshiko, M. S., and Berger, K. W. (1965) Sequence of respiratory muscle activity during varied vocal attack, *Speech Monographs,* **32,** 185–91.

Isshiki, N., Okamura, H., and Morimoto, M. (1967) Maximum phonation time and airflow rate during phonation, *Ann. Otol. (St. Louis),* **76,** 998–1007.

Klatt, D. H., Stevens, K. N., and Mead, J. (1968) Studies of Articulatory activity and airflow during speech, *Ann. N.Y. Acad. Sci.,* **155,** 42–55.

Konno, K., and Mead, J. (1967) Measurement of the separate volume changes of rib cage and abdomen during breathing, *J. appl. Physiol.,* **22,** 407–22.

Lindahl, J. W. S. H. (1970) Some thoughts and speculations on the voice, *Proc. roy. Soc. Med.,* **63,** 591–4.

Mead, J., Bouhuys, A., and Proctor, D. F. (1968) Mechanisms generating subglottic pressure, *Ann. N.Y. Acad. Sci.'* **155,** 177–81.

Negus, V. E. (1962) *The Comparative Anatomy and Physiology of the Larynx,* New York.

Perkins, W. H., and Koike, Y. (1969) Patterns of subglottal pressure variations during phonation, *Folio phoniat. (Basel),* **21,** 1–8.

Proctor, D. F. (1968) The physiologic basis of voice training, *Ann. N.Y. Acad. Sci.,* **155,** 208–28.

Rubin, H. J., Lecover, M., and Vennard, W. (1967) Vocal intensity, subglottic pressure and air-flow in singers, *Folia phoniat. (Basel),* **19,** 393–413.

Sonninen, A. A. (1956) The role of the external laryngeal muscles in length-adjustment of the vocal cords in singing, *Acta otolaryng.,* Suppl. **130,** pp. 1–102.

Subtelny, J. D., and Worth, J. H. (1966) Intraoral pressure and rate of flow during speech, *J. Speech Res.,* **9,** 498–518.

Yanagihara, N., Koike, Y., and von Leden, H. (1966) Phonation and respiration *Folia phoniat. (Basel),* **18,** 323–40.

5 The Configuration of the Chest Wall During Speech

J. MEAD

Department of Physiology, Harvard University School of Public Health, Boston, Massachusetts, U.S.A. and

T. J. HIXON

Department of Communication Disorders, Madison, Wisconsin, U.S.A.
with M. D. GOLDMAN

The contributions of our Laboratory in the area of speech physiology have largely been in methodology. I should like at this point in our programme to describe a method which grew out of our measurements on singers.

We were struck by the ability of singers to produce soft tones at relatively high lung volumes but with, as far as we could judge, completely relaxed diaphragms. We reasoned that they could achieve this only by substantial inspiratory activity of other muscles, which, through elevation of the rib-cage, could lower sub-diaphragmatic pressure sufficiently to compensate for diaphragmatic relaxation. To test this hypothesis we used pneumographs around the chest and abdomen; and noticed incidentally that when we closed off the airway and voluntarily shifted volumes from the rib-cage to the abdomen, the pathway traced by an X-Y plot of the signals from the pneumographs followed very nearly a single line, and that when the same manoeuvre was repeated at another fixed volume, another line was produced which quite closely paralleled the first. This suggested a graphic method for displaying the configuration of the chest wall that might be useful in assessing the separate volume contributions of the rib-cage and abdomen to respiratory volume changes.

Dr. Konno and I followed up these leads (Konno and Mead, 1967), and along the way developed something better than the simple pneumograph for measuring the movements. We found that the most prominent motions associated with breathing were to be seen in changes in the antero-posterior diameters of the rib-cage and abdomen. And we found that we could get electrical outputs proportional to these changes with magnetometers (Mead *et alii*, 1967). I will skip the details and get right on to the diagrams.

Rib-cage A-P diameter will be on the Y axis, increasing upward, and abdominal A-P diameter on the X axis, increasing to the right. Increasing lung volume will then be, in general, upward and to the right [Fig. 5.1]. But before we can interpret such a diagram in terms of volumes we need to adjust the relative gains of the two signals, and for this we make use of the iso-volume manoeuvre [Fig. 5.2]. If, at constant lung volume, I elevate my rib-cage and

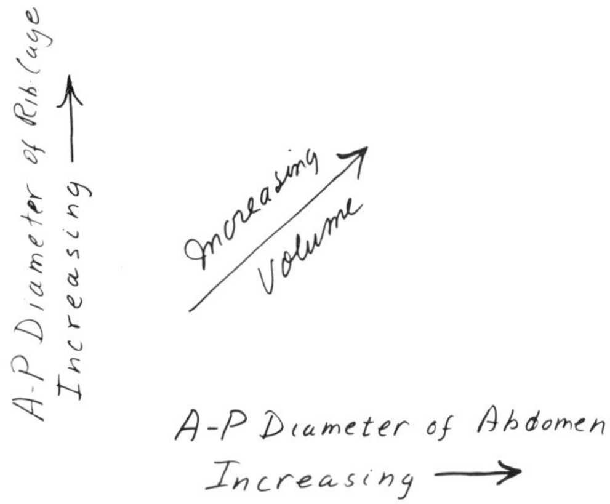

Fig. 5.1.

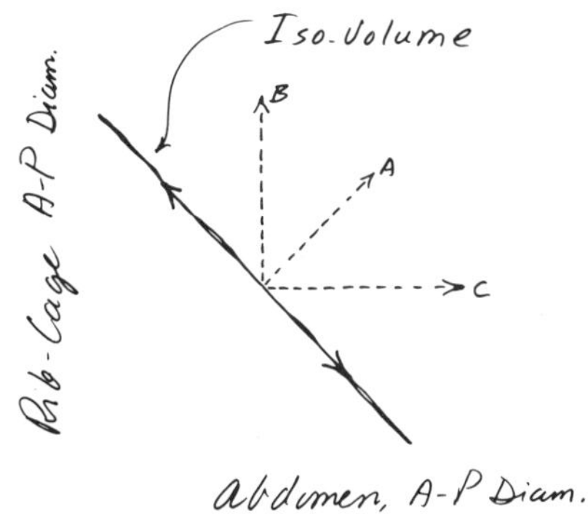

Fig. 5.2.

simultaneously contract my abdominal muscles so as to displace the abdominal wall inward,* since my total volume is constant the volume displaced by the surfaces of the rib-cage and abdomen must be equal and opposite. If, as I perform this manoeuvre, I adjust the signals so that the outputs are equal and opposite—that is, so that the pathway followed has a slope of -1 —then I know that the volumes displaced by the surfaces are equal for equal deflections.

* Demonstrated by Professor Mead on the lecture platform [*Ed.*].

I can now interpret the pathway taken during actual breathing in terms of the relative contributions of the rib-cage and abdomen. For example, I know that if the pathway during respiration has a slope of +1 (dashed line, A), the surface displacements of the rib-cage and abdomen are the same, and contribute equally to the volume change; whilst if the pathway is vertical (dashed line, B), all of the displacement is accounted for by the rib-cage and none by the abdomen, and *vice versa* (dashed line, C).

But I still have no idea of the actual volumes displaced. This I learnt by having the subject perform the iso-volume manoeuvre at different known lung volumes—for example, at 20 per cent increments of vital capacity. In this way I calibrate the diagram for volume [Fig. 5.3]. I can now estimate the total volume change between any two points and, from the slope of the line between the two points, the relative contributions of the rib-cage and abdomen to the total volume change.

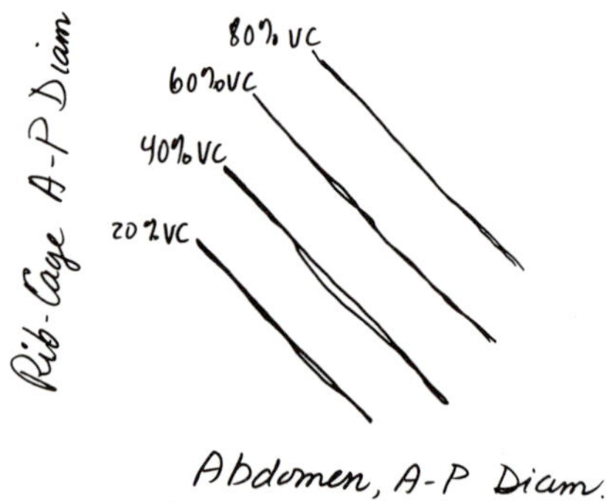

Fig. 5.3.

But the diagram is useful for more than estimating volumes and volume contributions. It can be used to provide a chart of the shape of the chest wall, and it is this aspect of the diagram I want to discuss with respect to speech. But first we need some landmarks. The particular landmark I want to point out is the so-called 'relaxation configuration'. When you voluntarily relax your respiratory muscles as best you can, your respiratory system comes to rest at a lung volume somewhere in the neighbourhood of 40 per cent of your vital capacity, if you are in the upright posture. On the diagram this corresponds to some point along the 40 per cent VC isopleth [Fig. 5.4]. You would find that this point was quite highly reproducible: you could relax from any other volume and you would tend to return to the near neighbourhood of this point. If you were then to breathe to various lung volumes and, before relaxing, were to close off your airway so that you would be relaxing at some volume other than the relaxation volume on the diagram, you would obtain a series of relaxation points which together would describe a curved line [Fig. 5.5]. Again you would find that this line was easily reproduced. This is the particular feature of the chart that I want to call to your attention. The hatched line describes the shape changes of the passively driven respiratory system—the pathway that would be followed if all of your respiratory muscles were paralyzed and if you were then passively driven by a mechanical respirator.

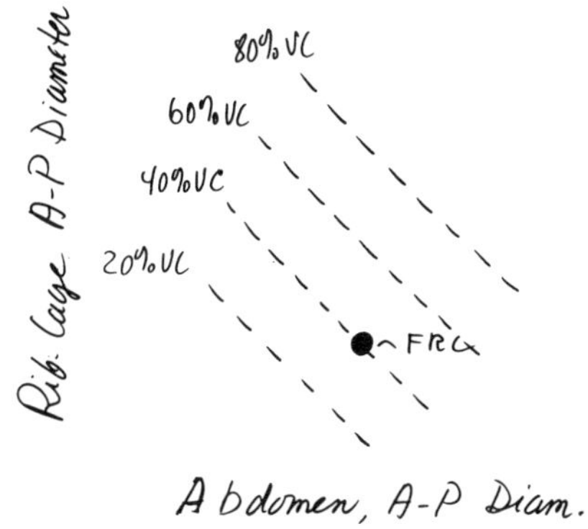

Fig. 5.4.

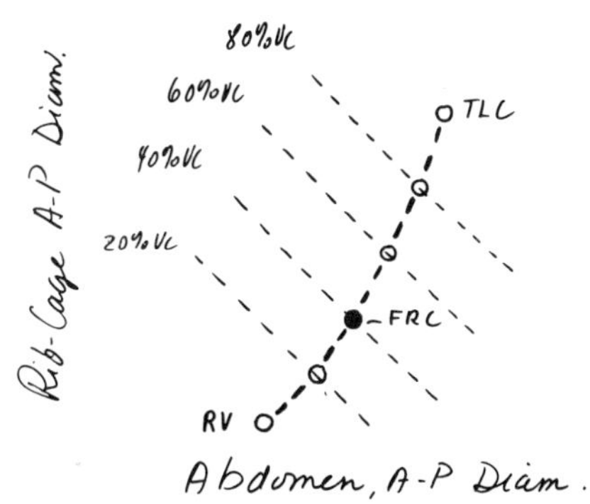

Fig. 5.5.

We can now return to the subject at hand. During spontaneous breathing, both at rest and during exercise, we stay quite close to the relaxation configuration. That is to say, during natural breathing the respiratory muscles do not distort the chest wall very far from the shape it would assume in the relaxed state. How we do this is an interesting question, and what it means in terms of energy expenditure is also interesting, but I want to tell you what happens during speech. Dr. Hixon has been looking into this question, and I will give a quick summary of some of his findings.

In the upright posture speech takes place well to the left of the relaxation configuration. This is true for conversation, reading aloud, singing and a variety of other vocal manoeuvres [Fig. 5.6]. Points to the left of the relaxation line indicate that the rib-cage is relatively more, and the abdomen correspondingly less expanded than they would be in the relaxed state at the same lung volume. What does this mean in terms of the activity of specific muscles? We cannot say on the basis of the diagram alone. This could represent either net inspiratory activity of inspiratory muscles acting on the rib-cage or preponderant expiratory activity of the abdominal muscles, or some combination of these.

But by measuring pressures at the same time, we can be somewhat more specific. Dr. Proctor has shown you [see p. 51] how it is possible to distinguish pressures developed by muscle activity from those developed by the passive properties of the respiratory system [see Fig. 4.15 in Dr. Proctor's Communication]. The same analysis can be accomplished separately for the rib-cage and abdomen. The pressure developed by the rib-cage is pleural pressure. The volume pressure characteristic of the rib-cage can be obtained by plotting rib-cage volume against pleural pressure. Rib-cage volume can be obtained directly from the rib-cage magnetometers, suitably calibrated by the method I described, and pleural pressure can be estimated from oesophageal pressure. The same can be done for the abdomen—for the pressure developed by the abdominal wall is intra-abdominal pressure, which can be estimated from gastric pressure.

The action of the muscles operating on the rib-cage can be assessed by comparing pleural pressure during activity with that at the same volume during relaxation [see dashed lines in Figs. 5.8 and 5.9]. Pressures more negative than the relaxation pressures indicate net inspiratory activity and pressures more positive, net expiratory activity. In the case of the abdomen, since all muscle activity of the abdominal wall is expiratory, all deviations from the relaxation characteristic are positive.

The remaining figures were obtained on myself just prior to this meeting, and summarize our approach. Figure 5.7 shows the representation we have already talked about: rib-cage *v.* abdominal A-P diameters. On the left are two iso-volume manoeuvres, 1 litre apart. On the right these are represented as dashed lines, along with a tracing during a relaxed expiration through pursed lips—indictating the relaxation configuration—and a tracing during ordinary talking. The speech was interrupted, and I relaxed immediately against a closed airway, producing the fine line parallel to the iso-volume isopleth. As in the previous examples, talking takes place in the neighbourhood of tidal breathing in terms of volume, but with the rib-cage relatively more expanded and the abdomen less expanded than during relaxation (and spontaneous breathing, not shown).

The diagrams in Fig. 5.8 are of rib-cage A-P diameter *v.* oesophageal pressure (Pes). The line sloping up to the right and extended with a dashed line is the relaxation characteristic, the heavy portion being a relaxed expiration through

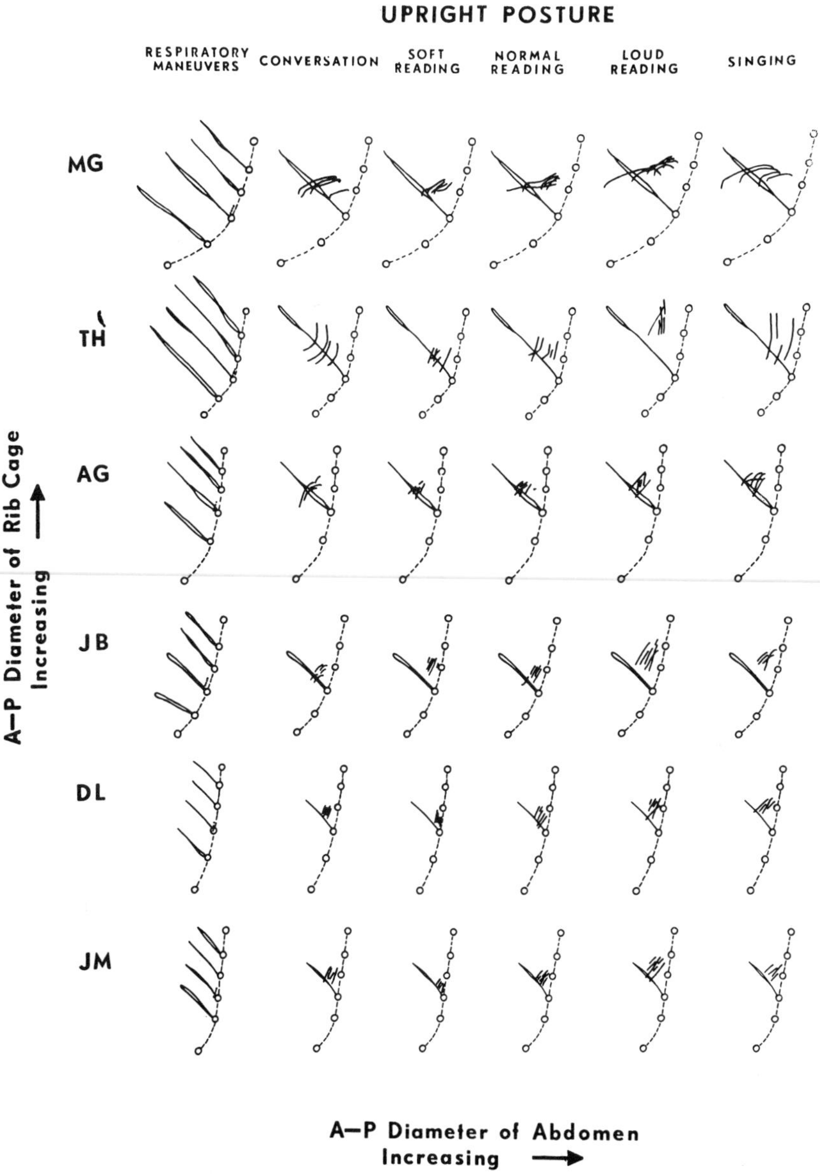
UPRIGHT POSTURE
RESPIRATORY MANEUVERS
CONVERSATION
SOFT READING
NORMAL READING
LOUD READING
SINGING
MG
TH
AG
JB
DL
JM
A–P Diameter of Rib Cage Increasing
A–P Diameter of Abdomen Increasing

Fig. 5.6.

pursed lips, comparable to that in Fig. 5.7. All pressure to the left of this line indicates net inspiratory muscle activity. The small loop on the upper left is a 'spontaneous' breath, and the larger one on the upper right a slow voluntary inspiration and expiration. On the lower left I was talking, and the pressure is to the right of the relaxation characteristic, indicating net expiratory activity of the muscles driving the rib-cage. When I voluntarily initiated a sentence at high lung volume, I had net inspiratory activity initially which decreased to zero and then proceeded into the positive range.

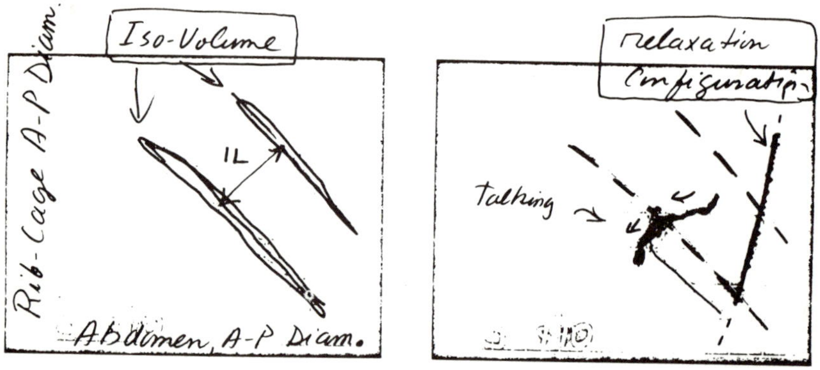

Fig. 5.7.

Figure 5.9 shows similar diagrams for the abdomen, pressure in this instance being sensed with a gastric balloon. Again, the dashed line indicates the relaxation characteristic. The heavy line represents a 'spontaneous breath' with conscious relaxation of the abdominal muscles; the abdomen proceeds along, or very close to its relaxation characteristic. Running speech takes place down and to the right of the relaxation characteristic, indicating activity of the abdominal muscles. Several sentences with interspersed inspirations are shown. During the inspirations my abdominal muscles remained active. When I interrupted speech and relaxed against a closed airway, I returned promptly to the relaxation characteristic.

In general we have found that during ordinary speech the rib-cage is either nearly relaxed or exhibits net expiratory activity, while the abdomen is invariably active. The shift from the relaxation configuration to relative expansion of the rib-cage is apparently the result of the dominant action of the abdominal muscles rather than of inspiratory activity of muscles operating on the rib-cage.

It is of interest that what appears to be a dominating influence of abdominal contraction has to do with the configuration at which speech takes place, and not the pressure generating the sound. If it were simply a matter of dominance, one might expect that during stressing, or what I refer to as 'emphatic speech', with each increase in pressure the abdomen might compress more than the rib-cage.

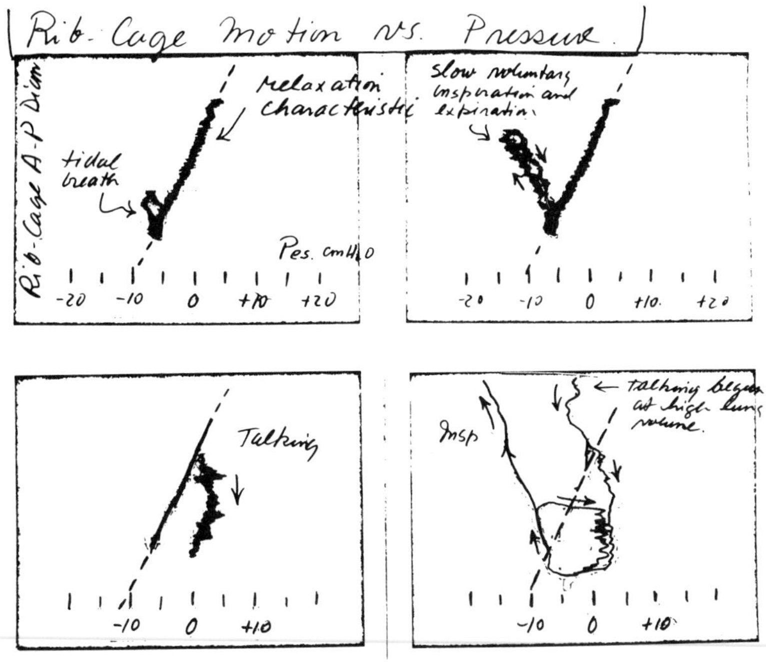

Fig. 5.8.

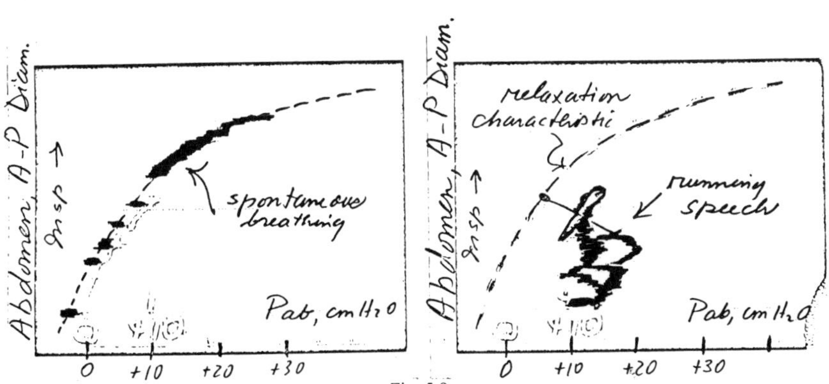

Fig. 5.9.

Figure 5.10 shows that in me, at least, this is not the case. On the left, gastric pressure is displayed against rib-cage A-P diameter. On the right is the simultaneous diameter-diameter graph superimposed on the iso-volume and relaxation lines, as in Fig. 5.8. The swings in pressure—which, the diaphragm being relaxed, are identical to the swings in pleural pressure—were quite large, but there are no synchronous ‘jogs’ in the configuration diagram.

Apparently we distort the chest wall from its relaxation configuration during speech, but we do not further distort it in developing pressure. In my earlier presentation (Communication 3), I suggested [see p. 35] that we appear to solve the bellows problem in two parts: we solve the volume problem and, in effect, use this solution as a platform from which to develop the required aleveolar pressure. I would now like to suggest that the platform is really better thought of in terms of the chest wall configuration. Apparently we can solve the configurational problem quite separately from the alveolar pressure problem. In Fig. 5.11 the

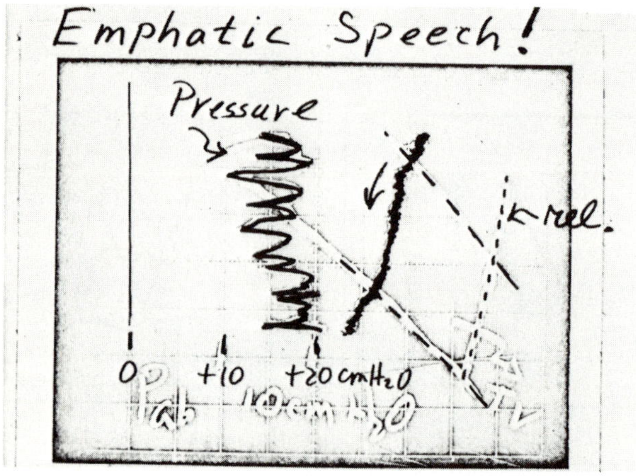

Fig. 5.10.

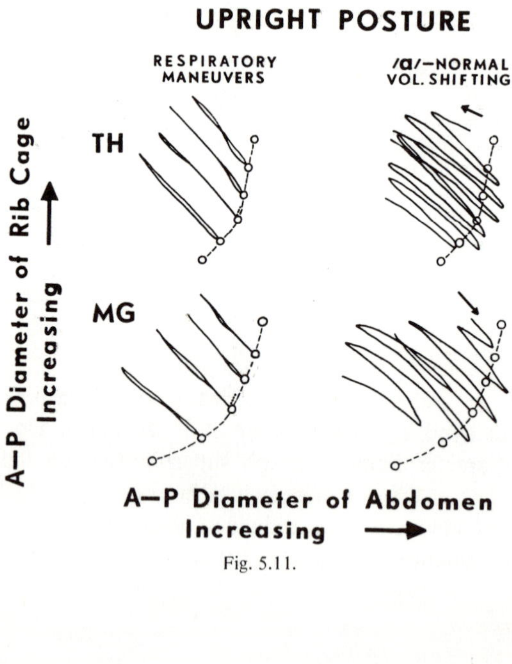

Fig. 5.11.

subjects produced a steady vowel sound starting from high lung volume, and as they produced the sound they simultaneously shifted volume back and forth between the rib-cage and abdomen. In effect they performed the 'iso-volume' manoeuvre as they phonated. This looks tricky; but try it—you will find it easy. I suppose that voice trainers take advantage of this separation, and try to get their students to use the configuration which is most advantageous from the standpoint of voice control; and I was interested to hear what Dr. Proctor had to say about this. My objective, in this talk, has been to show how some fairly simple measurements of the external dimensions of the chest wall, together with some moderately unpleasant measurements of pressure, can be used to interpret the mechanical action of the muscles on the respiratory pump.

REFERENCES

Konno, K., and Mead, J. (1967) Measurement of the separate volume changes of rib cage and abdomen during breathing, *J. appl. Physiol.*, **22,** 407–22.

Mead, J., Peterson, N., Grimby, G., and Mead, J. (1967) Pulmonary ventilation measured from body surface movements, *Science,* **156,** 1383–4.

DISCUSSION ON COMMUNICATIONS [3], [4], AND [5]

Klatt. In Dr. Proctor's movie, in which he showed cinéradiographs of himself singing, he indicated that the oral cavity was kept open to allow less airflow resistance to rapid inspiration. However, the movie gave me the impression that the tongue actually moved backwards in the pharynx and that there was therefore a pharyngeal obstruction. Would you care to comment on that?

Proctor. I have examined single frames from the ciné film. It is much easier to look at the details in single frames blown up than it is in the ciné film; and if you were to look at these single frames, I think you would agree that there is really a big increase in the over-all pharyngeal airway. The tongue is depressed, and in so doing it does move back towards the pharynx to a degree; but the over-all cross-sectional area of the pharynx is still far greater than it is during similar inspirations in ordinary conversation. You will have to take my word for that.

Brodnitz. Just two brief remarks in response to Dr. Proctor's paper. First of all, he has now proved with his beautiful and elegant instrumentation something that the older authors always instinctively postulated; and it may be appropriate, since we are in London, to quote something that was written in this city in 1886. Sir Morrell Mackenzie, the father of laryngology in the English-speaking countries, wrote a beautiful little book *Hygiene of the Vocal Organs,*[1] and in it there is a sentence that I constantly quote to my patients: 'in good speaking and singing, the minimum is always the optimum'—and this is exactly what Dr. Proctor has shown. There also came to mind an illustration that I have seen in another old book. In 1923, Nadoleczny[2] published a fascinating book that unfortunately has

[1] Mackenzie, M. (1886) *Hygiene of the Vocal Organs,* London.

[2] Nadoleczny, M. (1923) *Untersuchungen Ueber den Kuntsgesang,* Berlin.

never been translated into English, but which is a study of artistic singing containing dozens of pneumographs and spirographs of singers singing certain phrases; and in it there is one illustration where the author asked a very well trained singer to sing /*a*/ in a soft *piano,* and then at a given signal suddenly to change into an /*a*/ in a loud *forte.* The pneumographic and spirographic curves show what remarkable little difference there is between the two performances in this well-trained singer: you can hardly see a break in the pneumogram, and the expenditure of air rises only by one third at the change from a very soft *piano* to a very loud *forte.*

One more remark, Dr. Proctor: I was very happy when you said 'project your voice' and then added 'whatever that means'—for this phrase is an old pet hate of mine. A few years ago I wrote a little paper[1] in which I tried to show that the language we use in connexion with the description of voices is obsolete, and that all the traditional expressions that laymen and singers use lead to incorrect vocal behaviour. 'Project' is an exact example, as it implies the (incorrect) opposite of what you have demonstrated; but the trouble is, I have never found a really good expression as a substitute term to describe how the voice carries. I sometimes demonstrate to singers how, if one hits a tuning fork (which is a vibrator minus a resonator) you have to hold it very near to your ear to hear it; whereas if one puts it on a resonator (such as an empty box) the big tone suddenly erupts—and then I always say to them, 'this you get for free'. So it all fits in with your beautiful demonstration.

Proctor. May I say two things. First, I do not deserve the credit for the instrumentation and design of these studies—Dr. Mead's ingenuity is responsible for that. The other thing concerns this term 'project'—you are absolutely correct, Dr. Brodnitz, it is the wrong term. The trained singer learns to 'feel' the resonance very close to himself; but the way you tell, when singing to an audience, is by looking at someone in the back row—someone who has been reading the programme when you have been singing loudly. When you come to a soft passage and he looks up from his programme and pays attention, then you know your voice has 'carried'!

Kirchner. I would like to ask Dr. Proctor if he could give us his thoughts about the mechanism of the fine balance that he mentioned as being required to produce the proper subglottic pressure appropriate for a particular tone. First, I wonder how important are the abdominal muscles and the diaphragm as input sources, compared with any accurate information that might originate in the subglottic region of the larynx itself—from low-threshold pressure receptors in the subglottic mucosa, for example. Second, I wonder how accurate is the respiratory system or the larynx as a source of feed-back in monitoring the voice. We know, for instance, that a good singer can think of a tone and then open her mouth and hit $B^{\flat}$, or whatever is required—does this depend on some form of feed-back from the larynx or from the respiratory muscles; or is the tone immediately monitored by the auditory system, so that initially perhaps it is not the exact tone that the singer plans to produce, but within milliseconds it is immediately adjusted through the hearing system so that the singer slides into the right tone?

---

[1] Brodnitz, F. S. (1967) Semantics of the voice, *J. Speech Dis.*, **32,** 325.

Proctor. To my mind, what Dr. Mead has just told us probably casts the first clear light on your initial question. If I understand his studies properly, he is saying that the mechanism of adjusting the configuration of the chest and abdomen allows one to use elastic forces to their best efficiency; but superimposed upon this is a separate mechanism that is probably a major controller of the subglottic pressure—and this, I think, would be represented by the abdominal muscles, the diaphragm then being relaxed between chest and abdomen: I don't know whether Dr. Mead will agree with that or not. I think that probably there must be two mechanisms: for one thing, if you think of the opera singer who may be carrying out vigorous physical actions at the same time as he is singing, and yet is able to sing perfectly clearly, you visualize him as having one mechanism that clearly keeps his configuration appropriate in spite of the physical exercise as he moves about and another mechanism that controls the intensity of the tone he is emitting.

Another point about the feed-back system—most singing teachers use a term called 'musical intelligence', and they learn very quickly, when they have new pupils, who has this mysterious quality and who has not. If one listens to them sing in a concert one learns very quickly too, because they tend to hit tones just a little off the pitch at which they are aiming and then gradually slide into the tone they meant to hit—and I think the slowness with which they do this does not speak very well for an auditory feed-back mechanism. It is more likely that the person who is able to do this properly can, in his mind, set his vocal mechanism exactly where he wants it for a given tone and strike it quite clearly, with a very small variation from the actual pitch that he intends; but for this I have no proof—as you know, I am only stating my own impression.

Kirchner. Can a singer produce a designated note exactly, without hearing his own voice? Suppose one puts a masking system on the ears, and delivers enough masking noise such that a singer cannot hear his own voice, can a trained singer still hit the note that he intends?

Proctor. I think Dr. Mead did some studies on that subject.

Mead. I can't give the answer. I am sure others have done a better job on this matter. All we did was to use 'white' sound at the ear at an almost painful level and examine not only tones produced by the subjects's voice but also his ability to whistle (whistling is the easiest thing to damp out completely). We found that he could hit notes quite accurately (and quite accurately is the best I can say to you without quantitative data) in the absence of any acoustic awareness of the sound being emitted. I believe it is the experience of others here that people who have sudden onset of deafness are able to produce tones pretty effectively for a while, but in the course of months they gradually lose this ability. Such patients have no acoustic feed-back at all, so I think the statement can be made that some sort of proprioceptive reflex adjustment of vocal fold lengths can be quite adequate. I don't think that I have ever heard a good musician find it necessary to slide very far to hit a note—you 'think' the note, you can hear it 'in there', and you can produce it whether you hear the actual output yourself or not.

Winckel. To sing with minimum pressure, as Dr. Brodnitz mentioned, involves a process of voluntary control that has to be learned during training—and at first one uses the abdominal muscles. It is risky to use mainly the rib-cage muscles, as this may lead to over-pressure. What one has to learn is sensitive regulation of the abdominal muscles.

The next problem is the way that one acquires control of the resonance of singing in the vocal tract during performance, for increasing resonance absorbs more energy. This procedure must be fully controlled by the entire musculature, the activity of which does not depend exclusively on auditory feed-back—because if you are a stage singer in a Wagnerian opera, for example, you often cannot hear your own voice. Control in such circumstances depends on good training of the tactile afferent system that operates from receptors in the region of the pharynx and mouth. But to use this control accurately, you must have very good acoustics in the concert hall to provide coupling of the impedence between the mouth and the hall; and for this purpose, the reverberation time in the concert hall should be about 2 seconds—if it is less than this the singer loses control, as he does also if the reverberation time is too long. This is because he cannot balance the input from his oropharyngeal tactile receptors with his auditory feedback from the resonance effect—but we need more data on the mutual relations of these two processes.

Whitteridge. I should doubt very much, without direct evidence, that one can use the abdominal expiratory muscles without first bringing in the internal intercostals. As far as the normal human being is concerned, it is quite clear, in all the people we have tried, that you bring in the internal intercostals first above the ordinary FRC and then bring in the abdominal muscles, with the obliques before the recti—and I don't believe (and I won't believe it until somebody actually does the experiments) that even trained singers could bring in the external obliques and recti without using the intercostals.

As regards feed-back, there is a simple experiment you can do which at least shows there *are* some forms of auditory feed-back. If you take a tube from a manometer with a leak, and tell the chap to blow to keep up the pressure on the manometer, as long as the leak is making a noise he will not have the slightest difficulty holding the water or mercury steady; but if you then mask out the noise, he will oscillate a great deal more. Our suspicion was that wind instrument players were rather better than most people at performing without auditory feed-back.

The last point I would like to make is a matter of communication. We heard about Covent Garden opera singers singing *piano,* this being the total range that they produced. My experience of talking to singers is that *piano* in Covent Garden is what they would also call *mf* in chamber music: the way they use such words is entirely related to the musical effect produced, and is not in the least related to the physical effects that we would actually be measuring—and this does cause a bit of confusion.

Mead. Dr. Proctor has quoted me as likening the singer producing a steady tone to a violinist doing the same thing with springs attached to his bow. I should like to say that if violins came that way, violinists would probably soon learn to manage quite well; and, I would suppose, with mechanisms closely analogous to the ones we use to produce steady subglottic pressures in the face of an over-changing over-all muscular task. I like to think of the respiratory system as accomplishing the task in two steps. I think of it as solving the problem of providing zero alveolar pressure at any volume and in any body position or contortion. Then from this platform as it were, it produces the necessary squeeze to develop the alveolar pressure it wants in excess of zero. I should like to hear the neurophysiologists amongst us comment on how they envisage this as happening. Would it be worthwhile to model the situation with limb movements, where total muscle activity might more readily be monitored?

Anthony. I wonder if we could take the discussion away from the mechanics of muscle action and consider the organization of air flow on its own. I have been doing work over the last couple of years on this, devoted to three questions. One, of course, is the old dictum of phonetics that speech is modified breathing—but in phonetics teaching it is just left at that, and nobody goes any further. The second question is, if there is a conflict between speaking and breathing, how is it resolved—that is, how is the lung volume utilized in speech? The last question is whether a speaker in any way predicts how much air he is going to need for what he is going to say.

One thing I find odd is the use of non-normal speech in experimental studies of speech. On the one hand we have experimenters asking people to speak from 100 per cent of their vital capacity down to the bottom, when people just don't do this sort of thing in speaking; and on the phonetics side what we would really like to know is what happens in normal speech—that is, not speech that is all consonants or all vowels, but normal everyday speech. I am a little unhappy about some of the graphs that were presented, showing that people had performed a vital capacity manoeuvre and then concentrated on speaking. I think there is a possibility that the reason why they went so far down in their vital capacity during speaking was that they were still trying to recover from the vital capacity manoeuvre. In all the studies that I have done, I cannot find anyone going down below 28 to 30 per cent of their vital capacity—and it would be very informative if, in fact, the graph showing the features I mentioned were continued. In other words, after the man had stopped speaking it would be interesting to see how long it would be before he would, in fact, resume a quiet breathing pattern—and that would depend on whether he had been lying above or below his normal resting lung volume.

Mead. I think that the diagram showing the vital capacity and then the speech was really a convenience. The vital capacity data were an irrelevance, in that the vital capacity manoeuvre was not performed immediately before the particular speech episode. The speech performance that you refer to was reading passages aloud—indeed, a very artificial circumstance.

I also showed you a slide [Fig. 5.6] in which Hixon examined various forms and sorts of speech. Of course, it is quite difficult to study normal speech—just as it is almost impossible to study normal respiration. In fact, I don't think normal respiration in man has ever been studied since the invention of recording devices for measuring it, although before that perhaps they used to study it. But Hixon did include a fair amount of just plain talking, and the slide I showed indicated this. His subjects breathed near their functional residual capacity or slightly above it, and they breathed in a different configuration from the relaxed state. They certainly did not encompass a wide range of volumes. These people were talking in a steady state, more or less, and had not performed any vital capacity manoeuvre immediately before or immediately after the speech episode.

Anthony. I had great difficulty in making up my mind whether these were, in fact, continuous graphs; and in the slide you mention it is very difficult to see if they are. But if you look at the diagram on p. 106 of *The Respiratory Muscles*, by Dr. Campbell and his colleagues,[1] there is no doubt that there the speech follows on a vital capacity manoeuvre.

Mead. In that instance it was not really speech, but phonation of a tone. The subjects were not talking: they were merely producing a vocal tone, initiated at high lung volume and repeated at low lung volume. That does not show a single thing about the control of the muscles involved in phonation, other than in the production of a single tone. That is all that is intended.

Another slide I showed [Fig. 5.11] was of data obtained on Dr. Donald Proctor while he was singing a song. For these songs he indeed did a vital capacity manoeuvre before he started to sing, and he did another vital capacity after he had completed the song. I think he is fussy enough about his singing not to have let the vital capacity manoeuvre much influence the way he went about it—but he can speak about that himself.

Proctor. There always was a period of time between the two. In other words, we would do the vital capacity manoeuvre and then there might be a minute or two following before the song began, so that the two events did not immediately follow one another. But unless we had a vital capacity for the record we had no way of knowing exactly where we were on the same record during the singing.

Lindquist. I would like to come back to the question of the importance of mechanoreceptors in the control of the laryngeal muscles. We think there are no muscle spindles in these muscles, nor any in the facial muscles of the cat. Dr. Mårtensson and I[2, 3] have made some experiments in cats which show that mechanoreceptors in the mucosa of the mouth may be of importance in the control of the facial muscles; and more recently, Leanderson and Persson (1972)[4] have shown that blockade of trigeminal afferents in man results in changes in the pattern of electromyographic activity in the facial muscles during the pronunciation of simple syllables.

Ruth Bowden. The facial muscles in man and rabbit *do* contain classical muscle spindles, as well as a rather interesting type of spiral ending.[5, 6] Cutting the facial nerve below the stylomastoid foramen and preventing regeneration

---

[1] Campbell, E. J. M., Agostoni, E., and Newsom Davis, J. (1970) *The Respiratory Muscles: Mechanics and Neural Control,* 2nd ed., London.

[2] Lindquist, C., and Mårtensson, A. (1969) The possible role of exteroceptors in the control of facial muscles, *Acta physiol. scand.,* Suppl. 330, p. 116.

[3] Lindquist, C., and Mårtensson, A. (1970) Mechanisms involved in the cat's blink reflex, *Acta physiol. scand.,* **80,** 149–59.

[4] Leanderson, R., and Persson, A. (1972) The effect of trigeminal nerve block on the articulatory EMG activity of facial muscles, *Acta otolaryngol.,* **74,** 271–8.

[5] Bowden, R. E. M., and Mahran, Z. Y. (1956) The functional significance of the pattern of innervation of the muscle quadratus labii superioris of the rabbit, cat and rat, *J. Anat. (Lond.),* **90,** 217–27.

[6] Kadanoff, D. (1956) Die sensiblen Nervenendigungen in der mimischen Muskulatur des Menschen, *Z. mikr.-anat. Forsch.,* **62,** 1–15.

has shown that fibres from the 5th nerve enter the facial muscles[1]. We have yet to find the actual motor denervated spindles—perhaps it may be the 7th nerve that supplies motor fibres to the spindles. The 7th nerve does not contain any fibres of the calibre found supplying the annulospiral (primary) endings in the facial muscle spindles, whereas the 5th nerve does[1]; and if you combine both 5th and 7th nerve fibres in a fibre diameter distribution histogram, you get a classical bimodal distribution. So I am very interested to hear from Dr. Lindquist evidence suggesting that mucosal receptor endings play an important part in influencing the activity of the facial muscles.

Wyke. I just want to reinforce Dr. Bowden's comment about the little Stockholm misapprehension regarding the supposed absence of muscle spindles in the facial muscles, and add to it the statement that the idea that the intrinsic laryngeal muscles likewise do not contain stretch-sensitive receptors is also incorrect[2,3,4]. There will be a certain amount of evidence presented about this latter matter later in this Symposium [see pp. 382–6 and Communications 22 and 24] that should clear the air—at least I hope so.

Mårtensson. I know, of course, that we are going to hear much more about these things later on in the week. As I understand it, you must have very conclusive evidence for the presence of muscle spindles. What Dr. Bowden and Dr. Wyke have just said is very interesting, and we look forward to hearing more about the situation in the laryngeal muscles.

Malcolm. I was particularly interested in the chest volume distortion problems raised by Dr. Mead. If you consider the liver as producing mass and the chest wall elasticity, then the one can be balanced against the other in much the same way as in an electromagnetic analogue where you can shorten the lines of the magnetic field and raise the frequency, or shorten the lines of the electric field and lower the frequency—in the one case the cavity appears to be smaller, and in the other it appears to be larger. Another point is that I think the ribs may be acting as torsion bars, apart from being hinged on the costovertebral joints. Finally, I was a little worried about the term 'relaxation characteristic' because I find the term very confusing in the present context—perhaps Dr. Mead will explain?

Mead. In response to your last comment, I can accuse you perhaps of having dropped off during my Communication, because I didn't ever use the word 'characteristic'. I used the term 'relaxation configuration', just as I have done many times. With regard to the inertial impedence offered by the liver, everything that I talked about had to do with matters operating at frequencies below 5 Hz, and mostly well below 5 Hz. I think one can show that in the

---

1 Bowden, R. E. M., and Mahran, Z. Y. (1960) Experimental and histological studies of the extrapetrous portion of the facial nerve and its communications with the trigeminal nerve in the rabbit, *J. Anat. (Lond.)*, **94**, 375–86.

2 Abo-El-Enein, M. A., and Wyke, B. D. (1966) Myotatic reflex systems in the instrinsic muscles of the larynx, *J. Anat. (Lond.)*, **100**, 926–7.

3 Abo-El-Enein, M. A., and Wyke, B. D. (1966) Laryngeal myotatic reflexes, *Nature (Lond.)*, **209**, 682–6.

4 Wyke, B. D. (1973) Myotatic reflexogenic systems in the larynx, *Folia morphol. (Prague)*, **21**, 113–8.

respiratory system most of the inertia is (quite surprisingly) not in the liver but in the gas itself, because of the high velocities in the tracheobronchial tree. I know that sounds misleading, but that is where most of it is. I have no doubt that your model is appropriate to the fundamental frequencies of the voice, and perhaps that is a much more appropriate model to be concerned about. I don't think it applies to the aspect of mechanics that I was dealing with.

Martin. I wonder if I might steer the discussion in a rather different direction, as my own interest is in children's development. I think most of the work we have heard about—in fact, I suspect all the work described today—has been on adults. To me it is a matter of considerable clinical as well as theoretical importance to know more about what actually happens in children. For instance, the 3- or 4-year-old can phonate for only relatively short periods of time. When you consider the very young infant, the child of only a few days old, you find it has a respiratory cycle lasting perhaps 0.5 to one second; but even then such infants begin to introduce harsh respirations which, one suspects, are the beginning of phonation.

What I was wondering, in view of Dr. Mead's title 'Speech as Breathing', is whether what I have described might not almost be regarded in terms of breathing as speech. At what stage does this intermeshing of phonation and respiration occur, and when does respiratory modification occur as a result of phonation? I think this is of considerable linguistic significance. The length of the vocal utterance in children may be regulated perhaps not by psycholinguistic factors (as is always stated) but by much more mundane, but equally important, factors such as lung volume and the normal tidal air—a 3-month-old infant has perhaps only 40 ml, for instance. I would be grateful for information along these lines.

Proctor. I can only say that I do not know the answer to the question. We, too, are very much interested in the phonatory processes in infants and young children in connexion with hearing and speech problems as well as with normal development, and we do have some people working on it in our Laboratory—but I can't answer any of your questions as yet. I wish I could.

Mead. I want to clear up one point. I used the title 'Speech as Breathing', but I think it may be somewhat misleading. I like to confine the term 'breathing' to the natural process by which the lungs are ventilated, and it does not include speech. Speech is something else. It is not breathing: although incidentally for the act of speech the lungs are ventilated, I do not regard speech and breathing as being the same process at all, and I don't think any of us here does: I am sorry if my title implied that.

Plum. In so far as one can make relevant deductions from neurological observations in the adult patient, it is worth noting that brain stem lesions at the level of the upper pons that interrupt the corticobulbar pathway going to the lower cranial nerves completely interrupt the ability of the subject in any way to modify the respiratory act for any purpose, except for those effects induced by segmental reflexes coming from the hind brain (such as coughing, sneezing and so on). Such subjects are unable to cough and are unable to make any sound whatsoever during the period of obligate apnoea in periodic breathing; and can utilize the larynx only during the hyperpnoeic phase of obligate Cheyne-Stokes breathing. This would indicate that whatever these pathways may be, they originate higher than the pontine-midbrain junction—but more than that I cannot say.

Anthony. I can't talk about children's organization of air flow, although I agree this is something that we do need to know about. But I can talk about the rates at which people use up air when speaking, and there seems to be a considerable difference between how one organizes one's airflow in breathing and in spontaneous speech. What does emerge is that for each individual the rate at which one uses air while speaking is amazingly constant. If you take this in terms of time (that is, the air put out as a function of time) or as a function of the number of syllables uttered, there seems to be something in speech itself that determines that the rate in both sets of circumstances is constant.

6 Glottic Aerodynamics and Phonation

G. A. CAVAGNA AND E. M. CAMPORESI

Istituto di Fisiologia Umana
dell' Università di Milano and
Centro di Studio per la Fisiologia del
Lavoro Muscolare del C.N.R., Milan, Italy

INTRODUCTION

The mechanical behaviour of the glottis generator can be studied by analysing the conversion of the mechanical work done by the respiratory apparatus into sound energy. The fraction of the respiratory work, done in unit time, which may possibly be converted is given by the difference of pressure across the glottis [P] times the average flow of air through it [$\dot{V}$]. When P increases the sound intensity also increases; and a consistent linear relationship was found between log P and the number of decibels corresponding to the level of the sound emitted [db] (defined as 10 log (I_s/I_{so} , where $I_{so} = 10^{-16}$ watt/cm^2) (van den Berg, 1956; Margaria and Cavagna, 1959; Isshiki, 1964).

A similar relationship between db and the average air flow through the glottis cannot be observed in man: unlike the vibrating lamina of an accordion or a hooter, in which log $\dot{V}$ bears the expected linear relationship to db (Cavagna and Margaria, 1968), the data obtained in man are very scattered and do not allow us to determine a consistent relationship between $\dot{V}$ and db (Vogelsänger, 1954; van den Berg, 1956; Margaria and Cavagna, 1959; Isshiki, 1964; Bouhuys, Proctor and Mead, 1966; Rubin, LeCover and Vennard, 1967; Yanagihara and Koike, 1967). The great variability of the air flow through the glottis makes it difficult to determine the relationship between sound intensity and respiratory work.

However, by recording $\dot{V}$ and db simultaneously on an XY recorder [Fig. 6.1], it was noticed that the subject was not able to hold a tone of a given intensity with a flow smaller than a well-defined value, and that this 'minimum air flow' [$\dot{V}_{mn}$] was consistently related to sound intensity (Cavagna and Margaria, 1968). $\dot{V}_{mn}$ possibly indicates the significance of the flow necessary to drive the vocal folds without coarse losses. Flow rates greater than $\dot{V}_{mn}$ would indicate that part of the stream of air provided by the respiratory apparatus is being shunted at the glottis level, without participating in sound production. Therefore $\dot{V}_{mn}$ instead of $\dot{V}$ should be taken into consideration in order to study the function of the glottis generator. Log $\dot{V}_{mn}$ does not increase linearly with db as in the models (accordion and hooter)—except, possibly, in the range of high intensities [Fig. 6.2]. A line drawn through the high-loudness data only does not apply for the experimental data at lower intensity levels, for these are always appreciably above the line. At low sound levels, log $\dot{V}_{mn}$ also increases with increasing db in some cases, but at much lower rate; but in other cases an increase of the intensity is obtained with practically no change of $\dot{V}_{mn}$.

In order to explain this finding, the hypothesis was put forward that in the range within which $\dot{V}_{mn}$ is constant the loudness is increased by concentrating the rush of the air through the glottis at each cycle in a progressively shorter time, thus obtaining steeper and more effective rushes of air with about the same average air flow through the glottis (decreasing 'opening quotient' with increasing db). This mechanism of increasing the loudness of a tone was noticed by Fletcher

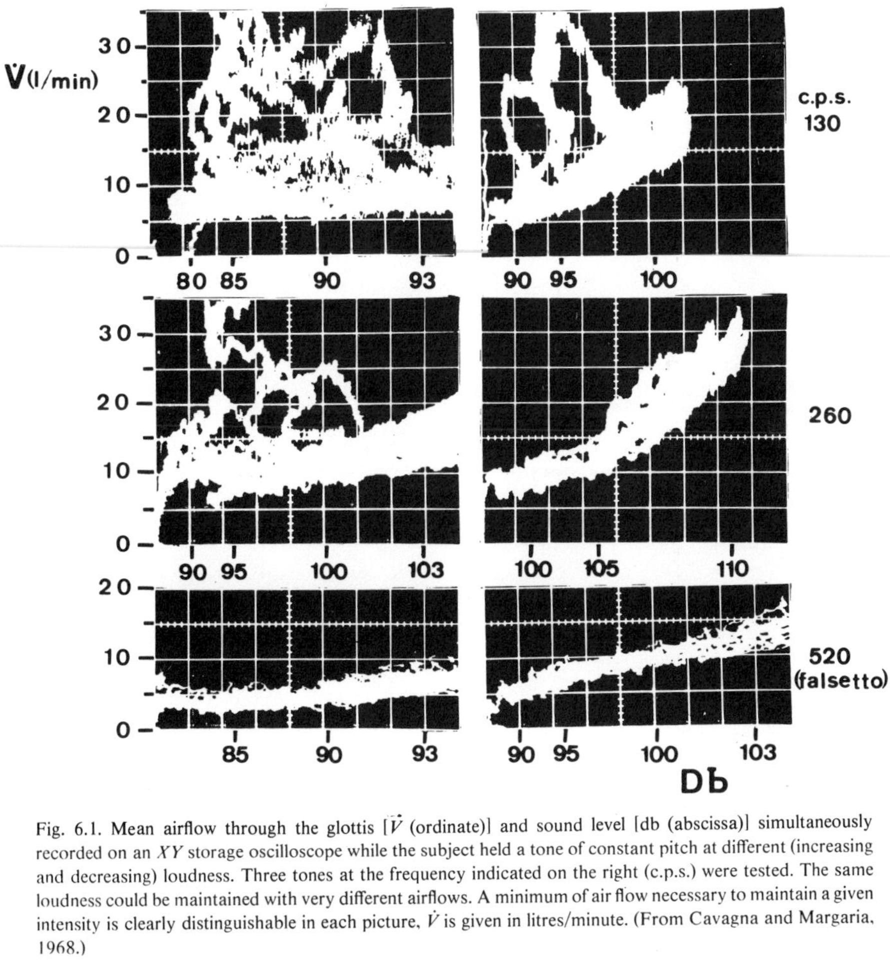

Fig. 6.1. Mean airflow through the glottis [$\dot{V}$ (ordinate)] and sound level [db (abscissa)] simultaneously recorded on an *XY* storage oscilloscope while the subject held a tone of constant pitch at different (increasing and decreasing) loudness. Three tones at the frequency indicated on the right (c.p.s.) were tested. The same loudness could be maintained with very different airflows. A minimum of air flow necessary to maintain a given intensity is clearly distinguishable in each picture. $\dot{V}$ is given in litres/minute. (From Cavagna and Margaria, 1968.)

(1950) in his analysis of high-speed motion pictures of the vibrating vocal folds (Farnsworth, 1940), and discussed by Flanagan (1958) and van den Berg (1956).

In the present experiments, high speed motion pictures of the vibrating vocal folds have been taken simultaneously with airflow and db measurements, in order: (1) to see how the shunt of air is attained when the flow through the glottis is greater than $\dot{V}_{mn}$; and (2) to determine the 'opening quotient' at different intensity levels when the airflow approaches $\dot{V}_{mn}$.

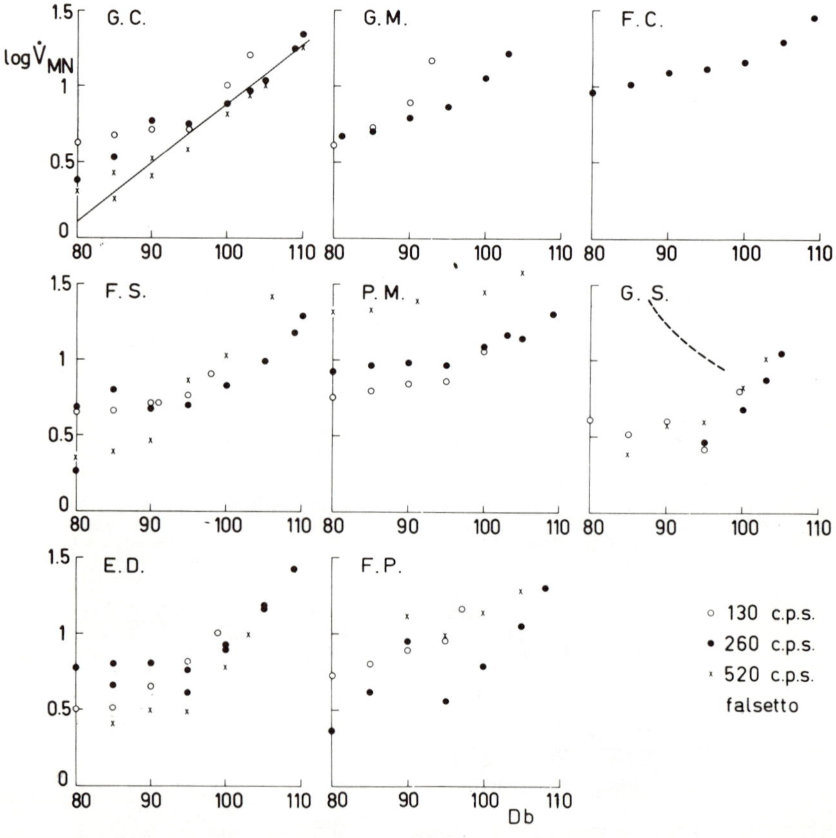

Fig. 6.2. The logarithm of the smallest airflow with which a given tone could be sustained |log $\dot{V}_{mn}$|, determined from tracings such as those in Fig. 6.1, is plotted as a function of sound level |db| for eight subjects. The broken line in the tracing G.S. indicates that low level tones were usually attained with a coarse waste of air flow; also in this subject, however, $\dot{V}_{mn}$ values are similar to those of the other subjects. The straight line in the tracing G.C. was calculated for the six closed circles (260 c.p.s.) above 95 decibels: the experimental data obtained below this sound level fall above this line, except possibly for the *falsetto* voice. A similar trend can be observed for the other subjects. F.P. is a trained singer. $\dot{V}_{mn}$ is given in litres/minute and P in cm H_2O. (Modified from Cavagna and Margaria, 1968.)

METHODS

The experiments were made on the senior author (G.C.), who has also served as a subject in previous work (Cavagna and Margaria, 1965, 1968). A schema of the experimental set-up is given in Fig. 6.3. The subject sat in a body plethysmograph [P], which was rigid and airtight around the neck: air was allowed to flow in and out of the plethysmograph during the respiratory movements through a hole (40 mm in diameter) covered with Monel gauze. The resistance offered by the gauze made it possible to measure the air flow [$\dot{V}$] from the changes of pressure inside the plethysmograph. The relationships between air flow and pressure changes were determined up to 30 l/min and found to be linear.

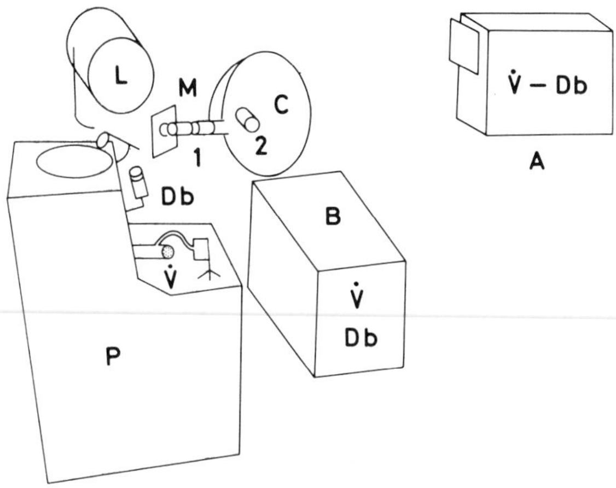

Fig. 6.3. Schematic representation of the set-up used during the experiments. The subject (not represented in the figure) sat in the body plethysmograph [P]. For other explanations see text.

The relatively great volume of air inside the plethysmograph introduced a time lag in the measurement of the air flow. A time constant of about the same order of magnitude also characterized the recording of the sound pressure level [db] by means of a phonometer whose characteristics are described elsewhere (Cavagna and Margaria, 1965). The slowness of the recording systems did not affect the results, since all measurements were made at steady state during the maintenance of constant tones.

The signals corresponding to the flow, $\dot{V}$, and to the sound level, db, were conveyed on the ordinate and on the abscissa respectively of an XY storage oscilloscope [A] which could be seen by the subject during phonation, thus allowing voluntary regulation of the flow through the glottis during the maintenance of the tone. $\dot{V}$ and db were also displayed on a second oscilloscope [B] whose spots were recorded on the film, through a second lens of the camera [C], simultaneously with the pictures of the vocal folds [see Figs. 6.4 and 6.5].

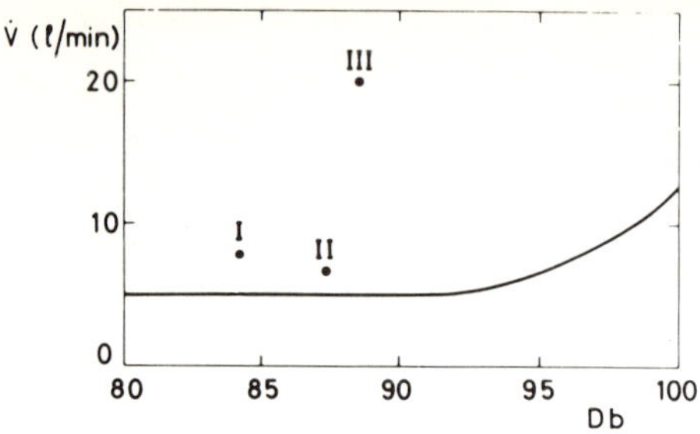

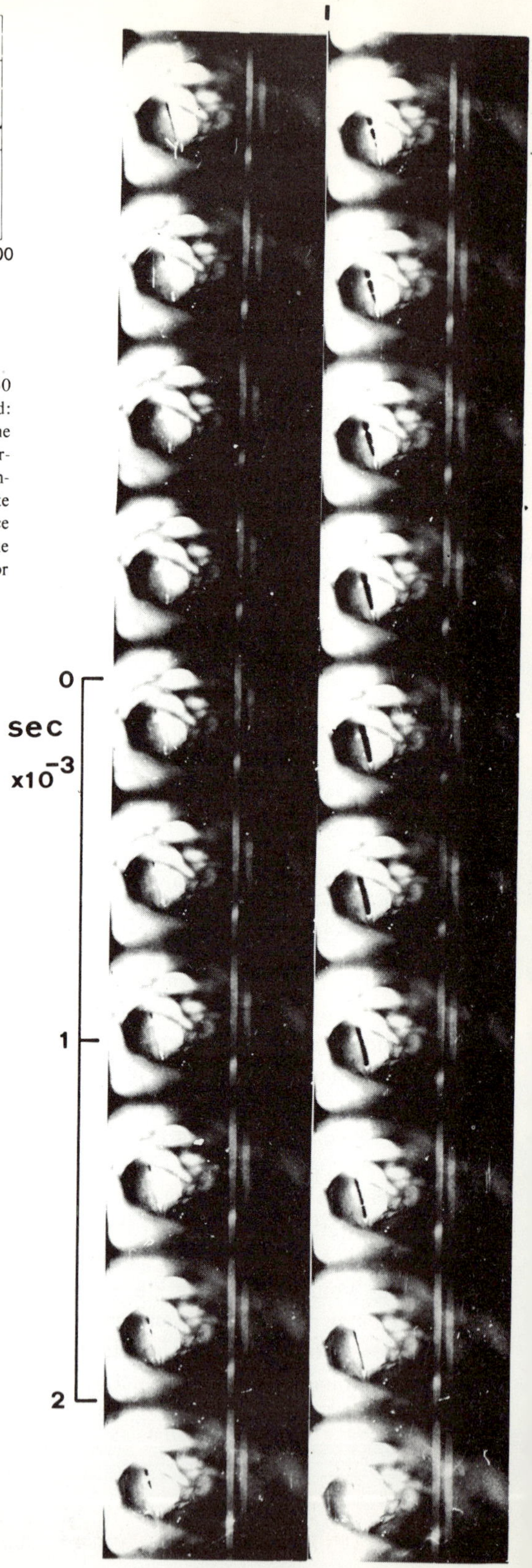

Fig. 6.4. Film sequence of the vocal folds vibrating at 130 Hz taken at 2400 frames/sec. Three cycles are illustrated: the flow through the glottis [$\dot{V}$] and the intensity of the sound emitted [db] during each cycle are given by the corresponding points in the inserted $\dot{V}$–*db* diagram. The continuous line in the diagram indicates the approximate trend of the minimal air flow for subject G.C. The trace on the films indicates the values of $\dot{V}$ and *db*. When the flow is appreciably greater than $\dot{V}_{min}$(III), the posterior portion of the glottis never closes during the cycle.

II III

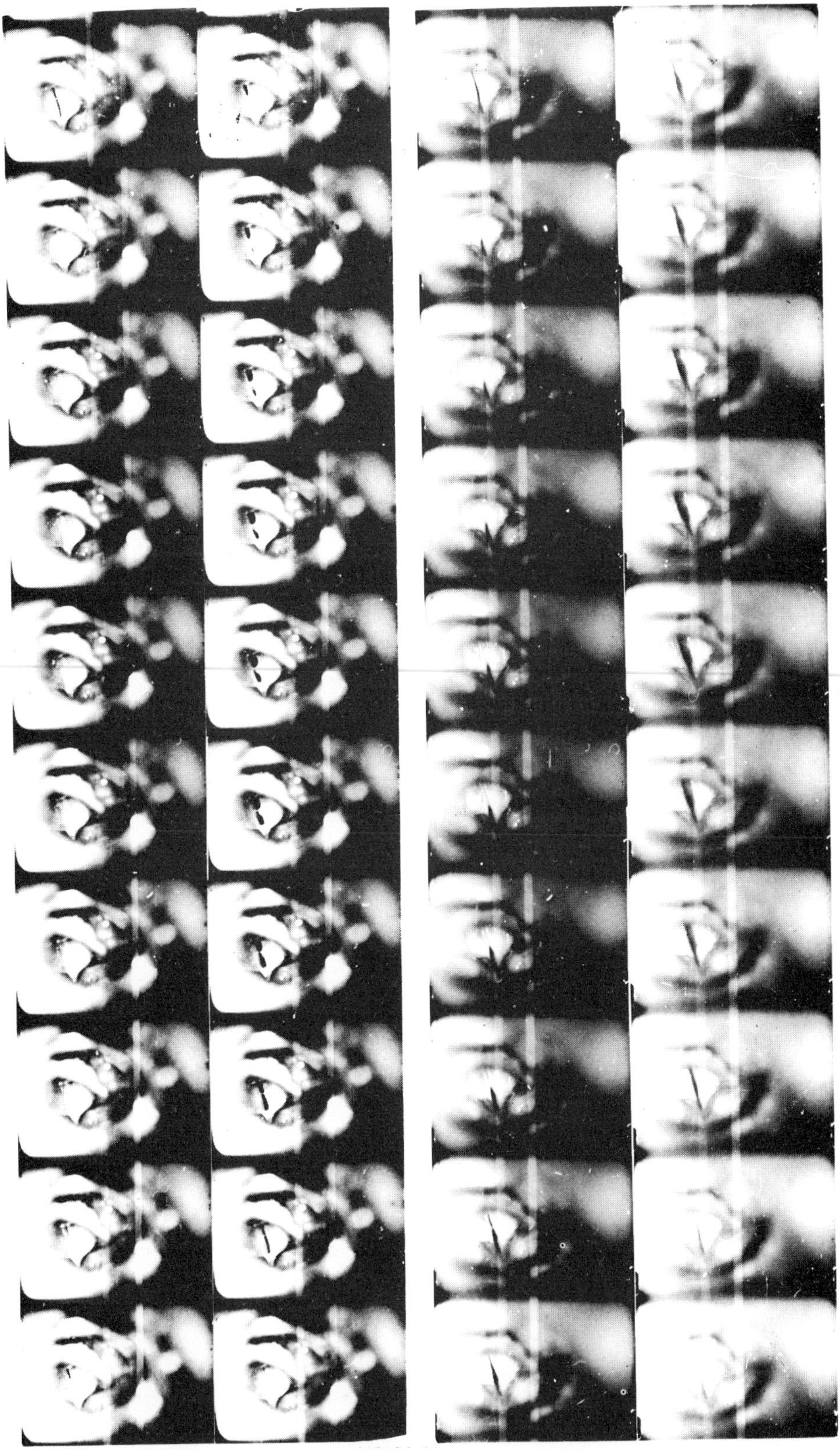

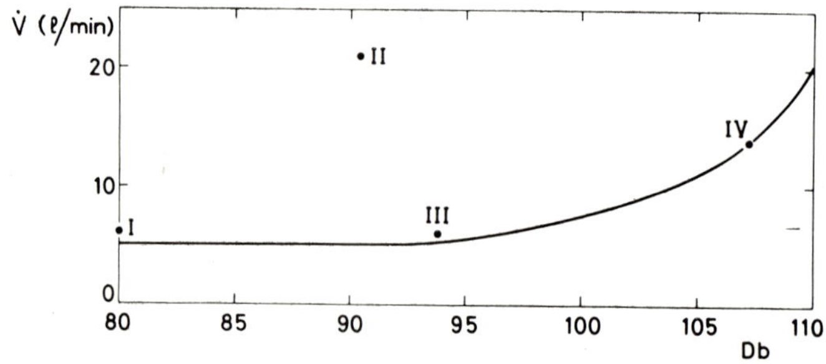

Fig. 6.5. Film sequences of four cycles at 260 Hz taken at about 2300 frames/sec. During cycle II the posterior portion of the glottis never closes, and the flow of air is much greater than $\dot{V}_{mn}$ The 'opening quotient' in cycles III and IV does not differ appreciably, in spite of the great difference in sound intensity. Other data as in Fig. 6.4.

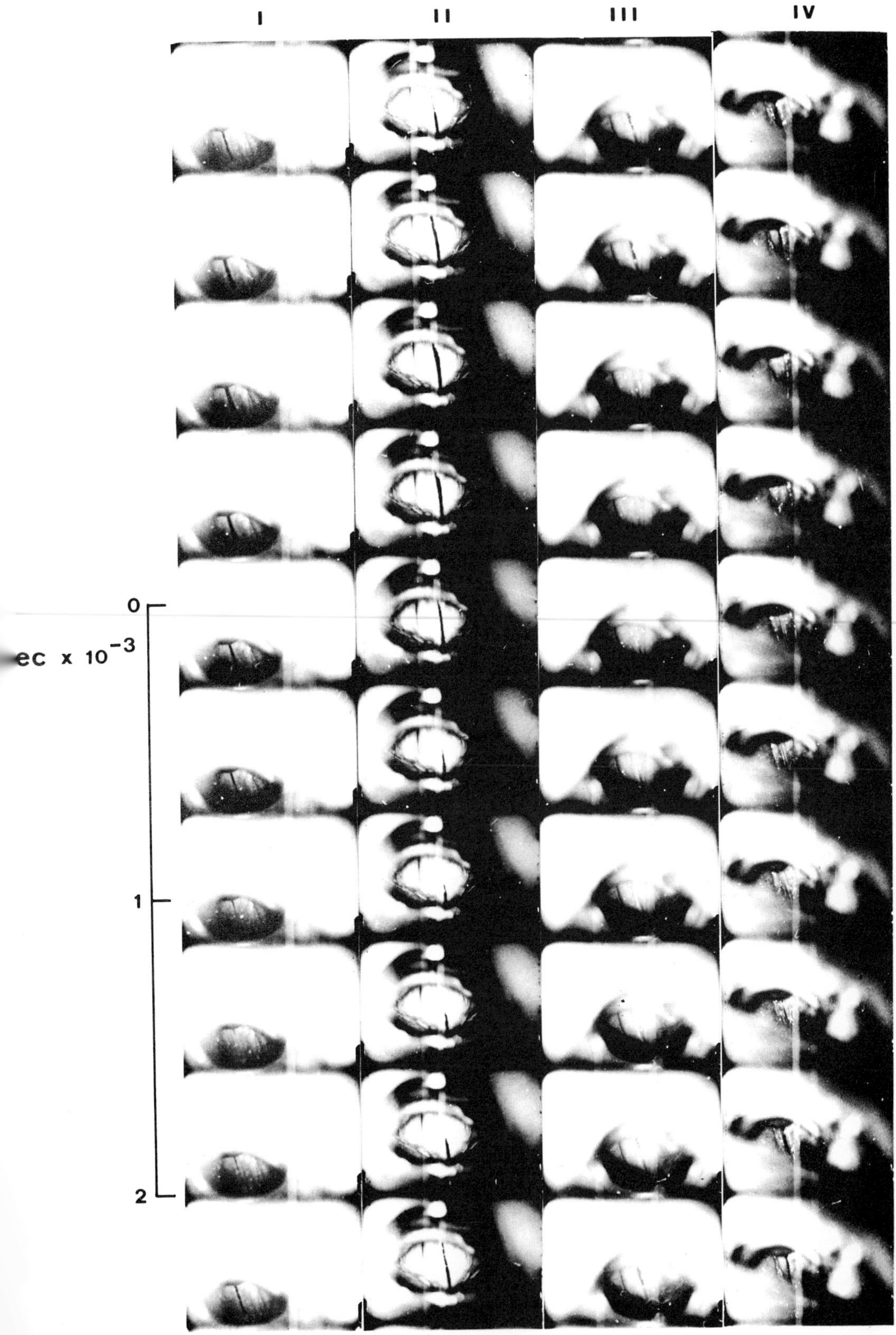

I
II
III
IV
0
ec x 10^{-3}
1
2

Calibrations were made on the film at the end of each experiment. By means of the high-speed camera the film was accelerated up to about 4000 frames/sec: samples at 2300–2400 frames/sec are given in Figs. 6.4 and 6.5. Timing marks were recorded on the film by a lamp within the camera, fed by a 50 Hz pulse generator.

Illumination of the vocal folds was supplied by a 7500 watt lamp [L]. In order to prevent 'sunburning' of the vocal folds, the red and the infra-red rays were in part reflected by a special glass inserted between the lamp and a holed mirror [M] placed in front of the camera: this reflected the light onto a small circular mirror placed in the pharynx of the subject—as is usually done in indirect laryngoscopy. The vocal folds were more clearly visible when pronouncing the vowel /*e*/ as in 'left'; and this phoneme was adopted throughout the experiments. Sustained tones of constant pitch (130 and 260 Hz, chest voice) were emitted. The intensity and the air flow through the glottis were chosen by the subject, according to the experimental procedure, by looking at the XY oscilloscope. Twenty films were taken: all were consistent with the results described below.

RESULTS

ACHIEVEMENT OF AIR SHUNTING THROUGH THE VOCAL FOLDS

In Figs. 6.4 and 6.5 the vocal folds are illustrated vibrating at a frequency of 130 and 260 Hz respectively, when the flow of air through the glottis was voluntarily maintained at a value (a) appreciably greater than, and (b) near to $\dot{V}_{mn}$. The values of air flow and of the loudness of the tones, recorded on the films, allowed us to define the points indicated on the $\dot{V}$-db diagrams represented in the Figures.

From the film sequences III in Fig. 6.4, and II in Fig. 6.5 it appears that the shunt of air is achieved by pulling apart the posterior portion of the vocal folds: in fact this section of the glottis never closed during the cycle, and its small changes of area, clearly visible on the films, are possibly not very effective in modulating the stream of air, therefore, in producing the sound. Even when the shunt is great, however, the anterior parts of the vocal folds come into contact, thus completely interrupting the flow of air through this section of the glottis.

As shown in Figs. 6.4 (I, II) and 6.5 (III, IV) the tone can be produced without coarse losses, with an air flow approaching $\dot{V}_{mn}$, by adduction of the vocal processes of the arytenoids. In this case, at middle and high intensity levels, the glottis seems to close completely during part of the cycle [Figs. 6.4, II and 6.5, III and IV].

At low intensity levels it is difficult to achieve complete adduction of the posterior portion of the vocal folds: these tones are usually emitted with a flow of air greater than $\dot{V}_{mn}$, and the intercartilaginous part of the glottis remains slightly open throughout the cycle. When, with some training, complete adduction of the posterior part of the vocal folds is attained even at such low intensities, the anterior part of the glottis may remain open throughout the cycle [Figs. 6.4, I and 6.5, I]. In fact it turned out that it is easier to close the glottis completely at high and also middle intensity levels (even with a flow appreciably greater than $\dot{V}_{mn}$) than at very low intensity levels with a flow near to $\dot{V}_{mn}$.

CHANGE OF THE OPENING QUOTIENT WITH VARIATIONS IN SOUND INTENSITY

The opening quotient (O.Q.) was determined by dividing the number of frames in which the glottis appeared open in some part of its length by the total number of frames composing one cycle. Measurements were generally made in the instances in which the air flow was kept voluntarily near to $\dot{V}_{mn}$. Since it was not usually possible to get pictures of the vocal folds throughout their whole length [see Figs. 6.4 and 6.5], the value of the O.Q. determined in this way may be somewhat smaller than the real one.

In Fig. 6.6 the opening quotient is given as a function of the loudness of the tone emitted for a frequency of 130 (open circles) and 260 Hz (closed circles) respectively. At 130 Hz it was not possible to determine the O.Q. beyond 91 db, because at higher intensities the vocal trace above the vocal folds was narrowed, thus hiding them. From Fig. 6.6 it appears that the O.Q. is higher at low than at high intensity levels; and this result is in agreement with the finding of Fletcher (1950), who reported that an increase in intensity was accompanied by an increase in the closed time of the vocal cycle. However, it also appears that a reduction of the O.Q. takes place only in the range of intensities in which $\dot{V}_{mn}$ is almost constant, From about 95 to 100 db, for the 260 Hz tone, the O.Q. is kept practically unchanged in spite of an appreciable increase in the loudness of the tone emitted. Furthermore, for the frequency of 130 Hz the opening quotient is decreasing in the range where $\dot{V}_{mn}$ is constant.

In conclusion, it seems that when the vocal folds are pressed together in order to vibrate without coarse losses: (1) in the range of high intensities the loudness of the tone emitted is increased without any change of the O.Q. simply by increasing the transglottic pressure; whereas (2) a progressive increase of the O.Q. up to 1 is observed down to low intensity levels. In addition, (3) when the flow through the glottis is greater than $\dot{V}_{mn}$, and/or at very low intensity levels, the glottis may stay open throughout the cycle. In this condition, the O.Q. of the vibrating part of the glottis may be calculated by considering the glottis 'open' when its area is greater than that of the shunt. This has been done in two instances at low intensity levels for a tone of 260 Hz (points between brackets in

Fig. 6.6). Even by making this assumption, however, the O.Q. is difficult to determine in these conditions since the shunt of air is not constant and the area continuously changes during the cycle.

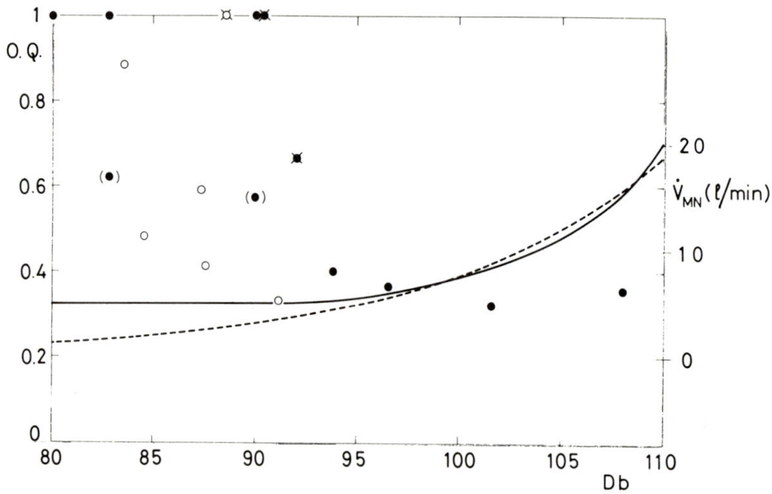

Fig. 6.6. The 'opening quotient' [O.Q., on the ordinate] is given as a function of the intensity of the sound emitted in db [abscissa] for the two frequencies of 130 (open circles) and 260 Hz (closed circles). The points enclosed in brackets refer to a condition in which the posterior portion of the glottis was always open during the cycle, and the O.Q. was calculated by considering only the anterior part of the glottis: the two points at the same db values and O.Q. = 1 were calculated, as were all the others, considering the glottis as a whole. The two curves refer to the right-hand ordinate: the broken one corresponds to the straight line in Fig. 6.2, and indicates how the minimal air flow through the glottis would change were log $\dot{V}_{mn}$ linearly related to db both at high and low sound intensity levels: the continuous line indicates the approximate trend of the relationship between $\dot{V}_{mn}$ and db actually found in the present experiments. It appears that above 95 db the two lines do not differ appreciably, and that in this range the O.Q. is kept about constant. At lower sound levels the actual flow is appreciably greater than the theoretical one and this is accompanied by greater values of the O.Q. The O.Q. was usually calculated for tones in which the flow through the glottis did not exceed the minimal flow of more than 5 l/min: higher flow rates are indicated by a cross on the point.

DISCUSSION

During phonation it is necessary to match the requirements of sound production with those of respiration. To increase the efficiency of sound production, the mechanical work performed by the respiratory apparatus should be converted totally, without shunts, into sound energy; but if this were done, the tone sustained during expiration might last for relatively too long a time, thus disturbing the regular succession of the respiratory movements. A shunt of air through the vibrating vocal folds (i.e. a continuous flow not participating in sound production) decreases the efficiency of the glottic generator, but it allows speeding up of expiration as required. Possibly this is the physiological significance of the variable flow of air through the glottis for a given sound intensity, described in Fig. 6.1: this may also explain the discordant results obtained in the literature on measurements of flow rates during phonation. The minimal flow necessary to vibrate the vocal folds may be adopted spontaneously by trained singers (Vogelsänger, 1954)—but certainly was not by our untrained subjects, particularly at low sound intensity.

The hypothesis put forward in our previous paper (Cavagna and Margaria, 1968) that in the range in which $\dot{V}_{mn}$ is about constant an increase of intensity is accompanied by a decrease of the opening quotient is confirmed by the present results [Fig. 6.6]. It is generally accepted in the literature that the O.Q. decreases with increasing db (see, for example, Fletcher, 1950). Timcke, von Leden and Moore (1958) reported that '. . . . the O.Q. is inversely proportional to the intensity of the sound produced'; but this is not confirmed by the present experiments, which show that in the range of high intensities sound pressure is increased with a constant O.Q.

In the range of high intensities, in which the O.Q. is constant, log $\dot{V}_{mn}$ seems to increase linearly with db, as in the accordion and in the hooter [Fig. 6.2]. When the intensity is reduced (below about 95 db in Figs. 6.2 and 6.6) the O.Q. increases, and log $\dot{V}_{mn}$ decreases less steeply, remains constant or even rises. This finding suggests that low sound levels (i.e. low transglottic pressures) can be reached only by decreasing the force pulling the vocal folds together, thus decreasing the pressure [P] necessary to open the glottis, and increasing the O.Q.

CONCLUSION

A tone of a constant pitch and loudness can be produced with very different flow rates through the glottis, but it has been shown that a minimal air flow [$\dot{V}_{mn}$] necessary to maintain vibration is consistently related to the sound level. High-speed motion pictures of the vibrating vocal folds, taken with simultaneous recordings on the film of the flow through the glottis and of the sound level, indicate: (1) that flow rates appreciably greater than $\dot{V}_{mn}$ can be attained by shunting the air through the posterior portion of the glottis; and (2) that at low sound levels a shunt of air through the glottis can be present even when the flow is voluntarily maintained as low as possible, near or equal to $\dot{V}_{mn}$.

The 'opening quotient' at first decreases (from 1 to about 1/3) with increasing intensity of the sound, and then it remains about constant while the intensity is appreciably increased further.

ACKNOWLEDGEMENTS

This research was partly supported by the Istituto di Cinematografia Scientifica del Politecnico di Milano. The authors wish to thank Ing. A. Berbenni and his collaborators for their technical assistance.

REFERENCES

Bouhuys, A., Proctor, D. F., and Mead. J. (1966) Kinetic aspects of singing, *J. appl. Physiol.,* **21,** 483–96.

Cavagna, G. A., and Margaria, R. (1965) An analysis of the mechanics of phonation, *J. appl. Physiol.,* **20,** 301–7.

Cavagna, G. A., and Margaria, R. (1968) Airflow rates and efficiency changes during phonation, *Ann. N.Y. Acad. Sci.,* **155,** 152–64.

Farnsworth, D. W. (1940) High speed motion pictures of human vocal cords, *Bell Laboratories Rec.,* **18,** 203–8.

Flanagan, J. L. (1958) Some properties of the glottal sound source, *J. Speech Res.,* **1,** 99–116.

Fletcher, W. W. (1950) A study of internal laryngeal activity in relation to vocal intensity, Ph. D. Thesis, Northwestern University, Evanston, Ill., U.S.A.

Isshiki, N. (1964) Regulatory mechanism of voice intensity variation, *J. Speech Res.,* **7,** 17–29.

Margaria, R., and Cavagna, G. A. (1959) Il meccanismo della fonazione, *Boll. Soc. ital. Biol. sper.,* **35,** 2075–7.

Rubin, H. J., LeCover, M., and Vennard, W. (1967) Vocal intensity, subglottic pressure and airflow relationship in singers, *Folia phoniat. (Basel.),* **19,** 393–413.

Timicke, R., von Leden, H., and Moore, P. (1958) Laryngeal vibrations: measurements of the glottic wave, *Arch. Otolaryng.,* **68,** 1–19.

Van den Berg, J. (1956) Direct and indirect determination of the mean subglottic pressure, *Folia phoniat. (Basel.),* **8,** 1–24.

Vogelsänger, T. G. (1954) Experimentelle Prüfung der Stimmleistung beim Singen, *Folia phoniat. (Basel.),* **6,** 193–227.

Yanagihara, N., and Koike, Y. (1967). The regulation of sustained phonation, *Folia phoniat. (Basel.),* **19,** 1–18.

DISCUSSION ON COMMUNICATION [6]

Proctor. I would like to compliment Dr. Cavagna on a beautiful piece of work. I think, perhaps, that Dr. Brodnitz and the other laryngologists here will agree with me that his photographs showing the open posterior portion of the larynx provide a beautiful picture of how patients develop contact ulcers. The portion that is open is the cartilaginous portion of the vocal folds; and when that is open continually and the rest of the folds are vibrating there is a continual banging together of the cartilaginous portions of the folds exactly where contact ulcers begin. Do you agree with that, Dr. Brodnitz?

Brodnitz. It's a new idea.

Malcolm. I should also like to congratulate Dr. Cavagna on a beautiful piece of work. From purely theoretical concepts, I have also been of the opinion that there must be a shunt during phonation and that the obvious place for the shunt was at the back, in relation to the arytenoids. I strongly suspect that the shunt is adjustable separately from the anterior portion of the vocal folds. One final point, one should not exclude the possibility that the shunt is also oscillating, but at a considerably lower frequency than the anterior portions of the vocal folds.

Cavagna. When the shunt of air is through the posterior portion of the glottis it appears that the width of that part of the glottis is changing a little during phonation. Obviously such changes are mainly in the anterior part of the glottis, but the posterior portion also changes—although not at a lower frequency, but with a lesser amplitude. We do, in fact, have amplitude changes in the width of the posterior part of the glottis during the operation of the shunt.

Ladefoged. I also liked the ideas you expressed—largely, of course, because they relate to some of my own work, which I will be talking about later on [see Communication 18]. But perhaps I can drop the seed in now of an idea that I would suggest, of characterizing the magnitude of the shunt (as you express it), or the magnitude of the amount of airflow through the non-vibrating part of the glottis, in terms of the distance between the arytenoid cartilages. I think, in fact, this is variable: I can produce different types of phonation where sometimes the arytenoid cartilages are held very close together and the posterior part of the glottis is not vibrating at all, only the anterior part—and in other types you can get them in a more normal position when the whole glottis is vibrating, and then further apart still when only the anterior part is vibrating because the posterior parts of the vocal folds are too far apart to vibrate.

Anthony. I would like to mention that we have seen a female patient who had this as a permanent feature. This was a woman in whom the posterior part of the glottis did not close at all. She had a very husky voice, and her voice broke very low down in the vocal range.

Cavagna. I have the original films here, and it may help if I now show them to you. [*Editor's note:* Dr. Cavagna showed his films at this point, with the following accompanying commentary].

The part of the glottis through which the shunt of air is produced during the utterance of sustained tones is shown by projecting high-speed films of the vocal folds vibrating at 260 Hz and 130 Hz, with a flow of air through the glottis of about 20 litres per minute at average sound intensity levels—i.e. with a flow appreciably greater than the minimal one for that degree of loudness (5 litres per minute).

It appears from the film that during each cycle of vocal fold vibration the area of the posterior shunt does not remain constant, but changes with the same frequency as and apparently in phase with the opening of the anterior part of the glottis. For this reason, it is not possible to measure the opening quotient when the shunt is present by considering the flow changes through the anterior part of the glottis as being merely superimposed on a continuous flow of air through the posterior portion.

The films were made with simultaneous recordings of air flow and sound level, and show that the shunt of air through the glottis is present not only in the so-called 'breathy voice', but also sometimes with a flow of air through the glottis near or equal to the minimal flow—particularly at a low intensity level.

In this condition the voice is not 'breathy'; i.e. the turbulence of the air through the glottic shunt is not noisy enough to be perceived. This explains why different flow rates through the glottis at the same sound intensity level, and (at least subjectively) with the same quality of voice during emission of sustained tones, have been reported in the literature. These observations indicate that when searching for a relationship between air flow rates through the glottis (or opening quotient changes) and sound intensity levels, it is necessary to look at a simultaneous recording of the flow in order to be sure that phonation is actually sustained with the minimal air flow through the glottis necessary to maintain vibration of the vocal folds.

Other films taken while singing tones of 130 Hz and 260 Hz at low intensity levels (with an airflow approaching the minimal value and the posterior portion of the glottis apparently closed during the cycle) show that the anterior part of the glottis never closes completely, even if it undergoes the greatest area changes during the vocal cycle. These pictures seem to indicate that at low intensity levels the force pressing the vocal folds together along their whole length decreases, in spite of the fact that the tone being emitted is voluntarily sustained with minimal air flow through the glottis.

Stevens. With regard to the changing opening quotients at lower sound pressures, a possible explanation could be that the vocal folds normally might be in a slightly open position without the application of any subglottic pressure: in other words, the normal static position would be slightly open. Then if you have just a small pressure across the vocal folds their amplitude of vibration will clearly be much smaller and therefore they will touch for only a small percentage of the time—or, perhaps, no percentage of the time.

Cavagna. Dr. Isshiki,[1] referring to the experiments of Furukawa,[2] has supported the idea that the opening quotient is not influenced by subglottic pressure if the closing force and tension of the vocal folds is constant. This is the conclusion from experiments on extirpated larynges. It may be that at high sound levels, when the opening quotient is constant, the force holding the vocal folds together is also constant and equal to its maximal value. When you want to decrease the sound intensity you need to develop less subglottic pressure [P] because of the linear relationship between log [P] and decibel output; and this reduced subglottic pressure may not be able to open the glottis if the force pressing the vocal folds together is kept unchanged. In order to open the glottis it is necessary to decrease the closing force, and this may lead to an increase in the opening quotient. What do you think about this explanation?

Stevens. It is reasonable to suppose that when you decrease the subglottic pressure you perhaps need to decrease the closing force, or spread the vocal folds slightly, in order to maintain vibration—otherwise, as you say, the vocal folds cannot be pushed apart.

Another couple of comments. When you open the posterior portion of the glottis, turbulence noise is generated and therefore is a source of 'breathiness' in

---

[1] Isshiki, N. (1969) Remarks on the mechanism for vocal intensity variation, *J. Speech Res.*, **12,** 665–72.

[2] Furukawa, M. (1967) Studies on the mechanism of phonation by means of extirpated larynx, *Pract. Otol., Kyoto*, **60,** 63–99.

the voice. Another point is that your data for the *falsetto* voice did not follow the same curve as the ordinary voice and had a curve that continued down all the way. Presumably, then, during *falsetto* the vocal folds do not come in contact during the vibration cycle at any point.

Cavagna. As to the first point: it is very easy to hear the turbulence when one utters something like this—*AAAAAAAAH* [Dr. Cavagna demonstrates—*Ed.*]—but sometimes it is not possible to understand from what you hear if there is a shunt or not. When you are near the minimal air flow (for example, when you are within 5 litres per minute more than the minimal air flow) you can have a shunt without producing any subjective sensation of turbulence.

Fourcin. There is one other factor that is related to the opening quotient ratio which probably has as great, if not a greater contribution to make towards the intensity of the sound that is emitted. This is the rate at which the vocal folds actually close when interrupting the air stream. Since the rate at which the vocal folds close is generally going to alter when the vocal quotient is changed, then the effect of rapidity of closure is in a sense going to be masked—but it may well be the dominant factor in determining the intensity of the sound uttered. The rate at which vocal fold closure occurs, rather than the length of time during which the folds are actually closed, is quite possibly the most important factor in determining the change of intensity as other factors (volume of flow, subglottic pressure) vary.

Wyke. Could I enter a caveat about the quotation from Isshiki's paper (*loc. cit.*) that has just been introduced by Dr. Cavagna, as I suggest that it might mislead us rather seriously unless it be emphasized that that work was done on isolated cadaver larynges. Certainly the observations quoted are absolutely true of such experiments, whose history goes back well beyond Dr. Isshiki: for example, some of you will recall that Dr. van den Berg[1,2] did many experiments to show that the size of the aperture of the glottis is not necessarily a direct function of subglottic pressure in the isolated larynx. But I think there is now good experimental evidence, part of it from Dr. Kirchner's Laboratory in Yale[3] and part from ours[4,5,6] (and other places as well[7,8,9,10]), to suggest that in the living, innervated larynx there is a reflex relationship during phonation between the aperture separating the vocal folds and the actual subglottic pressure obtaining

1 van den Berg, J. (1964) Some physical aspects of voice production, in *Research Potentials in Voice Physiology*, ed. Brewer, D. W., pp. 63–101, New York, State University of New York.

2 van den Berg, J. (1968) Sound production in isolated human larynges, *Ann. N.Y. Acad. Sci.*, **155**, 118–27.

3 Kirchner, J. A., and Suzuki, M. (1968) Laryngeal reflexes and voice production, *Ann. N.Y. Acad. Sci.*, **155**, 98–109.

4 Wyke, B. D. (1967) Recent advances in the neurology of phonation: phonatory reflex mechanisms in the larynx, *Brit. J. Disor. Commun.*, **2**, 2–14.

5 Wyke, B. D. (1970) Neurological mechanisms in stammering: an hypothesis, *Brit. J. Disor. Commun.*, **5**, 6–15.

6 Wyke, B. D. (1971) The neurology of stammering, *J. psychosom. Res.*, **15**, 423–32.

7 Lieberman, P. (1968) Vocal cord motion in man, *Ann. N.Y. Acad. Sci.*, **155**, 28–38.

8 Ladefoged, P. (1968) Linguistic aspects of respiratory phenomena, *Ann. N.Y. Acad. Sci.*, **155**, 141–50.

9 Dunker, E. (1968) The central control of laryngeal function, *Ann. N.Y. Acad. Sci.*, **155**, 112–21.

10 Koyama, T., Kawasaki, M., and Ogura, J. H. (1969) Mechanics of voice production. I. Regulation of vocal intensity, *Laryngoscope*, **79**, 337–54.

at the time. I therefore think that we should be careful about extrapolating from studies made on isolated larynges to what actually happened in the beautiful pictures of his own glottis shown to us by Dr. Cavagna.

Anthony. I'd like to support that. I have done some work with isolated larynges, and can state that it is a direct consequence of the procedure one uses with such larynges that the arytenoids are closed together.

Stevens. There has been a recent theory (which Dr. Klatt will mention in his paper later—see Communication 17) with regard to the vibration of the vocal folds; namely, the 'two-mass model' of the vocal folds. This theory predicts that as you decrease the pressure across the vocal folds, the optimum opening at which you can most easily get vibrations increases slightly.

Brodnitz. Dr. Cavagna, in what sense do you use the term *falsetto*—because you know there is a lot of confusion about it. I suppose you use it in the way singers use it, singers who have to use *falsetto* voice like that required in Verdi's 'Falstaff'. But some others now use *falsetto* for what we used to call 'head register'. Clearly there is a different mechanism of voice production in each case—which one is yours?

Cavagna. I don't know, but this is what I mean [Dr. Cavagna sings a note at 520 Hz in the manner referred to—*Ed.*]. I don't know if my vocal folds are operating in *falsetto* or not there.

Brodnitz. That's a high head tone.

Malcolm. From a rheologic point of view the vocal folds are mostly elastic at the front and mass at the back: therefore, it is resonable to expect the back to produce the lower notes and the front the high notes in the presence of the shunt you described. Regarding the intensity of the note (without considering the total energy output), the effect of raising the energy is to produce pulses; and shortening the pulse would certainly raise the energy level. One last matter—this question of growling—if the vocal folds oscillate in two modes and the shunt oscillates at a low frequency, it would act as a chopper (similar to the chopper in photo-electric devices).

SECTION II

Mechanisms Influencing Ventilatory Behaviour

CHAIRMAN

Professor E. J. M. Campbell

Hamilton, Ontario, Canada

7 Control of Depth and Rate of Respiratory Movements

C. VON EULER

Nobel Institute for Neurophysiology,
Karolinska Institutet, Stockholm, Sweden

Neither ablation experiments nor recording from respiratory neurones in the medulla and pons have so far given much understanding of how the demand for ventilation is converted into rhythmic breathing, or how the interrelation between tidal volume and respiratory frequency is chosen and controlled.

Analyses of the relationships between ventilation [V_E], tidal volume [V_T] and rate of breathing [f] in cats under various conditions with and without intact vagus nerves (von Euler, Herrero and Wexler, 1970) have revealed that the increase in rate normally seen in addition to an increase in tidal volume in response to an increased chemical drive is largely dependent on vagal mechanisms governed by the inflation receptors in the lungs (Scott, 1908; Naesland *et alii,* 1966; Richardson and Widdicombe, 1969). In the absence of any volume information feed-back from the lungs, the rate of breathing depends on the internal rate setting of the bulbopontine mechanisms which are modulated by temperature and other influences from more rostral central nervous structures. The rate of increase of inspiratory activity, and thus of volume expansion of the lungs in inspiration, was found to depend on the chemical drive but to be very little influenced by volume information from the lungs or by the setting of the internal bulbopontine frequency-determining mechanisms. Tidal volume thus seems to be largely determined by two factors: (1) the rate of volume increase; and (2) those parts of the frequency-governing mechanisms which determine inspiratory duration.

The internal bulbopontine 'pacemaker' alone seems to be the dominating frequency governor in eupnoeic breathing in man, and in thermal polypnoea and panting in animals. The vagal mechanism, on the other hand, seems to carry the main responsibility for the increase in rate of breathing in response to CO_2.

In further studies, Clark and von Euler (1970, 1972) have analysed the sequence of events in a breathing cycle in terms of inspiratory time, expiratory time and volume. Experimental data from cats (under pentobarbitone anaesthesia) and man were obtained from rebreathing trials, and from cats also by measuring the effects of brief inflations applied at various times during the respiratory cycle. The data from breath-by-breath measurements were

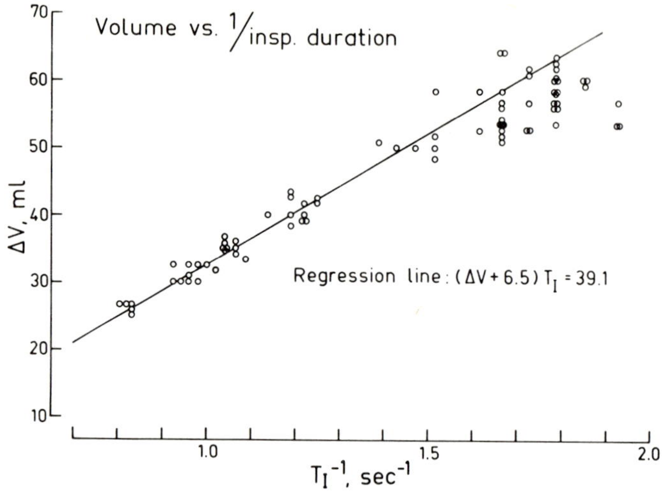

A

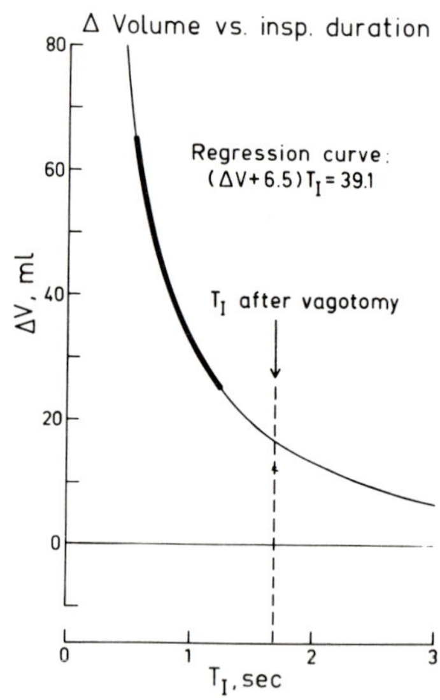

B

Fig. 7.1. From cat under pentobarbitone anaesthesia, showing [A] the approximately linear relationship between volume (ΔV) and the reciprocal of inspiratory duriation ($1/T_I$), and [B] the corresponding (hyperbolic) relationship between ΔV and T_I. (Clark and von Euler, unpublished data.)

analysed by digital computer (IBM 1800), and gave the following main results:

Both in rebreathing experiments and with artificially imposed inflations the inspiratory duration, T_I, was found to be dependent upon tidal volume, V_T. With both types of experiment this dependence could be closely approximated by a relationship of the form $(V_T - V_0)\,T_I = c$, where V_0 and c are constants, and V_0 can assume either positive or negative values. This approximately hyperbolic relationship between V_T and T_I (and linear relationship between V_T and $1/T_I$; see Fig. 7.1 A and B) can be regarded as the time course of the volume threshold for termination of inspiration which, thus, gets lower with time from the onset of the inspiratory phase. The time course of the 'Hering-Breuer' threshold curve was dependent on intact vagus nerves but seemed to be independent of the CO_2 level or inputs from muscle receptors. After vagotomy, the inspiratory duration remained essentially constant and did not change

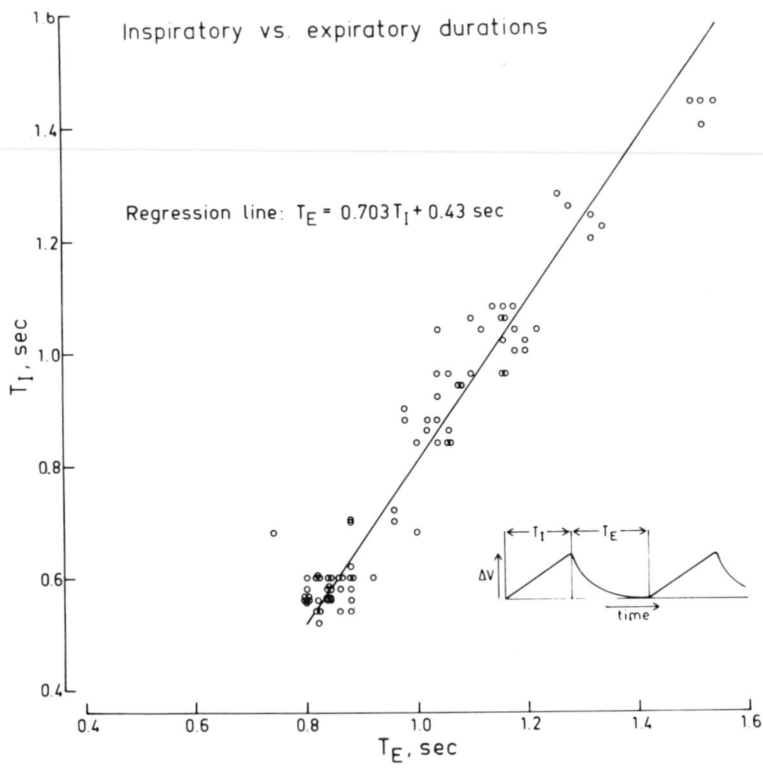

Fig. 7.2. From cat under pentobarbitone anaesthesia, showing the relationship between inspiratory duration (T_I) and expiratory duration (T_E). (Clark and von Euler, unpublished data.)

when tidal volume was altered either by artificial inflation or by the increased respiratory drive due to accumulation of CO_2 during rebreathing.

In man, during rebreathing, the relation between volume and inspiratory duration showed two different characteristics. (1) At tidal volumes up to about twice eupnoeic values, inspiratory duration did not change as tidal volume increased in response to increased PCO_2. At this volume range, designated *range 1*, the relationship was thus similar to that of the vagotomized cat. (2) As tidal volume increased above *range 1* a second range, designated *range 2*, was observed where inspiratory duration showed strong dependence on tidal volume in the same manner as in the cat with intact vagus nerves. The differences between cat and man appeared to be largely quantitative rather than qualitative in nature, as illustrated in Fig. 7.2. Vagotomy in cats abolished the *range 2* relationship, leaving only a *range 1* characteristic—i.e. a fairly constant value of T_I for all tidal volumes. In man, a similar effect has been reported (Guz *et alii*, 1966, 1970), where breathing rate no longer increased with increasing ventilation when the vagi were blocked with a local anaesthetic agent. The duration of the inspiratory phase of a breath thus appeared to be governed by two distinct mechanisms—i.e. with and without volume feed-back.

In both cat and man, expiratory duration was dependent on inspiratory duration, usually with a linear relationship which could be approximated by the expression $T_E = a \cdot T_I + b$, where a and b are constants [see Fig. 7.3].

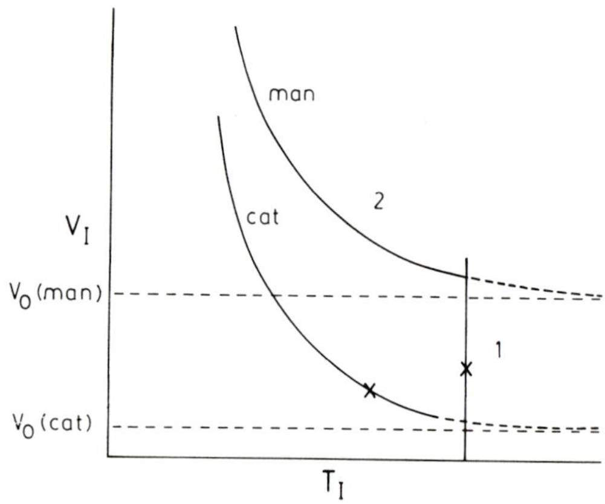

Fig. 7.3. Comparison of the relationships between inspiratory volume (V_I) and inspiratory duration (T_I) as often seen in man, and in the cat anaesthetized with pentobarbitone. In *range 1* [1], T_I remains fairly constant as V_I increases in response to increasing O_2 or imposed inflations. In *range 2* [2], T_I decreases with increasing V_I. The crosses indicate typical V_I, T_I values in eupnoeic breathing. (From Clark and von Euler, 1970.)

The experimental results were assembled in the form of a model consisting of the inspiratory volume-time characteristic and the inspiratory-expiratory timing relationship. The normal operation of the system can thus be described

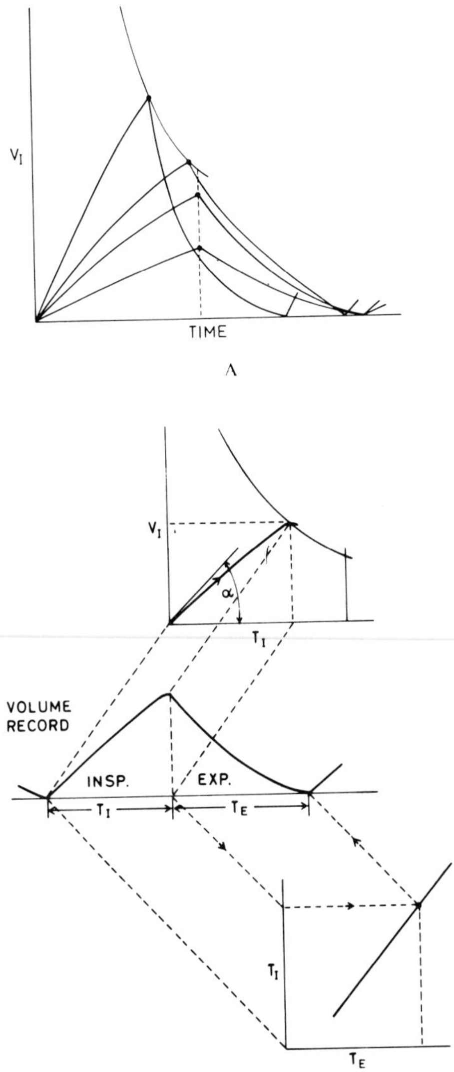

Fig. 7.4. Schematic illustrations referring to the control of respiratory rate and depth in *range 1* and *range 2* in man. [A] shows volume *v.* time for 4 breaths at different PCO_2 levels: the inspiratory phase of the two smallest breaths was terminated by the 'timing' mechanism of *range 1*, whereas in the two largest breaths inspiration was terminated by the 'Hering-Breuer mechanism' of *range 2*.

[B] is a schematic representation of the operation. A volume trace of a breath has been projected on to the inspiratory and timing characteristics to illustrate the main events in the cycle. The angle $[\alpha]$ indicates the rate of rise of inspiratory lung expansion which seems to be the main variable depending on the chemical drive to ventilation. (From Clark and von Euler, 1972.)

by referring to Fig. 7.4, which emphasizes the characteristics observed in man. The primary manner in which breathing is adjusted to meet the demand for ventilation is by changing the rate at which lung volume increases (von Euler, Herrero and Wexler, 1970). In eupnoeic breathing (man) volume increases relatively slowly, and the operation is in *range 1*. As respiratory drive increases, the motor output to the respiratory muscles increases faster and the lungs fill more rapidly. Thus the operation at first moves higher up in *range 1* without change in rate, and then into *range 2*—where larger volumes give shorter breath durations.

In order to study the neural mechanisms responsible for the *range 2* operation, i.e. the situation in which the volume threshold for inspiratory termination falls rapidly with increasing time after the start of the inspiratory phase we (von Euler, Hayward, Marttila and Wyman, to be published) have studied the electrical activity of single neurones close to the afferent vagal input in the solitary tract complex. The anatomical location and the morphological features of the neurones were determined by a fluorescent dye (Procion Yellow, ICI) injection technique (Stretton and Kravitz, 1968). The results suggest that the ventrolateral nucleus of the solitary tract contains neurones with functional properties and morphological features that make them likely to be the critical neuronal elements giving the characteristic inspiratory volume-time relationship in the 'Hering-Breuer' range.

It is entirely possible that inputs from skin, muscles and joints (especially the intercostal muscles and costovertebral joints) could influence breathing indirectly via changes in the neural correlates to the parameters of the system. The most interesting problem concerns the possibility that the whole threshold curve can move up and down along the volume axis.

REFERENCES

Clark, F. J., and von Euler, C. (1970) Regulation of depth and rate of breathing in cat and man, *Acta physiol. scand.*, **80,** 20–21A.

Clark, F. J., and von Euler, C. (1972) On the regulation of depth and rate of breathing, *J. Physiol. (Lond.),* **222,** 267–95.

von Euler, C., Herrero, F., and Wexler, I. (1970) Control mechanisms determining rate and depth of respiratory movements, *Resp. Physiol.,* **10,** 93–108.

Guz, A. M., Noble, M. I. M., Widdicombe, J. G., Trenchard, D., and Mushin, W. W. (1966) The effect of bilateral block of vagus and glossopharyngeal nerves on the ventilatory response to CO_2 of conscious man, *Resp. Physiol.,* **1,** 206–10.

Guz, A. M., and Widdicombe, J. G. (1970) Pattern of breathing during hypercapnia before and after vagal blockade in man, in *Breathing: Hering-Breuer Centenary Symposium,* pp. 41–44, ed. Porter, R., London.

Nesland, R. S., Plum, F., Nelson, J. R., and Stedler, H. D. (1966) The graded response to stimulation of medullary respiratory neurons, *Exp. Neurol.*, **14,** 57–76.

Richardson, P. S., and Widdicombe, J. G. (1969) The role of the vagus nerves in the ventilatory responses to hypercapnia and hypoxia in anaesthetized and unanaesthetized rabbits, *Resp. Physiol.*, **7,** 122–35.

Scott, F. H. (1908) On the relative parts played by nervous and chemical factors in the regulation of respiration, *J. Physiol. (Lond.),* **37,** 301–26.

Stretton, A. O. W., and Kravitz, E. A. (1968) Neuronal geometry: Determination with a technique of intracellular dye injection, *Science,* **162,** 132–4.

DISCUSSION ON COMMUNICATION [7]

Cohen. I wish to present some of my own results to complement those of Dr. von Euler with regard to two phenomena: (1) the relation between inspiratory and expiratory phase durations; and (2) the 'all-or-none' character of the cessation of inspiration.

The first slide [Fig. D.5] shows that there is a monotonic relation between inspiratory and expiratory phase duration, when inspiratory phases of different durations are produced by application of inspiration-inhibiting stimulation at

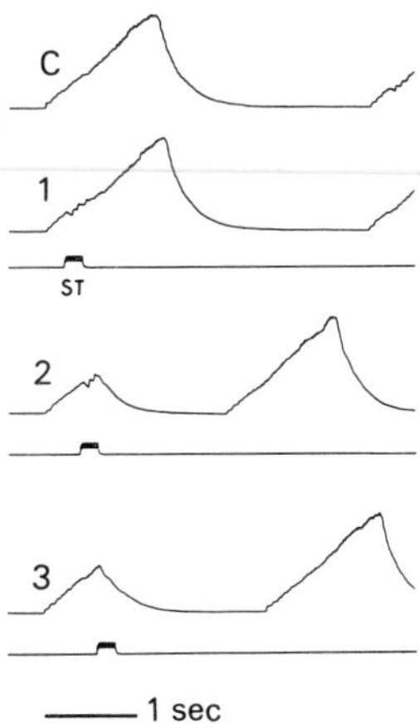

Fig. D.5. Dependence of inspiratory phase termination (i.e. switching to the expiratory phase) on delivery time of stimulus trains during the inspiratory phase (stimulation site: ventrolateral pontine inspiration-inhibiting point). Tracings show integrated phrenic signal. [C]: control inspiratory phase; [1–3]: inspiratory phases in which stimulus trains (ST, 20 pulses at 100/sec) were delivered, starting at the following delays from onset of the phase: [1] 200 msec; [2] 400 msec; [3] 600 msec.

different times in the phase. The traces show integrated phrenic potentials in a vagotomized cat (my favourite subject) during a control inspiratory phase [C] and during inspiratory phases [1–3] in which brief stimulus trains were applied at an inspiration-inhibiting point in the ventrolateral region of the rostral pons (Cohen, 1971[1]). When the train is delivered early in the inspiratory phase [1],

[1] Cohen, M. I. (1971) Switching of the respiratory phases and evoked phrenic responses produced by rostral pontine electrical stimulation, *J. Physiol. (Lond.),* **217,** 133–58.

there is only a transient change in phrenic discharge; but when it is delivered later in the phase an abrupt cessation of the inspiratory phase results, with switching to the expiratory phase. It can be seen that when the inspiratory phase is longer (due to later application of the inhibitory train), the succeeding expiratory phase is also longer (compare [2] and [3]). Thus, the relation of inspiratory to expiratory phase duration is similar to that found by Dr. von Euler when he varied inspiratory phase duration by other procedures.

The linkage between inspiratory and expiratory phase durations seems to be a general phenomenon. Although a detailed explanation of the effect cannot yet be given, I would place it in the category of 'post-inhibitory rebound'—during the inspiratory phase, expiratory neurones are being inhibited; this inhibition is greater during a longer inspiratory phase; and following such a phase there is a greater excitatory rebound of expiratory neurone discharge, leading to a longer expiratory phase.

The 'all-or-none' character of the cessation of inspiration is also apparent in Fig. D.5: stimulation applied early in the inspiratory phase has a minor effect, whilst stimulation applied later in the phase results in switching to the expiratory phase. This 'all-or-none' character is also apparent in experiments where diminution of the Hering-Breuer reflex (by airway occlusion) is used to lengthen the inspiratory phase.

The effect of airway occlusion at minimum lung volume on phrenic discharge is shown in Fig. D.6 (from an experiment in a spontaneously-breathing cat with intact vagi). This occlusion (imposed at the end of the expiratory pause) prevents the normally occurring afferent input due to lung expansion and thereby produces a lengthening of the inspiratory phase, due to disinhibition. The tracings in Fig. D.6 show computer-summed phrenic activity in 20

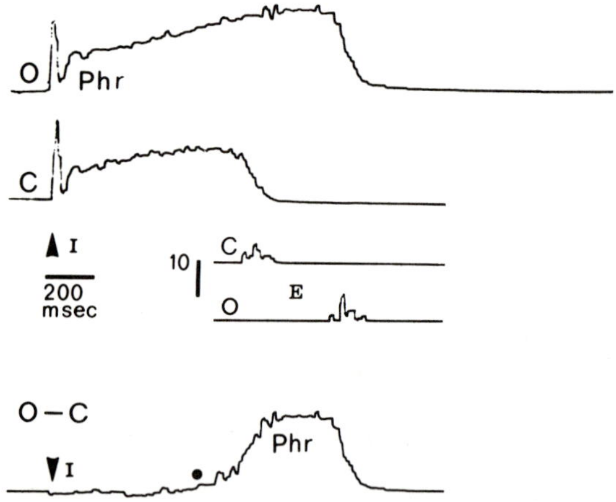

Fig. D.6. Effects of airway occlusion at minimal lung volume (occlusion started by I pulse, ended by E pulse) on phrenic discharge pattern. *Top tracings*: computer-summed phrenic activity (15 msec bins) in 20 control [C] and 20 occlusion [O] inspiratory phases. Computer sweep triggers were pulses locked to the I pulses (arrows). *Bottom tracing* [O-C]: computer-summed [O] activity minus [C] activity; dot indicates inflection point where O activity started to exceed C activity. *Middle tracings*: incidence of E pulses, which mark end of C and O phases (vertical bar, 10 pulses).

control [C] inspiratory phases and in 20 occlusion [O] inspiratory phases (occlusion was started at the beginning of the inspiratory phase and ended at the beginning of the expiratory phase). It can be seen that the over-all phrenic discharge pattern (initial burst, secondary depression, and then linearly increasing discharge) is not markedly changed by occlusion: the main effect of occlusion is to prolong the phrenic burst and the inspiratory phase. A more exact assessment of the activity change produced by occlusion was obtained by subtracting the summed control activity from the summed activity during occlusion ([O–C], bottom tracing). As shown by the initial flatness of this tracing, occlusion produced no change in phrenic activity for about 600 msec—a time equivalent to about two-thirds of the duration of the control inspiratory phase (cf. histogram of incidence of E pulses in middle tracing). Thereafter (after the dot indicating the inflection point of the bottom tracing), there was a moderate increase of activity for 240 msec, a time corresponding to the last third of the control inspiratory phase duration. Finally, the last part of the tracing shows activity in the prolonged portion of the inspiratory phase; i.e. in the time equivalent to the difference between the occlusion and the control inspiratory phase durations.

Thus airway occlusion—and therefore (by inference) pulmonary stretch receptor input—has no effect on phrenic discharge for a time equivalent to most of the control inspiratory phase duration. The major effect of occlusion is to prolong the duration of the phrenic burst, without changing its general pattern and in particular without changing the initial rate of increase of activity (indicated by the slope of the summed activity tracings). Therefore, the Hering-Breuer reflex seems to act by reducing the triggering level for cessation of the inspiratory burst. The inspiratory burst, once it has started, proceeds in a predetermined pattern; and the major effect of pulmonary stretch receptor input is to advance the onset of switching to the expiratory phase without, however, markedly affecting the discharge pattern (except for a relatively minor inhibition near the end of the inspiratory phase). This conclusion is in accord with the experiments of Clark and von Euler (1972)[1], who found that termination of inspiration by volume pulses depended on the time of their application during the inspiratory phase in an 'all-or-none' manner.

Plum. This is very lovely, Dr. von Euler. I have two questions for you. One of the curious things to us in our own[2] micro-electrode studies of brain stem respiratory neurones was that the number of discharges per burst seemed to be numbered (or limited) in the presence of an intact vagal afferent discharge. That is to say, given a single neurone in the medullary inspiratory population, the discharge would be 27 ($\pm$3) over wide ranges of tidal volume—the discharge being more condensed in the presence of a higher tidal volume and more spread out in the presence of a lower tidal volume, as if there were a numerical relationship to vagally-generated inhibition. I wonder if you have obtained similar data, and if you have an explanation for this?

The other question is whether you have any notions as to how the vagal afferent influence regulates the following expiratory cycle? This observation

[1] Clark, F. J., and von Euler, C. (1972) On the regulation of depth and rate of breathing, *J. Physiol. (Lond.)*, **222**, 267–95.

[2] Nesland, R. S., Plum, F., Nelson, J. R., and Siedler, H. D. (1966) The graded response to stimulation of medullary respiratory neurones, *Exp. Neurol.*, **14**, 57–76.

has been established since the early part of the century and conveniently ignored: it is not compatible with a simple inhibitory feedback, and requires a concept to explain how both expiration and subsequent inspiration are limited by the afferent discharge of the Hering-Breuer system.

von Euler. We have not yet done any numerical analysis of the data from that point of view.

Concerning your second question of how the phase of expiration is influenced by the preceding inspiratory phase the answer is that we don't know; but in our humble trials to model the system we think of integrative neural functions of the same type and time constant as are used during the phase of inspiration. We have, however, to take into account the fact that the expiratory duration can be subjected to reflex modulation over and above the basic setting determined by the mechanisms of the inspiratory phase.

Anthony. Gutzman (1928)[1] is the only reference I have found to the ratio of the times of inspiratory and expiratory periods. He found the ratio to lie between 1.5 and 1.8 to 1. I found it to be a bit less than that, somewhere between 1.2 and 1.5 to 1. When you plot T_I against T_E for each individual, then you find you get a cluster of points in a small area peculiar to each individual. There is some indication that for different types of lung disease this ratio changes, and I wonder if you would like to comment on the possibility of using it as a clinical measure.

von Euler. In trying to give some sort of answer to your question I should like to emphasize that the relation between the inspiratory duration and the expiratory duration can be disturbed by the volume information coming in during the expiratory phase, in that inflation of the lungs prolongs expiration. We have actually two phases of the Hering-Breuer mechanism—one is the arrest of the inspiratory phase, which is an 'all-or-nothing' phenomenon; and the other is the effect from the same receptors during the phase of expiration, where the prolongation of the expiratory phase is a graded reflex directly related to the amount of volume that remains in the lungs at specific times. This may be the reason, I think, why during some pulmonary diseases with increased expiratory resistance (so that the volume is not expelled at a normal rate) the expiratory duration is found to be relatively longer than in normal conditions.

Widdicombe. I imagine, Dr. von Euler, that you would agree that this kind of analysis may be applicable only to your particular experimental conditions, involving mainly inflation of the lungs and stimulation of the vagus nerves. Because if you look at other afferent inputs to the respiratory centres you get quite different patterns that don't fit with the diagrams you have shown—and there are two in particular which I think are worth mentioning. I am thinking now of the linear relationship you described between tidal volume and frequency—which probably applies in exercise and hypercapnia and hypoxia, where when we increase our tidal volume we increase our frequency. But if you get stimulation of lung receptors other than those operating the Hering-Breuer

---

[1] Gutzman, H. (1928) *Physiologie der Stimme und Sprache,* 2nd ed., Brunswick.

reflex (that is to say, either J receptors or irritant receptors) you get rapid shallow breathing which, of course, means that tidal volume has gone down and frequency has gone up—the opposite relationship; and the hyperbolic curve you drew of tidal volume against time of inspiration has, in these circumstances, to be displaced downwards. You might say this is a sensitization of the Hering-Breuer reflex; but as far as we know, Hering-Breuer reflex afferents are not involved.

The other reflex that has a completely different pattern of response, and one which is perhaps more relevant to this Symposium, is that produced when you stimulate the laryngeal mucosa with very weak solutions (or low concentrations) of irritant aerosols. These do not cause coughing; but they change the pattern of phrenic discharge and the pattern of respiration so that the phrenic discharge build-up is slower—and yet breathing becomes deeper and slower. This is way off what I might call the 'von Euler curve'. May I ask, have you studied any other afferent inputs, apart from Hering-Breuer inflation reflex, to see how they affect your relationships?

von Euler. This we regard, so far, as being a general scheme—a background against which other things could be studied. As I mentioned, we want to use this sort of analysis as a background for finding out something about the influence of proprioceptive information coming not only from the respiratory muscles but from all the muscles. Of course, there is also the possibility that these neurones receive impulses from other regions—but that is something we have not yet studied.

Ruth Bowden. Is it possible to tell how far the dendritic field spreads—whether it is entirely unilateral or bilateral—and whether the connexions are ipsilateral or bilateral, or is it too early at this stage?

von Euler. Yes, it is too early at this stage. We have to inject the dye electrophoretically for about 6 minutes to stain the cell body and the proximal parts of the dendrites, and it has proved difficult to stay inside a cell even for that length of time. So in most instances we have injected too little dye to be able to see the far ends of the dendrites. For this reason I can't tell you about bilateral distribution, either.

Otis. In the linear relationship between expiratory and inspiratory times you have an intercept. Do you have an interpretation of that intercept, as it implies that for zero inspiratory time you have a finite expiratory time?

von Euler. The relationship between inspiratory and expiratory durations was approximately linear over a fair signal range. However, some bending of the curve occurred at both extremes—especially with deep breaths and sighs, and when an inspiration-augmenting reflex was evoked. Concerning our interpretation of the intercept, we believe at present that it may represent a transition time between the phases of the breathing cycle.

Otis. What was the magnitude of the intercept?

von Euler. The magnitude of the intercept is, on the average, 0.4 sec.

Freedman. I was very intrigued, first of all, by your suggestion that this whole relationship may be shifted up and down the volume axis by inputs from other things. I don't know whether some of our data may be relevant in this regard, but I think they are.

This slide [Fig. D.7] describes a series of experiments in which subjects were breathing with elastic loads added to their breathing. If you concentrate on this particular subject (pointing), here is the increase in PCO_2 as he rebreathes and

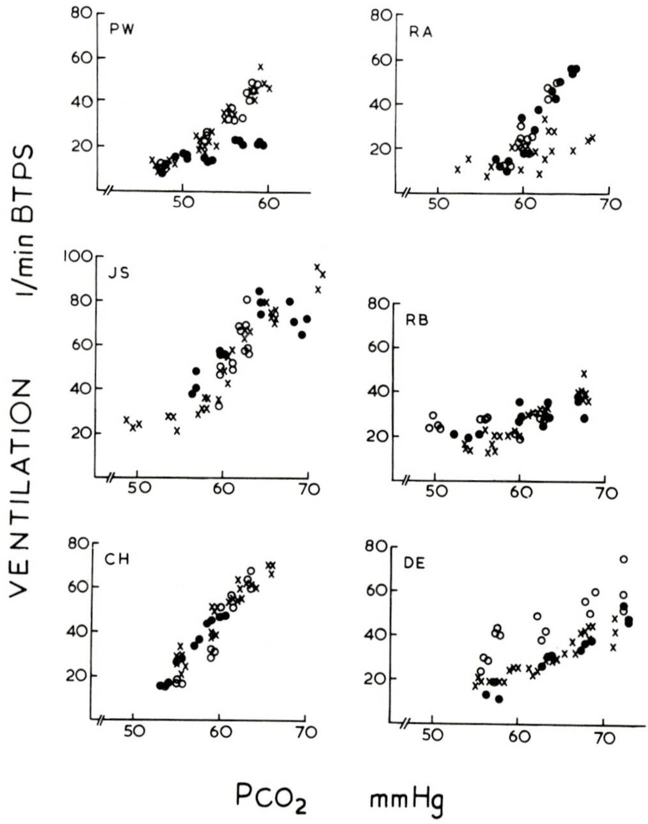

Fig. D.7. Relationship between ventilation and PCO_2 in six normal subjects breathing against added external elastic loads. [x = no load; o = load of 14 cm H_2O/l; ● = load of 23 cm H_2O/l]. (From Freedman, S., Dalton, K. J., Holland, Denise, and Patton, J. M. S. (1972). The effects of added elastic loads on the respiratory response to CO_2 in man, *Resp. Physiol.*, **14**, 237–50.)

here is his ventilation. The crosses are with no load and the dots are with various degrees of increased elastic load; and here we have a region in which (for a given PCO_2) he always keeps the same ventilation irrespective of the load. In the next slide [Fig. D.8], if you look at the pattern of breathing, however, you can see that without a load here (pointing) is his ventilation and here his tidal volume. But with a load, to achieve the same ventilation we have a much smaller tidal volume—there is hardly any increase in tidal volume at all, the effect being almost entirely an increase in frequency. I think the intriguing thing about this is that with this method of adding a load you don't affect in any way the pulmonary or transpulmonary pressure-volume relationships. For a given degree of lung inflation the transpulmonary pressure

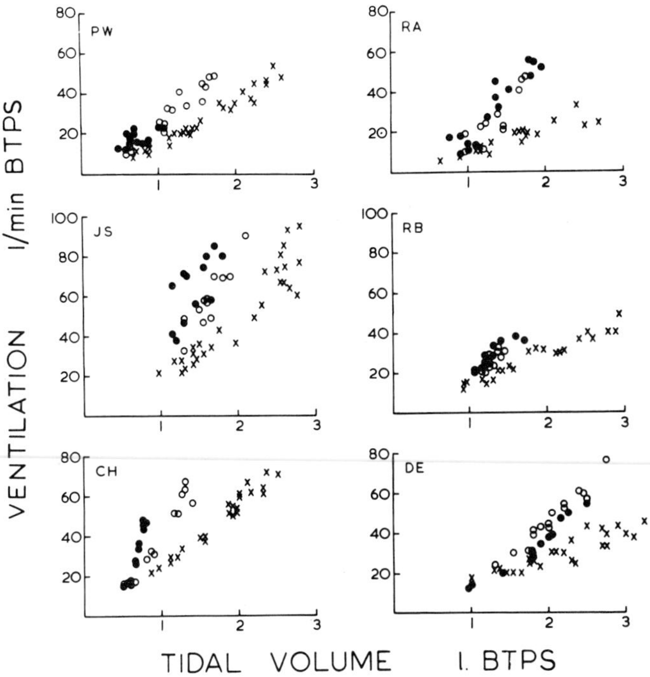

Fig. D.8. Relationship between tidal volume and ventilation drawn from same data as in Fig. D.7. Note that with added elastic loads a given ventilation is achieved by a higher respiratory frequency and smaller tidal volume than without a load.

remains the same, the added load being taken entirely by the large extra-thoracic airways and the chest wall. I would suggest that this might support your idea of other afferent influences moving the responses up and down the volume threshold.

Campbell. I think it is only fair to point out that this is in conscious subjects, and might not necessarily be true in unconscious subjects. Is that not right?

Freedman. I would guess, from the results of Bland *et alii* (1967)[1] with elastic loads in anaesthetized dogs, that you might find exactly the same thing in anaesthetized human subjects.

[1] Bland, S., Lazerou, L., Dyck, G., and Cherniack, R. M. (1967) The influence of the 'chest wall' on respiratory rate and depth, *Resp. Physiol.*, **3**, 47–54.

von Euler. I am intrigued to see this.

Mead. I notice that your volumes are all referred to relative tidal volume. I am curious to know whether you have information on whether we should be thinking in terms of absolute lung volume as well. There is this paradox that a man exposed to positive pressure may shift his volume quite substantially, well into the threshold that you mention, without any appreciable change of frequency.

von Euler. We have tried to keep track of the absolute volume; and in man we have actually used the method that you described yesterday [see Communication 5] by having one pneumograph around the thorax and another around the abdomen, connecting them up and adjusting their relative sensitivities so that the iso-volume recording with a closed glottis would give zero output. We have then used the pneumograph record as an 'absolute' volume reference for the spirogram.

Your question, I guess, concerns the problem of how rapidly we may adapt to a new volume situation in the lungs and I can't answer that precisely; but we have in some experiments added some positive pressures (this is especially in cats) and found an adaptation to the new situation which was somewhat quicker than the receptor adaptation should permit. On the other hand, the prolongation of the expiratory phase did not adapt in the same way as did the effect on the inspiratory phase. It thus seems possible that here we are concerned with some sort of central adaptation, rather than receptor adaptation.

Cohen. I wish to make some additional comments on the relation between inspiratory and expiratory phase durations. There is suggestive evidence that the expiratory phase is divided into two portions, related respectively to 'early-expiratory' and to 'late-expiratory' neurone discharge. In some of my experiments (unpublished) I have found that when the inspiratory phase was lengthened (by stimulation, or spontaneously as in a gasp), there was a lengthening of the immediately succeeding burst from an early-expiratory neurone: in contrast, there was no marked change in the burst duration of a late-expiratory neurone during the expiratory phase succeeding the lengthened inspiratory phase. These observations suggest that the lengthening of the expiratory phase is caused mainly by changes in activity of the early-expiratory neuronal population.

Another important observation is that the inspiratory-expiratory duration relationship is not symmetrical. That is to say, the expiratory phase duration is related to the duration of the preceding inspiratory phase, but the inspiratory phase duration is independent of the duration of the preceding expiratory phase. The latter effect, which is observed in expiratory-inspiratory phase-switching produced by pontine electrical stimulation (Cohen, 1961[1], Fig. 4), indicates that the characteristics of inspiratory discharge are programmed in a relatively rigid manner.

von Euler. Yes, I agree with you—all our evidence points to the fact that the inspiratory phase and the following expiratory phase form an entity that is not

---

[1] Cohen, M. I. (1971) Switching of the respiratory phases and evoked phrenic responses produced by rostral pontine electrical stimulation, *J. Physiol. (Lond.)*, **217**, 133–58.

dependent on what happened in the cycle before or on the characteristics of the cycle following it—unless you go to high volumes and provoke augmented inspiration or a 'deep breath'. When we have sighs, or 'deep breaths', then something happens to the whole breathing mechanism which upsets it for a while with a few breaths of smaller tidal volume and shorter duration in a way suggestive of a downward shift of the volume threshold.

Wyke. I want to ask about a possible implication of the idea that there are two populations of inspiratory neurones, one of which is vagally modulated and one which apparently is not. Presumably, the assumption is that the population which is not vagally modulated may contain neurones that are modulated by somatic (for example, myotatic or articular) afferent inputs from the thorax (or elsewhere)—and these, as I understand it, are the ones that have spinal projections. Would you care to speculate on that, in the light of the fact that what evidence there is from studies of somatic afferent influences on brain stem nuclei is that many neurones to which such inputs project are neurones that have spinal projections to fusimotor neurones. I am especially interested in this particular population of inspiratory neurones in view of the possibility of inputs to them from costo-vertebral articular receptors, and from the various receptor systems in the larynx.

von Euler. I don't know that I can comment very much on that. I think that the neurones with vagal modulation, and which are non-spinally projecting neurones, relay into mechanisms that Dr. Cohen has already mentioned and will perhaps report on more fully this afternoon [see Communication 12]—that is to say, the switch mechanism. It is also possible that other afferents such as joint afferents may feed on to these same neurones by relaying from one neurone to another in sequence in the brain stem—but I don't know. Nor is it known whether the second group of inspiratory neurones (that is, the 'non-vagal' but spinally-projecting neurones) may receive inputs from afferent systems such as those of myotatic or articular origin.

8 Co-ordination of Neuromuscular Efferent Systems in Ventilation

A. A. VILJANEN

Department of Pulmonary Diseases, University Central Hospital, Helsinki, Finland

Our present studies are for the most part direct continuations of the work presented at the Hering-Breuer Centenary Symposium in 1969 (Viljanen, 1970). The respiratory centres produce the desired ventilation by programming the activity of the respiratory muscles. The neural efferent activity of the respiratory centres generates the mechanical activity of the respiratory muscles. The relationship between the mechanical and electrical activities of the respiratory muscles is relatively well understood (Campbell and Green, 1953 a and b; Fink *et alii,* 1958; Delhez *et alii,* 1959; Agostoni *et alii,* 1960; Petit *et alii,* 1965; Viljanen *et alii,* 1965; Viljanen, 1967; Sears and Newsom Davis, 1968). Assuming a constant relationship of the total neural efferent activity of the respiratory centres to the electrical activity of the respiratory muscles, an idea can be gained of the functional activity of the respiratory centres by registering the electrical activity of the respiratory muscles. This also provides an opportunity to estimate, for instance, the total work performed by the respiratory muscles. It is further possible to utilize the said electrical activity, for instance, for monitoring the respiration of patients in an intensive care unit.

While we have to treat the respiratory centres as a 'black box', we have the possibility of varying the afferent inputs, whereby we may gain information concerning the functional principles of the respiratory centres by recording their output. Information is also obtained on the distribution of the output between different respiratory muscles in different situations. It is well known that the gradation of contraction of skeletal muscle is based upon the recruitment of motor units, upon their discharge frequency and upon their duration of action (Adrian and Bronk, 1929; Smith, 1934; Bronk and Ferguson, 1935; Lindsley, 1935; Gilson and Mills, 1941; Gesell *et alii,* 1941; Pitts, 1942; Gesell *et alii,* 1949; Tokizane *et alii,* 1952; Bigland and Lippold, 1954).

Our studies aim to clarify, in addition to the electrical-mechanical relationships in the respiratory muscles, the relationships between the electrical activities of the nerves and the corresponding respiratory muscles, as well as the functional principles of the respiratory centres. In order to facilitate the treatment of the results, we have devised a mathematical model of the regulation of breathing.

METHODS

Our measurements of mechanical quantities were made by means of a pneumotachograph and electrical pressure transducers. The electrical activities were recorded from the external intercostal muscles with needle or surface electrodes, and from the diaphragm with oesophagus electrodes in our human studies and with needle electrodes in animals. The data were subsequently analysed by a Linc computer from tape recordings. The analysis of the electrical activity of the respiratory muscles was considerably facilitated by the use of logical units conditioning the discharges for computer analysis (Viljanen and Raitanen, 1970). A Univac 1108 computer was employed to work out the mathematical model.

In the experiments on cats, anaesthesia was produced by administering pentobarbitone (40 mg/kg) or fluothane (1.0–1.5 per cent).

RESULTS

As was mentioned above, the electrical and mechanical activities of the respiratory muscles have been shown to be closely related. Recording the electrical activity of motor units in the cat diaphragm as a function of the pressure impulse produced by the respiratory muscles, we have found a comparatively good linear relationship between these variables when the number of motor units monitored exceeded four or five.

We have tried to find out whether the impulse activity recorded from a phrenic nerve multifibre preparation is linearly related to the diaphragmatic EMG activity (cf. Dittler, 1910; Dittler and Garten, 1912; Gasser and Newcomer, 1921). Figure 8.1, referring to a cat, shows that a linear relationship seems to exist between activity in the right phrenic multifibre preparation and the left diaphragmatic EMG impulse activity (the right phrenic nerve was isolated from the diaphragm). Our preliminary findings suggested a linear relationship between the electrical activity of the diaphragm and intercostal muscles during one inspiration, both in man (Viljanen and Halttunen, 1967) and in the cat (Viljanen *et alii,* 1968). We have now checked these findings in a more extensive series of experiments, and find the assumption to be correct in the cat [Fig. 8.2] and also during quiet breathing in man.

It appears that under stable conditions the respiratory centres decide in advance the nature of the next inspiration, and programme it to reach a predetermined end-state. It is well known that the vagal afferent drive has an inhibitory effect on the output of the respiratory centres. By injecting mepivacaine (*Carbocaine,* 5 mg) into a femoral vein, a marked increase in the efferent impulse activity of the respiratory centres was produced during the next inspiration. This

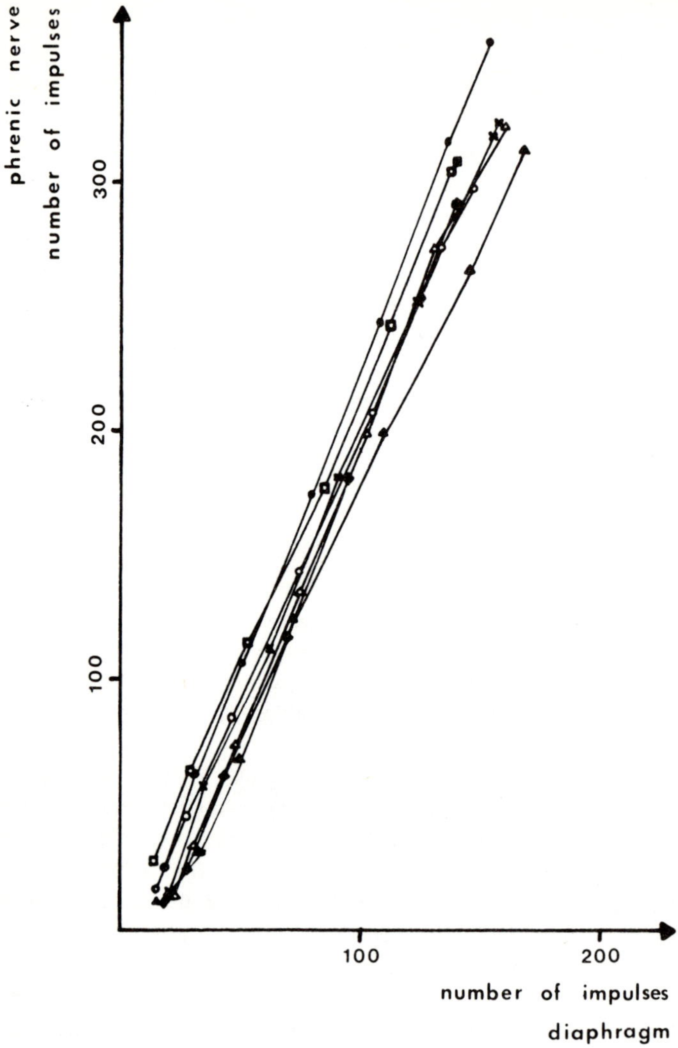

Fig. 8.1. Number of impulses from a right phrenic nerve multifibre preparation plotted against the number of discharges from the left hemidiaphragm during eight inspirations in a cat.

effect was seen mainly as an increase in the number of impulses, rather than in their frequency.

It seems that the vagal afferent drive continuously participates in the programming of the respiratory centres in the cat. In several experiments, the increase in number of efferent impulses when both vagal nerves were blocked was almost twice that observed when only one vagus was blocked [Fig. 8.3]. The effect seems to vary between individual subjects, and with different anaesthetic agents.

Furthermore, we wish to call attention to the occurrence of rapid changes in the activity of the respiratory centres following changes in the oxygen content of the inspired air. On administration of 100 per cent or 6 per cent oxygen after normal

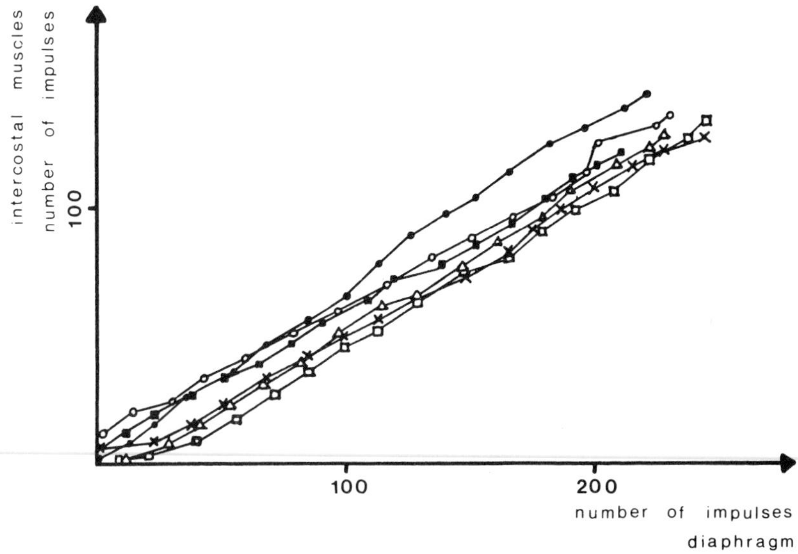

Fig. 8.2. Number of discharges from the external intercostal muscles plotted against the number of discharges from the diaphragm during six inspirations in a cat.

air to a cat, an effect was regularly observed during the second or third inspiration thereafter.

We have also studied the effect of a voluntarily produced efferent drive on the output of the respiratory centres. A category production study was carried out in which the subjects were instructed to perform first a normal inspiration followed by inspirations two, three, four, five and six times as large. Hyperventilation and fatigue were avoided. We found the number of motor unit discharges to increase, perhaps progressively, with the subjectively produced category of breathing [Fig. 8.4]. The discharge frequency may be more nearly linear with reference to category.

On the basis of previous findings, and assuming that the respiratory centres decide in advance the nature of the next breath and that they possess a certain adaptive capacity, we have devised a mathematical model of the respiratory centres (Viljanen and Hämäläinen, unpublished data; cf. Rohrer, 1915; Otis *et alii,* 1950; Mead, 1960; Widdicombe and Nadel, 1963; Priban and Fincham, 1965; Horgan and Lange, 1968; Wald *et alii,* 1968; Yamashiro and Grodins, 1971).

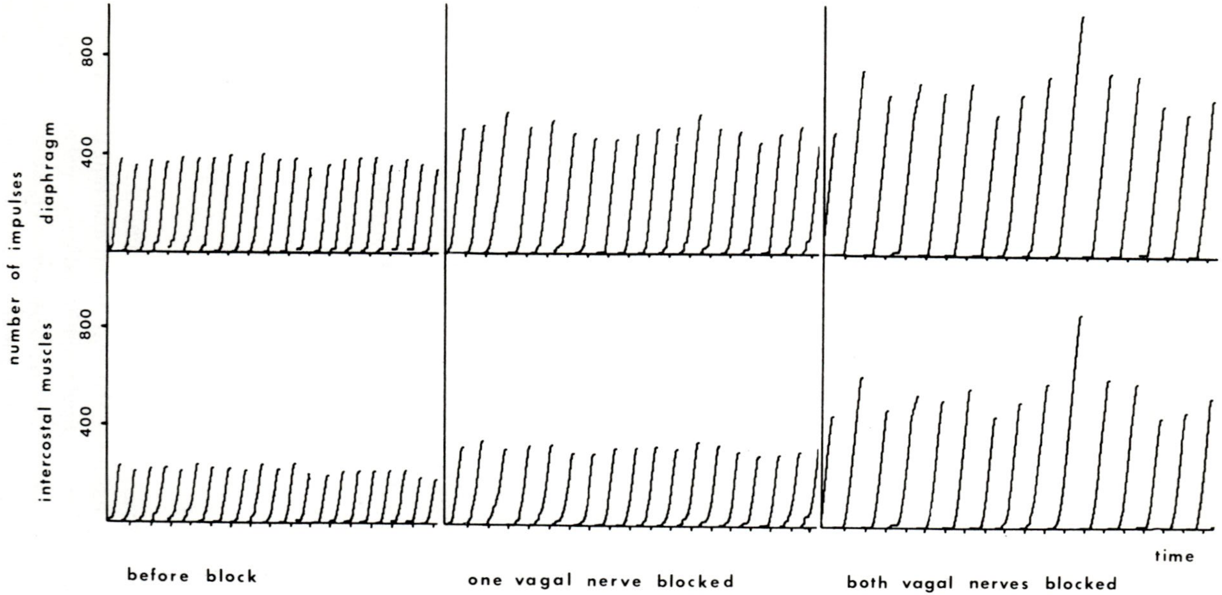

Fig. 8.3. Effect of blocking one or both vagus nerves on the motor unit activity recorded from the diaphragm and external intercostal muscles in a cat.

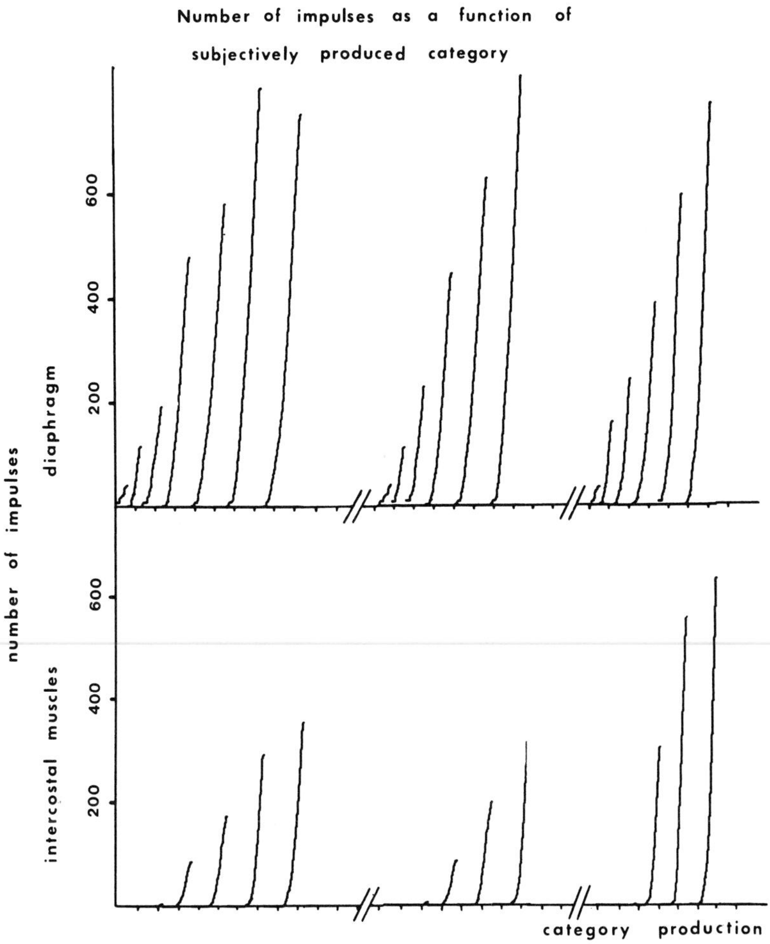

Fig. 8.4. Number of discharges from the external intercostal muscles and diaphragm recorded in three category production tests, involving a normal inspiration followed by inspirations two, three, four, five and six times as large.

We divided our model into two parts: lower level and higher level. The system equation used by us is:

$$P_{\text{mus}}(t) = KV(t) + R\,V(t),$$

where P_{mus} = pressure produced by respiratory muscles
K = elastance
V = displacement from the relaxation volume
$\dot{V}$ = rate of gas flow
R = resistance

First we shall consider the lower level, where the execution of one breath is programmed. During inspiration the following performance index will be minimized:

$$J_{I,1} = \int_0^{T_I} \left(\ddot{V}^2(t) + \alpha_1 P_{\text{mus}}(t) \dot{V}(t) \right) dt = \min_{V(.)}!$$

where T_I = inspiratory time
α_1 = an individual constant

The boundary conditions are:

$$t = 0;\ V = V_0;\ \dot{V} = 0$$

$$t = T_I;\ V = V_0 + V_T;\ \dot{V} = 0,$$

where V_0 = displacement from relaxation volume
V_T = tidal volume

During expiration the following performance index will be minimized:

$$J_{I,2} = \int_0^{T_E} \left(\ddot{V}^2(t) + \alpha_2 P^2_{\text{mus}}(t) \right) dt = \min_{V(.)}!$$

where T_E = expiratory time
α_2 = an individual constant

The boundary conditions are:

$$t = 0;\ V = V_0 + V_T;\ V = 0$$

$$t = T_E;\ V = V_0$$

Secondly, the higher level part is to be considered. It regulates the entire mechanical ventilatory procedure to meet the alveolar ventilation demand. The following performance index will be minimized:

$$J_{II} = \frac{1}{T_I + T_E + T_3} \left[\beta_1 \int P^2_{\text{mus}_i}(t) dt + \beta_5 \int P^2_{\text{mus}_e}(t) dt + \right.$$

$$\left. \beta_3 \int_0^{T_I} \dot{V}_i^2(t) dt + \beta_4 \int_0^{T_E} \dot{V}_e^2(t) dt + \beta_5 \int_0^{T_I + T_E} dt \right] = \min_{T_I, T_E, T_3, V_0, V_D}!$$

where β_1–β_5 = individual constant weighting factors
P_{mus_i} = pressure produced by inspiratory muscles
P_{mus_e} = pressure produced by expiratory muscles
T_3 = non-active period

Constraints:

$$V(t) = V^*(t)$$

$$\dot{V}_A = \frac{V_T - V_D}{T_I + T_E + T_3}$$

$$R_A(V_D) = \frac{0.082}{V_D - 0.0825} + 0.4 \text{ (Widdicombe and Nadel, 1963)}$$

where V^* = the value yielding optimum lower level performance
$\dot{V}_A$ = alveolar ventilation
R_1 = airway resistance
V_D = airway volume

To date, we have completed the testing of the lower level model in comparisons of calculated and experimental data. Fig. 8.6 shows theoretical flow rate graphs as calculated from the model for given values of tidal volume and inspiration and expiration times, which have been superimposed on experimental recordings. The imperfect fit observed in the left top figure at the inspiration and expiration peaks can be explained by the circumstance that a merely linear systems equation is employed in our model. A better fit at the expiratory flow peak was achieved when an artificial additional linear expiratory resistance was added (in the bottom left figure). In the figures on the right, an additional in-

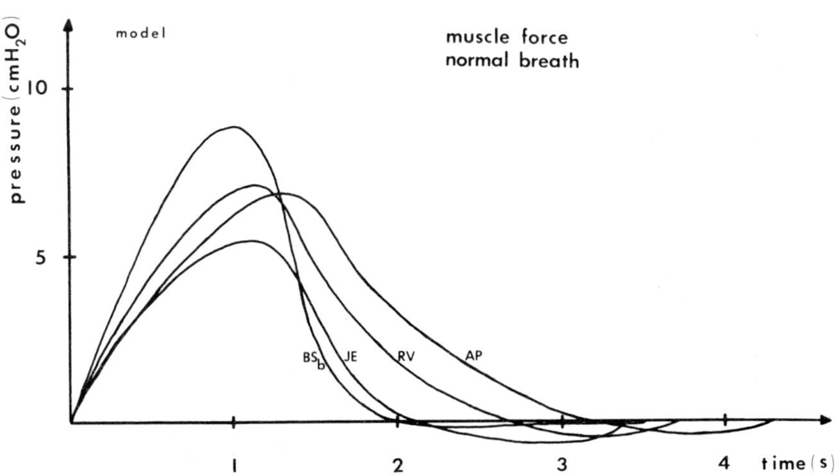

Fig. 8.5. Graphs showing the force produced by the respiratory muscles during one respiratory cycle, as calculated by computer according to the model described in the text, for four different subjects.

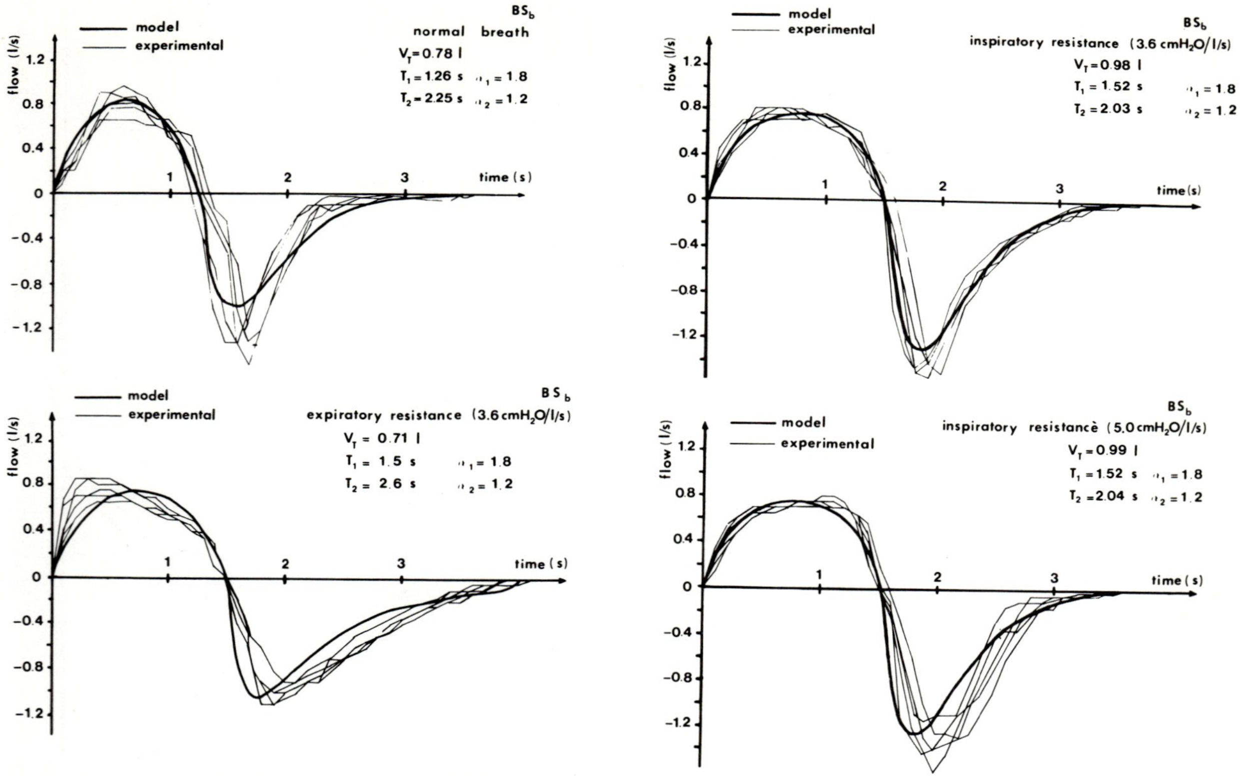

Fig. 8.6. Flow rate graphs determined by computer, in accordance with the model described in the text, for given values of tidal volume and inspiration and expiration time, superimposed on experimental recordings of one and the same subject. In the experiments represented by the lower left and both right figures, artificial expiratory/inspiratory resistances had been added, as indicated in the top corners of these figures.

spiratory resistance was interposed. It is to be noted that in all these instances, in which one and the same subject was involved, the fit is very good.

The theoretical force produced by the respiratory muscles during one respiratory cycle has been calculated from the model for four different subjects in Fig. 8.5.

The appropriateness of the model is being tested in a similar manner with regard to the higher level model. Such results as have emerged so far indicate a good reproduction of experimental data by the computer runs.

DISCUSSION

Although the control of the function of the diaphragm and intercostal muscles by the respiratory centres seems to come from one and the same functional site, their profunctional contributions change in certain conditions, and presumably the change has functional significance. It should be kept in mind, when the functional activity of the respiratory centres is considered in terms of the electrical activity recorded from the respiratory muscles, that the integration is effected at the spinal level. As a consequence, the recorded electrical activity possibly gives a distorted picture of the impulse discharge from the respiratory centres. However, far more detailed information is obtained concerning respiratory events than if mechanical events alone are studied. According to previous studies the neural discharge from the spinal level, at least, bears a linear relationship to the electrical activity of the corresponding muscle, and this in its turn is linearly related to the force developed by the respiratory muscles with regard to motor unit frequency, for instance. In steady state conditions the efferent discharge from the respiratory centres probably goes to the α-motoneurones in adherence to certain specific laws. An interesting phenomenon is also encountered in the relationship between respiratory category production and the corresponding number of discharges. This finding seems to be in agreement with observations by Halttunen *et alii* (1971), who examined the tidal volume as a function of category production.

It is conceivable that the respiratory centres effect a required inspiration by transmitting to the respiratory muscles a given number of impulses, which could be an index of neural energy; on arrival at the respiratory muscles, these impulses cause a certain physical action. The character of the breath is determined by peripheral circumstances. Even though human studies suggest that the respiratory centres determine in advance the nature and end-state of the inspiration, it has been shown in animal experiments, for instance, that the vagal afferent drive has an immediate effect on the activity of the expiratory and inspiratory motoneurones (Sears, 1964). A phenomenon of the same kind is produced by injecting mepivacaine into the femoral vein of a cat, whereby the next efferent discharge becomes considerably greater than those preceding. This should also be demonstrable in man. It is further noted that the quantity of vagal afferent discharge seems to have a definite quantitative effect on the output of the respiratory centres. The vagal afferent drive seems to participate constantly in the programming of the respiratory centres.

From the results which the mathematical model of the regulation of breathing yielded, the inference could be drawn that the character of the respiratory cycle and its frequency, for a given level of ventilation, are determined by structural characteristics (in the model: $\alpha_1, \alpha_2, \beta_1$–$\beta_5$). On the other hand, such regulation of breathing takes place at the neural level, possibly with reference being made to experience. It might be possible to assign each individual to a given category, whereby perhaps the changes of respiratory events in different situations could be anticipated.

No far-reaching conclusions are as yet justified from the results obtained with the aid of our model, as long as the testing of its reliability in the higher level mode is still in progress. We believe that the use of a model of this kind is the sole means by which quantitative respiratory studies can be carried further. One of the arguments pointing in this direction is the great number of variables which influence the regulation of breathing: it is indeed wellnigh hopeless to strive to achieve any over-all picture of the compound behaviour unless one is able to apply a computer model re-programmed over and over again. Provided that we succeed in constructing an appropriate mathematical model of the regulation of breathing, it will be possible to examine the co-ordination of neuromuscular efferent systems in a more coherent manner.

SUMMARY

The relationships between the electrical activities of the intercostal muscles and of the diaphragm, and between those of the phrenic nerve multifibre preparation and the diaphragm, have been shown to be linear in the anaesthetized cat.

A comparatively good linear relationship exists between the activity of motor units and the pressure impulse produced by the inspiratory muscles during a single inspiration in the cat, when the number of active motor units exceeds four or five.

A specific quantitative relationship may exist between the vagal afferent input and the output of the respiratory centres.

A rapid effect on the output of the respiratory centres is noted following changes in the concentration of oxygen and carbon dioxide in the inspired air in the cat.

The relationship between category production of inspiratory volume and the electrical activity of the respiratory muscles has been examined in man.

A mathematical model of the regulation of breathing has been devised, and its reliability has been partially tested.

ACKNOWLEDGEMENTS

These studies have been aided by grants from the Finnish Anti-Tuberculosis Association, the National Research Council for Medical Sciences, Finland, and the Paulo Foundation, Helsinki.

REFERENCES

Adrian, E. D., and Bronk, D. W. (1929) The discharge of impulses in motor nerve fibres, *J. Physiol. (Lond.),* **67,** 119–51.

Agostoni, E., Sant'Ambrogio, G., and del Portillo Garrasco, H. (1960) Electromyography of the diaphragm in man and transdiaphragmatic pressure, *J. appl. Physiol.,* **15,** 1093–7.

Bergström, R. M., Halttunen, P. K., and Viljanen, A. A. (1972) The voluntary regulation of breathing in man, *Acta physiol. scand.,* **84,** 428–9.

Bergström, R. M., and Kerttula, Y. (1961) On the neural control of breathing as studied by electromyography of the intercostal muscles of the rat, *Ann. Acad. Sci. fenn., A5,* 79.

Bigland, B., and Lippold, O. C. J. (1954) Motor unit activity in the voluntary contraction of human muscle, *J. Physiol. (Lond.),* **125,** 322–35.

Bronk, D. W., and Ferguson, L. K. (1935) The nervous control of intercostal respiration, *Amer. J. Physiol.,* **110,** 700–7.

Campbell, E. J. M., and Green, J. H. (1953a) The expiratory function of abdominal muscles in man, *J. Physiol. (Lond.),* **120,** 409–18.

Campbell, E. J. M., and Green, J. H. (1953b) The variations in intra-abdominal pressure and the activity of the abdominal muscles during breathing: a study in man, *J. Physiol. (Lond.),* **122,** 282–90.

Delhez, L., Petit, J. M., Sneppe, E. T., and Bottin, R. (1965) Particularités méthodologiques et intérêt de l'électromyographie intégrée des piliers du diaphragme, *Rev. Electrodiagn. Therap.,* **2,** 303–17.

Dittler, R. (1910) Über die Aktionströme des Nervus phrenicus bei natürlicher Innervation, *Arch. ges. Physiol.,* **131,** 581–8.

Dittler, R., and Garten, S. (1912) Die zeitliche Folge der Aktionströme in phrenicus und Zwerchfell bei der natürlichen Innervation, *Z. Biol.,* **58,** 420–50.

Fink, B. R., Ngai, S., and Duncan, H. A. (1958) Effect of air flow resistance on ventilation and respiratory muscle activity, *J. Amer. med. Ass.,* **168,** 2245–9.

Gasser, H. S., and Newcomer, H. S. (1921) Action currents in the phrenic nerve, *Amer. J. Physiol.,* **57,** 1–26.

Gesell, R., Atkinson, A. K., and Brown, R. G. (1941) The gradation of the intensity of inspiratory contraction, *Amer. J. Physiol.,* **131,** 659–73.

Gesell, R., Hunter, J., and Lillie, R. H. (1949) Electrical and functional activity of motor neurons, *Amer. J. Physiol.,* **159,** 15–28.

Gilson, A. S. Jr., and Mills, W. B. (1941) Activities of single motor units in man during slight voluntary efforts, *Amer. J. Physiol.,* **133,** 658–68.

Horgan, J. D., and Lange, R. L. (1968) Chemical control in the respiratory system, *IEEE Trans. Bio-Medical Engineering, BME-15,* 119–27.

Lindsley, D. B. (1935) Electrical activity of human motor units during voluntary contraction, *Amer. J. Physiol.,* **114,** 90–99.

Mead, J. (1960) Control of respiratory frequency, *J. appl. Physiol.,* **15,** 325–36.

Otis, A. B., Fenn, W. O., and Rahn, H. (1950) Mechanics of breathing in man, *J. appl. Physiol.,* **2,** 592–607.

Petit, J. M., Milic-Emili, J., and Delhez, L. (1960) Role of the diaphragm in breathing in conscious normal man: an electromyographic study, *J. appl. Physiol.,* **15,** 1101–6.

Priban, I. P., and Fincham, W. F. (1965) Self-adaptive control and the respiratory system, *Nature, (Lond.),* **23,** 339–43.

Rohrer, F. (1915) Der Strömungswiderstand in den menschlichen Atemwegen und der Einfluss der unregelmässigen Verzweigung des Bronchialsystems auf den Atmungsverlauf in verschiedenen Lungenbezirken, *Arch. ges. Physiol.,* **162,** 225.

Sears, T. A. (1964) The slow potentials of thoracic respiratory motoneurones and their relation to breathing, *J. Physiol. (Lond.),* **175,** 404–24.

Smith, O. C. (1934) Action potentials from single motor units in voluntary contraction, *Amer. J. Physiol.,* **108,** 629–38.

Tokizane, T., Kawamata, J., and Tokizane, H. (1952) Electromyographic studies on the human respiratory muscles. Studies on the activity pattern of neuro-muscular units, *Jap. J. Physiol.,* **2,** 232–47.

Viljanen, A. A. (1967) The relation between the electrical and mechanical activity of human intercostal muscles during voluntary inspiration, *Acta physiol. scand.,* **296,** (Suppl.), 1–61.

Viljanen, A. A. (1970) Patterns of motor activity in breathing in response to varying sensory inputs, in *Breathing: Hering-Breuer Centenary Symposium,* pp. 183–93. ed. Porter, R., London.

Viljanen, A. A., Poppius, H., Bergström, R. M., and Hakumäki, M. (1965) Electrical and mechanical activity in human respiratory muscles, *Acta neurol. scand.,* **296,** (Suppl.), 237–9.

Viljanen, A. A., and Raitanen, E. Y. (1970) A method for conditioning the EMG signals for computer analysis, *Proc. First Nordic Meeting on Medical and Biological Engineering,* pp. 149–51, ed. Spring, E., Jauhiainen, T., and Honkavaara, T., Helsinki.

Viljanen, A. A., Salorinne, Y., Laitinen, L. A., and Paloheimo, M. (1968) The effect of ether anesthesia on electrical activity of the diaphragm and intercostal muscles during inspiration in the cat, *Scand. J. clin. Lab. Invest.,* **21,** Suppl. 101, 64.

Wald, A. A., Murphy, T. W., and Mazzia, V. D. B. (1968) A theoretical study of controlled ventilation, *IEEE Trans. Bio-Medical Engineering, BME-15,* 237–48.

Widdicombe, J. G., and Nadel, J. A. (1963) Airway volume, airway resistance, and work and force of breathing: theory, *J. appl. Physiol.,* **18,** 863–8.

Yamashiro, S. M., and Grodins, F. S. (1971) Optimal regulation of respiratory airflow, *J. appl. Physiol.,* **30,** 597–602.

DISCUSSION ON COMMUNICATION [8]

Plum. The remarkable thing here is the fit between the empirically obtained curves and the predictions, when so many assumptions were made in the collection of the data. What I am talking about is the assumption that when one measures impulse frequency in *an* intercostal muscle or *a* slip of diaphragm, one can extrapolate from such data on single motor units to a prediction of the behaviour of the whole muscle population. This is extraordinary, in the light of the changes that take place in the activity of the intercostal muscles in relation to posture and to diseases of the chest (as I understand the work of von Euler, Sears and others), and in terms of the direct recordings that you yourself have made. I wonder if Dr. Viljanen has brought any of his primary recordings with him, so that we might look directly at the correlations between muscle activity and nerve output? For example, one question is how did he record impulse frequency in the phrenic nerve—did he dissect one or two fibres: that is, are we talking about the activity of one or two fibres? Because if so, his data would imply that there is no recruitment of new units during increased activity. Perhaps he can help us with this.

Viljanen. We recorded the electrical activity from a multifibre preparation of the phrenic nerve; but the model is not based mainly on the electrical recordings. We have just designed a theoretical model, and now we are testing it. When we compare the pressure predicted by the model with the frequency of the electrical activity of the respiratory muscles there seems to be rather a good correlation between them.

Plum. I think Adrian and Bronk[1] long ago showed that phrenic nerve discharge was related to volume of air flow—but are your data such that you can predict air flow quantitatively from, say, intercostal activity?

Viljanen. I am sorry, I must admit I do not understand.

Plum. As I understand it, you were making the inference that you could predict ventilation quantitatively from *either* intercostal *or* diaphragmatic muscle potential frequency.

Campbell. I didn't understand that to be the case. I understood from your earlier slides that there was a relationship between phrenic nerve activity and the diaphragmatic electromyogram, and between intercostal and diaphragmatic electromyograms, both of which were linear—and I wonder were there any sampling problems here? Also, within what range of conditions do these very nice linear correlations apply—do they have to be under very carefully controlled conditions of posture and respiratory status? Could you tell us a bit more about the relationships between phrenic neuronal activity and diaphragmatic electromyogram and intercostal electromyogram? How far can you generalize from observations on one nerve or muscle?

[1] Adrian, E. D., and Bronk, D. W. (1928) The discharge of impulses in motor nerve fibres. I. Impulses in single fibres of the phrenic nerve, *J. Physiol. (Lond.)*, **66**, 81–101.

Viljanen. That is difficult to answer. But there is a close relationship between activity recorded from phrenic motor nerve fibres and electromyographic activity recorded from the diaphragm. There is also a linear relationship between activity recorded from the diaphragm and activity recorded from the intercostal muscles.

Campbell. Well, let me ask a more specific question. If you change your electrode position on the phrenic nerve but leave the diaphragm electromyogram alone, or if you move the EMG electrode and leave the phrenic electrode alone, would you still get the same numerical relationship—or if you don't get the same numerical relationship, is it at least always linear?

Viljanen. The electrodes are not always at the same place. But it is very well known that there is no difference between different parts of the diaphragm, and I too have not found any difference. Of course, the number of impulses depends on the number of nerve fibres from which we have recorded the electrical activity.

Wyke. Could I put the problem another way, because from a neuromuscular point of view this is critical for your thesis—on this, at least, I think we would all agree. When you record the activity of an intercostal muscle, do you put *one* needle in the intercostal muscle and reinsert it in the same place every time, or do you put it in and leave it there? Or do you put *two* electrodes in the same intercostal muscle and record the gross motor unit activity between them: because some of us would *not* agree that motor unit structure and behaviour are the same everywhere in every muscle—especially in the diaphragm and intercostals, neither of which are all of a piece in these respects[1,2].

Viljanen. The electrodes employed are concentric bipolar needles. Their reinsertion at absolutely the same point in different experiments cannot be guaranteed, but each was left in place for the duration of one experiment.

Wyke. All your recordings then are bipolar concentric needle recordings, both for the diaphragm and for the intercostals?

Viljanen. Yes.

Campbell. Do you just record from one side?

Viljanen. From one side only, in one experiment.

Campbell. Some years ago, Bigland and Lippold (1954)[3] (using gross surface electrodes) showed that under very specific conditions there was a correlation between the integrated electromyogram and muscle force, but you were showing

---

[1] Agostoni, E., and Sant'Ambrogio, G. (1970) The diaphragm, in *The Respiratory Muscles: Mechanics and Neural Control,* 2nd ed., ed. Campbell, E. J. M., Agostoni, E., and Newsom Davis, J., pp. 145–60., London.

[2] Campbell, E. J. M., and Newsom Davis, J. (1970) The intercostal muscles and other muscles of the rib cage, in *The Respiratory Muscles: Mechanics and Neural Control,* 2nd ed., ed. Campbell, E. J. M., Agostoni, E., and Newsom Davis, J., pp. 161–174, London.

[3] Bigland, B., and Lippold, O. C. J. (1954) The relation between force, velocity and integrated electrical activity in human muscles, *J. Physiol. (Lond.),* **123,** 214–24.

correlations between nerve and muscle activity. It was the degree of reproducibility of that observation, and the correlation of the activity of one muscle with that of another muscle, that I was asking about.

Viljanen. I have also measured the activity from the intercostal muscle on the contralateral side, and have found a close correlation between the activity of the muscles on the two sides.

Mead. A simple question. What is the correlation between nerve impulse frequency and integrated electromyographic activity?

Viljanen. Integrated nerve impulse frequency correlates well with the electrical activity recorded from the respiratory muscles.

Mead. So you would say that measurement of nerve impulse frequency is as good and reliable as measurement of integrated electromyographic activity. Is that generally agreed?

Cohen. I believe I can help with that. There have been several recent studies[1,2,3] which show that either phrenic or diaphragmatic discharge is a good indicator of tidal volume.

Mead. That is not an answer to my question. I want to know the relationship between integrated output and impulse frequency, whether it be in nerve or muscle.

Cohen. I am getting around to that. In a recent detailed study (Eldridge, 1971)[3] it was shown that peak integrated phrenic activity (i.e. integrated activity in the last 0.1 sec of inspiration) is linearly related to tidal volume over a wide range. Other measures of activity were less adequate: (1) Total integrated activity per breath was not suitable as an indicator of tidal volume, because changes of inspiratory duration affected this measure. (2) Impulse frequency in the phrenic nerve was sometimes well correlated with tidal volume, but was often less adequate than integrated activity, presumably because action potentials of different amplitudes (from fibres of different sizes) were counted equally. In short, peak integrated phrenic nerve activity, which has a linear relation to tidal volume, is the most accurate indicator of respiratory centre output.

Widdicombe. I think I can also help to answer Dr. Mead's point. The best work on this, I believe, has been done by Wiemer[4]. I am not sure if the work has yet been published, but he has looked very critically at impulse discharges in the phrenic nerve and in the respiratory muscles and at their integrals (using a large

---

[1] Fink, B. R., Hanks, E. C., Holaday, D. A., and Ngai, S. H. (1960) Monitoring of ventilation by integrated diaphragmatic electromyogram. Determination of carbon dioxide (CO_2) threshold in anesthetized man. *J. Amer. med. Ass.*, **172**, 1367–71.

[2] Lourenço, R. V., Cherniack, N. S., Malm, J. R., and Fishman, A. P. (1966) Nervous output from the respiratory centre during obstructed breathing. *J. appl. Physiol.*, **21**, 527–33.

[3] Eldridge, F. L. (1971) Relationship between phrenic nerve activity and ventilation. *Amer. J. Physiol.*, **221**, 535–43.

[4] Wiemer, W. Personal communication.

number of different ways of integrating them). Of course, it depends what you mean by 'integrating' when you ask your question, as there are several ways of doing it; but if you adjust all your electronic integrators very carefully indeed, then you get an integral output that corresponds fairly well to the number of impulses. But even with the most careful adjustment in Wiemer's hands, there was still about a 20 per cent error—and if you don't bother about such adjustments and simply plug in the integrator hoping to get the right answer, the error could be far larger.

Could I also say something further on Dr. Wyke's point about the diaphragm, because I think it is worth emphasizing, as he said, that it is not just a simple unimodal organ. There are several pieces of research now which show that the diaphragm contains both fast and slow motor units—including, in fact, 'twitch' motor units. This inhomogeneity is supported by the histological work of Gauthier and Padykula (1966)[1], who showed that the histological appearances of diaphragmatic muscle fibres vary enormously from fibre to fibre; and both Decandia *et alii* (1966)[2] and Sant'Ambrogio *et alii* (1970)[3] (and more recently Nail, Sterling and myself)[4] have studied single phrenic motor units, and have shown that many of them are what might be called 'twitch units' that only become active either in vigorous breathing or in rather complicated respiratory actions like coughing or sneezing. Therefore I think that the kind of electromyographic pattern that you get, especially from integrators, may be distorted by large fibre activity that comes in with deep breaths.

von Euler. In addition to what Drs. Wyke and Widdicombe have said, I could add that motor units in different portions of the diaphragm have slightly different inputs to them, so that different parts of the diaphragm can work in slightly different ways from one another, depending on the afferent influences—especially from the intercostals.

Another point worth mentioning is that when you relate the electrical activity of a muscle to the force or tension of that muscle you must realize that the well-known length/tension relation is not at all reflected in the electrical activity, or any 'integral' of it. The electrical activity of a muscle is dependent only on the efferent activity in the nerve fibres innervating the muscle, and is completely independent of any changes in the length of the muscle—whereas the contractile force of a muscle, as you know, is very much dependent on its length. In the performance of the intercostals and especially of the diaphragm in the act of breathing, the length/tension relationship is a factor of great importance—which makes it fairly difficult to use the 'integrated' electromyogram as a quantitative index of the force developed by the respiratory muscles in unrestricted breathing.

---

[1] Gauthier, G. F., and Padykula, H. A. (1966) Cytological studies of fibre types in skeletal muscle: a comparative study of the mammalian diaphragm, *J. Cell Biol.*, **28,** 333–54.

[2] Decandia, M., Gantchev, G. N., and Sant'Ambrogio, G. (1966) Correlation between intrapulmonary pressure and rate of firing of diaphragmatic motor units during inspiratory efforts, *J. Physiol. (Lond.)*, **186,** 3–4P.

[3] Sant'Ambrogio, G., Decandia, M., and Gantchev, G. N. (1969–70) The composite structure of the diaphragm of the rabbit, *Arch. Fisiol.*, **67,** 27–39.

[4] Nail, B. S., Sterling, G. M., and Widdicombe, J. G. (1972) Patterns of spontaneous and reflexly-induced activity in phrenic and intercostal motoneurones, *Exp. Brain Res.*, **15,** 318–32.

Campbell. I think I should remind you that in Dr. Viljanen's earlier work[1] he did actually explore these relationships between total number of impulses, frequency of impulses and things like total volume inspired and rate of inspiration. I think you published that, didn't you, as an extension of the work of your mentor in Helsinki?

Viljanen. Yes. For instance, we have studied the relation between the total number of impulses recorded from the respiratory muscles and their physical action [see General Discussion, at p. 459].

Campbell. I want to try and get some order into this discussion. We are spending a lot of time on the fundamental question of the extent to which one can argue from the registration of either muscular or neuronal events to interpretation of either mechanical events or central behaviour; and it seems to me that the sense of the discussion so far implies a certain caveat. It may be that one can argue mechanically from electrical events, and one may be able to argue about central controllers from peripheral neural events; but I think many of you would like to have the pedigree of Dr. Viljanen's work in a little bit more detail before accepting the fact that he met the various precautions that have to be taken into account. If so, I would like to give an opportunity of speaking to those who wish to consider the conclusions he is drawing from his work—or would you like to spend some more time on the fundamental question? [*Editor's note*: A further explanatory statement from Dr. Viljanen may be found in the General Discussion, at p. 459].

Wyke. Would you allow me further to belabour the fundamental point that you have just made, because not only is it cogent in relation to the conclusions we have just heard from Dr. Viljanen but also it is going to crop up again and again throughout this Symposium. This is because the problem that we neurologically-orientated people have in trying to fit our data to the data generated by what I might call our mechanically-orientated colleagues (measuring forces, pressures, flow rates and so on) is to correlate electrophysiological data obtained with electromyograms or electroneurograms with the measurements that are obtained with electromechanical devices such as pressure transducers and flow rate recorders. One of the fundamental reasons for this difficulty (as Dr. von Euler has said) is the lack of direct match between the integrated electromyographic output of muscles and the tension they develop, especially because the relationships are completely different in isometric and isotonic circumstances[2,3,4,5]. Therefore, without more direct technical description of what is being done here, I think we should regard Dr. Viljanen's matching of his measured curves to his predicted curves with a certain amount of reserve, as one cannot keep on multiplying caveats to the *n*th power.

---

1 Viljanen, A. A. (1966) Patterns of motor activity in breathing in response to varying sensory inputs, *Acta physiol. scand.*, **68,** (Suppl. 277), 1–212.

2 Lippold, O. C. J. (1954) The relation between integrated action potentials in a human muscle and its isometric tension, *J. Physiol. (Lond.),* **117,** 492–9.

3 Bigland, B., and Lippold, O. C. J. (1954) The relation between force, velocity and integrated electrical activity in human muscles, *J. Physiol. (Lond.),* **123,** 214–24.

4 Basmajian, J. V. (1967) *Muscles Alive: Their Functions Revealed by Electromyography,* 2nd ed., Baltimore.

5 Lenman, J. A. R. (1969) Integration and analysis of the electromyogram and related techniques, in *Disorders of voluntary Muscle,* 2nd ed., ed. Walton, J. N., pp. 843–76, London.

Beverly Bishop. I feel that the use of the electromyogram or the electroneurogram is really an index of what the respiratory 'centre' is putting out, so I think we can justify Dr. Viljanen's data on that score, at least.

I would like to call attention to the record he showed in which the number of impulses recorded from the diaphragm increased in response to inhalation of an hypoxic gas mixture. In my own work I have found that the diaphragm responds very quickly and dramatically to changes in the oxygen tension in an inspired gas mixture without any measurable change in minute ventilation—which quite disturbs me. I could even demonstrate that there could be a fall in the alveolar CO_2 measured at the airway (end-tidal CO_2) without a change in minute ventilation. Thus the total output of the system appeared not to be changing, yet there was apparently some change in alveolar ventilation. Using the over-all measure of minute ventilation would thus have distorted our interpretation of the response to this kind of chemostimulation—but using the electrical output signals from the muscles showed that there had been a change in the *pattern* of muscular activity, even though the total output remained the same. I was glad to see that Dr. Viljanen and I agree on this response of the diaphragm to minimal hypoxic stimulation.

Campbell. The intercostal activity was also shown to increase in Dr. Viljanen's recordings, if I remember rightly—that is, both intercostal and diaphragmatic activity increased.

Beverly Bishop. This is another point I wanted to ask about. I understood from Dr. Viljanen's presentation that hypoxia increased the total number of impulses recorded from the diaphragm without changing their frequency of discharge, whereas the intercostals changed their discharge frequency. This bothered me, in that one muscle was changing its frequency of discharge while the other was changing its total number of discharges. Maybe you would clarify that for me?

Viljanen. The relations are not at all clear between frequency of discharge and total number of impulses in different muscles during hypoxia.

Campbell. I feel that we must note, Dr. Bishop, that you have just presented us with a very difficult question—if, in fact, there is an increase in the activity of the muscles with a change in the blood gases but no change in ventilation.

Beverly Bishop. I used a 12 per cent oxygen mixture, with no measurable change in minute ventilation. These experiments have bothered me, and so are not yet published.

Campbell. I think you are right to be bothered!

Freedman. Can I ask for some clarification regarding your model, Dr. Viljanen? You used the model to predict what the subject's flow pattern ought to be, but on what data do you base these predictions? Do you specify the model to produce a given volume in a given time—I am particularly interested in whether you specify the time?

Viljanen. [Dr. Viljanen shows a slide [Fig. D.9] of his mathematical

MATHEMATICAL MODEL

SYSTEMS EQUATION

$$P_{mus}(t) = KV(t) + R\dot{V}(t)$$

I LOWER LEVEL PERFORMANCE CRITERIA

1. INSPIRATION

$$J_{I,1} = \int_0^{T_1} \left(\ddot{V}^2(t) + \alpha_1 P_{mus}(t)\,\dot{V}(t) \right) dt = \min_{V(\cdot)}!$$

BOUNDARY CONDITIONS:

$$t = 0\ ;\ V = V_0 \qquad \dot{V} = 0$$
$$t = T_1\ ;\ V = V_0 + V_T\ ,\ \dot{V} = 0$$

2. EXPIRATION

$$J_{I,2} = \int_0^{T_2} \left(\ddot{V}^2(t) + \alpha_2 P^2_{mus}(t) \right) dt = \min_{V(\cdot)}!$$

BOUNDARY CONDITIONS:

$$t = 0\ ;\ V = V_0 + V_T\ ,\ \dot{V} = 0$$
$$t = T_2\ ;\ V = V_0 \qquad ,\ \dot{V} = 0$$

$\Rightarrow$ OPTIMAL FLOWPATTERN $\dot{V}^*(t)$

II HIGHER LEVEL PERFORMANCE CRITERIA

$$J_{II} = \frac{1}{T_1+T_2+T_3} \left[\beta_1 \int P^2_{mus,i}(t)\,dt + \beta_2 \int P^2_{mus,e}(t)\,dt + \beta_3 \int_0^{T_1} \ddot{V}_i^2(t)\,dt + \beta_4 \int_0^{T_2} \ddot{V}_e^2(t)\,dt + \beta_5 \int_0^{T_1+T_2} dt \right] = \min_{T_1, T_2, T_3, V_0, V_D}!$$

CONSTRAINTS:

$$V(t) = V^*(t)$$

$$\dot{V}_A = \frac{V_T - V_D}{T_1+T_2+T_3}$$

$$R_A(V_D) = \frac{0.082}{V_D - 0.0825} + 0.4$$

$\Rightarrow$ OPTIMAL $T_1^*, T_2^*, T_3^*, V_0^*, V_D^*$

Fig. D.9.

model—*Ed.*] You ask something about this. In the case of low-level performance [Fig. D.9I] we have to specify tidal volume and inspiratory time and expiratory time. For computing a high-level performance run [Fig. D.9II], only the minute ventilation is given. The other individual data you see are constants.

Campbell. As your Chairman, could I ask you to get together with Dr. Cohen and Dr. Wyke and find a form of words that will convey the significance of this mathematical model to the readers of the Proceedings of this Symposium—it is obviously rather difficult to understand. [*Editor's note:* see Dr. Viljanen's supplementary explanation in the General Discussion, at p. 459].

9 Pulmonary Reflex Mechanisms in Ventilatory Regulation

J. G. WIDDICOMBE *

University Laboratory of Physiology, Oxford, England

The best-known reflex from the lungs is the Hering-Breuer inflation reflex, which is mediated by 'pulmonary stretch receptors' and is responsible for adjusting the pattern of breathing in response to changes in respiratory drive and in lung mechanical conditions (Widdicombe, 1964; Fillenz and Widdicombe, 1971; von Euler *et alii,* 1970). The lung receptors for this reflex are relatively insensitive to drugs and chemical irritants, and to pathological changes in the lungs. Although the rôle of the reflex has not been assessed quantitatively in conditions other than those with healthy lungs, there is little reason to doubt that it is qualitatively different in disease; in physiological and pathological conditions it will influence the pattern of breathing for any given minute volume, but it is unlikely to provide a significant tonic drive or tonic inhibition to the respiratory brain stem complex. In this sense it is a modulating rather than a controlling influence. The reflex has been discussed elsewhere in this volume [see von Euler, 1974, at p. 95] and will not be dealt with here.

Over the past 100 years other vagal reflexes from the lungs have been described, of which the characteristics do not fit well with the inflation reflex and the properties of pulmonary stretch receptors. These include the reflex responses to deflation and collapse of the lungs (Breuer, 1868) [Fig. 9.1, B], the 'paradoxical' (inspiratory) response to large lung inflations (Head, 1889; Reynolds and Hilgeson, 1965) [Fig. 9.1, A, C], the stimulation of breathing caused by inhalation of irritant gases, aerosols and dusts (Banister *et alii,* 1950; Widdicombe *et alii,* 1962), the rapid shallow breathing of multiple pulmonary micro-embolism (Whitteridge, 1950), the respiratory changes associated with increased pressures in the pulmonary vascular bed (Churchill and Cope, 1929; Downing, 1957) and the hyperpnoea that occurs with drug-induced bronchoconstrictions (Karczewski and Widdicombe, 1969a). Many of these responses are associated with reflex changes in bronchomotor tone, usually bronchoconstriction. In the last few years the lung receptor mechanisms that mediate these reflexes have been analysed, chiefly by recording nerve impulse traffic in single vagal fibres from the receptors, and two afferent pathways have

* Present address: Department of Physiology, St. George's Hospital Medical School, Tooting, London, England.

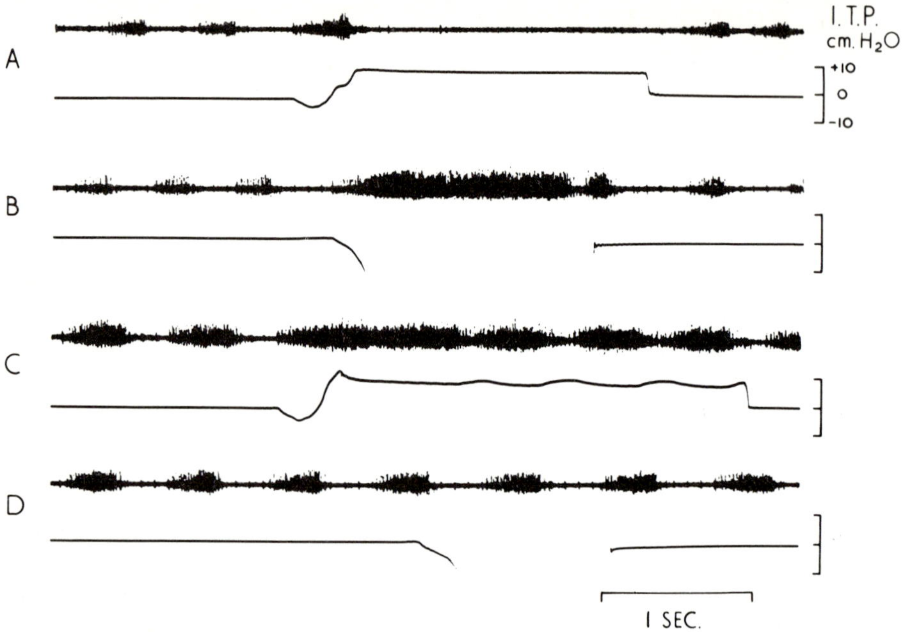

Fig. 9.1. Effect of inflation [A and C] and deflation [B and D] of the lungs of a rabbit on a diaphragmatic electromyogram (upper traces) and intratracheal pressure (lower traces). In [A] and [B] the vagus nerves were not cooled; inflation causes a transient increase in diaphragmatic activity followed by inhibition, whereas deflation greatly increases activity. In [C] and [D] the vagus nerves were cooled to 5°C: inflation now increases diaphragmatic activity (Head's paradoxical reflex) whilst deflation has no effect. (From Widdicombe, 1964.)

been described, studied and shown to be responsible for much or all of the reflex activity. Firstly there are 'lung irritant receptors', which lie in and under the epithelium of the intrapulmonary airways and have small diameter vagal myelinated afferent nerve fibres (Mills *et alii,* 1969); and secondly there are J-receptors, which occur in the alveolar wall and have non-myelinated vagal afferent fibres (Paintal, 1970). In many conditions of the lungs these two groups of receptors interact to drive respiration, and they also presumbly interact with the pulmonary stretch receptors (Adrian, 1933) to determine the pattern of breathing.

LUNG IRRITANT RECEPTORS

RECEPTOR RESPONSES

These have vagal myelinated fibres (conduction velocity, 3.6–25.8 m/sec in the rabbit), and respond to stimuli with a characteristically irregular discharge.

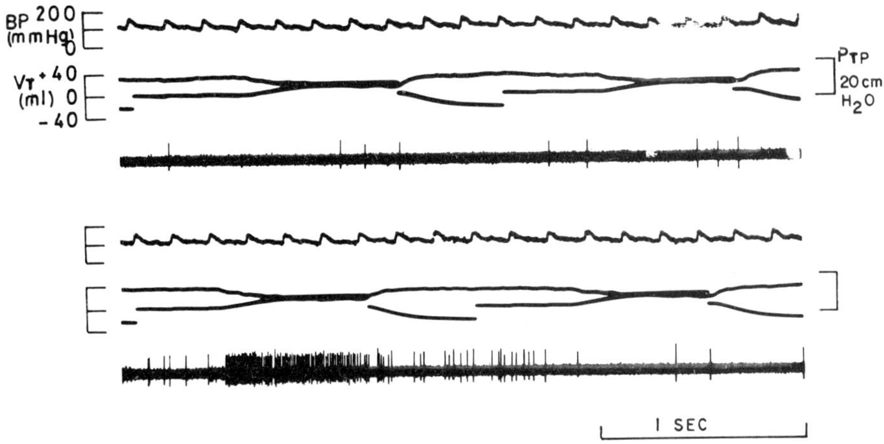

Fig. 9.2. Response of a lung irritant receptor to inhalation of carbon dust. Traces from above down: systemic arterial blood pressure [BP], transpulmonary pressure [P_{TP}], tidal volume [V_T], (zeroing at points of zero air flow) and action potentials in a single vagal nerve fibre from a lung irritant receptor. Upper record, control showing slow spontaneous discharge; lower record, during inhalation of dust, showing maximum stimulation of the receptor. The rabbit was paralysed and artificially ventilated and vagotomized. (From Sellick and Widdicombe, 1971.)

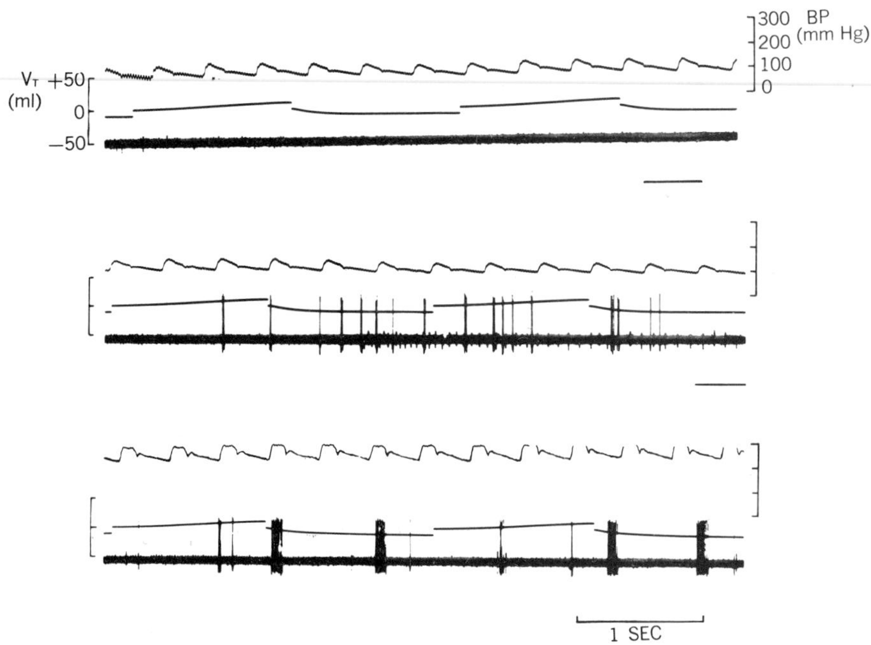

Fig. 9.3. Response of a pulmonary irritant receptor to intra-right atrial injection of histamine acid phosphate, 100 μg/kg (at signal in uppermost record) in a vagotomized, paralysed, artificially-ventilated rabbit: 5.5 sec between upper two traces, 20.5 sec between lower two. Histamine caused an increase in blood pressure and a receptor discharge without clear relation to respiratory phase. (From Mills *et alii,* 1969.)

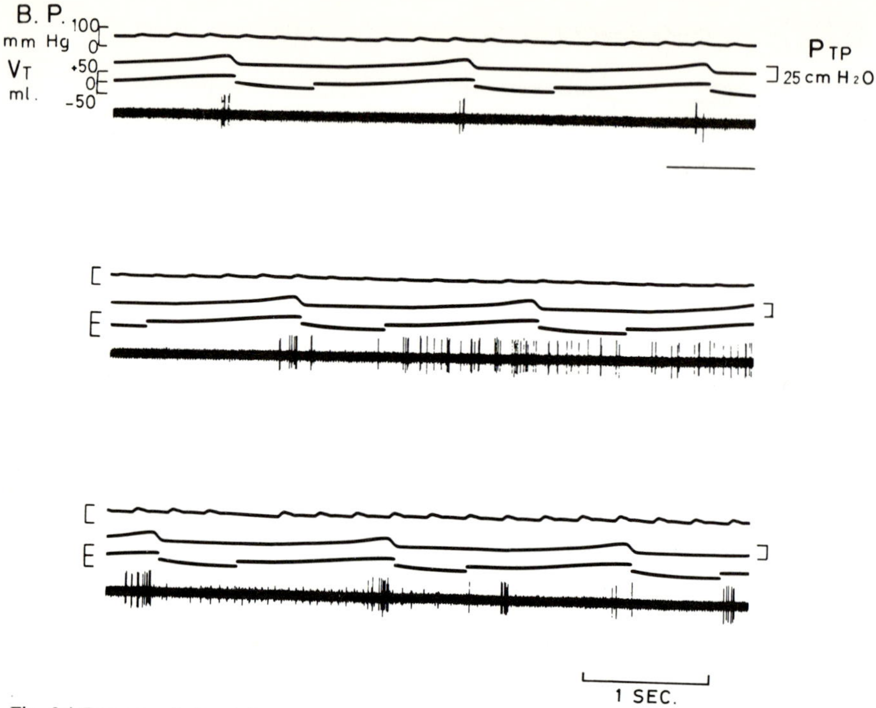

Fig. 9.4. Response of a lung irritant receptor to pulmonary micro-embolism. Traces from above down: blood pressure, transpulmonary pressure, tidal volume, action potentials and signal. The rabbit was paralysed and artificially ventilated. (From Mills *et alii*, 1969.)

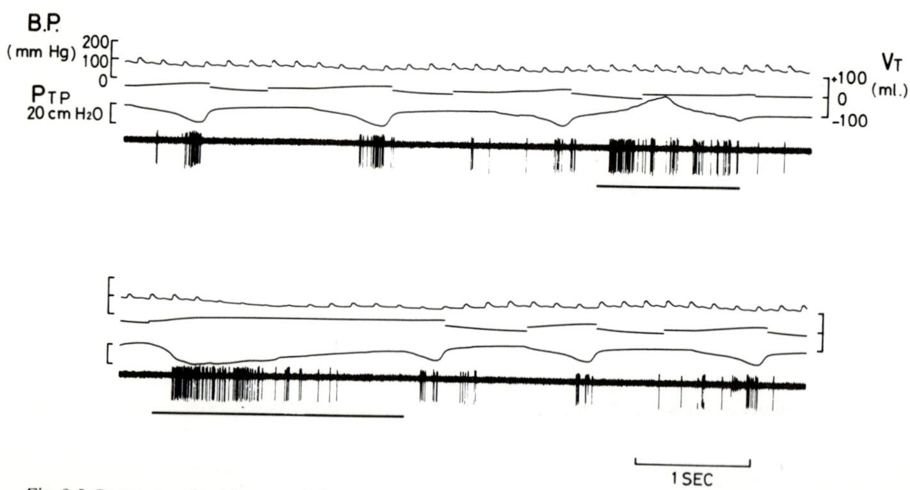

Fig. 9.5. Responses of a pulmonary irritant receptor to deflation (upper record) and inflation (lower record) of the lungs. From above down: systemic arterial blood pressure [BP], tidal volume changes [V_T, trace zeroing at points of zero air flow; inflation upwards], transpulmonary pressure [P_{TP}] and action potentials in a single vagal fibre. Deflation and inflation were during the horizontal signal bars. Note the rapidly-adapting irregular discharges. (From Mills *et alii*, 1969.)

They are stimulated by three groups of stimuli (Mills *et alii,* 1969; Sellick and Widdicome, 1969, 1971): (1) Inhalation of chemical irritants such as ethyl ether, ammonia vapour and cigarette smoke; and mechanical irritants such as 'inert' carbon dust [Fig. 9.2]. Presumably these stimuli act from the lumen of the airways. The receptor terminals reach up to the ciliary layer of the epithelium (see below), and would therefore be sensitive to chemical or mechanical irritation. It is not known whether the chemicals work by inducing mechanical deformation of the epithelium, or whether the mechanical irritants act by release of tissue chemicals; the responses appear similar in pattern. (2) Contraction of the smooth muscle of the airways by bronchoconstrictor drugs such as histamine, whether given as an aerosol or intravenously [Fig. 9.3]. Since these receptor responses are greatly inhibited by isoprenaline, presumably acting on the airway smooth muscle, the discharge seems to be secondary to smooth muscle contraction rather than a direct action of histamine. (3) Mechanical changes in the lungs which lead to a greater pull on the airways and distortion of the airway wall. This group includes pneumothorax, atelectasis, anaphylaxis, pulmonary congestion and multiple micro-embolism [Fig. 9.4]. In conditions like anaphylaxis and micro-embolism, there could also be release of active chemicals from damaged lung. In this group also is included large inflations and deflations of the lung [Fig. 9.5], and the passage of an endobronchial catheter. Thus the receptors seem to be stimulated by any sudden or severe mechanical change in their environment, and this is consistent with the rapidly-adapting discharge they exhibit. It now seems certain that these endings are the same as those studied by Knowlton and Larrabee (1946) and called 'rapidly adapting pulmonary stretch receptors'—a group which may also include tracheal cough receptors.

HISTOLOGY

Nerve fibres entering and ramifying in the epithelium of the airways from the trachea to the respiratory bronchioles can be demonstrated with silver and methylene blue staining methods (Larsell, 1921; Gaylor, 1934; Elftman, 1943). In the rabbit fine processes, staining for non-specific cholinesterase and continuous with nerve bundles, can also be seen to enter the surface epithelium. In some instances the epithelial terminations can be seen to arise from a myelinated axon. Electronmicroscopy of the tracheal epithelium shows the presence of varicose axones running towards the surface of the epithelium. The varicosities lack the vesicles characteristic of motor fibres (Fillenz and Woods, 1970).

The histological appearance of the receptors is consistent with their response to the various stimuli described above. Thus one would expect them to be stimulated by intraluminal irritants since they terminate so near the surface, and to be excited by an endobronchial catheter.

REFLEX ACTION OF LUNG IRRITANT RECEPTORS

Since lung irritant receptors are stimulated by deflation of the lungs and by pneumothorax, and since their reflex action is to cause hyperventilation and hyperpnoea, it has been concluded that they play a major rôle in the Hering-Breuer deflation reflex (Luck, 1970; Sellick and Widdicombe, 1970). The inhibition of pulmonary stretch receptor discharge by deflation of the lungs will also exert an excitatory action on breathing, except in man where the inflation reflex seems to play little part in eupnoea (Widdicombe, 1961).

Large inflations of the lungs can cause an inspiratory effort, or gasp, by a vagal reflex (Head, 1889; Widdicombe, 1964; Reynolds and Hilgeson, 1965). This is seen with intact vagus nerves, and when the vagus nerves of the rabbit are partially blocked by cold, inflation causes a maintained inspiratory effort—'Head's paradoxical reflex'. Lung irritant receptors are a likely candidate for this reflex. They are stimulated by inflation of the lungs and their reflex action is to augment breathing. In addition, the 'gasp reflex' is enhanced by collapse of the lungs and inhibited by over-inflation and increased compliance (Reynolds and Hilgeson, 1965): exactly similar properties have been described for lung irritant receptors (Sellick and Widdicombe, 1970).

In conditions like pulmonary anaphylaxis and histamine-induced bronchoconstriction there is vagal reflex hyperpnoea and bronchoconstriction (Karczewski and Widdicombe, 1969a, b; Koller, 1973) and stimulation of lung irritant receptors. Hyperpnoea also stimulates the endings (Sellick and Widdicombe, 1969). Thus there would be a tendency for augmentation of the reflex responses by positive feed-back; on the other hand this would be limited by the opposite action of pulmonary stretch receptors and blood gas changes.

Recent studies show that lung irritant receptors have a reflex action on the intrinsic muscles of the larynx. Injections or inhalations of histamine, which preferentially stimulates lung irritant receptors, increase the phasic discharge in inspiratory (abductor) motoneurones to the larynx (Szereda-Przestaszewska, 1971). The drug also increases the discharge in expiratory (adductor) motoneurones and increases the resistance to airflow through the larynx in expiration (Stransky, Szereda-Przestaszewska and Widdicombe, unpublished). Thus in the presence of histamine the larynx constricts more in expiration and probably dilates more in inspiration, at the same time as there is reflex bronchoconstriction (Karczewski and Widdicombe, 1969a; Mills *et alii,* 1970).

Finally, there is evidence that lung irritant receptors can cause unpleasant respiratory sensations. If the lung is first collapsed by breathing oxygen or by centrifugation of subjects, on re-inflation of the collapsed lung there is instantaneous pain or an unpleasant 'tearing sensation' (Ernsting, 1960; Burger and Macklem, 1968). The nature and timing of this sensation can only be explained by the action of lung irritant receptors, which are stimulated in these conditions

(Sellick and Widdicombe, 1970). J-receptors are not excited, and pulmonary stretch receptors no more than on inflation of normal lung. Passage of an endobronchial catheter causes an immediate painful or burning sensation in paralysed patients (Prys-Roberts, 1970). This excites only epithelial receptors. Inhalation of irritant gases and aerosols causes a painful or distressing sensation, or a sense of tightness over the chest. This stimulus would also excite J-receptors [see below]. It is possible that both lung irritant and J-receptors can give rise to unpleasant respiratory sensations, just as in the skin there are pain receptors with non-myelinated fibres and others with δ-myelinated fibres. The evidence that the dyspnoea of lung disease in patients may be mediated by lung afferent fibres has been summarized elsewhere (Guz *et alii,* 1970b). It fits with the involvement of lung irritant and/or J-receptors.

J-RECEPTORS

RECEPTOR RESPONSES

Paintal (1955) described the activity of 'specific deflation receptors'. He recorded from vagal non-myelinated fibres in the cat; but since their response to lung deflation and to pneumothorax was transient and weak, they have now been renamed 'juxta-pulmonary capillary receptors' or J-receptors. They are thought to lie close to the pulmonary capillaries. Some are weakly excited by large lung inflations (Coleridge *et alii,* 1965; Sellick and Widdicombe, 1970). Stimulation is caused by pulmonary congestion, pulmonary oedema, multiple micro-embolism, inhalation of strong irritants such as chlorine, and halothane (Paintal, 1969). Since they are stimulated within 2.5 sec by drugs such as phenyl diguanide injected into the right atrium, and within 0.3 sec by inhalation of ethyl ether or halothane vapour [Fig. 9.6], and since these response times are unaffected by circulatory arrest, Paintal concluded that the endings lie in the alveolar wall. The afferent vagal fibres have conduction velocities of 0.8–7.0 m/sec, so presumably some of the fibres are myelinated, but the majority conduct at less than 3.0 m/sec. Impulse frequencies during strong stimulation reach peak values of 20–50 impulses/sec, with average values of 7.5 impulses/sec.

HISTOLOGY

Non-myelinated fibres can be seen in the alveolar wall with silver staining (Fillenz, 1970; Fillenz and Widdicombe, 1971), but this method does not distinguish between afferent and efferent fibres. Electronmicroscopy has shown a few fibres which lack the vesicles characteristic of cholinergic or monoaminergic motor fibre terminals (Fillenz and Woods, 1970), but the fibres are rare and need not be nerve terminals or receptors. Merrick and Reid (1971) have observed

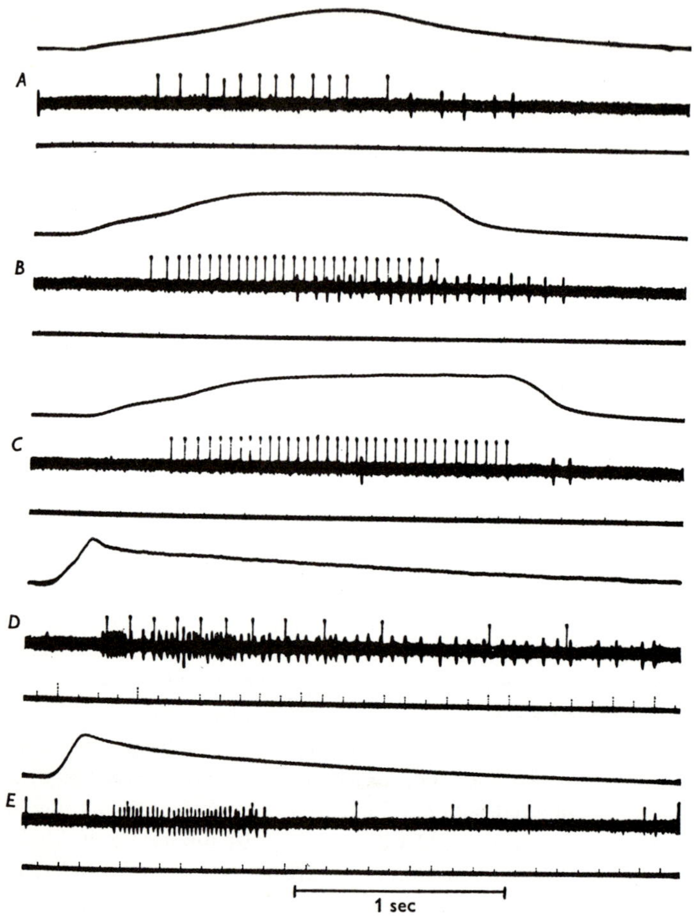

Fig. 9.6. Responses of type-J receptor (small diphasic spike) to inflation of the lung [A, B and C] with open chest and insufflation with halothane (60 ml) at [D] and [E]. The large monophasic spikes in A, B, C and D are those of a pulmonary stretch fibre and these show the consistent response to inflation of the lung with 60 ml in A and about 150 ml in B and C, in contrast to the variable response in the type-J fibre which is excited during the deflation phase in A and during inflation in B (but after a significant delay); there is comparatively little effect in C (again about 150 ml inflation). Insufflation of halothane had no excitatory effect on the pulmonary stretch fibre in D (normal circulation) in contrast to the marked excitation of the type-J receptor, which is excited with a similar latency in E after cutting the great vessels and removing the ventricles. From above downwards in each record: intratracheal pressure, impulses in filament and 0.1 sec time marks. Gain of amplifier for intratracheal pressure in A, D and E is twice that in B and C. (From Paintal, 1969.)

fibres in the alveolar wall of the rat which appear to be afferent and, on one occasion, a fibre which could be seen terminating in a cluster of receptor cells. Whether these fibres are connected to J-receptors has not been established.

REFLEX ACTION OF J-RECEPTORS

Phenyl diguanide is a convenient drug to use for the study of J-receptors. It is not a specific stimulant for them, since it can also excite lung irritant receptors and peripheral chemoreceptors (Dawes and Comroe, 1954; Mills *et alii,* 1969) and has central nervous actions (Glagowska and Widdicombe, 1973); but when injected in threshold doses into the right atrium, the short latency of its response indicates that its main reflex action is via J-receptors. This has been confirmed by the observation that the drug still produces reflex actions on breathing when the vagal myelinated fibres are blocked by anodal current or by cold (Guz *et alii,* 1970a). If vagal conduction is unimpaired, phenyl diguanide causes an expiratory apnoea followed by rapid shallow breathing in the cat, and an inspiratory apnoea followed by rapid shallow breathing in the rabbit. In both species there is hypotension and bradycardia.

Injections of phenyl diguanide also inhibit the monosynaptic hindlimb stretch reflex in the cat via a vagal afferent pathway (Deshpande and Devanandan, 1970). However, although it was concluded that this response was a primary reflex action of J-receptors, the inhibition of the reflex was greatly reduced by artificial ventilation of the cat; and this, and the fact that there was a close correlation between the response and the timing of the reflex hypotension, suggests that the spinal reflex changes may be partly secondary to blood gas and blood flow changes.

Stimulation of J-receptors by phenyl diguanide causes a reflex constriction of the larynx (amounting virtually to closure of the larynx) coincident with the reflex apnoea, and a vigorous discharge in laryngeal expiratory (adductor) motoneurones (Stransky, Szereda-Przestaszewska and Widdicombe, unpublished), as well as a reflex bronchoconstriction (Karczewski and Widdicombe, 1969a).

It seems probable that the J-receptors are part of a non-myelinated nociceptive system distributed throughout the lungs. This would be consistent with their weak mechanosensitivity; with the fact that they are stimulated by lung tissue damage caused by injection of hot water (Frankstein and Sergeeva, 1966) and by inhalation of chlorine; and with their response to pulmonary congestion and oedema (Paintal, 1970). Paintal has suggested that the true stimulus to the receptors is an increase in interstitial fluid volume in the alveolar wall secondary to an increase in pulmonary capillary pressure or permeability, and that the receptors may be stimulated in exercise. Their response to halothane is interesting, as this gas is not usually regarded as a strong irritant; however, halothane has been shown to cause tachypnoea by stimulation of lung receptors with non-myelinated vagal fibres (Kaminski *et alii,* 1968), and this is consistent with a J-receptor reflex.

INTERACTION OF REFLEXES

It is clear, from studies of action potential traffic in single vagal nerve fibres, that a number of stimuli in the lungs increase the discharge of both lung irritant receptors and J-receptors. These include inhalation of irritant gases, multiple micro-embolism and pulmonary congestion. Other stimuli may be more specific. For example, halothane and small doses of phenyl diguanide seem to act on J-receptors rather than on irritant receptors, and histamine aerosol and probably 'inert' dust seem to act on irritant receptors rather than on J-receptors.

There have been two attempts to differentiate the rôles of the two reflex systems in lung conditions. Guz and Trenchard (1971) developed the 'anodal block' technique whereby a direct current applied to the vagus nerve or nerves can be adjusted to block conduction in all myelinated fibres (e.g. from irritant receptors) while leaving conduction in non-myelinated fibres (e.g. from J-receptors) intact (Trenchard and Widdicombe, 1973). Using this method, they confirmed that J-receptors were chiefly responsible for the respiratory changes due to intravenously injected phenyl diguanide and to multiple micro-embolism in rabbits. Unfortunately they did not show that the reflex response to stimulation of irritant receptors was present in their experimental conditions; and it is at least possible that the irritant receptor reflex had been abolished by anaesthesia, since irritant receptors are known to be stimulated by micro-embolism and by appropriate doses of phenyl diguanide (Mills *et alii,* 1969).

Widdicombe and Glogowska (1973) adopted Guz and Trenchard's anodal block technique and showed that the respiratory responses to inhalation of histamine aerosol and to pulmonary oedema in rabbits were chiefly mediated by myelinated vagal fibres, presumably from lung irritant receptors; whereas the respiratory response to inhaled halothane was chiefly mediated by non-myelinated fibres, presumably from J-receptors. Again, a differential action of anaesthesia may have been present, although reflexes from both groups of receptor were qualitatively intact.

CONCLUSIONS

Both lung irritant and J-receptors are stimulated by a variety of physiological and pathological conditions in the lungs. Their reflex actions on breathing, laryngeal calibre and bronchomotor tone are superficially similar: J-receptors cause reflex bradycardia, hypotension and inhibition of the spinal muscle stretch reflex, but these three responses have not been determined for the irritant receptors. Presumably the two reflexes interplay in many lung conditions, and the action of pulmonary stretch receptors will also modify the reflex responses, but the research to establish quantitatively the relative rôles of the three reflexes is at a very early stage. The indications are that 'specific' reflexes from the groups of receptors are improbable, compared with an involvement and interaction of the three groups in most lung conditions.

ACKNOWLEDGEMENTS

I am grateful to Drs. Stransky and Szereda-Prezestaszewska for permission to describe their unpublished research, and to them and Drs. Guz and Trenchard for much valuable discussion.

REFERENCES

Adrian, E. D. (1933) Afferent impulses in the vagus and their effect on respiration, *J. Physiol. (Lond.),* **79,** 332–58.

Banister, J., Fegler, G., and Hebb, C. (1950) Initial respiratory responses to the intratracheal inhalation of phosgene or ammonia, *Quart. J. exp. Physiol.,* **35,** 233–50.

Breuer, J. (1868) Die Selbststeuerung der Athmung durch den Nervus vagus, *Akad. Wiss. Wien.,* **58** (II): 909–37.

Burger, E. J., and Macklem, P. (1968) Airway closure: demonstration by breathing 100 per cent O_2 at low lung volumes and by N_2 washout, *J. appl. Physiol.,* **25,** 139–48.

Churchill, E. D., and Cope, O. (1929) The rapid shallow breathing resulting from pulmonary congestion and oedema, *J. exp. Med.,* **49,** 531–7.

Coleridge, H. M., Coleridge, J. C. G., and Luck, J. C. (1965) Pulmonary afferent fibres of small diameter stimulated by capsaicin and by hyperinflation of the lungs, *J. Physiol. (Lond.),* **179,** 248–63.

Dawes, G. S., and Comroe, Jr., J. H. (1954) Chemoreflexes from the heart and lungs, *Physiol. Rev.,* **34,** 167–201.

Deshpande, S. S., and Devanandan, M. S. (1970) Reflex inhibition of monosynaptic reflexes by stimulation of type-J pulmonary endings, *J. Physiol. (Lond.),* **206,** 345–58.

Downing, S. E. (1957) Reflex effects of acute hypertension in the pulmonary vascular bed of the dog, *Yale J. Biol. Med.,* **30,** 43–56.

Elftman, A. G. (1943) The afferent and parasympathetic innervation of the lungs and trachea of the dog, *Amer. J. Anat.,* **72,** 2–28.

Ernsting, J. (1960) Some effects of oxygen breathing on man, *Proc. roy. Soc. Med.,* **53,** 96–98.

von Euler, C. (1974) Control of depth and rate of respiratory movements. This volume [p. 95].

von Euler, C., Herrero, F., and Wexler, I. (1970) Control mechanisms determining rate and depth of respiratory movements, *Resp. Physiol.,* **10,** 93–108.

Fillenz, M. (1970) Innervation of pulmonary capillaries, *Experientia (Basel),* **25,** 842.

Fillenz, M., and Widdicombe, J. G. (1971) Receptors of the lungs and airways, in *Handbook of Sensory Physiology,* Vol. 3, pp. 82–112, ed. Neil, E.; Heidelberg.

Fillenz, M., and Woods, R. I. (1970) Sensory innervation of the airways, in *Breathing: Hering–Breuer Centenary Symposium,* pp. 101–7, ed. Porter, R., London.

Frankstein, S. I., and Sergeeva, Z. N. (1966) Tonic activity of lung receptors in normal and pathological states, *Nature (Lond.),* **210,** 1054–5.

Gaylor, J.B. (1934) The intrinisic nervous mechanism of the human lung, *Brain,* **57,** 143–60.

Glogowska, M., and Widdicombe, J. G. (1973) The respiratory action of phenyl diguanide in vagotomised rabbits, *Acta neurobiol. exp.,* **33,** 81–7.

Guz, A., Noble, M. I. M., Eisele, J. H., and Trenchard, D. (1970a) The role of vagal inflation reflexes in man and other animals, in *Breathing: Hering-Breuer Centenary Symposium,* pp. 17–40, ed. Porter, R., London.

Guz, A., Noble, M. I. M., Eisele, J. H., and Trenchard, D. (1970b) Experimental results of vagal block in cardiopulmonary disease, in *Breathing: Hering–Breuer Centenary Symposium,* pp. 315–28, ed. Porter, R., London.

Guz, A., and Trenchard, D. W. (1971) The role of nonmyelinated vagal afferent fibres from the lungs in the genesis of tachypnoea in the rabbit, *J. Physiol. (Lond.),* **213,** 345–72.

Head, H. (1889) On the regulation of respiration, *J. Physiol. (Lond.),* **10,** 1–70; 279–90.

Kaminski, B., Karczewski, W., and Gorzynski, J. (1968) Influence of certain inhalation anaesthetics on the function of pulmonary stretch receptors, *Acta physiol. pol.,* **19,** 533–41.

Karczewski, W., and Widdicombe, J. G. (1969a) The role of the vagus nerves in the respiratory and circulatory responses to intravenous histamine and phenyl diguanide in rabbits, *J. Physiol. (Lond.),* **201,** 271–91.

Karczewski, W., and Widdicombe, J. G. (1969b) The role of the vagus nerves in the respiratory and circulatory reaction to anaphylaxis in rabbits, *J. Physiol. (Lond.),* **201,** 293–304.

Koller, E. A. (1973) Afferent vagal impulses in anaphylactic bronchial asthma, *Acta neurobiol. exp.,* **33,** 51–6.

Knowlton, G. C., and Larrabee, M. G. (1946) A unitary analysis of pulmonary volume receptors, *Amer. J. Physiol.,* **151,** 547–53.

Larsell, O. (1921) Nerve terminations in the lung of the rabbit, *J. comp. Neurol.,* **31,** 105–31.

Luck, J. C. (1970) Afferent vagal fibres with an expiratory discharge in the rabbit, *J. Physiol (Lond.),* **211,** 63–71.

Meyrick, B., and Reid, L. (1971) Nerves in rat intra-acinar alveoli: an electronmicroscopic study, *Resp. Physiol.,* **11,** 367–77.

Mills, J., Sellick, H., and Widdicombe, J. G. (1969) The role of lung irritant receptors in respiratory responses to multiple pulmonary embolism, anaphylaxis and histamine-induced bronchoconstriction, *J. Physiol. (Lond.)*, **203**, 337–57.

Mills, J. E., Sellick, H., and Widdicombe, J. G. (1970) Epithelial irritant receptors in the lungs, in *Breathing: Hering-Breuer Centenary Symposium*, pp. 77–92, ed. Porter, R., London.

Paintal, A. S. (1955) Impulses in vagal afferent fibres from specific pulmonary deflation receptors. The response of these receptors to phenyl diguanide, potato starch, 5-hydroxytryptamine and nicotine, and their role in respiratory and cardiovascular reflexes, *Quart. J. exp. Physiol.*, **40**, 89–111.

Paintal, A. S. (1969) Mechanism of stimulation of type-J pulmonary receptors, *J. Physiol. (Lond.)*, **203**, 511–32.

Paintal, A. S. (1970) The mechanism of excitation of type-J receptors and the J-reflex, in *Breathing: Hering-Breuer Centenary Symposium*. pp. 59–70, ed. Porter, R., London.

Prys-Roberts, C. (1970) In *Breathing: Hering-Breuer Centenary Symposium*, p. 249, ed. Porter, R., London.

Reynolds, L. D., and Hilgeson, M. D. (1965) Increase in breathing frequency following the reflex deep breath in anesthetised cats, *J. appl. Physiol.*, **20**, 491–9.

Sellick, H., and Widdicombe, J. G. (1969) The activity of lung irritant receptors during pneumothorax, hyperpnoea and pulmonary vascular congestion, *J. Physiol. (Lond.)*, **203**, 359–82.

Sellick, H., and Widdicombe, J. G. (1970) Vagal deflation and inflation reflexes mediated by lung irritant receptors, *Quart. J. exp. Physiol.*, **55**, 153–63.

Sellick, H., and Widdicombe, J. G. (1971) Stimulation of lung irritant receptors by cigarette smoke, carbon dust, and histamine aerosol, *J. appl. Physiol.*, **31**, 15–19.

Szereda-Przestaszewska, M. (1971) Activity of vagal respiratory motoneurones in anaphylactic shock in rabbits, *Acta allergologica*, **26**, 17–38.

Trenchard, D., and Widdicombe, J. G. (1973) Assessment of differential block of conduction by direct current applied to the cervical vagus nerve, *Acta neurobiol. exp.*, **33**, 89–96.

Whitteridge, D. (1950) Multiple embolism of lung and rapid shallow breathing, *Physiol. Rev.*, **30**, 475–87.

Widdicombe, J. G. (1961) Respiratory reflexes in man and other mammalian species, *Clin. Sci.*, **21**, 163–70.

Widdicombe, J. G. (1964) Respiratory reflexes, in *Handbook of Physiology*, Section 3: Respiration, Vol. I, pp. 585–630, Washington, D.C., American Physiological Society.

Widdicombe, J. G., and Glogowska, M. (1973) Relative roles of irritant, type-J and pulmonary stretch receptors in lung reflexes, *Acta neurobiol. exp.*, **33**, 21–31.

Widdicombe, J. G., Kent, D. C., and Nadel, J. A. (1962) Mechanism of bronchoconstriction during inhalation of dust, *J. appl. Physiol.*, **17**, 613–16.

DISCUSSION ON COMMUNICATION [9]

Mead. Could we ask you to fill in the Table that you started to construct on the blackboard?

Widdicombe. I will, if you like. [*Editor's note:* Dr. Widdicombe constructed Table D.1 on the blackboard.]

TABLE D.1

REFLEX INFLUENCES OF PULMONARY RECEPTORS

MUSCLES	PULMONARY STRETCH RECEPTORS	J RECEPTORS	LUNG IRRITANT RECEPTORS
Inspiratory muscles	Inhibition	Inhibition	Stimulation
Expiratory muscles	?	Inhibition	Inhibition
Laryngeal abductors (inspiratory)	Inhibition (? stimulation)	Stimulation	Stimulation
Laryngeal adductors (expiratory)	Inhibition	Stimulation	Stimulation
Laryngeal resistance	Decreased	Increased (may be sustained)	Increased (never sustained)

I should really also show a slide here, because the action of the inflation reflex of the abductor (inspiratory) type of motoneurone is complicated [see Fig. D. 10]. The inflation was made at the break in the signal line [Fig. D.10, A]; and you notice that this results in a continuous discharge in the recurrent laryngeal nerve fibre—so one would say that the inflation reflex stimulates the abductor nerve fibres. However, between run [A] and run [B] we gave the animal 18 mg of pentobarbitone (a fairly small dose for these cats); and now exactly the same stimulus (at the break in the signal line in Fig. D.10, B) causes no activity, but instead complete inhibition. So I would certainly put inhibition here [see Table D.1., column 2], but query the possibility of some stimulation if the anaesthetic conditions are appropriate. The adductors are also inhibited [Table D.1., column 2] and, as you would expect from that, you get a decrease in laryngeal resistance. In face of this, it seems that the pulmonary stretch receptors switch off all laryngeal muscular activity, or sometimes may bring in a bit of abductor activity.

Cohen. I am afraid Dr. Widdicombe has complicated things by bringing out a new observation; namely, that many laryngeal expiratory fibres have 'late-expiratory' discharge patterns. This conflicts with my own observations, as well

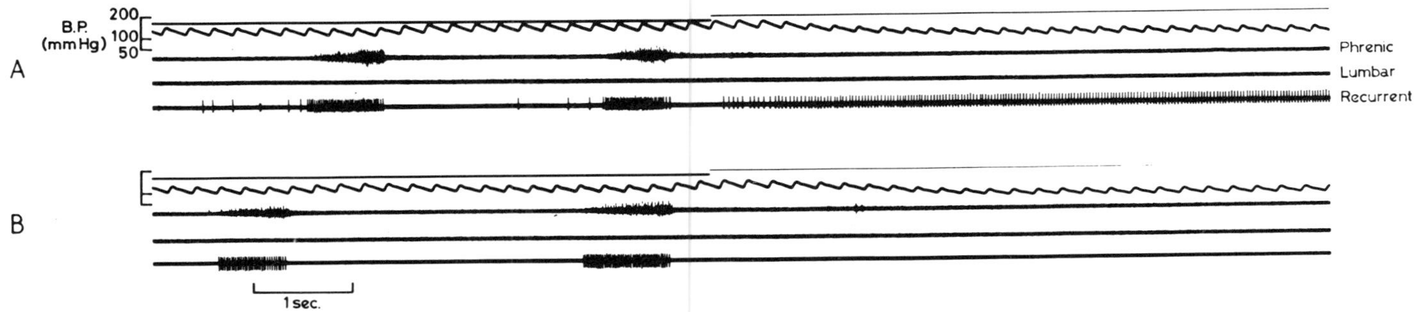

Fig. D.10. Response of an inspiratory recurrent laryngeal motoneurone to lung inflation. Traces from above down: signal, blood pressure, phrenic electroneurogram, lumbar electroneurogram, recurrent laryneal nerve fibre activity. In [A] inflation inhibits the phrenic discharge and causes a continuous discharge in the laryngeal nerve fibre. [B] is after an intravenous injection of 18 mg of pentobarbitone sodium: inflation now inhibits both phrenic and recurrent laryneal nervous activity.

as with several in the literature (Eyzaguirre and Taylor, 1963[1]; Barillot and Bianchi, 1971[2]), to the effect that laryngeal expiratory fibres have early-expiratory discharge patterns. From the teleological point of view, if one thinks in terms of laryngeal resistance it seems reasonable that these fibres should have early-expiratory patterns, since their discharge tends to close the glottis and thus prolong the time of lung expansion by exerting a 'braking' action on lung collapse. Conversely, the late-expiratory discharge patterns observed in thoracic and abdominal expiratory muscles (Floersheim, 1960)[3] act to empty the lungs below the level of FRC. The latter effect would be opposed by late-expiratory laryngeal discharge—which seems to lead to a teleological puzzle.

Widdicombe. I quite agree: I am aware of that problem. Why we get mainly end-expiratory responses and other people get early expiratory ones I don't know: it might well be an effect of the anaesthetic state, or of the animal's general condition. We have seen very few early expiratory responses when recording from the recurrent laryngeal nerve.

Cohen. To elaborate the point further, my own (unpublished) recordings from the whole recurrent laryngeal nerve (in decerebrate cats) show that expiratory activity is maximal at the start of the expiratory phase, and that the minimum activity in the nerve occurs in the late expiratory phase. So possibly the type of response is, to some extent, a function of the anaesthetic or the decerebrate situation.

Widdicombe. Our studies were all made on cats under pentobarbitone anaesthesia, and not on decerebrate preparations.

Plum. To what degree is the pattern (or the phase) of the activity actually dependent on the anaesthetic situation and the type of preparation used? I ask this because of the common clinical observation in hyperpnoeic subjects with congested lungs [see also Communication 13] that expiratory abdominal muscular activity is obvious—one can palpate it: but you have described only three reflex systems, none of which facilitates expiratory muscular activity [see Table D.1].

Another question is, how much hyperpnoea is evoked by stimulation of the pulmonary reflex system alone, without chemoreceptor enhancement?

Widdicombe. How much hyperpnoea does it cause? Well, I am sorry to be obvious, but it depends on the size of the stimulus. If you stimulate (in cats) the irritant receptors by (say) half a minute of inhalation of histamine aerosol—which is probably what most human subjects would require to cause fairly mild bronchoconstriction—and you measure blood gas tensions, then you find that the PCO_2 may drop from something like 30 mm Hg (which you find in most nor-

---

[1] Eyzaguirre, C., and Taylor, J. R. (1963) Respiratory discharge of some vagal motoneurones, *J. Neurophysiol.*, **26**, 61–78.

[2] Barillot, J.-C., and Bianchi, A.-L. (1971) Activité des motoneurones laryngés pendant les reflexes de Hering-Breuer, *J. Physiol. (Paris)*, **63**, 783–92.

[3] Floersheim, G. L. (1960) Zur elektromyographischen Charakterisierung der Atmungsactivität, *Helv. physiol. pharmacol. Acta*, **18**, 17–24.

mal cats) to 20 mm Hg. PO_2 also decreases, suggesting that there are ventilation/perfusion disturbances in the lungs as well—which complicates things. But the tensions of both blood gases will drop an appreciable amount with fairly mild stimulation of irritant receptors. As for the J-reflex, I cannot really answer the question because it is more complicated, as there is first a period of apnoea that allows the blood gases to become asphyxial and then you may have some relative hyperventilation.

With regard to the reflexes not causing expiratory abdominal muscle contraction, we have not studied these reflexes in decerebrate animals (apart from one or two). All I can say is that in animals anaesthetized with pentobarbitone we don't get any expiratory muscle contractions, even when we deliberately let them get rather light: so presumably your clinical patients must have some other mechanism coming in—or they are different from cats, quite rightly!

Wyke. I want to suggest that this problem of different patterns of activity in relation to anaesthesia might, in fact, be a non-problem, in view of the fact that a few years ago we reported[1] some studies on the differential effects of changing monitored states of barbiturate anaesthesia on the reflex activity of the individual muscles of the larynx. I am going to say a little about this at the end of this week [see Communication 24]; but perhaps I should point out now that in actual fact the mutually related patterns of contraction and relaxation of the intrinsic laryngeal muscles in relation to the respiratory cycle are very much a function of the stage of anaesthesia obtaining at the time. So I would suggest that we do not get too worked up about the apparent discrepancies, providing we remember that the degree of anaesthesia is a critical variable in this type of experiment (and should therefore always be assessed in the conduct of such experiments).

von Euler. I should like to ask Dr. Widdicombe whether he has any information from this work about changes in the responsiveness of the various reflexes with time during the different phases of the respiratory cycle. We know from Knox's's[2] work in our Laboratory that the shortening of the expiratory phase in response to deflations is much more marked in the early part of expiration, and that there is a period of complete non-responsiveness at the end of the expiratory phase. The last period of the expiratory phase also shows a complete lack of responsiveness to inflation of the lungs.

Widdicombe. I have not done any experiments to test this point—although, knowing of your work, I have thought about how the experiments could be done. They are not quite as easy as they might seem because, as far as I know, there is no very easy way in which one can stimulate selectively either J-receptors or irritant receptors briefly (say, for a quarter of a second) to see what they do in a particular phase of respiration.

You certainly cannot do that with J-receptors—there is no possible way of doing it. In theory you could do it with the irritant receptors if, for example, you could cause a sudden deflation (just for a quarter of a second) in a certain phase of respiration. Up to now I have not tried this experiment, because I had assumed

[1] Wyke, B. D. (1968) Effects of anaesthesia upon intrinsic laryngeal reflexes: an experimental study, *J. Laryngol.*, **82,** 603–12.

[2] Knox, C. K. (1973) Characteristics of inflation and deflation reflexes during expiration in the cat. *J. Neurophysiol.*, **36,** 284–95.

that if you did you would get an interaction between the effects of the irritant receptor afferent volley and (say) the inhibition of pulmonary stretch receptor activity that accompanies lung deflation. But I think, in view of your observation that at the beginning of the expiratory pause the pulmonary stretch receptors are ineffective on the respiratory cycle, that this might be a useful time at which to test the effect of a volley of irritant receptor discharges—I think that is the only time in the respiratory cycle at which one could put in a volley of irritant receptor discharge without it interacting with the pulmonary stretch receptor effect.

von Euler. If the deflation effect is caused by the irritant receptors, then this period at the end of expiration should also lack responsiveness to these receptors.

Widdicombe. Yes, I agree. To get an irritant receptor response you have got to have a fairly large inflation (quite possibly larger than you have used), or a fairly vigorous, quick deflation. Have you tried that? And if nothing happened, then this would indicate that the irritant receptors have no action when firing off in that phase.

Campbell. Maybe something slipped by me there: are there no deflation receptors now?

Widdicombe. Deflation receptors don't exist, except in Switzerland.

Campbell. Now you have thoroughly confused us!

Widdicombe. I will give you the historical background. Paintal (1955)[1] originally called the type J receptors 'specific deflation receptors'; but they do not respond to lung deflation at all, and that is one reason why he changed the name to 'type J'. A few years ago some research students working with me at Oxford[2,3,4], and some workers in Basle[5,6], recorded from afferents from lung receptors that the Oxford group called 'irritant receptors' and the Basle group 'collapse receptors'. We have stuck to 'irritant receptors'—but there are *no* deflation receptors; and no matter whether you call them irritant or collapse receptors, they are certainly different from J-receptors.

Campbell. I suspect that one of the most 'irritating' things you can do to a lung is to deflate it!

---

[1] Paintal, A. S. (1955) Impulses in vagal afferent fibres from specific pulmonary deflation receptors, *Quart. J. exp. Physiol.*, **40,** 89–111.

[2] Mills, J., Sellick, H., and Widdicombe, J. G. (1969) The role of lung irritant receptors in respiratory responses to multiple pulmonary embolism, anaphylaxis and histamine-induced bronchoconstriction, *J. Physiol. (Lond.),* **203,** 337–57.

[3] Sellick, H., and Widdicombe, J. G. (1969) The activity of lung irritant receptors during pneumothorax, hyperpnoea and pulmonary vascular congestion, *J. Physiol. (Lond.),* **203,** 359–82.

[4] Sellick, H., and Widdicombe, J. G. (1970) Vagal deflation and inflation reflexes mediated by lung irritant receptors, *Quart. J. exp. Physiol.*, **55,** 153–63.

[5] Koller, E. A. (1967) Atmung und Kreislauf in anaphylaktischen Asthma bronchiale des Meerschweinchens. III. Die lungen Veränderungen in Asthmaanfall und die inspiratorische Reaktion, *Helv. physiol. pharmacol. Acta,* **26,** 153–70.

[6] Homberger, A. C. (1968) Beitrag zur Nachweis von Kollapsafferenzen im Lungenvagus des Kaninchens, *Helv. physiol. pharm. Acta,* **26,** 97–118.

Moran. I should like to ask Dr. Widdicombe something about the effects of pulmonary emboli. In about 100 per cent, I think, of some kinds of animal experiments micro-emboli result in bronchoconstriction; and there are about nine or ten cases in the literature in which pulmonary emboli have been described as presenting in human subjects with clinical evidence of so-called bronchoconstriction. We have come across about ten such cases presenting like asthma, in our experience of about four hundred people investigated from the point of view of their pulmonary function. Now you specifically mentioned *micro*-emboli—and the question I am asking is, why 'micro' in view of the plight of the human subject? Is this the general clinical experience—and what about large emboli? Finally, what about the field of the lung, as it were, which is reflexly affected by this bronchoconstriction: is it only the part containing the emboli, or does the effect spread?

Widdicombe. There are several difficult questions tucked away there. In animal work, as I am sure you know, large emboli do not cause much change in the pattern of breathing or in bronchomotor tone—they have to be micro-emboli of 50 to 100 μ diameter; and most animal workers believe that these do cause reflex bronchoconstriction. I can only speculate as to why this might not be apparent in man (that is, in your patients), but it may be because various compensatory mechanisms would tend to cancel each other out. For example, if the patients were hyperventilating and their PCO_2 were going down, this would have a bronchodilator action. If they were taking deep breaths, again this would have a bronchodilator action—which might not occur in animals. But when I say that micro-emboli cause bronchoconstriction, this is a primary reflex which can either be abolished or even reversed by secondary actions due to responses like changes in blood gas tensions and changes in lung volume—but that is speculation.

With regard to where in the lung the bronchoconstriction occurs, I don't think anybody has tested this very carefully. You certainly get it in both lungs if the emboli go into one lung—this is a well-established effect in animals. Most physiologists assume (and I think it quite plausible) that there is a generalized constriction of the innervated airways (in other words, the rather large airways) in both lungs—but I would not like to be asked to give evidence to back that up.

Proctor. From the point of view of teleology (which has been brought up), one possible usefulness of constriction of the laryngeal lumen would be to increase the deposition of inhaled particulate matter in the upper trachea and the larynx itself—which does not occur in the normally open glottis.

A question now: have you ever studied electrical activity in the superior laryngeal nerve, the external branch of which supplies the cricothyroid muscle—which is a fairly potent adductor of the vocal folds?

Widdicombe. Drs. Stransky and Szereda-Przestaszewska, working with me, have done this recently—but not more than half a dozen times. The results of these half-dozen experiments are not extensive enough for conclusions to be drawn—but the suggestion is a good one.

Godfrey. To revert back to the pulmonary embolism story, surely this is just a question of the time at which the patients are studied. Most people who have studied patients in the acute phase of pulmonary embolism have found airway obstruction—and we certainly have. But if you study them later on you might not find this effect going along with the hypoxia, which is universal in acute pulmonary embolism but which passes off after some hours. I think that might explain the apparent discrepancy.

Could I ask Dr. Widdicombe one question: do you now know any effective way of blocking J- and irritant receptors?

Widdicombe. I wish I did. People are trying hard to find ways of doing it, as I am sure you know. I think the most promising is the way in which Guz and Trenchard[1] are looking at the problem at the moment—which is to try to block the irritant receptors with local anaesthetic aerosols while leaving the J-receptors unaffected.

Godfrey. Might I carry on? The reason I am asking this is particularly because of asthma—and especially because of the effects of exercise in asthmatics, which have all the characteristics (or at least some of the characteristics, anyway) of these reflexes. Both irritant and J-receptors might be involved in the bronchoconstriction that comes on in such patients within a minute of stopping exercise, perhaps because the receptors in the asthmatic are unduly sensitized in response to exercise.

Widdicombe. I think that is very likely. We know the irritant receptors are stimulated by bronchoconstriction, and by stimuli like mucus in the respiratory tract. It is more difficult to see how the J-receptors could come in under these conditions, unless there is pulmonary congestion after exercise. One problem about exercise-induced asthma—if it is not completely irrelevant, Mr. Chairman—is that there is some difference of opinion as to whether or not it is atropine-sensitive. If the effect is due to reflexes, it should be atropine-sensitive: on the other hand, nearly all the doses of atropine that have been given to test this in man have really been homeopathic.

Murakami. Dr. Widdicombe showed a very beautiful recording of the activity in a single expiratory fibre in the recurrent laryngeal nerve—but how did you distinguish this so-called expiratory activity from the expiratory activity associated with whining or straining or phonation? It depends on the degree of anaesthesia, surely?

Widdicombe. Naturally we made certain assumptions. We called the fibres expiratory because they fired in the pause between phrenic discharges, or while the abdominal muscles were contracting in expiration in a period of respiratory activity that looked as though it was spontaneous and eupnoeic. One can't say, of course, that there wouldn't have been some kind of noise coming from the larynx if it had been connected in the circuit—except that in the control situation the fluctuations are fairly small (and are only just measurable as resistance changes, in fact). The glottal resistance tracing in the control preparations showed little measurable resistance variation, although these expiratory fibres

---

[1] Guz, A., and Trenchard, D. W. (1971) The role of non-myelinated vagal afferent fibres from the lungs in the genesis of tachypnoea in the rabbit, *J. Physiol. (Lond.)*, **213**, 345–72.

were found in all the preparations. So whilst I agree that we do not know that activity in these fibres might not be associated with some kind of vocalization in the respiratory tract were the larynx in series with the lungs, we simply have to assume that if they fire on expiration they are expiratory neurones.

Murakami. In my experience of animal experiments, I have never succeeded in making a recording of expiratory activity in the recurrent laryngeal nerve—either with electromyographic recordings from the adductor muscles, or with neurograms recorded directly from the nerve branch innervating an adductor muscle.

Widdicombe. I can check the figures, but speaking from memory of what is more or less random sampling (assuming the fibre sizes are the same) we have had about 80 inspiratory and about 30 or 40 expiratory fibres. It is just possible that on occasions my colleagues have only looked for the expiratory fibres, so it may not be a completely random sample—but certainly there has been no problem finding them. It is also true, though, that if you stimulate laryngeal activity with reflexes like coughing, then the majority of the silent fibres that come in are expiratory ones. There is very little recruitment of inspiratory fibres—they are mainly firing off all the time—but you get a lot of recruitment of expiratory fibres in these circumstances.

Mårtensson. I think it is the experience of some workers (and perhaps this relates to what Dr. Wyke and Dr. Murakami have already mentioned) that activity in the expiratory nerve fibres is easily wiped out by general anaesthesia, whereas that in the inspiratory fibres is much more resistant to general anaesthetic depression.

Campbell. You are referring to the recurrent laryngeal nerve, are you not?

Mårtensson. Yes.

Murakami. I don't know whether this question is relevant to your Communication or not, but although I have never studied the afferent nerve fibres from the thorax in the recurrent laryngeal nerve that respond to deflation or inflation of the lungs, all of us know that the Hering-Breuer reflex is easily disturbed after bilateral dissection of the vagus nerves. I really want to know, what is the physiological significance of that afferent discharge in the recurrent laryngeal nerve?

Widdicombe. I looked at this some years ago,[1] and it is quite true that there is a number of afferent nerve fibres coming up the recurrent laryngeal nerve from the chest. I am not sure, quantitatively, how many there are compared with the vagus—but I would have thought only a very small percentage. You certainly get a number of cardiovascular afferents, a few pulmonary stretch afferents and a few cough receptor fibres coming up there; but I have always assumed that these were a minority of fibres that happened to have lost their way and so go up the

---

[1] Widdicombe, J. G. (1966) Action potentials in parasympathetic and sympathetic afferent fibres to the trachea and lungs of dogs and cats, *J. Physiol. (Lond.)*, **186,** 56–88.

recurrent laryngeal nerve, join the superior laryngeal nerve, and then pass back into the vagus that way. These pathways were mapped out to some extent by Jewett (1964)[1] a few years ago; and Dr. Kirchner and his colleagues[2] have also found cardiovascular afferent fibres traversing the recurrent laryngeal nerves. I have always suspected that the total number of these fibres is so small that they have very little physiological significance because, as you say, if you cut the vagus nerves and leave the recurrent laryngeal nerves intact then virtually all the lung reflexes go.

Campbell. But what about the vagal block experiments now? In a normal man breathing quietly, vagal blockade has no effect on his breathing but it does affect his response to CO_2. In man with various diseases of the lungs, blockade of the vagi does affect the pattern of breathing—and in normal and in diseased states there is prolongation of the breath-holding time. Can you put this together in a story that correlates the different receptors and their respective actions?

Widdicombe. No. I think it is a very complicated matter, as you well know. I think it very likely that in the conditions that obtain in a man with hyperventilation and dyspnoea and short breath-holding time (and in whom vagal block prolongs the breath-holding time and relieves the sensation of dyspnoea) the afferent pathway coming up from the lungs is from irritant or J-receptors, mainly because the pulmonary stretch receptors don't fit into this picture at all: I think the evidence is in favour of the irritant receptors—but I may be biased. I don't know of any direct evidence on this, until the current experiments with lignocaine aerosol are extended: specific block of the irritant receptors in man will answer this problem, and preliminary experiments by Petit (1970)[3] indicate that irritant receptors in the airways may be involved.

Godfrey. The relevant experiments have already been done and I am going to tell you about them this afternoon [see Communication 11]—so I will leave it for now.

Murakami. One very short question. Do you think the laryngeal abductor is an inspiratory muscle or a tonic muscle?

Widdicombe. All I can talk about is what I have found in the anaesthetized cat, in which there are various differing patterns of inspiratory abductor fibre discharge. The firing of the majority of these fibres is limited to inspiration and is simultaneous with a phrenic discharge; but you do find a number (one of which I showed) in which the discharge carries on into expiration—and you even find some (which may not be going to the larynx at all) where you have a continuous discharge that is slightly accentuated in inspiration, but these are a small minority. The great majority of the fibres we found are clearly inspiratory, and we have

---

[1] Jewett, D. L. (1964) Direct nervous pathways between the cervical vagus and the aortic or superior laryngeal nerves traversed by baroreceptor, 'pulmonary stretch' and other fibres in the cat, *J. Physiol. (Lond.)*, **175**, 358–71.

[2] Strauss, S. G., Fukuda, H., and Kirchner, J. A. (1972). Personal communication.

[3] Petit, J. M. (1970) General discussion, in *Breathing: Hering–Breuer Centenary Symposium*, pp. 111–15, ed. Porter, R., London.

simply assumed that they are abductor in function—but, of course, we have no direct evidence to prove that.

Wyke. May I say something à propos of that question? Again it depends on the precise stage of general anaesthesia, or whether you are looking at an un-anaesthetized human being electromyographically, as to what muscles you would regard as 'tonic' and what you would regard as respiratory abductors and adductors. I don't want to go into this now, because I am going to deal with it in detail in my Communication (No. 24) later in the week.

Campbell. I would just like to remind you—in the general spirit of a consideration of control systems—that Sir Victor Negus stated[1] that there was contraction of the laryngeal adductor muscles during expiration, and that this might have some effect on the distribution of air in the lungs or on the pulmonary circulation. But I don't know whether these observations have good support, and if they do have good support whether these may be lung-laryngeal interactions or centrally organized control mechanisms.

Widdicombe. A very good idea. What we are going to do—in fact, the week after next—is to close up that loop between larynx and trachea and see whether or not these laryngeal constrictions feed back to the lung and change tidal volume and ventilation. We have not yet done that.

Godfrey. Bearing precisely on that, of course, is the absolutely beautiful example provided by the baby with the respiratory distress syndrome. The respiratory 'grunt' heard in this syndrome is due to laryngeal occlusion, which represents (if you like) an attempt to maintain lung inflation, to prevent collapse. This is just the situation where your irritant and J-receptors are both being stimulated like anything in the baby's lungs—so here is your laryngeal adaptation. In fact, if you intubate such a baby and don't put it on positive airway pressure ventilation you do it harm.

---

[1] Negus, V. E. (1949) *The Comparative Anatomy and Physiology of the Larynx*, London.

10 Respiratory Control Mechanisms in Loaded Breathing

S. FREEDMAN, E. R. KAFER*
AND D. J. C. READ†

Departments of Medicine and Anaesthetics,
Royal Postgraduate Medical School,
London, England

During the normal course of life the respiratory muscles maintain ventilation and pulmonary gas exchange in the face of a frequently changing mechanical load due, for example, to speech and changes of posture and to external factors such as clothing and furniture. Changes in mechanical load also occur in disease. It is often difficult to measure these changes in load; but in order to study how the respiratory system adapts to them we can crudely imitate the natural situation and examine the changes consequent upon the imposition of added external loads. Such loads can be characterized in the same fluid mechanical terms as we use to measure the mechanical properties of the respiratory system; i.e. as resistance and elastance. This account of respiratory control mechanisms in loaded breathing reflects our interest in certain aspects of the subject, and we shall describe mainly recent work.

In normal man, the main response to an added load is due to the inspiratory muscles, whether the load is added during inspiration or expiration. If added during expiration, there is usually an increase in end-expiratory level due to the action of inspiratory muscles, and Campbell (1958) has shown that there may be no electrical activity in expiratory muscles with expiratory resistive loads as big as 10 cm H_2O/l/sec. Further, from the control point of view, respiratory centre output appears to be directly related to inspiratory, but not to expiratory, muscle activity (Milic-Emili and Tyler, 1963). For these reasons we shall concentrate on the effects of inspiratory loading.

THE RESPONSE TO CONTINUOUS LOADING

When inspiratory loads were imposed on normal conscious human subjects for periods of several minutes there was remarkably little change in ventilation

* Present address: Cardiovascular Research Institute and Department of Anesthesia, University of California, San Francisco, California, U.S.A.

† Present address: Department of Physiology, University of Sydney, Sydney, New South Wales, Australia.

(Freedman and Campbell, 1970). Added resistance produced a small reduction in ventilation, but there was a lot of individual variation and the falls were only significant with loads greater than 70 cm H_2O/l/sec—which is about 18 times greater than the normal total resistance of lungs and chest wall. At this level, ventilation was on average reduced by 30 per cent. In the same subjects, with added elastic loads (produced by breathing from closed jars of various sizes), there was, on average, an increase in ventilation which amounted to 22 per cent with a load of 100 cm H_2O/l, but these changes were not statistically significant.

Changes in pattern of breathing were much more striking. The same levels of resistive and elastic loading quoted above produced respectively a 64 per cent fall and a 131 per cent increase in respiratory frequency, tidal volume being essentially unaltered by resistive, and much reduced by elastic, loading. These changes in pattern of breathing are well recognized and have been described previously with added resistances and elastances in man by many authors, from Marey (1865) onwards.

There are obvious advantages in these breathing patterns. They enable ventilation to be maintained with minimal increases in inspiratory muscle force and work at each breath, and with a minimal increase of pressure across the extrathoracic airways and chest wall. They are not seen when anaesthetized subjects are loaded (Nunn and Ezi-Ashi, 1961; Howell, 1966; Freedman and Campbell, 1970), and for this reason it has been suggested that they serve mainly to reduce subjective discomfort. However, in these experiments inhalational anaesthetic agents were used, which are known to increase the sensitivity of pulmonary stretch receptors (Whitteridge and Bülbring, 1944; Coleridge *et alii,* 1968) and which may also have affected other reflexes. Bland *et alii* (1967) found, in dogs anaesthetized with barbiturates, that elastic loading did cause an increase in frequency and a fall in tidal volume that persisted after vagotomy.

On exercise, ventilation at a given work load is reduced by resistive loading, but oxygen uptake is unaffected (Gee *et alii,* 1968; Cerretelli *et alii,* 1969).

STABILIZING MECHANISMS WITH ADDED LOADING

There are three groups of mechanisms which may serve to stabilize ventilation in the face of a change of mechanical load: chemical, neurological and the intrinsic impedance of the respiratory system. All three groups could have been operative in the experiments so far described. Chemical factors are of primary importance. Any tendency for ventilation to fall would be wholly or partially counteracted by the effects of the consequent increase in arterial PCO_2 and fall of PO_2 (e.g. see Zechman *et alii,* 1957; Freedman, 1969).

EFFECTS OF LOADING ON THE VENTILATORY RESPONSE TO CO_2 AND HYPOXIA

One way of examining the interaction of chemical and other mechanisms as they affect loading, is to study the ventilatory response to CO_2. This has been shown to be depressed by added resistances (Cherniack and Snidal, 1956; Eldridge and Davis, 1959). Milic-Emili and Tyler (1963) showed that although the response to CO_2 was depressed when expressed in terms of ventilation, it was unaffected in terms of the work rate of the inspiratory muscles. There was a linear relationship between PCO_2 and the power output of the inspiratory muscles, but not of the expiratory muscles, suggesting that the chemoreceptor control system exerts its effects only via the inspiratory muscles.

Added elastic loads do not depress the ventilatory response to CO_2. This is true whether the load is produced by chest strapping (Thompson and Read, 1967), or by breathing from closed jars (Freedman *et alii,* 1972). In both these experiments there was the previously described increase in frequency and fall in tidal volume with added elastic loads. Freedman *et alii (loc. cit.)* found that inspiratory and total muscle power outputs were also unaffected by added elastic loads, and this was achieved by increasing respiratory frequency and reducing end-expiratory lung volume so that less elastic work was done on each breath; and the load, although inspiratory, was effectively shared by both inspiratory and expiratory muscles.

Barnett and Rasmussen (1970) found that an added resistance of about 35 cm H_2O/l/sec (presumably present on inspiration and expiration) depressed the ventilatory response to hypoxia as well as the CO_2.

REFLEX STABILIZING MECHANISMS

There are clearly many possible neurological pathways and reflexes that can affect the response to added loads.

At the spinal level, the intercostal muscles have been shown to possess a stretch reflex system (Glebovskii, 1966); and in the cat, tracheal occlusion at the onset of inspiration (the equivalent of a very large elastic load) causes an increase in discharge from inspiratory muscle spindles (Critchlow and von Euler, 1963), and also an increased rate of discharge in alpha motoneurones to these muscles which is abolished by dorsal root section (Corda *et alii,* 1965a). The cat diaphragm, in contrast to the intercostals, is comparatively poorly supplied with muscle spindles (Corda *et alii,* 1965b), and tracheal occlusion does not lead to a reflex increase in phrenic alpha motoneurone discharge nor to an increased rate of rise of the integrated diaphragmatic EMG. In unanaesthetized rabbits, Sant'Ambrogio and Widdicombe (1965) found an increased rate and duration of EMG discharge from a single intercostal motor unit in response to an added resistance, but not from a diaphragmatic motor unit. In man, Newsom Davis and Sears (1970) have demonstrated increased intercostal EMG activity following loading, which by its timing must have been reflexly mediated.

In general it is true to say that most studies of changes in reflex activity after loading do not provide any quantitative correlation with changes in ventilation.

Von Euler and Fritts (1963) did, however, quantify the effects of spinal and vagal reflexes on the respiratory muscles of cats by measuring static pressure and volume changes developed during inspiratory efforts against a clamped trachea. They found that vagotomy decreased the apparent elastance (i.e. there was more volume change for a given muscle pressure), whereas section of the dorsal roots (C.3 to T.12) had the opposite effect. They concluded that vagal and dorsal root afferents exerted reflex inhibitory and excitatory effects respectively on the respiratory motoneurones.

Many respiratory reflexes are mediated via the vagus; but in man, vagal afferents appear to exert very little effect on breathing within the tidal range (Guz *et alii,* 1964; 1966). The effect of vagal block on the response to added loads has not been studied in man. In cats, Lynne-Davies *et alii* (1971a, b) found that about one-third of the compensatory change on the first breath following addition of an elastic load could be attributed to increased drive to the intercostal muscles, dependent on intact vagi. The remainder of the compensatory change in respiratory muscle activity in these experiments was attributed to the intrinsic mechanical properties of the muscles themselves.

STABILIZING EFFECTS OF THE INTRINSIC PROPERTIES OF THE RESPIRATORY SYSTEM

Recently attention has been focused on the rôle of the intrinsic mechanical properties of the respiratory system in general, and of the respiratory muscles in particular, in the stabilization of ventilation following a change in mechanical load. The availability of such mechanisms for load compensation was probably first suggested by von Niekerk and ter Braak (1935).

In general, the greater the elastance and resistance of the respiratory system, the smaller will be the effects of any imposed change in elastance or resistance, whether internal or added externally.

Pengelly *et alii* (1971) showed that in cats and in four conscious men during electrical stimulation of the phrenic nerves, total elastance was greater than when measured with relaxed muscles. The same was true for resistance. In the men, on average, the elastance was 13 cm H_2O/l and the resistance 6 cm H_2O/l/sec. They termed these values the 'effective' elastance and resistance. The conventional average values of total elastance and resistance are 8 cm H_2O/l and 3 cm H_2O/l/sec (Agostoni, 1970).

The difference between the 'effective' and conventional (or 'passive') values was attributed to the intrinsic force-length and force-velocity relationships of the diaphragm. In the respiratory system the equivalents of force, length and velocity are pressure, volume and flow—so that the force-length relationship can be quantified as elastance and the force-velocity component as resistance. The force-velocity relationship of the human respiratory system has been shown to be similar to that of isolated muscle (Agostoni and Fenn, 1960). Pengelly *et alii* (1971) developed the equations of Otis *et alii* (1950) to predict the change in tidal volume after addition of an external load, assuming no change in ventilating pressure—i.e. no change in input to the respiratory muscles. This model assumes that the waveform of the volume changes during breathing is sinusoidal.

Milic-Emili and Pengelly (1970) compared these predictions with the observed tidal volumes on the first breath after addition of an elastic load in man. They studied the first breath, to avoid changes in chemical drive. They found that tidal volume fell with increasing elastic load, and that at each level of loading the observed tidal volumes exceeded those predicted from their model when using 'passive' values of elastance and resistance, but that when they used 'effective' elastance and resistance, predictions and observations agreed exactly.

They concluded that on the first loaded breath, such stabilization of tidal volume as occurred was entirely due to the intrinsic impedance of the respiratory muscles and that there was no increase in neural drive to these muscles. During added loading the force-velocity relationship has the result that for a given stimulus, as inspiratory flow rate is decreased, the muscles will develop a greater force; and similarly, as tidal volume is reduced the force-length relationship will result in greater force of contraction. Another property of the respiratory system that has a similar result to the force-length relationship is the curvature of the diaphragm—which will develop greater force when the radius of curvature is smaller; i.e. when lung volume or tidal volume is smaller (Marshall, 1962).

We have recently done some similar experiments using added inspiratory resistances in normal subjects breathing different gas mixtures (Read *et alii*, 1972a). We found that for a given resistance the percentage reduction in tidal volume was the same, irrespective of total ventilation or PCO_2 [Fig. 10.1A]. This was true in all but one of 19 experiments on 5 subjects. Tidal volume, expressed as a percentage of tidal volume without an added load, was also the same whether the resistance was added, or used as a continuous background and changed to a lower or higher value for a single breath [Fig. 10.1B].

However, the observed tidal volume of the first breath with a resistive load always exceeded the prediction of Pengelly and Milic-Emili's model.

The discrepancy appears to be due to two defects in the model when applied to resistive loads. In the simple analysis proposed by Pengelly and Milic-Emili it is assumed that: (1) a steady-state equation can be applied to the first loaded breath; and (2) sinusoidal functions can replace the actual respiratory waveforms without introducing a significant error in the prediction of peak-to-peak volume change (i.e. tidal volume).

Read *et alii* (1972b) have shown that these two assumptions lead to erroneous

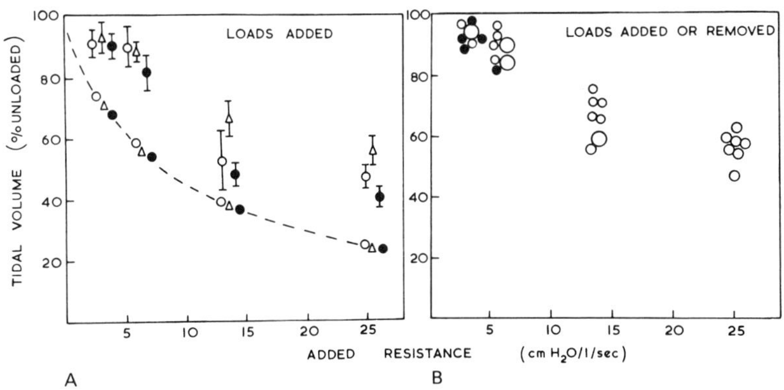

Fig. 10.1. Effect of added resistances on the first loaded breath while breathing different gas mixtures. Two experiments on one subject. Left-hand panel [A] shows tidal volume of loaded breath (as percentage of immediately preceding control breath) in an experiment where the loads were added for a single breath. Points and bars represent mean ±2SEM of 10 observations. Open circles—breathing oxygen; triangles—5 per cent CO_2 in oxygen; closed circles—7 per cent CO_2 in oxygen. Dashed line joins values predicted from theory of Pengelly and Milic-Emili, using 'effective' values of elastance and resistance of the respiratory system. Right-hand panel [B] shows results of an experiment breathing oxygen or 5 per cent CO_2 in oxygen where there was a continuous background of resistance and this was (for a single breath) decreased, or increased, or removed altogether. Tidal volumes expressed as percentage of value achieved when resistance was removed altogether. Large symbols represent steady-state values with continuous resistance, open symbols values when resistance was increased, and closed symbols when decreased. Each point is mean of 5–10 observations. Confidence limits omitted for clarity.

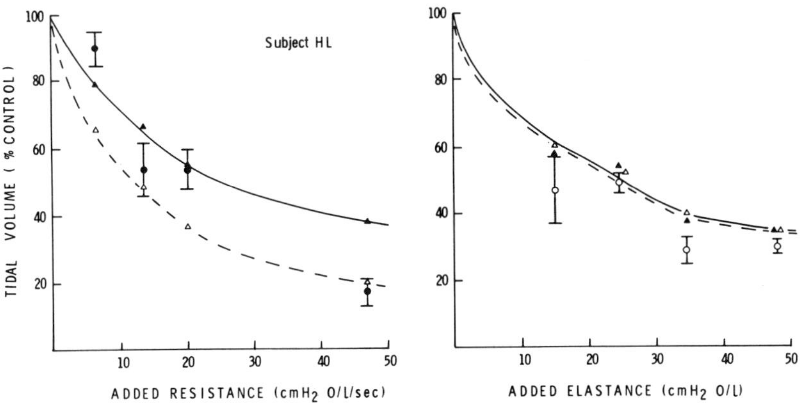

Fig. 10.2. Effects of added elastic and resistive loads on the tidal volume of the first loaded breath in one conscious seated subject. Points and bars represent mean ±2SEM of 10 observations with each load. Continuous line represents values predicted from non-steady state analysis of Read, and dotted line predictions from theory of Pengelly and Milic-Emili. Both predictions use 'effective' values of elastance and resistance.

predictions for the first breath with resistive loading. This was established by comparing the simple predictions with a non-steady state analysis based on the actual waveform of breathing.

The non-steady state analysis was performed using digital computer techniques:

1. To obtain volume and air flow every 0.1 sec in a series of control breaths.
2. To compute average waveforms from these pooled data.
3. To calculate the waveform of driving pressure generated by the respiratory muscles (P_{mus}), using either 'passive' or 'effective' values of intrinsic elastance and resistance.
4. To calculate the waveforms of volume and air flow for successive breaths after the introduction of either resistive or elastic external loads.
5. To compare the predicted response on the first loaded breath with the average waveform calculated from data obtained every 0.1 sec in repeated tests.

Using this method of non-steady state analysis, we have compared the predicted and experimental responses of the first breath with various resistive or elastic loads. In nine conscious subjects tested with 4 resistive and 4 elastic loads, the agreement found between predicted and experimental tidal volumes (as percentage of control) was reasonably good [Fig. 10.2]. The refined analysis therefore seemed to remove the major discrepancy found when our resistive loading data had been compared with the simple prediction of Pengelly and Milic-Emili. The elastic loading data obtained in the same subjects confirmed Pengelly and Milic-Emili's findings, agreeing with both simple and refined predictions.

The refined analysis, however, drew attention to another problem. In the conscious subjects, the inspiratory period was frequently prolonged by resistive loading. This sometimes resulted in a tidal volume greater than predicted, despite good matching of the early waveforms [Fig. 10.3]: minor modification of the constants used for 'effective' elastance and resistance could produce a perfect match of peak-to-peak volume at the price of poor matching in waveforms. This indicates that despite a major improvement in the accuracy of its predictions, the refined model still suffers serious limitations [Fig. 10.4].

This problem was not found when the results of resistive loading during light thiopentone anaesthesia were analysed (Kafer *et alii,* 1971). In these subjects, the resistive load did not alter the inspiratory time. Remarkably good matching of both the peak-to-peak volume changes (tidal volume) and the waveforms was found [Figs. 10.5 and 10.6].

At present the basis for the difference between responses in conscious and anaesthetized subjects is not clear.

We used light thiopentone anaesthesia because barbiturates cause no interference with muscle contraction, neuromuscular transmission or pulmonary stretch receptors (Whitteridge and Bülbring, 1944), and in low doses there is

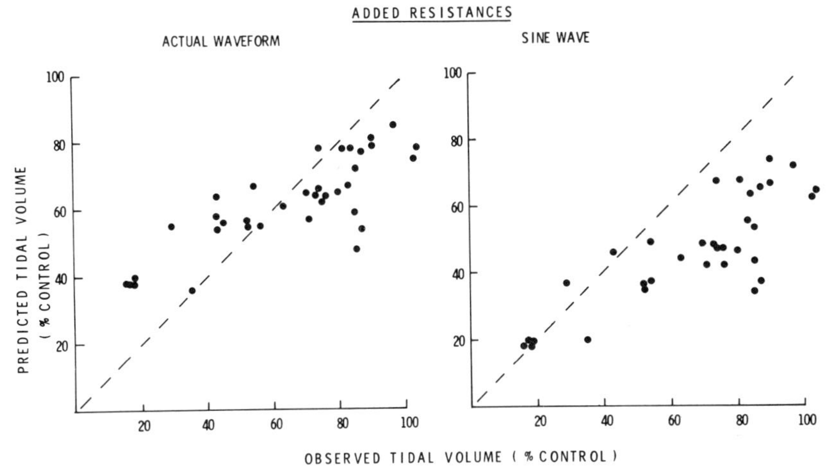

Fig. 10.3. Predicted and observed tidal volumes with added resistances in nine conscious seated subjects. Each point is mean of 10 observations. Left hand panel shows predictions from non-steady state analysis of Read and right hand panel those of steady state ('sine wave') analysis of Pengelly and Milic-Emili. Dashed line is line of identity.

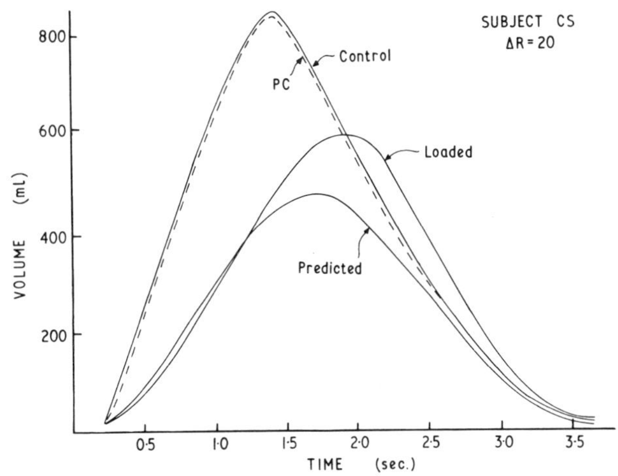

Fig. 10.4. Average waveform of control, loaded and predicted loaded tidal volumes in one conscious seated subject with an added resistance of 20 cm H_2O/l/sec. Data derived from 10 observations. PC = control breath predicted from waveform of ventilating pressure, as a check on mathematical procedures (see text).

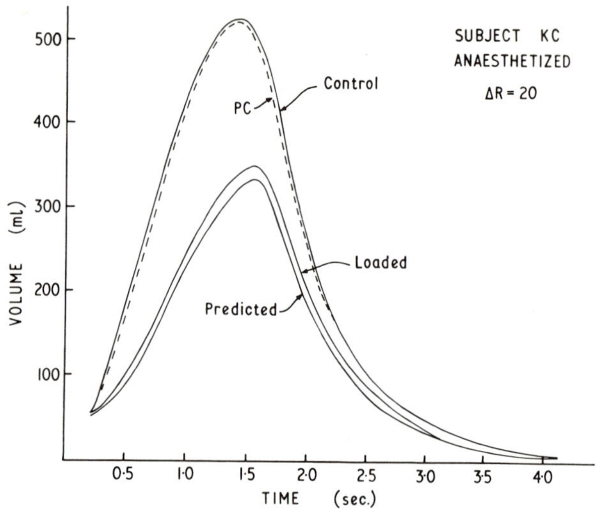

Fig. 10.5. Average waveform of control, loaded, predicted control (PC) and predicted loaded tidal volumes in one supine subject anaesthetized with thiopentone, with an added resistance of 20 cm H_2O/l/sec.

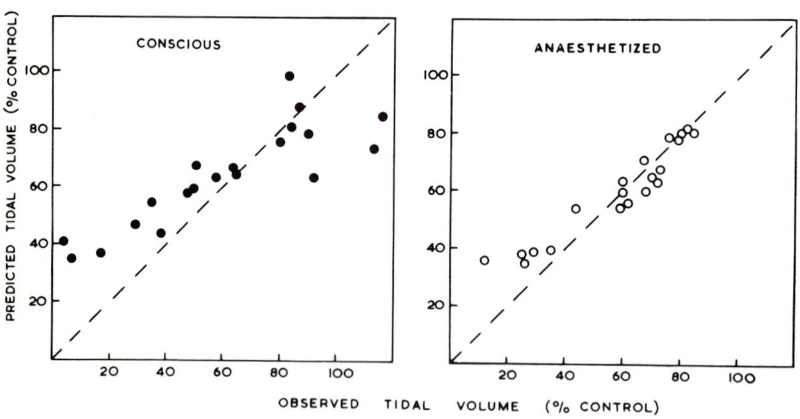

Fig. 10.6. Predicted and observed tidal volumes with added resistances in five supine subjects when conscious and anaesthetized. Predictions from non-steady state analysis of Read. Each point is mean of 4–10 observations.

minimal interference with monosynaptic spinal reflexes. Despite this, pharmacological effects of anaesthesia cannot be excluded completely—although we believe that the elimination of psychogenic factors (Freedman and Weinstein, 1965) offers a more likely explanation.

It is tempting to interpret all these results as showing that the intrinsic properties of the respiratory muscles are entirely responsible for the observed changes immediately after loading, and that there is no change in neural drive to the respiratory muscles; but there are several reasons why our interpretation should be more cautious.

First is the extrapolation of values of 'effective' elastance and resistance, obtained during electrophrenic respiration, to spontaneous breathing. During electrophrenic respiration, the rib cage undergoes paradoxical movement (i.e. in an expiratory direction) and volume changes recorded at the mouth will differ from the true volume change of the respiratory system.

Secondly, an apparent increase in elastance and resistance over 'passive' values need not necessarily be due to the intrinsic properties of the respiratory muscles. Konno and Mead (1968) have suggested that the impedance of the respiratory system may be greater during spontaneous breathing than as conventionally measured, because it departs from its configuration in the relaxed state; whilst Agostoni and Mognoni (1966) have demonstrated large phase differences between dorsoventral and lateral chest movements when breathing against added loads.

Thirdly, as von Euler and Fritts (1963) have shown [see above], the effects of reflexes can also be expressed in terms of an increased impedance.

For these reasons we prefer to interpret the values of 'effective' elastance and resistance as empirical constants embodying a variety of intrinsic mechanical and reflex mechanisms which contribute to the stability of ventilation after a sudden change of load. Finally, whatever the contribution of intrinsic mechanisms to the immediate changes, it is apparent that in the longer term the stability of ventilation is also dependent on changes in pattern of breathing that are due to nervous mechanisms.

ACKNOWLEDGEMENTS

D.J.C.R. acknowledges support by the North Foundation, California, U.S.A., E.K. was supported by a Travel Grant from the University of Sydney, N.S.W., Australia, and by the Medical Research Council.

REFERENCES

Agostoni, E. (1970) Dynamics, in *The Respiratory Muscles: Mechanics and Neural Control,* 2nd ed., pp. 81–84, ed. Campbell, E. J. M., Agostoni, E., and Newsom Davis, J. M., London.

Agostoni, E., and Fenn, W. O. (1960) Velocity of muscle shortening as a limiting factor in respiratory air flow, *J. appl. Physiol.,* **15,** 349–53.

Agostoni, E., and Mognoni, P. (1966) Deformation of the chest wall during breathing efforts, *J. appl. Physiol.,* **21,** 1827–32.

Barnett, T. B., and Rasmussen, B. (1970) Ventilatory responses to hypoxia and hypercapnia with external airway resistance, *Acta physiol. scand.,* **80,** 538–51.

Bland, S., Lazerou, L., Dyck, G., and Cherniack, R. M. (1967) The influence of the 'chest wall' on respiratory rate and depth, *Resp. Physiol.,* **3,** 47–54.

Campbell, E. J. M. (1958) *The Respiratory Muscles and the Mechanics of Breathing,* p. 31, London.

Cerretelli, P., Sikand, R. S., and Farhi, L. E. (1969) Effect of increased airway resistance on ventilation and gas exchange during exercise, *J. appl. Physiol.,* **27,** 597–600.

Cherniack, R. M., and Snidal, D. P. (1956) The effect of obstruction to breathing on the ventilatory response to CO_2, *J. clin. Invest.,* **35,** 1286–90.

Coleridge, H. M., Coleridge, J. C. G., Luck, J. C., and Norman, J. (1968) The effect of four volatile anaesthetic agents on the impulse activity of two types of pulmonary receptor, *Brit. J. Anaesth.,* **40,** 484–92.

Corda, M., Eklund, G., and von Euler, C. (1965a) External intercostal and phrenic α motor response to changes in respiratory load, *Acta physiol. scand.,* **63,** 391–400.

Corda, M., von Euler, C., and Lennerstrand, G. (1965b) Proprioceptive innervation of the diaphragm, *J. Physiol. (Lond.),* **178,** 161–77.

Critchlow, V., and von Euler, C. (1963) Intercostal muscle spindle activity and its γ motor control, *J. Physiol. (Lond.),* **168,** 820–47.

Eldridge, F., and Davis, J. M. (1959) Effect of mechanical factors on respiratory work and ventilatory responses to CO_2, *J. appl. Physiol.,* **14,** 721–6.

von Euler, C., and Fritts, H. W. (1963) Quantitative aspects of respiratory reflexes from the lungs and chest wall of cats, *Acta physiol. scand.,* **57,** 284–300.

Freedman, S., and Weinstein, S. A. (1965) Effects of external elastic and threshold loading on breathing in man, *J. appl. Physiol.,* **20,** 469–72.

Freedman, S. (1969) The functional capacity of the respiratory muscles in man, Ph.D. Thesis, University of London.

Freedman, S., and Campbell, E. J. M. (1970) The ability of normal subjects to tolerate added inspiratory loads, *Resp. Physiol.,* **10,** 213–35.

Freedman, S., Dalton, K. J., Holland, Denise, and Patton, J. M. S. (1972) The effects of added elastic loads on the respiratory response to CO_2 in man, *Resp. Physiol.,* **14,** 237–50.

Gee, J. B. L., Burton, G., Vassallo, C., and Gregg, J. (1968) Effects of external airway obstruction on work capacity and pulmonary gas exchange, *Amer. Rev. resp. Dis.,* **98,** 1003–12.

Glebovskii, V. D. (1966) Stretch reflexes of intercostal muscles, *Fed. Proc. Trans.,* Suppl. **25,** T937–43.

Guz, A., Noble, M. I. M., Trenchard, D., Cochrane, H. L., and Makey, A. R.

(1964) Studies on the vagus nerves in man: their role in respiratory and circulatory control, *Clin. Sci.*, **27**, 293–304.

Guz, A., Noble, M. I. M., Widdicombe, J. G., Trenchard, D., and Mushin, W. W. (1966) The effect of bilateral block of vagus and glossopharyngeal nerves on the ventilatory response to CO_2 of conscious man, *Resp. Physiol.*, **1**, 206–10.

Howell, J. B. L. (1966) Breathlessness in pulmonary disease, in *Breathlessness*, pp. 165–77, ed. Howell, J. B. L., and Campbell, E. J. M., Oxford.

Kafer, E. R., Read, D. J. C., and Freedman, S. (1971) Tidal volume stabilization to resistive loading during anaesthesia, *Brit. J. Anaesth.*, **43**, 718.

Konno, K., and Mead, J. (1968) Static volume-pressure characteristics of the rib-cage and abdomen, *J. appl. Physiol.*, **24**, 544–8.

Lynne-Davies, P., Couture, J., Pengelly, L. D., and Milic-Emili, J. (1971a) Immediate ventilatory response to added inspiratory elastic loads in cats, *J. appl. Physiol.*, **30**, 512–16.

Lynne-Davies, P., Couture, J., Pengelly, L. D., West, D., Bromage, P. R., and Milic-Emili, J. (1971b) Partitioning of immediate ventilatory stability to added elastic loads in cats, *J. appl. Physiol.*, **30**, 814–19.

Marey, E. (1865) Pneumographie: étude graphique des mouvements respiratories et des influences qui les modifient, *J. anat. (Paris)*, **2**, 425–53.

Marshall, R. (1962) Relationship between stimulus and work of breathing at different lung volumes, *J. appl. Physiol.*, **17**, 917–21.

Milic-Emili, J., and Pengelly, L. D. (1970) Ventilatory effects of mechanical loading, in *The Respiratory Muscles: Mechanics and Neural Control*, 2nd ed., pp. 271–90, ed. Campbell, E. J. M., Agostoni, E., and Newsom Davis, J. M., London.

Milic-Emili, J., and Tyler, J. M. (1963) Relation between work output of respiratory muscles and end-tidal CO_2 tension, *J. appl. Physiol.*, **18**, 497–504.

Newsom Davis, J., and Sears, T. A. (1970) The proprioceptive reflex control of the intercostal muscles during their voluntary activation, *J. Physiol. (Lond.)*, **209**, 711–38.

von Niekerk, J., and ter Braak, J. W. G. (1935) Die Anpassung des Atmungsvorganges an Widerstandsanderungen in den Atmungswegen, *Arch. ges. Physiol.*, **236**, 44–51.

Nunn, J. F., and Ezi-Ashi, T. I. (1961) The respiratory effects of resistance to breathing in anaesthetized man, *Anesthesiology*, **22**, 174–85.

Otis, A. B., Fenn, W. O., and Rahn, H. (1950) Mechanics of breathing in man, *J. appl. Physiol.*, **2**, 592–607.

Pengelly, L. D., Alderson, A. M., and Milic-Emili, J. (1971) Mechanics of the diaphragm, *J. appl. Physiol.*, **30**, 797–805.

Read, D. J. C., Milic-Emili, J., and Freedman, S. (1972a) The effect of added resistance on the first loaded breath in man. [In preparation].

Read, D. J. C., Kafer, E. R., and Freedman, S. (1972b) An analysis of some factors influencing the immediate response to added loads in man. [In preparation.]

Sant'Ambrogio, G., and Widdicombe, J. G. (1965) Respiratory reflexes acting on the diaphragm and inspiratory intercostal muscles of the rabbit, *J. Physiol. (Lond.)*, **180**, 766–79.

Thompson, J. F., and Read, D. J. C. (1967) Ventilatory responses to chest strapping and negative intrapulmonary pressures during progressive hypercapnia in conscious man, *Aust. J. exp. Biol. med. Sci.*, **45,** 18P.

Whitteridge, D., and Bülbring, E. (1944) Changes in activity of pulmonary receptors in anaesthesia and their influence on respiratory behaviour, *J. Pharmacol. exp. Ther.*, **81,** 340–59.

Zechman, F., Hall, F. G., and Hull, W. E. (1957) Effects of graded resistance to tracheal air flow in man, *J. appl. Physiol.*, **10,** 356–62.

11 Respiratory Sensation and Respiratory Muscle Activity

S. GODFREY

Institute of Child Health,
The Hammersmith Hospital, London, England

The study of respiratory sensation is beset with problems, chief of which is that breathing in normal subjects is usually an unconscious act. Probably the only occasions on which the normal subject becomes aware of respiratory sensation are during exhausting exercise and during breath holding. These two situations have one feature in common; namely, that there is an imbalance between what the respiratory muscles are actually achieving and what they ought to be achieving. The concept of inappropriateness as a source of respiratory sensation was originally proposed by Campbell and Howell (1963), and it could apply equally well to normal subjects or patients with pulmonary disease.

Conventional neurophysiology relates sensation to the application of stimuli in a fairly direct manner, and in the past it has generally been assumed that the same applies to sensations generated in the respiratory system—for example, the unpleasant sensation produced by breathing high concentrations of CO_2. In similar terms, the usual response to an unpleasant sensation is to move the part directly affected away from the stimulus. These ideas have to be abandoned when studying respiratory sensations, because it has become apparent from the studies to be described that such sensations may arise and be relieved by very indirect pathways.

Respiratory physiologists have been fascinated for many years by the study of breath holding because it is a source of very potent respiratory sensations, and because it is an excellent model for studying the drive to breathe. Such studies have shown that the maximum tolerable duration of a breath-hold—the breath holding time (B.H.T.)—is controlled by a combination of mechanical and chemical factors. There is a direct relationship between the B.H.T. and the lung volume at which the breath is held (Muxworthy, 1951); and prior hyperventilation with O_2 (by lowering the PCO_2) results in a gross prolongation of B.H.T. (Klocke and Rahn, 1959). The essentially linear nature of these relationships can be seen in Figs. 1 and 4 of the earlier review by Godfrey (1969).The effect of O_2 has usually been studied by noting the B.H.T. with various combinations of PO_2 and PCO_2 (Hill and Flack, 1908; Douglas and Haldane, 1909; Ferris *et alii,* 1946; Otis *et alii,* 1948). It was found that there was a curved relationship, such that the higher the PO_2 the higher the PCO_2 that could be tolerated. As a result of these studies, various groups postulated that the sensation which caused the subject to break his breath-hold was due to a simple combination of stimuli related to lung volume, PCO_2 and PO_2 (Klocke and Rahn, 1959; Mithoefer, 1959). This idea was very similar to that about the factors causing the breaking point during rebreathing with restricted tidal volume and various combinations of gas tensions (Mithoefer *et alii,* 1953).

The suggestion that the sensation during breath holding arises from a simple summation of lung volume and chemical stimuli cannot, however, explain why rebreathing can be continued for much longer than breath holding despite a greater degree of asphyxia (Hill and Flack, 1908), or why it is possible to resume a breath-hold, having been forced to break previously, by taking a few breaths of gas which do not improve the asphyxia (Fowler, 1954).

By a modification of the latter type of experiment, Godfrey *et alii* (1969) have shown that it is possible to begin a breath-hold at a higher PCO_2 and smaller lung volume than existed at the breaking point of the preceding breath-hold. Because neither absolute levels of chemical stimuli nor simple summation of chemical and mechanical factors could explain these findings, a new model was proposed by Godfrey and Campbell (1968) which envisaged the stimuli as summating over a period of time. A version of this model is shown in Fig. 11.1. The chemical stimulus is taken as equivalent to PCO_2 in the presence of high concentrations of O_2, and it rises steadily during rebreathing or breath holding along the line ADB. With the onset of breath holding at point A, mechanical stimuli are added and the total stimulus then rises along AC. For simplicity at this time it is taken that the

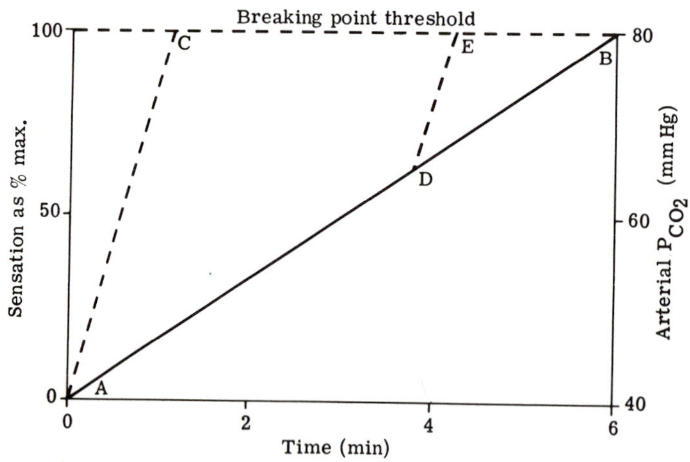

Fig. 11.1. Diagram to illustrate the summation of stimuli during breath holding. The solid line represents chemical stimuli related to PCO_2 and the dotted lines AC and DE represent the sum of chemical and mechanical stimuli during breath holding. (After Godfrey and Campbell, 1968.)

sensations due to chemical and mechanical stimuli follow the same pathways as the stimuli with respect to time, and the subject reaches the breaking point for breath holding at point C or for rebreathing at point B. When rebreathing is resumed at the breaking point of the first breath-hold, it was postulated that some of the stimulus would be removed by the act of moving the lungs or chest wall so that it is therefore possible to continue on to point D, and begin a second breath-hold. The stimuli rise along DE and the breaking point is reached after a shorter time than in the first breath-hold. It should be noted that the second breath-hold begins at a PCO_2 of 65 mm Hg, which is higher than that existing at the breaking point of the first breath-hold (48 mm Hg). This type of model can explain the observations above about breath holding and rebreathing, but begs the very important question of the relationship between stimuli and sensation.

A very striking feature about the breaking point of breath holding is the very rapid relief that occurs when breathing is resumed, at a time well before any chemical changes could have occurred in the blood. This is so, even if the relieving breath consists of nitrogen (Wright and Branscombe, 1954). A more formal study by Kobayasi and Sasaki (1967) confirmed that a single breath at breaking point could relieve the unpleasant sensation. Godfrey and Campbell (1969) found that a single breath was as effective as five breaths when taken at the breaking point of a breath-hold, in terms of the duration of the subsequent breath-hold [Fig. 11.2A].

Further work by Godfrey and Zeidifard (unpublished data) has attempted to define whether half a breath is as good as a whole breath in relieving the sensation at breaking point. This was done by performing breath-holds separated by rebreathing, starting with 7 per cent CO_2 in O_2 (as described fully by Godfrey and Campbell, 1969). This enabled the construction of a B.H.T.—PCO_2 response curve for breath-holds carried out at a total lung capacity (TLC), and another curve for breath-holds at residual volume (RV). In each case only a single breath from the rebreathing bag was permitted between breath-holds. A third curve was then constructed for breath-holds that were made alternatively at TLC and RV so that there was only half a breath between any pair of breath-holds. The whole set of experiments was then repeated with either two and a half or three whole breaths between breath-holds. The results are shown in Fig. 11.2B for 3 normal subjects, in terms of the B.H.T. for an initial PCO_2 of 55 cm Hg. Again there was no difference between 1 and 3 breaths, but there was also no difference between $\frac{1}{2}$, 1, $2\frac{1}{2}$ or 3 breaths in terms of the duration of the following breath-hold.

From the above work it would appear that movement of the chest wall or lungs can permit the resumption of a breath-hold even if it represents only half a breath; but there is no clear information about the effect of the size of this half breath. This question has been explored in terms of the tolerable duration of rebreathing by Mithoefer *et alii* (1953) and Kobayasi and Sasaki (1967), who

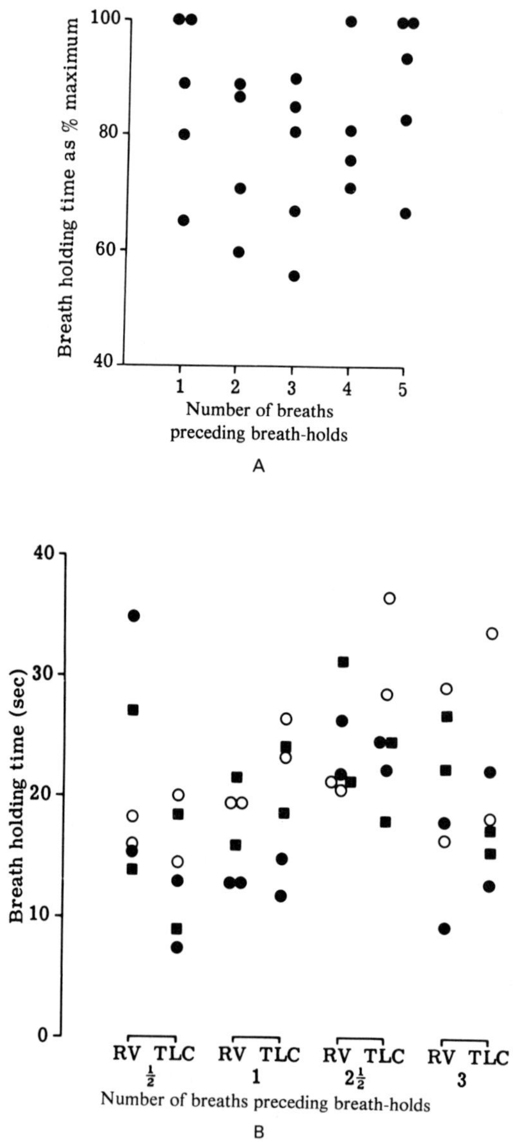

Fig. 11.2[A]. Breath holding time in relation to the number of preceding breaths. Each symbol represents one subject who performed 3–5 of the experiments. The number of breaths had no significant effect. (From data of Godfrey and Campbell, 1969.) [B]. Breath holding in three subjects at residual volume (RV) and total lung capacity (TLC) after the number of breaths indicated. Each symbol represents one subject who performed each experiment twice. There is no difference between the half breath and whole breath experiments. (Godfrey and Zeidifard, unpublished data.)

showed that the larger the tidal volume the longer the tolerated rebreathing time and the higher the tolerated PCO_2 at breaking point. In a very elegant series of experiments, Remmers *et alii* (1968) found that there was a close interrelationship between frequency of breathing, tidal volume and the maximum tolerated PCO_2 [Fig. 11.3A]. They plotted the tolerated level of PCO_2 against the product of tidal volume and frequency and obtained essentially straight lines. I have replotted their data with the axes reversed to give ventilation in terms of PCO_2—i.e. in conventional ventilation-PCO_2 response curve format [Fig. 11.3B]—and I have added the mean relationship for the ventilatory response to CO_2 during free breathing found by Godfrey *et alii* (1971). There is a striking similarity of intercept and slope in two cases, and the slope in the other two would also lie within the normal range. The similarity of these two kinds of ventilation-PCO_2 response curves implies that the naturally selected ventilation at any given level of PCO_2 is very close to the ventilation needed to tolerate this level of PCO_2, and that any restriction of free ventilation should rapidly generate an intolerable sensation.

The evidence presented so far suggests that the stimuli or sensations generated during rebreathing or breath holding can be wholly or partly removed by moving the lungs or chest wall. Two conflicting views have been held on the way in which such movement could affect sensation. The conventional approach (Mithoefer *et alii,* 1953; Wright and Branscombe, 1954) was that Hering-Breuer impulses from pulmonary stretch receptors somehow 'neutralized' the stimuli reaching the central nervous system from the chemoreceptors. An alternative view (Fowler, 1954) suggested that the sensation (during breath holding) arose from the voluntary prevention of shortening of the respiratory muscles and that relief of the sensation by movement was because 'the resulting motor activities are more appropriate to the prevailing level of stimulation at the respiratory centre'. Fowler believed that the origin of the sensation was the involuntary twitching of the diaphragm that he noted during breath holding.

Agostoni (1963) made a formal study of diaphragmatic contractions during breath-holding, and found that twitching began some time after the start of the breath-hold at a constant level of PCO_2—provided the subject was not hypoxic. His records clearly show that the electrical activity of the diaphragm increased progressively up to breaking point. The diaphragmatic activity was not related to the lung volume at which the breath was held; and although rebreathing at the breaking point of a breath-hold relieved the sensation and permitted the resumption of breath holding, the diaphragmatic activity returned almost immediately. Agostoni (1963) thought that the diaphragmatic activity reflected chemical stimuli only, but he did not quantitate the electrical signal. However, the data of Kobayasi and Sasaki (1967) show that the PCO_2 at the onset of diaphragmatic twitching was not fixed and depended on the concentration of CO_2 (in O_2) inspired at the beginning of the breath-hold. They produced an illustration showing

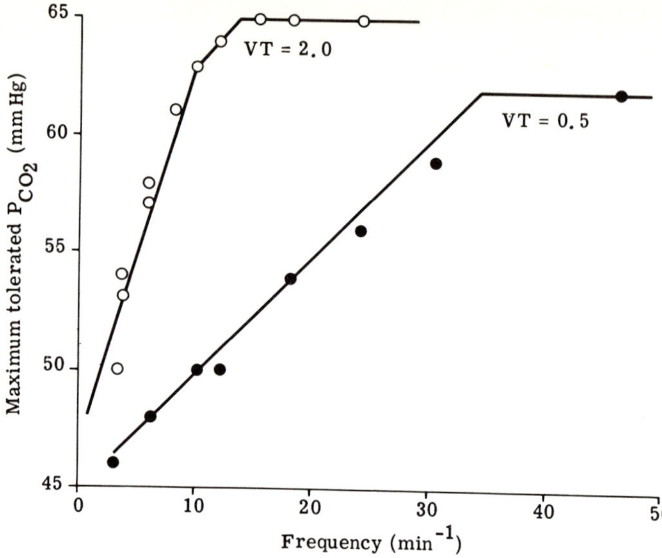

A

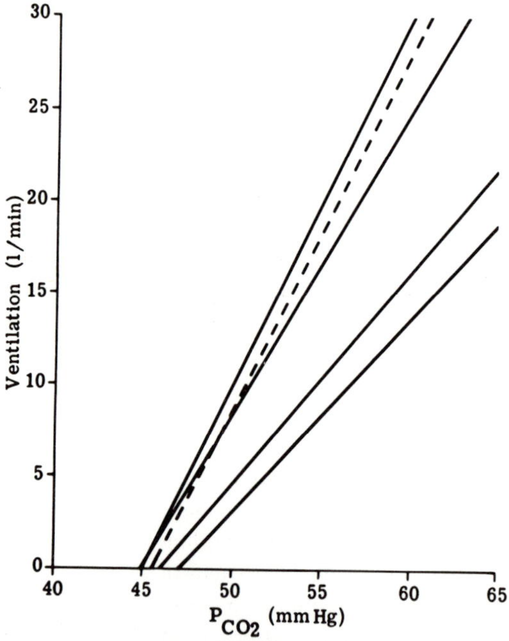

B

Fig. 11.3 [A]. Effect of frequency of breathing and tidal volume on the maximum tolerated PCO_2 during rebreathing. (From Remmers *et alii*, 1968.) [B]. Relationship between ventilation and maximum tolerated PCO_2 (solid lines) from the data of Remmers *et alii* (1968), redrawn on conventional ventilation-CO_2 axes. The interrupted line is the ventilatory response to CO_2 during unrestricted breathing found by Godfrey *et alii* (1971).

a hyperbolic relationship between the PCO_2 at the onset of the twitching and the duration of the preceding part of the breath-hold. Moreover, from another illustration in their paper it would appear that rebreathing between breath-holds did actually reduce the quantity of diaphragmatic activity which then built up again. This whole subject is still open, but it seems quite possible that the diaphragmatic twitching reflects the total stimuli arising during the breath-hold, rather like the interrupted lines in Fig. 11.1.

Recent studies have attempted to define the origin of the sensation during breath holding by blocking various neural pathways. Guz *et alii* (1966a) produced temporary blockade of both vagus nerves in normal subjects, and found that this had little effect on normal breathing but grossly prolonged the B.H.T. The unpleasant sensation of breath holding was virtually absent, and diaphragmatic twitching did not occur. However, the B.H.T. was still longer at larger lung volumes and the ability to detect added elastic loads to breathing was unimpaired. In another study, Guz *et alii* (1966b) found that vagal blockade reduced the ventilatory response to inhaled CO_2 and their subjects did not then feel distressed by breathing the CO_2. As a result of these studies they believed that the sensation during breath-holding and rebreathing arose from within the lung and travelled by way of vagal afferents. They later found that blockade of the chest wall up to the first thoracic segment had no effect on breath holding, rebreathing, or the detection of added loads to breathing (Eisele *et alii*, 1968); but the last result differed from that of Newsom Davis (1967), who found impaired detection of mechanical loads in patients with high spinal cord lesions.

The very prominent diaphragmatic twitching during breath holding led Campbell and his colleagues to speculate on the effect of muscular paralysis on breath holding. A method was designed whereby it was possible to paralyse conscious normal subjects and maintain communication through hand signals by excluding one limb from the effect of the tubocurarine used (Campbell *et alii*, 1967). They found that paralysis resulted in a gross prolongation of B.H.T. (the time between the cessation of artificial ventilation and its resumption at the request of the subject or observers). The unpleasant sensation was abolished, as during vagal blockade. This type of experiment was repeated by Campbell *et alii* (1969) to study the effect of paralysis on breath holding at high levels of PCO_2. The results of this study are shown in Fig. 11.4, where the B.H.T. values of the subject are expressed as percentages of his predicted B.H.T. based on either the PCO_2 at the start or the end of control breath-holds, obtained from B.H.T.-PCO_2 response curves. The B.H.T. after atropine is also shown, because this drug was always given before the tubocurarine. It can be seen that the B.H.T. was little affected by atropine but very considerably prolonged by paralysis, even when the starting PCO_2 was as high as 61 mm Hg. When considering the B.H.T. in relation to the breaking point PCO_2, the prolongation with paralysis is even more striking; and in one case it was infinitely longer than the control, because it had proved impossible for the subject to continue any breath-hold above a PCO_2 value of 69 mm Hg in the control experiments. Once again the paralysed breath-holds were devoid of unpleasant sensation, and were actually terminated in response to the anxiety of the observers rather than the wish of the subject.

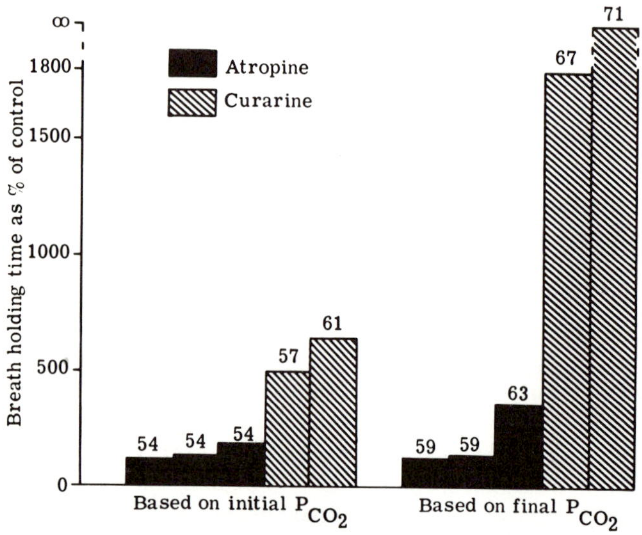

Fig. 11.4. Effect of total paralysis on breath holding time compared with the predicted times based on the initial or final PCO_2 of the breath-hold. The solid columns represent three experiments after atropine alone and the hatched columns represents two experiments after paralysis, all in the same subject. The numbers over the columns indicate the PCO_2. (After Campbell *et alii*, 1969.)

These experiments with tubocurarine emphasize that all sensations, including those due to very high levels of PCO_2, must be mediated by contraction of the respiratory muscles. In fact, Noble *et alii* (1971) showed that the B.H.T. could be prolonged by blockade of the phrenic nerves alone, and suggested that the diaphragm was the most important source of the sensation. The results of the three types of blocking experiment described are summarized in Fig. 11.5. One subject, M.N., took part in all three types of experiment, and it is interesting to compare the efficacy of the various methods of prolonging B.H.T. The absolute B.H.T. was longest after total paralysis and almost as long after vagal blockade, whilst phrenic blockade was not so effective; but when considered as percentages of the control B.H.T. measured at the same session, there was little to choose between the three types of experiment.

The results of all the experiments discussed above have been incorporated into a model of the genesis of sensation during breath holding or rebreathing that is shown in Fig. 11.6 and which updates the earlier model of Godfrey and

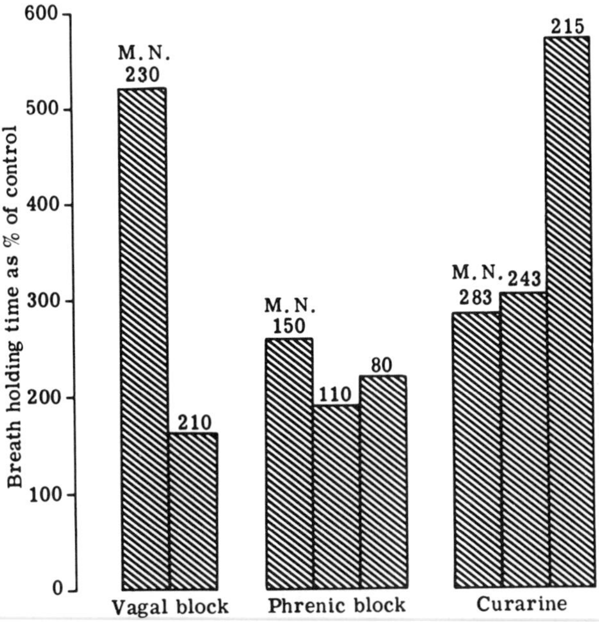

Fig. 11.5. Prolongation of breath holding time after blocking the vagus or phrenic nerves and after total paralysis by curarine. M.N. indicates one subject who performed all three types of experiment. The numbers above the columns indicate the breath holding time in seconds. (Extracted from the data of Campbell *et alii*, 1967; Guz *et alii*, 1966b; and Noble *et alii*, 1971.)

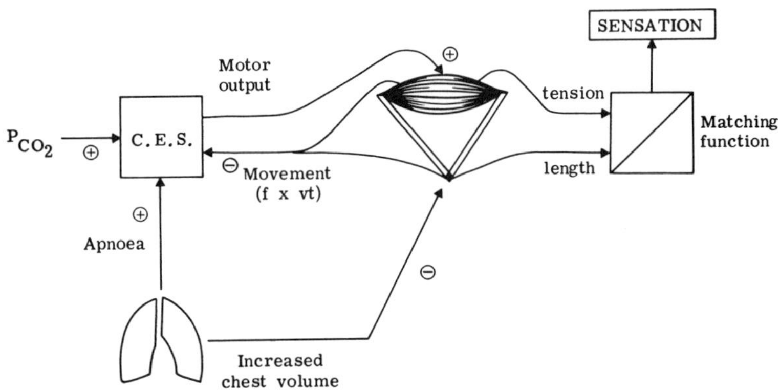

Fig. 11.6. Model of the genesis of sensation during breath holding or rebreathing. Chemical and mehanical factors summate to cause tension in the respiratory muscles, and the imbalance between the tension and length of these muscles generates the sensation. The derivation of the model is explained in the text.

Campbell (1968, 1970). In the model, a central excitatory state (C.E.S.) is generated as a result of chemical stimuli (PCO_2 in the absence of hypoxia) summating with non-chemical stimuli arising from the lung and travelling via the vagi. During breath holding these pulmonary stimuli signal apnoea, but they do not arise from conventional Hering-Breuer receptors because an increased rate of lung shrinkage does not affect B.H.T. (Godfrey *et alii,* 1969) and because the lung volume effect on B.H.T. persists after vagal blockade. The increase in C.E.S. results in an efferent drive to the respiratory muscles (principally to the diaphragm) which are voluntarily prevented from shortening during breath holding, or whose ability to contract may be limiting as the drive becomes excessive during rebreathing. This efferent drive is the sum of chemical and mechanical factors and can be partially inhibited by even half a breath at the breaking point of breath holding, or by ventilation during rebreathing that is just appropriate to the level of PCO_2.

When the efferent drive is voluntarily resisted during breath holding, or when ventilation cannot keep up with the drive in rebreathing, there is imbalance between the length and the tension of the respiratory muscles. This imbalance is sensed by muscle and joint receptors which provide information for a matching system and generates an unpleasant sensation proportional to the imbalance. It is this sensation which is felt and which causes the subject to reach breaking point. The effect of static lung volume on B.H.T. is probably mediated through the setting of the muscle and/or joint receptors, because its effect is independent of vagal blockade and it does not affect efferent phrenic disharge.

On the basis of this model, B.H.T. is prolonged after vagal blockade because 'apnoea' stimuli are removed and the extent of the prolongation of B.H.T. emphasizes how important they are—but it must be admitted that they cannot be further identified at present. Total paralysis removes all the tension in the respiratory muscles and therefore there cannot be any length-tension mismatching, irrespective of the level of PCO_2 or apnoeic stimuli. Phrenic blockade removes a large part of the sensation and therefore the diaphragm must be the chief site of length-tension matching.

In patients with lung disease, the disproportionate tension in the respiratory muscles needed to produce ventilation, or the inability to keep ventilation up to the demand, could well result in the sensation of dyspnoea through the pathways described above.

REFERENCES

Agostoni, E. (1963) Diaphragm activity during breath holding: factors related to its onset, *J. appl. Physiol.,* **18,** 30–36.

Campbell, E. J. M., and Howell, J. B. L. (1963) The sensation of breathlessness, *Brit. med. Bull.,* **19,** 36–40.

Campbell, E. J. M., Freedman, S., Clark, T. J. H., Robson, J. G., and Norman, J. (1967) The effect of muscular paralysis induced by tubocurarine on the duration and sensation of breath-holding, *Clin. Sci.,* **32,** 425–32.

Campbell, E. J. M., Godfrey, S., Clark, T. J. H., Freedman, S., and Norman, J. (1969) The effect of muscular paralysis induced by tubocurarine on the duration and sensation of breath-holding during hypercapnia, *Clin. Sci.,* **36,** 323–8.

Douglas, C. G., and Haldane, J. S. (1909) The regulation of normal breathing, *J. Physiol. (Lond.),* **38,** 420–40.

Eisele, J., Trenchard, D., Burki, N., and Guz, A. (1968) The effect of chest wall block on respiratory sensation and control in man, *Clin. Sci.,* **35,** 23–33.

Ferris, E. B., Engel, G. L., Stevens, C. D., and Webb, J. (1946) Voluntary breath holding. III. The relation of the maximum time of breath holding to the oxygen and carbon dioxide tension of arterial blood with a note on its clinical and physiological significance. *J. clin. Invest.,* **25,** 734–43.

Fowler, W. S. (1954) Breaking point of breath holding, *J. appl. Physiol.,* **6,** 539–45.

Godfrey, S. (1969) Breath holding and the sensations due to chemical and mechanical stimuli to breathing, *Brit. J. Dis. Chest.,* **63,** 177–92.

Godfrey, S., and Campbell, E. J. M. (1968) The control of breath holding, *Resp. Physiol.,* **5,** 385–400.

Godfrey, S., and Campbell, E. J. M. (1969) Mechanical and chemical control of breath holding, *Quart. J. exp. Physiol.,* **54,** 117–28.

Godfrey, S., and Campbell, E. J. M. (1970) The role of afferent impulses from the lung and chest wall in respiratory control and sensation, in *Breathing: Hering-Breuer Centenary Symposium,* pp. 219–32, ed. Porter, R., London.

Godfrey, S., Edwards, R. H. T., Copland, G. M., and Gross, P. L. (1971) Chemo-sensitivity in normal subjects, athletes and patients with chronic airways obstruction, *J. appl. Physiol.,* **30,** 193–9.

Godfrey, S., Edwards, R. H. T., and Warrell, D. A. (1969). The influence of lung shrinkage on breath holding time, *Quart. J. exp. Physiol.,* **54,** 129–40.

Guz, A., Noble, M. I. M., Widdicombe, J. G., Trenchard, D. I., Mushin, W. W., and Makey, A. R., (1966a) The role of vagal and glossopharyngeal afferent nerves in respiratory sensation, control of breathing and arterial pressure regulation in conscious man, *Clin. Sci.,* **30,** 161–70.

Guz, A., Noble, M. I. M., Widdicombe, J. G., Trenchard, D., and Mushin, W. W. (1966b) The effect of bilateral block of vagus and glossopharyngeal nerves on the ventilatory response to CO_2 of conscious man, *Resp. Physiol.,* **1,** 206–10.

Hill, L., and Flack, M. (1908) The effect of excess of carbon dioxide and of want of oxygen upon the respiration and the circulation, *J. Physiol. (Lond.),* **37,** 77–111.

Klocke, F. J., and Rahn, H. (1959) Breath holding after breathing oxygen, *J. appl. Physiol.,* **14,** 689–93.

Kobayasi, S., and Sasaki, C. (1967) Breaking point of breath holding and tolerance time in rebreathing, *Jap. J. Physiol.,* **17,** 43–56.

Mithoefer, J. C. (1959) Lung volume restriction as a ventilatory stimulus during breath holding, *J. appl. Physiol.,* **14,** 701–5.

Mithoefer, J. C., Stevens, C. D., Ryder, H. W., and McGuire, J. (1953) Lung volume restriction, hypoxia and hypercapnia as interrelated respiratory stimuli in normal man, *J. appl. Physiol.*, **5,** 797–802.

Muxworthy, J. F., Jr. (1951) Breath holding studies: relationship to lung volume, *A.F. Tech. Rep. 6528, Wright-Patterson Air Force Base,* pp. 452–6, Dayton, Ohio.

Newsom Davis, J. (1967) Contribution of somatic receptors in the chest wall to detection of added inspiratory airway resistance, *Clin. Sci.,* **33,** 249–60.

Noble, M. I. M., Eisele, J. H., Frankel, H. L., Else, W., and Guz, A. (1971) The role of the diaphragm in the sensation of holding the breath, *Clin. Sci.,* **41,** 257–83.

Otis, A. B., Rahn, H., and Fenn, W. O. (1948) Alveolar gas changes during breath holding, *Amer. J. Physiol.,* **152,** 674–86.

Remmers, J. E., Brooks, J. G., and Tenney, S. M. (1968) The effect of controlled ventilation on the tolerable limit of breath holding, *Resp. Physiol.,* **4,** 78–90.

Wright, G. W., and Branscombe, B. V. (1954) The origin of the sensation of dyspnoea, *Trans. Amer. Climatological Assoc.,* **66,** 116–25.

DISCUSSION ON COMMUNICATIONS [10] AND [11]

Opened by E. J. M. Campbell. It says in the programme that I am to open this discussion, but I am not going to take very long. First, I would like to make one point. In the study of the ability to detect mechanical loading that Dr. Freedman and I (with others) did 10 years ago[1,2,3], we found that you can detect a load of about 20 per cent or so added to the existing internal normal load. Recently we tackled the question of what happens if you just have a volume load—that is to say, the breathing is just increased without there being any mechanical distortion—and we found[4] that the tidal volume has to be increased by about 700 ml before you are aware that your breathing has increased. In a resting subject the tidal volume is about 500 ml, and therefore the tidal volume can be doubled before you know your breathing has increased. The point I wish to make is that provided the mechanical events are *appropriate* then you don't feel very much, compared with what you feel if a mechanical distortion is introduced into the system.

While I have the privilege, I would like to direct a question to each of the speakers. First of all to Dr. Godfrey: you said that a half-breath when you reach breaking point causes the subsequent breath-hold to be as long as if you had had a full breath, and you said that this was true whether you went in the inspiratory

[1] Campbell, E. J. M., Freedman, S., Smith, P. S., and Taylor, M. E. (1961) The ability of man to detect added elastic loads to breathing, *Clin. Sci.,* **20,** 223–31.

[2] Bennett, E. D., Jayson, M. I. V., Rubenstein, D., and Campbell, E. J. M. (1962) The ability of man to detect added non-elastic loads to breathing, *Clin. Sci.,* **23,** 155–62.

[3] Campbell, E. J. M., Bennett, E. D., and Rubenstein, D. (1963) The ability to distinguish between added elastic and resistive loads to breathing, *Clin. Sci.,* **24,** 201–7.

[4] West, D. W. M., Ellis, C. G., and Campbell, E. J. M. (1971) Detection of increased ventilation in man, *Fed. Proc.,* **30,** 270.

direction or the expiratory direction. I take it, from what you said, that this was a complete half-breath from residual volume to total lung capacity, and then from total lung capacity to residual volume. That would make it difficult to compare the effects of each breath, because the lung volumes at the end of the relieving half-breath would be different. Can you therefore say whether inspiration is better than expiration at relief?

Godfrey. May I show a slide?

Campbell. While we are waiting for it, maybe Dr. Freedman would like to answer my next question, or make a comment. Everything you were saying related to the first loaded breath; and I think for the record we should say that there were some scattered observations some 10 years ago or so suggesting that there is a progressive breath-by-breath response which, as far as I am concerned, has never been adequately documented or adequately explained. Is that a fair summary of the position?

Freedman. Yes, that is fair. We did go on in the conscious and in the anaesthetized subjects to look at subsequent breaths. With resistive loads there is not very often a gradual increase with subsequent breaths: with elastic loads, as you described, this is much more consistent, and I have done a few experiments on conscious subjects at different levels of PCO_2 to try and sort this out—but it sometimes happens and it sometimes doesn't. If one looks at dogs anaesthetized with barbiturates, again it sometimes happens and sometimes doesn't. All I can say is that I cannot relate it to change in blood gases or to the state of the blood gases when the load is added. I don't know what it can be related to; but one cannot explain it all as a change in end-expiratory lung volume, as you showed originally.

Campbell. Thank you. Dr. Godfrey?

Godfrey. [Commenting on slide—*Ed.*]. This [Fig. D.11] is the protocol for the experiment where we had the subject rebreathing, and then breath holding at residual volume with one breath in between (a); or performing a total lung capacity breath-hold when he breaks with one breath (b), or with two or three breaths. These (c) are half-breath experiments when he rebreathes—at first out, holds his breath, and then breathes right in. By plotting breath-holding times and PCO_2 response curves we could compare the half-breath experiments with the predicted total lung capacity breath-holding times after one breath, or the residual volume breath-holds after one breath.

Widdicombe. Could I, first of all, correct a slight error that Dr. Godfrey made which I am sure he would like me to correct; namely, that Guz *et alii* (1966),[1] when they blocked the vagus nerves in human subjects, never claimed in their paper that during vagal blockade the sensation of breath holding was abolished. Their paper was worded very carefully to say that the unpleasant sensation of

---

[1] Guz, A. Noble, M. I. M., Widdicombe, J. G., Trenchard, D., Mushin, W. W., and Makey, A. R. (1966) The role of vagal and glossopharyngeal afferent nerves in respiratory sensation, control of breathing and arterial pressure regulation in conscious man, *Clin. Sci.*, **30**, 161–70.

breath holding was alleviated in their subjects, in that it was far less acute and far less severe—but the sensation was still there.

I think you gave very convincing evidence that the unpleasant sensation of breath holding is due to a large number of different afferent inputs. But you did not convince me—and I have never been convinced by reading the literature—that there is any experimental evidence for the idea of length/tension inappropriateness or mismatch being the basis of this sensation. Everything a respiratory physiologist does to a subject introduces mismatch and makes tension

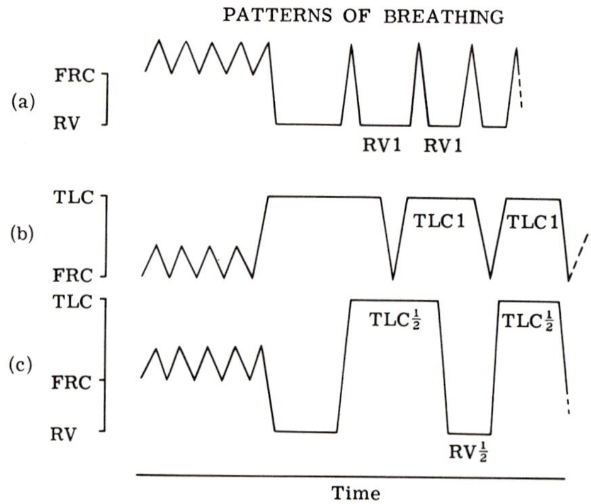

Fig. D.11. Experimental protocol for breath-holding experiments for diagramatic illustration. Above (a): the subject takes one breath between residual volume breath-holds. In the middle illustration (b) he takes one breath between total lung capacity breath-holds. In the lower illustration (c) he takes half a breath between alternating residual volume and total lung capacity breath-holds.

inappropriate. Every patient with respiratory or cardiovascular disease has this, so the fact that you can correlate length/tension inappropriateness with unpleasant respiratory sensation is no evidence at all of a causal relationship. My question is, have you any experimental evidence that length/tension inappropriateness feeds in as a separate component, apart from all the many inputs that we know cause sensation? Can you show us experimentally that length/tension inappropriateness results in an unpleasant sensation?

Godfrey. How would you account for the human curare experiments[1] in other terms?

Widdicombe. I would say that if you are curarized you are full of length/tension inappropriateness! Buckets of it! None of the muscles have any active tension at any length.

[1] Campbell, E. J. M., Clark, T. J. H., Freedman, S., Norman, J., and Robson, J. G. (1967) The effect of muscular paralysis induced by tubocurarine on the duration and sensation of breath-holding. *Clin. Sci.*, **32**, 425–32.

Godfrey. But you aren't. How could you say you are inappropriate in these circumstances—the muscles presumably have no tension and are at their relaxed length: so they are not inappropriate.

Widdicombe. This is why I think that length/tension inappropriateness is not a good concept to put forward as an explanation of the unpleasant sensation. You can have it in paralysis (and you can produce it in other ways) and not have unpleasant respiratory sensations. I think what one does with curare is to suppress the activity of a large number of afferent inputs—from tendon receptors and joint receptors and muscle receptors—and therefore you don't have such an unpleasant sensation when you hold your breath.

Godfrey. Well, I think the curare experiments certainly show one thing, and that is that all the other sensory mechanisms (including those related to the high CO_2, the apnoea and everything else) are finally channelled through the chest wall. But in terms of inappropriateness, I don't see what is inappropriate in having slack muscles with no tension in them—it seems to me that they are not then inappropriate. If, however, you try to breathe when you are paralysed and conscious then you immediately feel very inappropriate; and it was notable in the curarization experiment (and perhaps Dr. Campbell, who was the subject, will speak of this himself) that the subject wanted to be artificially ventilated during progressive paralysis at a time when his tidal volume was actually quite adequate. If he was trying to breathe and feeling conscious of his breathing he felt uncomfortable; but if he relaxed and just lay back, he did not.

Widdicombe. I think that supports my argument, because his trying to breathe is not introducing any more length/tension inappropriateness in muscles that already are not contracting. What it is doing is putting another afferent input into the cortex.

Godfrey. That is inappropriateness! It is like putting in a drive to which he cannot respond. But perhaps we should turn to the originator of the idea.

Campbell. I don't think we should let this discussion go on. I think that there is a long historical story to be unravelled here. First of all, when we introduced the idea 10 years ago[1,2] we spoke of length and tension—and we should really have spoken only of 'length'. I would also quarrel with the suggestion that we should use the term 'input'. This may be metaphysical or metaphysiological; but we are really concerned with the nature of the sensation, rather than the nature of the afferent information giving rise to the sensation.

We have discussed this at length on other occasions, and I have never seen any need to postulate specific length and tension information from the periphery as being the essence of this sensation. Our suggestions were concerned with the nature of the sensation itself, rather than afferent information. We need better information regarding the neurophysiological basis of the perception of such things as weight and elasticity, and in such activities as playing games (like ten-

---

1 Campbell, E. J. M., Freedman, S., Smith, P. S., and Taylor, M. E. (1961) The ability of man to detect added elastic loads to breathing, *Clin. Sci.*, **20,** 223–31.

2 Campbell, E. J. M., and Howell, J. B. L. (1963) The sensation of breathlessness, *Brit. med. Bull.*, **19,** 36–40.

nis) in which you need very active proprioceptive information for the accurate programming required to achieve given tasks. One would then, I think, be in a better position to specify the wiring diagram underlying the respiratory sensation. We, unfortunately, found ourselves using words like 'length' and 'tension'. 'Tension' I am quite prepared to take out now; but I do not think the basic idea stands or falls by whether or not there are specific length and tension inputs, or a given peripheral site in which length and tension are compared. I think, frankly, that our original idea is not very helpful because it is so difficult to test, and I am rather sorry that it got the attention that it did.

von Euler. I don't think that Dr. Campbell should be sorry that the idea about length/tension inappropriateness got the attention that it did: it has certainly proved a very fruitful hypothesis giving rise to much new and informative experimentation.

With regard to the effects on respiratory sensation of vagal block and of curarization, I think that the results suggest some central interaction: both vagal information and afferent information via the phrenic nerve seem to be of importance, but apparently only if the other pathway is also in functional operation. That suggests to me that a crucial point might be the central convergence of the afferents from these two channels, perhaps in an additive fashion.

In this connexion, I should like to put in a word of warning against the belief that one should always expect a close correlation between specific afferent inputs and conscious perception. The brain is able to take all sorts of information into account. For instance, when some respiratory loading is introduced at the onset of some speech activity I presume the brain (at least in most cases) is aware of the talking and is able to take this into account, and to tell the respiratory control mechanism that the incoming information about sudden change in respiratory impedance does *not* mean that there is an immediate threat of suffocation and should *not* give rise to alarming sensations: the body is just busy talking.

Beverly Bishop. I want to ask specifically about the doses of curare, and just how much they were affecting. The reason I am asking is that some time ago an experimental psychologist used curare in animal experiments[1] in an attempt to abolish the functions of the cerebral cortex to see if lower motor centres could 'learn'; and I seem to remember that the curare did affect the cortex in relatively low doses. When we are speaking about conscious sensation I think we are certainly talking about functions in which the cerebral cortex is involved.

Godfrey. The experiment I referred to wasn't done in the cat but in a conscious, apnoeic respiratory physiologist. I can't remember the exact doses of curare that were used, but they were given in graduated doses according to the subject's tidal volume all the time. Just enough was given to abolish his tidal volume; but I think, perhaps, we should ask the subject himself—who is our Chairman.

Campbell. The dose was about 50 mg[2]—but I think the point is well taken. It is still quite possible that the sensation was abolished because of some other

---

[1] Culler, E., Coakley, J. P. Shurrager, P. S., and Ades, H. W. (1939) Differential effects of curare upon higher and lower levels of the central nervous system, *Amer. J. Psychol.*, **52,** 266–73.

[2] Campbell, E. J. M., Clark, T. J. H., Freedman, S., Norman, J., and Robson, J. G. (1967) The effect of muscular paralysis induced by tubocurarine on the duration and sensation of breath-holding, *Clin. Sci.*, **32,** 425–32.

effect of the curare; but note has to be taken of the observation that phrenic block does it as well. We actually used *d*-tubocurarine, and looked at its pharmacology at the time and came to the conclusion, as far as we could then, that the doses were insufficient to affect cortical perceptual functions—but it is still a possibility.

Wyke. The answer to this particular problem is, in fact, now available from assessments of the transfer of *d*-tubocurarine from blood into brain. Without citing the references at length here [see Wyke, 1969][1], there is substantial evidence in the literature that when systemic doses of *d*-tubocurarine of the order that were used in the circumstances in which our Chairman was involved[2] are given, they do not have any significant direct effect on any electrophysiological criterion of brain function (and presumably, therefore, on perceptual functions).

Can I also use this opportunity to monopolize the floor for one moment more, because Dr. Bishop said in a moment of pithy inspiration yesterday that one of the functions of this Symposium might have been achieved if we could establish some kind of common vocabulary; and one of the things that has always worried the neurologists (who are in a minority here today) is that the neurological vocabulary used by respiratory physiologists sometimes seems to us to be a slightly foreign language. I was therefore very gratified to hear Dr. Godfrey admit (albeit indirectly) that the respiratory physiologists have at last discovered that all that is afferent is not sensory—as Dr. von Euler has already indicated. I would like to suggest, therefore, that in any consideration of where this perceptual experience of respiratory discomfort comes from, it has to be shown anatomically and physiologically that the postulated pathways actually project from identified peripheral receptors ultimately to the cerebral cortex.

Now this, of course, relates especially to the postulate that muscle spindles might contribute to awareness of the breath-holding sensation. It might well be that some part of the brain short of the cerebral cortex can generate such a perceptual experience—I don't think we know, although (instinctively, I must confess) I don't believe this to be the case. But I think what *has* to be taken into account here is that there is no good evidence that muscle spindle afferents project to the cerebral cortex in higher animals (that is, in primates) in a way capable of evoking perceptual experience [see Matthews, 1972][3]. I will be interested to hear what Dr. Godfrey and Dr. Widdicombe (who has already got his finger in the air) might have to say in reaction to that.

Campbell. I would not like to waste time in chasing windmills: but I don't think any of us associated with this idea originally would now suggest that the muscle spindle is a 'sensory' endorgan, nor would we implicate the muscle spindle in providing the sensation of inappropriateness, or the sensation underlying breath holding, or the sensation of dyspnoea.

---

[1] Wyke, B. D. (1969) *Principles of General Neurology: An Introduction to the Basic Principles of Medical and Surgical Neurology*, pp. 343, 730, London.

[2] Campbell, E. J. M., Clark, T. J. H., Freedman, S., Norman, J., and Robson, J. G. (1967) The effect of muscular paralysis induced by tubocurarine on the duration and sensation of breath-holding, *Clin. Sci.*, **32,** 425–32.

[3] Matthews, P. B. C. (1972) *Mammalian Muscle Receptors and Their Central Actions*, London.

Widdicombe. Just two very brief comments. First, I don't think one is entitled to assume that sensations of this sort are localized in the cortex—they might well be below it (as Dr. Wyke suggested). Second, even if it is shown that muscle spindles do not send their afferents to the cerebral cortex (or wherever this sensation comes from) they might very well interact with other pathways at lower levels and therefore contribute to the unpleasant sensation. I don't think one can entirely rule out a contribution from muscle spindles because of the arguments that have just been raised.

Plum. I think one should not lose sight of the fact that these extraordinarily stimulating and generative observations have attracted a lot of thought about the experience of the individual, not least because of the individuals involved—they were distinguished physiologists and professors of medicine.

Nevertheless, the cortex is not necessary for the organism to respond to respiratory afferent inputs. If one occludes the airway in a decerebrate animal, it struggles in such a way as to suggest that the breath-holding experience is extremely 'unpleasant'; and the same is true of human subjects during surgical anaesthesia. This behaviour is in marked contrast to what Reis and McHugh (1968)[1] found when they induced apnoea by amygdaloid stimulation; the apnoea then lasted indefinitely, to a point where hypoxia disrupted the cardiovascular system, and yet there was no struggling whatsoever. The same principle is true in man—decerebrate patients struggle against resistances to the airway, and one finds no evidence that this response requires the cerebral cortex.

Campbell. I haven't seen it personally, but I am told by paediatricians who have seen children with anencephaly that obstruction of their airway causes them to display plenty of evidence of distress. I think this leads us into the problem of whereabouts in the brain one feels this distress—which is difficult. It is a very intriguing problem, and could keep us busy for many hours—but I wonder does anybody have any questions now for Dr. Freedman, because he has some really hard information about passive loading and what it implies for the control systems.

Wyke. May I interpolate the remark that one must distinguish between brain stem reflex responses to airways obstruction and the perceptual experience of respiratory distress—which latter (as far as I am concerned) involves the participation of the cerebral cortex.

von Euler. I think Dr. Freedman's Communication was very interesting, and I certainly found it challenging because I have been in the game of myotatic reflex controls and have advocated their importance for load compensation. But as I stated in my introduction this morning [see Communication 7], we were fairly soon provoked to regard the effects of load compensation in response to added external loads as a minor task for the myotatic reflex system and the muscle spindles. The myotatic reflexes seem to be of greater importance for the co-ordination of posture and breathing, whereby an undisturbed function of the volume pump can go on while we are adjusting posture or moving the trunk.

---

[1] Reis, D. J., and McHugh, P. R. (1968) Hypoxia as a cause of bradycardia during amygdala stimulation in the monkey, *Amer. J. Physiol.*, **214**, 601–10.

More challenging perhaps are the relations between volume and time that you showed. You have already suggested yourself that a bridge between your results and ours might be found in the hypothesis about up and down shifts in the Hering-Breuer threshold curve.

With respect to the responses to added loads, you seem to agree that there is some reflex increase in the motor output. If there is an increase—or decrease in the case of the diaphragm—of the efferent output to the muscles you would, in consequence, get a magnified change in the intrinsic mechanism of the muscles since the length/tension relationship changes its slope fairly steeply when the efferent activity changes. With reflex changes in efferent activity to the diaphragm it reacts differently along the length/tension diagram, and probably in a direction opposite to the intercostals. I am now hinting at one possible reason why you see so little effect on the diaphragm when the intercostals are changing in an opposite direction.

Mead. I have a question, although not related to that last point. You improved the fit of your prediction by taking into account the fact that the waveform is not a sine wave. I would like to know if you maintain the assumption that the sine wave was appropriate, and if you have adjusted the frequency for the duration of inspiration. Would you go as far toward a better prediction? Have you done that?

Freedman. That is the first thing we did. We just looked at the sine wave predictions using only the inspiratory time, and we moved further away from the observed values; because if you just set up a theoretical case, the slower the frequency the closer you get to the prediction—and the duration of inspiration is usually less than the duration of expiration. If we just looked at the frequency based on inspiratory duration, we moved further away from the observed facts.

If I could comment on the remarks of Dr. von Euler, I would say I completely agree with him. I think his work and that done with Fritts[1] showed that you could quantify the effects of the vagus and of the dorsal root afferents on the whole system in terms of a pressure/volume diagram which changed its slope; and this is what I was trying to imply, perhaps rather badly, in my last comments.

Mead. May I ask Dr. Godfrey a question? I am interested to know whether or not you have taken advantage of the iso-volume manoeuvre to produce substantial motions of rib-cage and abdomen at a fixed lung volume in breath-holding.

Godfrey. No.

Mead. You can put the rib-cage through a good 60 per cent of its vital capacity, and the abdomen through a full 100 per cent of its vital capacity, at iso-volume quite readily—and that would be an interesting thing to examine.

Godfrey. What I have done (although I haven't any data with me to show here) is to try getting subjects to perform forced inspiration and expiration during breath holding against an absolute obstruction (raising mercury columns up and

---

[1] von Euler, C., and Fritts, H. W. (1963) Quantitative aspects of respiratory reflexes from the lungs and chest walls of cats, *Acta physiol. scand.*, **57**, 284–300.

down) to see whether this kind of method of generating pressure would have any effect on breath-holding time, and whether breath holding against a sustained contraction of the respiratory muscles to sustain a higher pressure (or to sustain a negative pressure) would have any influence on breath-holding time. The answer is that none of these manoeuvres altered the breath-holding time at all—but that's not quite the same thing.

Mead. Not at all the same thing!

Godfrey. What were you getting at then, with your iso-volume manoeuvre?

Mead. That you can put the rib-cage through a wide range of its potential motion without changing lung volume, and similarly the abdomen and diaphragm. When you made your maximal pressure effort, your presumably did not change the configuration of the chest wall much.

Godfrey. I did not measure it, but it probably did change quite a bit.

Beverly Bishop. Dr. Mead's suggestion appeals to me, because it is a way of sorting out the difference between mere configurational changes in the chest and abdomen and actual air flow movements.

Campbell. I don't wish to muddy the water but I have done that informally, and it does produce a slight prolongation of breath-holding time—but nothing comparable with what you get with curare. In fact, we thought that maybe curarization might shorten breath-holding time, because one's ability to do it was abolished—but it didn't turn out that way.

Stevens. This comment may not be too appropriate to these papers, but during speech production the resistance of the upper airways changes by a factor of ten or so over a period of perhaps 100 msec, and then back again. I wonder if anyone could make a comment about what sort of adjustments go on during such a short time—as when one makes a sound like 'H' or 'T'. The resistance during such sounds is then very low, and the volume flow is very high.

Freedman. The only information I know that relates to time is that of Newsom Davis and Sears (1970).[1] They applied negative or positive pressures to the mouth with very rapid onset and removal; and their results did show changes in the intercostal electromyogram (which is what they were looking at), so presumably the system can respond down to or below 100 msec.

Moran. My question is directed to Dr. Godfrey, and relates to that part of his model of which he said (or implied) that certain respiratory muscle movements might put an inhibitory bias on the central excitatory state. Some years ago we did some rebreathing experiments, the results of which could be described like this: at different levels of exercise we produced a signal at the point at which we just recognized an additional sensation of some discomfort in breathing, while

---

[1] Newsom Davis, J., and Sears, T. A. (1970) The proprioceptive reflex control of the intercostal muscles during their voluntary activation, *J. Physiol. (Lond.)*, **209**, 711–38.

breathing with a rebreathing bag (the PCO_2 in the rebreathing bag being recorded at the same time); and the result of this was that the point at which this sensation was detected corresponded to a higher PCO_2 in the bag for higher levels of exercise. This surprised us a bit; but just about that time Dr. Campbell and his colleagues started to get interested in the downstream relationships between the bag PCO_2 and the alveolar PCO_2, and we started to do the same sort of thing in relation to our experiments but never finished it, which is why I haven't published it. My question now is, does this seem to you to support the idea that at high levels of exercise—at which ventilation and respiratory muscle movement are increased—there might be a central inhibitory effect of the muscular activity (or of some other aspect of the movement) counterbalancing the possible CO_2 effect? Or do you think it is all just a time-delayed sort of phenomenon?

Godfrey. I think that observation is very nice, because it is exactly what we found as well a lot later on, when we did rebreathing on exercise while looking at the ventilatory response and the breath-holding response to CO_2. I think the simple explanation is that when you have high levels of exercise the rate of rise of PCO_2 in the rebreathing bag is very much greater than at lower levels of exercise—so it is not really surprising that you hit higher levels of PCO_2 before you quit. In fact, if you put all the facts together that are known about the ventilatory response to CO_2 during exercise and the breath-holding response, you come to the conclusion that during exercise there is an increased rate of rise of chemical drive because the CO_2 is going up faster during rebreathing or breath holding, and there is also an increase in the rate of rise of the non-chemical drive. I don't think it has got anything to do with the downstream effect.

Moran. Just one further point. It seemed to us that the time scale of the rates of rise, and the possible rates of centripetal signal transfer in this respect, don't fit in with that sort of explanation.

Campbell. Before we close this discussion, may I use the Chairman's privilege to make a final statement. I do not believe that the idea of length/tension can get any further until we have a better understanding of the control of motor acts in general; and this, all my reading suggests, is in a very confused situation. It follows that our ability to test our ideas in terms of the specific situation of the act of breathing cannot make much progress until we know more about the control of movement in general. I will go on record as saying I believe that understanding of the sensation of dyspnoea and of the situation in conditions when breathing is mechanically difficult resides in the act of breathing, not in the stimuli that increase the breathing. That is my position, and I crave your indulgence to make it clear once more.

12 Activity of Respiratory Control Neurones in the Brain Stem

MORTON I. COHEN

Department of Physiology,
Albert Einstein College of Medicine,
New York, N.Y., U.S.A.

This paper summarizes the results of several related projects dealing with the origin of respiratory periodicity. The general aim has been to explain how the respiratory act originates from the interactions of populations of neurones in the brain-stem respiratory centres. The principle of analysis was to produce respiratory changes by various experimental procedures, and to compare the changes in activity of individual respiratory neurones with changes in the over-all respiratory cycle and in respiratory motor discharge.

Experiments were conducted in a simplified preparation where the major portion of ventilation-related afferent input was eliminated; namely, the vagotomized, paralysed, thoracotomized and artificially-ventilated cat. (In studies specifically dealing with vagal afferent influences, the vagi were left intact.) These experimental conditions resulted in isolation of the central oscillatory system, whose over-all state was indicated by efferent phrenic discharge.

The responses to three types of experimental input were studied: (1) carbon dioxide tension (Cohen, 1968); (2) lung inflation, in animals with intact vagi (Cohen, 1969); and (3) brain-stem electrical stimulation (Cohen, 1971). Several types of response of respiratory units were observed, on the basis of which a functional classification of neurones was proposed, and a preliminary schematic circuit of the respiratory oscillator was presented (Cohen, 1970).

In another type of analysis, the short-term interactions between inspiratory neurones were studied (Cohen, 1973, in press). These interactions are manifested by synchronization of inspiratory discharge in the form of high-frequency (60–110/sec) oscillations in phrenic and in medullary inspiratory activity. This kind of oscillation is an electrical sign of inspiratory activity which, studied by auto- and cross-correlation analysis, has furnished information about the connexions within the medullary inspiratory centres and about the efferent inspiratory pathways from medulla to spinal cord.

METHODS

Experiments were performed on cats, decerebrated or anaesthetized (with urethane or Dial-urethane), which were paralysed by succinylcholine or gallamine, thoracotomized, and artificially ventilated. The animals were usually vagotomized, except in one study on vagal afferent influences (Cohen, 1969). Further details of methods can be found in Cohen (1968, 1969, 1971).

Recordings were taken on magnetic tape of: (1) efferent phrenic discharge; (2) brain-stem neural activity (both unit and population) as seen by microelectrodes (NaCl pipettes or micropointed insulated tungsten rods); (3) standard pulses marking (a) time points in the respiratory cycle, and (b) electrical stimuli. By processing phrenic potentials, standard pulses marking the start of the inspiratory (I) and expiratory (E) phases were obtained; and standard pulses corresponding to unit spikes were obtained by use of an amplitude discriminator (Cohen (1968), App. 1).

Electrical stimuli were delivered to brain-stem points by bipolar nichrome electrodes. The parameters of stimulation (including time of stimulus delivery in the respiratory cycle) were set by digital logic circuits (Cohen, 1971).

In experiments on the effects of afferent input from pulmonary stretch receptors (Cohen, 1969), the test of airway occlusion was used: a solenoid-operated valve was switched electrically by pulses time-related to the respiratory cycle, so that the lungs could be maintained at any chosen volume level.

Data Processing

Patterns of both spontaneous and evoked activity were analysed with an average-response computer (CAT 1000), using as sweep triggers the I pulses, E pulses, and stimulus-related pulses. Periodicities in phrenic and brain-stem activity were analysed with a correlation computer (Mnemotron 258). The computer memory contents were read out in digital form to a printer, and in analogue form to an ink-writer or an oscilloscope.

RESULTS AND DISCUSSION

DISCHARGE PATTERNS OF RESPIRATORY NEURONES

The first step in the analysis of function of respiratory neurones was classification by discharge pattern (spike incidence distribution within the respiratory cycle). A schematization of the major discharge patterns is shown in Fig. 12.1. These patterns fall into two major categories.

1. Patterns sharply locked to a phase transition of the respiratory cycle (inspiratory or expiratory neurones [Fig. 12.1, left] are found predominantly in

the lateral medulla. Most investigators agree that within the medulla there is an anatomical localization of discharge types: inspiratory neurones are more numerous rostral to the obex, whilst expiratory neurones are more numerous caudal to the obex (Baumgarten *et alii,* 1957; Haber *et alii,* 1957; Merrill, 1970). Furthermore, the inspiratory neurones are found in two anatomically distinct regions: (a) the 'dorsal respiratory nucleus', located ventral to the nucleus and tractus solitarius; and (b) the 'ventral respiratory nucleus' or nucleus retro-ambigualis, located more ventrolaterally at the same level (Merrill, 1970; Bianchi, 1971; Cohen, unpublished observations).

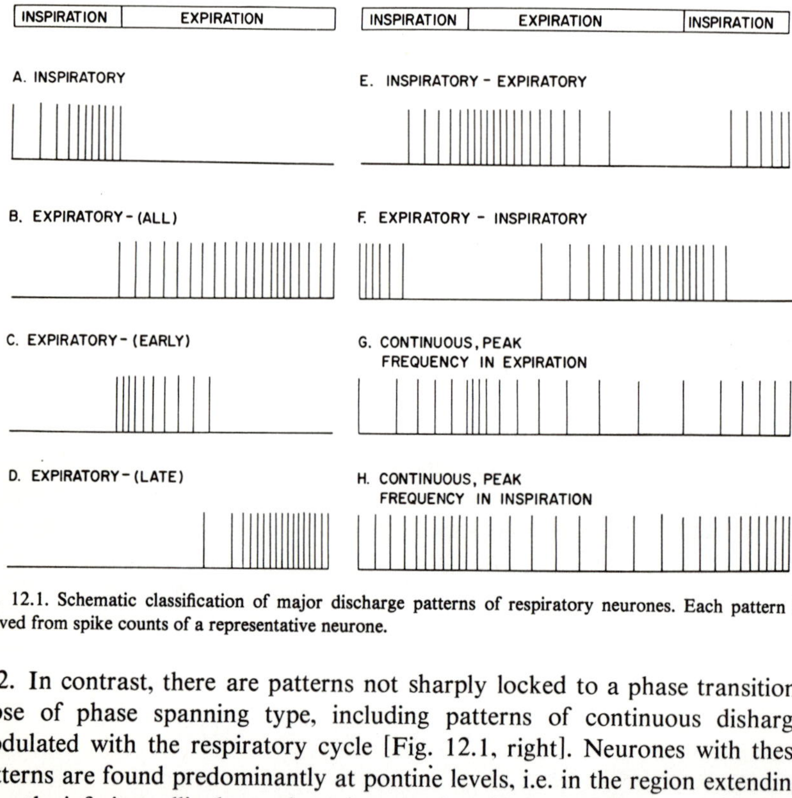

Fig. 12.1. Schematic classification of major discharge patterns of respiratory neurones. Each pattern is derived from spike counts of a representative neurone.

2. In contrast, there are patterns not sharply locked to a phase transition: those of phase spanning type, including patterns of continuous disharge modulated with the respiratory cycle [Fig. 12.1, right]. Neurones with these patterns are found predominantly at pontine levels, i.e. in the region extending from the inferior colliculus to the striae acusticae (Cohen and Wang, 1959; Bertrand and Hugelin, 1971).

The discharge patterns of medullary inspiratory and expiratory neurones are similar to those found in peripheral respiratory nerves and muscles: (1) Inspiratory patterns are similar to those found in the phrenic nerve, diaphragm, and external intercostals (Gill, 1963; Eklund *et alii,* 1964; Sears, 1964). (2) The late-expiratory and expiratory-(all) patterns resemble those found in the internal intercostal and abdominal muscles (Floersheim, 1960; Sears, 1964). (3) The early-expiratory patterns resemble those of recurrent laryngeal expiratory fibres (Eyzaguirre and Taylor, 1963). These similarities suggest that the inspiratory and expiratory neurones in the medulla are involved in the systems generating the detailed pattern of peripheral respiratory discharge.

In contrast, the phase-spanning discharge patterns of pontine neurones (inspiratory-expiratory and expiratory-inspiratory) are not found in respiratory nerves and muscles. This observation, as well as observations obtained by stimulation and lesion-making procedures, have led to the suggestion that pontine respiratory neurones are involved in the generation of the system's oscillation, and in particular in the generation of the transitions between the inspiratory and expiratory phases (Wang *et alii,* 1957; Cohen and Wang, 1959; Bertrand and Hugelin, 1971; Cohen, 1971).

EFFECTS OF CARBON DIOXIDE TENSION ON DISCHARGE PATTERNS

In a study of the effect of hypocapnia on the discharge patterns of respiratory neurones (Cohen, 1968), several types of response to lowered alveolar CO_2 tension were observed; these are shown schematically in Fig. 12.2. In the 'type-1' response, a neurone which normally fires in bursts (e.g. inspiratory or expiratory) continues to exhibit this phasic pattern as CO_2 level is lowered, until at a sufficiently low level the discharge becomes sporadic and disappears. In the contrasting 'type-2' response, a neurone's discharge never disappears during

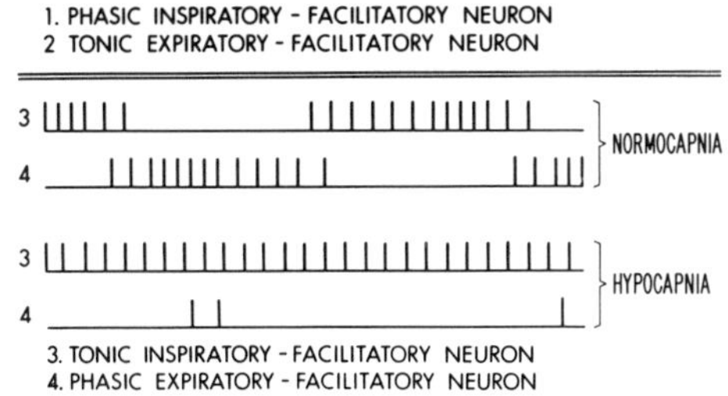

Fig. 12.2. Schematic representation of contrasting responses to hypocapnia, shown by two pairs of reciprocally firing neurones. *Top pair*: 1, inspiratory neurone with type-1 response; 2, expiratory neurone with type-2 response. *Bottom pair*: 3, expiratory-inspiratory neurone with type-2 response; 4, inspiratory-expiratory neurone with type-1 response.

hypocapnia, but rather at sufficiently low CO_2 levels the discharge becomes continuous or 'tonic'.

In the schematic representation of Fig. 12.2, the patterns of response typical for some common classes of neurone are shown. The patterns of two pairs of neurones are shown; for each pair, the two members fire in reciprocal portions of the cycle and also have reciprocal responses to hypocapnia. The first reciprocal pair consists of: (1) an inspiratory neurone whose discharge disappears during hypocapnia; and (2) an expiratory neurone whose discharge becomes continuous during hypocapnia, the silent period being eliminated. The second reciprocal pair consists of: (3) an inspiratory-expiratory neurone which becomes silent during hypocapnia; and (4) an expiratory-inspiratory neurone whose discharge becomes continuous (tonic) during hypocapnia. These responses were typical for neurones of the classes shown.

Two major conclusions can be reached: (1) The fact that certain pairs of neurones firing in reciprocal portions of the cycle also have reciprocal responses to change of CO_2 level supports the idea that there are mutually inhibitory connexions between populations typified by each member of a pair (e.g. mutual inhibition between inspiratory and expiratory neurones). (2) The existence of a tonic discharge in certain types of neurone during hypocapnia supports the idea that respiratory periodicity arises on a background of continuous activity that is made periodic by feed-back inhibitory connexions. Two major types of tonic activity were observed: (a) the expiratory neurones with 'type-2' responses tend to fire continuously (as in expiratory apnoea), if not periodically inhibited by inspiratory neurones; and (b) the expiratory-inspiratory neurones, which predominate in the lower pons (Cohen and Wang, 1959; Cohen, 1968) and which probably form the substrate of the apneustic mechanism, tend to fire continuously (as in apneusis) if not periodically inhibited by pneumotaxic and vagal input.

EFFECTS OF LUNG INFLATION ON DISCHARGE PATTERNS

In order to analyse some of the mechanisms of the Hering-Breuer reflex in cats with intact vagi, the level of pulmonary stretch receptor input was manipulated by means of airway occlusion, and the responses in phrenic discharge and in brain-stem respiratory unit activity were observed (Cohen, 1969).

In the classical Hering-Breuer reflex, lung inflation exerts inspiratory-inhibitory and expiratory-facilitatory effects, with respect to respiratory phase durations as well as respiratory motor discharge. Therefore it might be expected, and indeed it was observed, that brain-stem respiratory neurones would show responses in this direction: many inspiratory neurones were inhibited, and many expiratory neurones were excited, by inflation. However, other neurones showed 'paradoxical' responses, i.e., responses in a direction opposite to those in peripheral discharge: some inspiratory neurones were excited, and some expiratory neurones were inhibited, by inflation.

To explain such paradoxical responses, it has been suggested (Baumgarten and Kanzow, 1958; Cohen, 1969) that neurones having these responses are part of interneuronal inhibitory networks subserving the Hering-Breuer reflex. According to this view, those inspiratory neurones excited by inflation act in turn to inhibit the inspiratory neurones which excite inspiratory motor discharge. Similarly, those expiratory neurones inhibited by inflation are inhibitory to the expiratory neurones that excite expiratory motor discharge, so that inflation causes expiratory facilitation by means of disinhibition.

EFFECTS OF BRAIN-STEM ELECTRICAL STIMULATION ON DISCHARGE PATTERNS

Electrical stimulation in the 'pneumotaxic centre' region of the rostrolateral pons produces powerful respiratory effects (Bertrand and Hugelin, 1971; Cohen, 1971). Stimulation in the dorsolateral region of the rostral pons (nucleus parabrachialis) produces inspiratory facilitation, whilst stimulation more ventrolaterally at the same level produces expiratory facilitation. Furthermore, the short latency (4–7 msec) of the phrenic responses to such stimuli indicates that rostral pontine systems are closely connected to the systems that generate inspiratory discharge.

The emphasis in my own studies (Cohen, 1971) has been on the mechanisms producing the abrupt transitions between the inspiratory and expiratory phases. Thus, a brief train of stimuli delivered in the nucleus parabrachialis during the latter part of the expiratory phase produces switching to the inspiratory phase, i.e. premature onset of a complete inspiratory phase. Similarly, switching from the inspiratory to the expiratory phase can be produced by stimulus trains delivered at ventrolateral points during the inspiratory phase.

In order to explore the mechanisms of phase transition, micro-electrode recordings were made of brain-stem unit and population activity during these stimulus-evoked phase transitions. An example [Fig. 12.3] shows the average response of a lower pontine neurone during stimulus trains (applied in the nucleus parabrachialis during the latter part of the expiratory phase) which produced premature termination of the expiratory phase and initiation of the inspiratory phase in 80–100 msec. The spontaneous discharge pattern of the neurone (top histogram) was continuous, with gradually increasing frequency during the expiratory phase, and more rapidly increasing frequency during the inspiratory phase.

During the stimulus trains, the neurone's discharge frequency increased more than sevenfold (middle histogram). Furthermore, the evoked spikes of the neurone were locked to individual stimuli (bottom histogram): the spikes tended to follow each shock by about 6 msec, which was also the latency of the response to the first shock. Since the onset of the evoked phrenic burst occurred considerably later (80–100 msec after the first shock), it is apparent that the neurone must have been located relatively near, synaptically speaking, to the upper pontine centre being stimulated. The observations suggest that this neurone is part of a transmission chain between the pontine pneumotaxic centre and the medullary inspiratory centre.

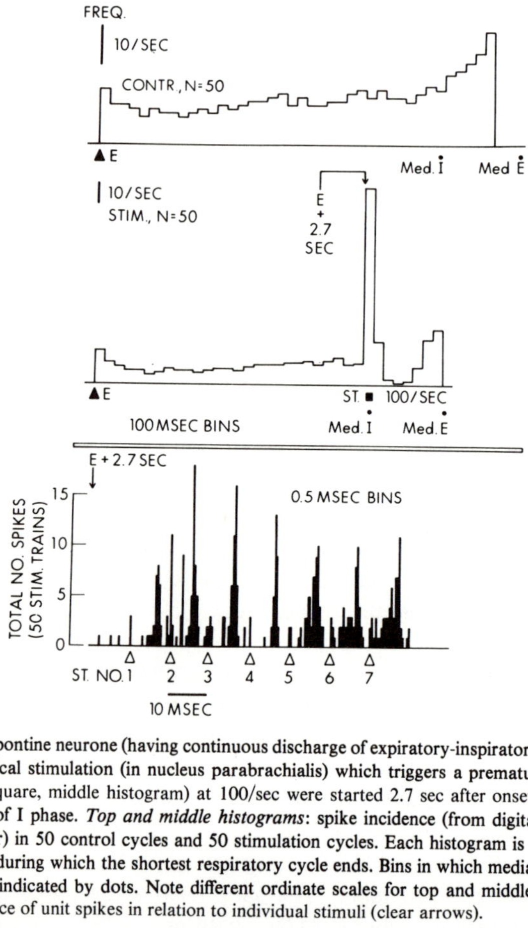

Fig. 12.3. Lower pontine neurone (having continuous discharge of expiratory-inspiratory type) excited by upper pontine electrical stimulation (in nucleus parabrachialis) which triggers a premature inspiratory phase. Stimulus trains (square, middle histogram) at 100/sec were started 2.7 sec after onset of E phase and terminated at onset of I phase. ***Top and middle histograms***: spike incidence (from digital readout of average response computer) in 50 control cycles and 50 stimulation cycles. Each histogram is terminated at the bin preceding the bin during which the shortest respiratory cycle ends. Bins in which median incidence of I or E pulses occurs are indicated by dots. Note different ordinate scales for top and middle histograms. ***Bottom histogram***: incidence of unit spikes in relation to individual stimuli (clear arrows).

Further observations on the mechanisms of phase transition were obtained from recordings of medullary inspiratory wave activity during stimulus-evoked switching from the expiratory to the inspiratory phase. Oscillatory wave activity (frequency 60–110/sec), occurring during inspiration, is found in the ventrolateral region of the medulla medial to the nucleus ambiguus, and is probably a reflection of synchronized synaptic potentials in inspiratory neurones [see below]. The spontaneous oscillations in both medullary and phrenic activity during the early part of the inspiratory phase are shown in Fig. 12.4 (SPONT), where it can also be seen that the oscillations in the two signals had a definite phase relation. When stimulus trains were applied in the nucleus parabrachialis during the latter part of the expiratory phase [Fig. 12.4, STIM], a premature inspiratory phase was evoked; further, as the train was continued into the inspiratory phase, evoked waves of activity were produced in both the phrenic and medullary recordings.

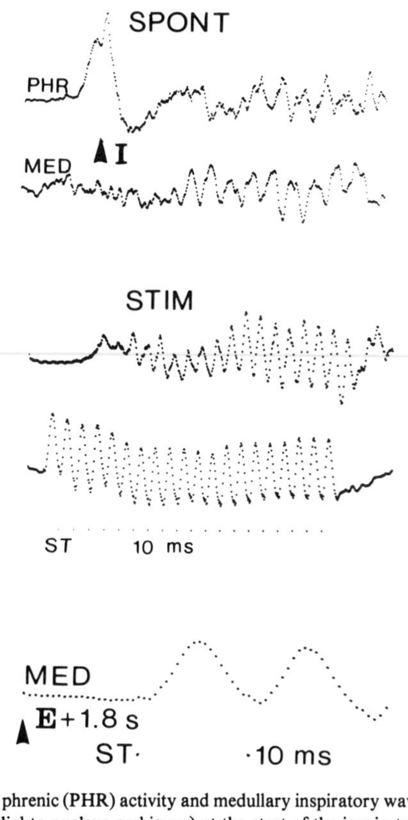

Fig. 12.4. Relation between phrenic (PHR) activity and medullary inspiratory wave (MED) activity (recorded in ventrolateral medulla medial to nucleus ambiguus) at the start of the inspiratory phase in two conditions: SPONT, spontaneously occurring inspiratory phase. STIM, inspiratory phase prematurely evoked by electrical stimulation in nucleus parabrachialis during the expiratory phase. Negativity upwards for both signals. Each trace is the analogue readout from the average-response computer of summed activity in 48 analysis sweeps (0.5 msec bins). *Computer sweep triggers*: SPONT, pulses delayed 180 msec from the I pulses (I, arrow), with the tape played backward (the readout display has the correct time and voltage polarity). STIM, pulses occurring 1.8 sec after start of E phase (preceding first stimulus of each train by 10 msec). Stimulus trains consisted of 20 stimuli at 100/sec; timing indicated by dots (ST). The dots are also time markers (10 msec) for the SPONT traces. *Lowest trace*: expanded readout of medullary (MED) summed evoked activity, showing time relation to first three stimuli (ST, dots). *Gain ratio* (STIM/SPONT) of readouts: PHR, 0.6; MED, 0.2.

The most significant observation shown in Fig. 12.4 (STIM) is that the evoked medullary waves occurred well before the onset of the evoked phrenic burst (which had a latency of *ca.* 30 msec); three such evoked waves occurred before the triggering of the premature inspiratory phase. Further (lowest trace), the latency from the first shock to the start of the first evoked medullary wave was quite short: 1.0–1.5 msec. Apparently, then, neurones in the ventrolateral medulla were being excited synaptically well before phase-switching was produced. It may be suggested that depolarization in these neurones was being successively increased by each stimulus, till levels were reached for mass discharge in the population of inspiratory neurones (phase-switching).

Further analyses similar to those shown in Figs. 12.3 and 12.4 are being conducted, with the aim of identifying functionally the classes of neurone involved in switching from the expiratory to the inspiratory phase. In addition, this type of analysis is being applied to the mechanisms of switching from the inspiratory to the expiratory phase.

SYNCHRONIZATION OF DISCHARGE BETWEEN INSPIRATORY NEURONES

In another type of analysis, the short-term time relations between the activity of different inspiratory neurones were studied. These short-term interactions are manifested in the final common pathway by high-frequency oscillation (60–110/sec) of phrenic discharge, a phenomenon reported by early investigators (Gasser, 1928; Wyss, 1939). In my own study (Cohen 1973, in press) the phrenic oscillations, as well as the simultaneously occurring oscillations in medullary inspiratory unit and inspiratory wave activity, were analysed by auto- and cross-correlation. Two types of medullary inspiratory activity which have high-frequency oscillation are shown in Fig. 12.5.

1. *Inspiratory wave activity* [Fig. 12.5, top]. This activity is found in the ventrolateral region of the medulla, medial to the nucleus ambiguus. The simultaneous recordings of phrenic and medullary activity indicate that both signals were oscillatory; and inspection of the expanded traces (B) suggests that the two oscillations were locked to each other. These impressions were confirmed by correlation analysis (right): the autocorrelograms (ACG) of both signals had the same oscillation period, 11 msec; and the crosscorrelogram (CCG) shows that peak phrenic activity tended to occur about 3 msec after peak negativity of the medullary signal. This oscillatory wave activity is probably a reflection of synchronized synaptic potentials in neighbouring inspiratory neurones of the ventrolateral medulla.

2. *Inspiratory multi-unit activity* [Fig. 12.5, bottom]. The traces are typical of recordings in the dorsal respiratory nucleus, located ventral to the nucleus and tractus solitarius: in this region, dense aggregates of inspiratory neurones are found. The correlograms (right) show that the oscillation of this multi-unit activi-

ty was synchronized to phrenic oscillation—both signals had the same oscillation period, 12 msec, as shown in the ACG's; and peak phrenic activity tended to occur about 4 msec after peak multi-unit activity, as shown in the CCG. In addition, recordings of many isolated inspiratory units showed relationships similar to those illustrated in Fig. 12.5, bottom.

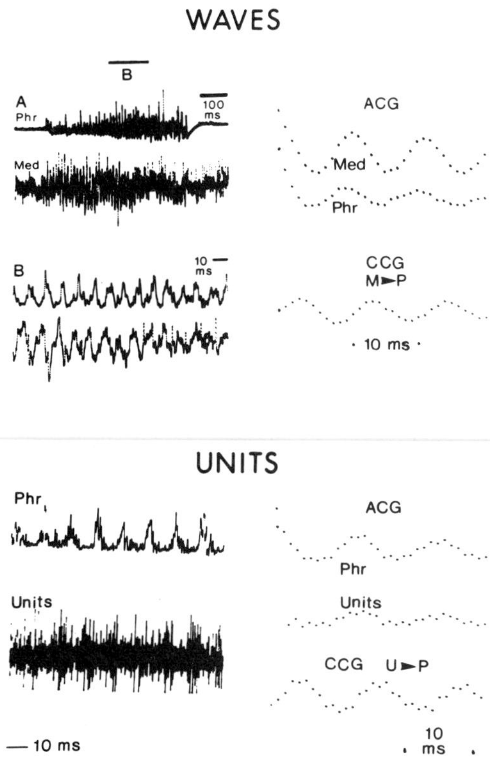

Fig. 12.5. Two types of medullary oscillatory activity occurring during the inspiratory phase and their time relations to phrenic oscillation. Negativity upwards for all signals. *WAVES*: inspiratory wave activity recorded in the ventrolateral medulla medial to the nucleus ambiguus (3.5 mm rostral to obex, 3.5 mm lateral to midline, 3.8 mm ventral to surface of medulla). [A] phrenic (PHR) and medullary (MED) activity during one inspiratory phase. [B] expanded portion of trace A. *UNITS*: activity of several inspiratory units recorded in dorsal respiratory nucleus (region ventral to nucleus and tractus solitarius: 2.5 mm rostral to obex, 3.5 mm lateral to midline, 1.3 mm ventral to surface of medulla); the traces of phrenic and unit activity start 310 msec after onset of an inspiratory phase. *Correlograms* (right) were computed from samples of inspiratory phases (*WAVES*, $N=100$; *UNITS*, $N=45$); traces are computer analogue readouts (32 1-msec bins). *ACG*: autocorrelograms of phrenic, wave, and multi-unit activity. *CCG*: crosscorrelograms between phrenic and wave activity (M→*P*), and between phrenic and multi-unit activity (*U*→*P*). The designation *A*→*B* means that the *A* signal was correlated with the subsequently occurring *B* signal. For the unit correlograms, spikes from each of 3 or 4 units triggered pulses (±2.5 volts, 1-msec duration) which served as input to the correlator. For further details of correlation methods see Cohen (1973).

The high-frequency oscillation of inspiratory discharge is useful as an electrical sign of inspiratory activity which furnishes information about inspiratory centre connexions as descending pathways. The majority of inspiratory neurones had phase relations to phrenic discharge that were similar to those shown in Fig. 12.5 (bottom): peak unit activity tended to precede peak phrenic activity by 3–5 msec. Also, similar lags were usually observed between medullary inspiratory wave and phrenic activity [as in Fig. 12.5, top]. These lags are of the same order of magnitude as the latencies of phrenic responses to electrical stimuli in the medullary inspiratory regions (Rijlant, 1942; Pitts, 1943). In one of my own experiments (previously unpublished), a stimulus delivered in the contralateral dorsal respiratory nucleus (through an electrode which previously had recorded multi-unit inspiratory activity) produced a phrenic response with a latency of 2.5 msec. This observation, together with the finding from antidromic stimulation that most neurones in the dorsal respiratory nucleus project to the cervical cord (Nakayama and Baumgarten, 1964; Bianchi, 1971), indicates that this nucleus is a portion of the descending system exciting phrenic motoneurone discharge.

Oscillatory wave activity (60–110/sec), occurring during the inspiratory phase, has also been recorded near expiratory cells in the caudal medullary 'expiratory centre' (Cohen, 1973). This type of activity reflects synaptic potentials in expiratory cells. Such oscillatory potentials have been found in intracellular recordings of expiratory neurones, where they are superimposed on the steady or slowly changing hyperpolarization that exists during the inspiratory phase (Salmoiraghi and Baumgarten, 1961, Fig. 1; Mitchell and Herbert, 1971; Cohen and Mitchell, unpublished observations). The occurrence of oscillatory phenomena during the inspiratory phase in both expiratory and inspiratory neurones (oscillating synaptic potentials in the former, oscillating spike activity in the latter) is direct evidence of synaptic connexions between neurones of the two classes. The data suggest that the oscillating synaptic potentials of expiratory neurones are produced by input from inspiratory neurones which have oscillating spike activity.

The existence of short-term synchronization of inspiratory discharge suggests that most elements of the inspiratory population are interconnected in a positive-feed-back (re-excitant) manner, so that they are caused to fire together within a short time span. Further studies, using the method of correlation analysis, are in progress with the aim of elucidating these mechanisms of interaction between inspiratory neurones.

SUMMARY

The mechanisms that generate respiratory periodicity are being studied by analysis of responses of brain-stem respiratory neurones to various experimental manipulations which change the over-all respiratory cycle, as monitored by efferent phrenic discharge.

1. During hypocapnia, several types of neurone (mainly expiratory-inspiratory and expiratory) display continuous or 'tonic' discharge. This observation supports the hypothesis that respiratory periodicity arises from the action of inhibitory feed-back connexions on tonic inspiratory-facilitatory and expiratory-facilitatory systems.

2. During the Hering-Breuer reflex evoked by lung inflation, various types of neuronal response are observed. It is suggested that neurones having 'paradoxical' responses (i.e. responses in a direction opposite to the peripheral reflex effects) are involved in the interneuronal inhibitory pathways of the Hering-Breuer reflex.

3. By electrical stimulation of the rostral pontine 'pneumotaxic centre' (nucleus parabrachialis) it is possible to trigger the transition from one respiratory phase to the other. By recording respiratory unit and population activity during these stimulus-induced phase transitions, information is being gathered on the ponto-medullary transmission chains for respiratory phase-switching.

4. The short-term time relations between activity of inspiratory neurones are being studied by correlation analysis of the synchronization of inspiratory discharge, which is manifested by high-frequency (60–110/sec) oscillations in phrenic discharge and in medullary inspiratory unit and inspiratory wave activity. These oscillations furnish an electrical sign of inspiratory activity which is useful in tracing inspiratory centre connexions and descending pathways.

ACKNOWLEDGEMENT

This research was supported by USPHS Grant NS 03970. The author is a Career Scientist of the Health Research Council of the City of New York.

REFERENCES

von Baumgarten, R., von Baumgarten, A., and Schaefer, K.-P. (1957) Beitrag zur Lokalisationsfrage bulboreticulärer respiratorischer Neurone der Katze, *Arch. ges. Physiol.*, **264**, 217–27.

von Baumgarten, R., and Kanzow, E. (1958) The interaction of two types of inspiratory neurons in the region of the tractus solitarius of the cat, *Arch. ital. Biol.*, **96**, 361–73.

Bertrand, F., and Hugelin, A. (1971) Respiratory synchronizing function of nucleus parabrachialis medialis: pneumotaxic mechanisms, *J. Neurophysiol.*, **34**, 189–207.

Bianchi, A. L. (1971) Localisation et étude des neurones respiratoires bulbaires. Mise en jeu antidromique par stimulation spinale ou vagale, *J. Physiol. (Paris)*, **63**, 5–40.

Cohen, M. I. (1968) Discharge patterns of brain-stem respiratory neurons in relation to carbon dioxide tension, *J. Neurophysiol.*, **31,** 142–65.

Cohen, M. I. (1969) Discharge patterns of brain-stem respiratory neurons during Hering-Breuer reflex evoked by lung inflation, *J. Neurophysiol.*, **32,** 356–74.

Cohen, M. I. (1970) How respiratory rhythm originates: evidence from discharge patterns of brain stem respiratory neurones, in *Breathing: Hering-Breuer Centenary Symposium,* pp. 125–50, ed. Porter, R., London.

Cohen, M. I. (1971) Switching of the respiratory phases and evoked phrenic responses produced by rostral pontine electrical stimulation, *J. Physiol. (Lond.),* **217,** 133–58.

Cohen, M. I. (1973) Synchronization of discharge, spontaneous and evoked, between inspiratory neurons, *Acta neurobiol. exp.,* **33,** 189–218.

Cohen, M. I., and Wang, S. C. (1959) Respiratory neuronal activity in pons of cat, *J. Neurophysiol.,* **22,** 33–50.

Eklund, G., von Euler, C., and Rutkowski, S. (1964) Spontaneous and reflex activity of intercostal gamma motoneurones, *J. Physiol. (Lond.),* **171,** 139–63.

Eyzaguirre, C., and Taylor, J. R. (1963) Respiratory discharge of some vagal motoneurons, *J. Neurophysiol.,* **26,** 61–78.

Floersheim, G. L. (1960) Zur elektromyographischen Charakterisierung der Atmungsaktivität, *Helv. physiol. pharmacol. Acta,* **18,** 17–24.

Gasser, H. S. (1928) The analysis of individual waves in the phrenic electroneurogram, *Amer. J. Physiol.,* **85,** 569–76.

Gill, P. K. (1963) The effects of end-tidal CO_2 on the disharge of individual phrenic motoneurones, *J. Physiol. (Lond.),* **168,** 239–57.

Haber, E., Kohn, K. W., Ngai, S. H., Holaday, D. A., and Wang, S. C. (1957) Localization of spontaneous respiratory neuronal activities in the medulla oblongata of the cat: a new location of the expiratory center, *Amer. J. Physiol.,* **190,** 350–5.

Merrill, E. G. (1970) The lateral respiratory neurones of the medulla: their associations with nucleus ambiguus, nucleus retroambigualis, the spinal accessory nucleus and the spinal cord, *Brain Res.,* **24,** 11–28.

Mitchell, R. A., and Herbert, D. A. (1971) Intracellular potentials from medullary respiratory neurones in the cat, *Physiologist,* **14,** 196.

Nakayama, S., and von Baumgarten, R. (1964) Loklisierung absteigender Atmungsbahnen im Rückenmark der Katze mittels antidromer Reizung, *Arch. ges., Physiol.,* **281,** 231–44.

Pitts, R. F. (1943) The basis for repetitive activity in phrenic motoneurons, *J. Neurophysiol.,* **6,** 439–54.

Rijlant, P. (1942) L'excitation du centre respiratoire bulbaire, *Mém. Acad. roy. Méd. Belg.,* **1,** (fasc. 10), 1–131.

Salmoiraghi, G. C., and von Baumgarten, R. (1961) Intracellular potentials from respiratory neurones in brain stem of cat and mechanism of rhythmic respiration, *J. Neurophysiol.,* **24,** 203–18.

Sears, T. A. (1964) Efferent discharges in alpha and fusimotor fibres of intercostal nerves of the cat, *J. Physiol. (Lond.),* **174,** 295–315.

Wang, S. C., Ngai, S. H., and Frumin, M. J. (1957) Organization of central respiratory mechanisms in the brain stem of the cat: genesis of normal respiratory rhythmicity, *Amer. J. Physiol.*, **190,** 333–42.

Wyss, O. A. M. (1939) Impulssynchronisierung im Atmungszentrum, *Arch., ges. Physiol.*, **241,** 524–38.

DISCUSSION ON COMMUNICATION [12]

Plum. Is 'AN' in your diagram the nucleus ambiguus?

Cohen. That's right. The wave activity is found medial to the nucleus ambiguus but not in the nucleus itself [see Fig. 5 in Cohen, 1973][1]. I haven't any good explanation of what neural structures give rise to the waves, but I suspect they may originate in dendrites or axons connecting the dorsal and ventral respiratory nuclei.

Plum. Can you tell us anything about the over-all energy output of this system when respiration is no longer phasic? The question is based on the observation that the expiratory motoneurone pool fires continuously during hypocapnia: also, some of its members fire continuously as one progressively planes off the brain stem and finally creates surgical apnoea. Do you have any idea about the quantitative relationship of the tonic activity to the activity of the phasically firing brain stem? One of the big problems we had in our studies, for example, was that we could find very little evidence of recruitment during increased activity in the medullary respiratory centre. In cell after cell after cell, we got (during CO_2 breathing, for example) an increase in the discharge frequency in the non-vagotomized animal, or an increase in the number of discharges in the vagotomized animal—but never could we recruit new units. I wonder, in your non-phasic animals, whether you could 'derecruit' them, as it were?

Cohen. The problem you are raising concerns the functional relation between two populations of expiratory neurones, which are distinguished on the basis of their responses to hypocapnia (Cohen, 1968)[2]. (1) The first group show phasic (type-1) responses; and at sufficiently low CO_2 levels they cease firing—i.e., they show 'derecruitment'. In agreement with your results, I found that most neurones of this group were 'derecruited' at rather low CO_2 levels: this was also true of most inspiratory neurones. The second group of expiratory neurones have tonic (type-2) responses to hypocapnia: at sufficiently low CO_2 levels, they fire without respiratory modulation. Although one cannot apply the term 'derecruitment' to the type-2 responses, it is of interest that the discharge gap (silent period) in the discharge of these neurones disappears only at very low CO_2 levels: this effect may be attributed to the 'derecruitment' of inspiratory neurones, whose discharge is presumably inhibiting these expiratory neurones.

With respect to the functional rôles of these two types of expiratory neurone, I had originally thought that the group with type-1 responses is concerned with

---

[1] Cohen, M. I. (1973) Synchronization of discharge, spontaneous and evoked, between inspiratory neurons, *Acta neurobiol. exp.*, **33,** 189–218.

[2] Cohen, M. I. (1968) Discharge patterns of brain-stem respiratory neurons in relation to carbon dioxide tension, *J. Neurophysiol.*, **31,** 142–65.

formation of peripheral discharge patterns, since the type-1 responses resemble those of expiratory motoneurones (abdominal and internal intercostal); and since type-2 responses had not been observed in the periphery, I thought that neurones with such responses were involved in formation of the basic rhythmicity. Recently, however, I was informed by Dr. F. L. Eldridge (personal communication) that he had found type-2 responses in the internal intercostal electromyogram—which, of course, complicates the story. Nevertheless it is perfectly conceivable, and indeed likely, that neurones that are involved in setting the basic rhythmicity also project to spinal motoneurones.

A related question concerns the observation of Dr. Bishop that abdominal expiratory motor units do not respond to CO_2, which seems to conflict with my own observations on the type-1 responses of medullary late-expiratory neurones (whose discharge patterns are similar to those of the abdominal motor units). This discrepancy might be partially explained, perhaps, by differences of experimental preparation—i.e. the spontaneously breathing cat with intact vagi (Dr. Bishop's preparations) versus the paralysed, vagotomized cat (my own preparations)—or else it may be that somewhere between the brain stem and the abdominal expiratory neurones you get some shift of the response curve. The problem clearly requires further investigation.

von Euler. I agree that it is entirely possible that there are two completely different types of expiratory neurones. One type may be working, together with an inspiratory population, to form some sort of a 'flip-flop' system—a term that has some application to a description of the phases of respiration. These neurones may have something to do with inhibition of the inspiratory neurones and may have very little to do with the production of expiratory efforts. The other type, which may be controlled quite differently, may be more directly concerned with managing the expiratory motor pathways.

Cohen. Yes: I think that is, for the moment, the most reasonable hypothesis which has to be tested.

von Euler. Concerning recruitment, we have seen that among the inspiratory neurones, some of them which start a bit late in inspiration often start a little earlier if you increase the CO_2. That is a sort of recruitment, in a sense.

Plum. These were in a vagal afferent sensitive pool?

von Euler. No—the non-vagally sensitive ones.

Plum. I ask, because the more rapid the lung expansion the earlier comes the vagal stimulus that may activate that neuronal population.

Cohen. I think recruitment is really a statistical matter. In other words, there is a distribution of response thresholds for the neurones in a given population, and every neurone must be recruited at some CO_2 level (albeit at a very low level). But I think your point is well taken that at normocapnic levels the great majority of inspiratory neurones have been recruited and are already firing.

Plum. You may remember Pitts[1] showed recruitment as a progressive phenomenon during increased activity in the phrenic neurogram; and for this reason we looked particularly carefully (as did you), but it is hard *not* to find units with a microelectrode, so that we reproduced the experience that you had—which was finding cells at either hypercarbia or normocarbia, and then hyperventilating so as to drop the PCO_2 level below what might be considered to be the threshold for a lower level of ventilation. We never could turn cells off when moving from hypercarbia to normocarbia; and only in occasional instances were we able to turn them off as we approached the breathing threshold—that is, to PCO_2 values of about 26–28 mm Hg. That was the basis for our conclusion.

Cohen. I agree; but I would like to add that the recruitment thresholds in the phrenic motoneurone population may be different from those in the medullary inspiratory neuronal population.

von Euler. I think it is interesting to find in the literature that the spinal projections of the inspiratory neurones of the solitary tract complex on the whole don't go down to the thoracic region, but stop in the lower cervical region—indicating that they project almost entirely on to the phrenic motoneurone pool (Bianchi, 1971)[2]. This is in agreement with the work of Torvik (1957)[3], showing (with retrograde degeneration) that the descending spinal path actually comes from the ventrolateral nucleus of the solitary tract where we found the inspiratory neurones I was talking about this morning [see Communication 7]. On the other hand, Merrill (1971)[4] has found that the other main group of inspiratory neurones (i.e. those of the nucleus retro-ambigualis) mainly project down to the thoracic region. This suggests that we have two different pools of inspiratory neurones, one of which goes to the phrenic nucleus and one of which projects to the intercostal motoneurone pools. This seems very interesting to me, and I wonder whether there are any indications that these two pools can be controlled separately; and with regard to the phase relationships, I wonder whether you have observed any phase differences between these two groups in your experiments?

Cohen. I have observed no differences (on the average) between inspiratory neurones in the two nuclei, with respect to the phase relations of their discharges to phrenic oscillations. It seems that the phase shifts between members of the two groups are within the range of variation of shifts within each group. This is understandable, because the two groups are located near each other at the same medullary level, and therefore the transmission pathway is short. However, one might expect to find a phase relation between medullary and thoracic inspiratory neurones different from that between medullary and phrenic neurones, because

---

[1] Pitts, R. F. (1943) The basis for repetitive activity in phrenic motoneurones, *J. Neurophysiol.*, **6**, 438–54.

[2] Bianchi, A. L. (1971) Localisation et étude des neurones respiratoires bulbaires. Mise en jeu antidromique par stimulation spinale ou vagale, *J. Physiol. (Paris)*, **63**, 5–40.

[3] Torvik, A. (1957) The spinal projection from the nucleus of the solitary tract: an experimental study in the cat, *J. Anat. (Lond.)*, **91**, 314–22.

[4] Merrill, E. G. (1971) The descending pathways from the lateral respiratory neurones in cats, *J. Physiol. (Lond.)*, **218**, 82–83P.

of the longer conduction pathway in the former case. But this comparison has not yet been examined experimentally.

The only differences between the two groupings of inspiratory neurones that I have found are: (1) The dorsal neurones are more densely packed, since one can usually record simultaneously from many neurones in the dorsal region, whilst in the ventral region it is easier to find isolated neurones. (2) In the dorsal region, one finds a pure population of inspiratory neurones, whilst in the ventral region inspiratory and expiratory neurones are somewhat intermingled.

I might add that the results of the cross-correlation analysis of inspiratory activity were somewhat disappointing, since almost all activities (both unit and wave) having high-frequency (60–110/sec) oscillations showed similar phase relations to the oscillations in phrenic discharge.

von Euler. I didn't get that completely. Can you reset the oscillations with a stimulus delivered into the pontine centre, and can you reset it in one population only and not in the other?

Cohen. I haven't got that far. I have made only a limited number of observations of the effects of stimulation on this type of oscillation, as follows:

1. Electrical stimulation in the dorsolateral rostral pons, when delivered during the inspiratory phase, can reset the oscillation in both phrenic and medullary inspiratory wave activity. During the stimulus train, medullary and phrenic activity preserves the same relative phase relation that exists spontaneously, but the stimulation produces a greater degree of synchronization, which takes the form of evoked activity locked to individual stimuli. Further [see Cohen, 1973,[1] Fig. 2], stimulation could synchronize inspiratory activity at a frequency different from that of spontaneous activity, but the degree of synchronization was greatest when the forcing (stimulus) frequency was close to the frequency of spontaneous oscillation (Cohen, 1973,[1] Fig. 3).

2. Electrical stimulation in the above-mentioned pontine region, delivered during the expiratory phase, produces switching to the inspiratory phase, and also produces evoked wave activity in the ventrolateral medulla which precedes by a considerable time (30–40 msec) the evoked phrenic burst [c.f. Fig. 12.4, at p. 195]. Thus, the oscillatory inspiratory activity can be evoked and modified by stimulation of the pontine 'pneumotaxic centre'.

Campbell. I am not very familiar with the intimacies of the medulla, I am afraid: but as to the path you are implicitly following, how does it relate to the one that Burns (1963)[2] was developing some years ago? Is there any conflict or match or support between the two views?

Cohen. There is no conflict, but I am trying to make some of the general ideas more specific.

Beverly Bishop. From a teacher's point of view, I would like to inquire what you tell the medical students about the control of respiration. If one goes back

---

[1] Cohen, M. I. (1973) Synchronization of discharge, spontaneous and evoked, between inspiratory neurons, *Acta neurobiol. exp.*, **33**, 189–218.

[2] Burns, B. D. (1963) The central control of respiratory movements, *Brit. med. Bull.*, **19**, 7–9.

through the physiology textbooks of the last 15 or 20 years, one can see that they run periodically through different hypotheses. Wherever we stand, it seems to me that you are back to the stage of saying that we have both inspiratory neurones that are tonically active and must be inhibited, and expiratory neurones that are tonically active and must be inhibited. You get the best of both worlds!

Cohen. In terms of medical students' needs, I would consider this material too recondite. Anyway, I advocate as the latest appropriate story (on an operational level) the version of Wang (Wang and Ngai, 1964)[1], which emphasizes the importance of the tonic inspiratory-facilitatory action of the lower pontine 'apneustic centre', an action which is unmasked during the state of apneustic respiration. In addition, I have suggested the importance of a tonic expiratory-facilitatory system in the medulla (Cohen, 1968)[2]. I have also emphasized the mutual 'flip-flop' action of the respiratory centres, which—by analogy with a multivibrator circuit—occurs when a continuous change of circuit variables has led to a threshold state for switching.

Widdicombe. Of course, although normally we do flip-flop, there are conditions when we do not. I am not thinking of actions like vomiting, but of responses like coughing and sometimes sneezing—where you get a simultaneous, vigorous discharge of inspiratory and expiratory neurones. We don't know whether these reflex acts feed in through the neurone pools that you have been describing: they could by-pass them, as Dr. Plum has shown, in other pathways. But if they were to feed into your neurone pools, then you have switched off the flip-flop.

Cohen. Conceivably: though I would question how simultaneous the activities are. In other words, if you looked at an inspiratory and expiratory discharge both on a short time scale you might find that they were alternating rapidly.

Widdicombe. I can show you records of diaphragmatic and abdominal electromyograms (which, admittedly, are a bit different) showing considerable overlap on a very fast time scale—of course, I don't know what is happening in the medulla.

Cohen. In that case I think that you have by-passed the oscillator. As you know, you can do this with voluntary movement.

Campbell. What about speech? Do you know what would happen if speech were suddenly superimposed on this?

Cohen. No, I'm afraid not—I try to avoid this by decerebrating!

---

[1] Wang, S. C., and Ngai, S. H. (1964) General organization of central respiratory mechanisms, in *Handbook of Physiology, Section 3. Respiration.* Vol. 1, ed. Fenn, W. O., and Rahn, H., Washington, American Physiological Society.

[2] Cohen, M. I. (1968) Discharge patterns of brain-stem respiratory neurons in relation to carbon dioxide tension, *J. Neurophysiol.*, **31**, 142–65.

Campbell. I was wondering about an interruption of your pattern—such as by speech. Could there be a switch-off at this level, or does the cycle go on and be interrupted downstream if you do something like cough or speak?

Cohen. I think there are probably pathways which by-pass the medullary respiratory centres. Thus, Sears (1966)[1] has stimulated the pyramidal decussation in the medulla and has produced short-latency responses in intercostal motoneurones. However, I believe that these supramedullary pathways which by-pass the oscillator must in addition produce switching of the oscillator (perhaps via collaterals); otherwise you would get great confusion due to opposing inputs arriving at the motoneurones.

Wyke. Following on that question from Dr. Campbell and moving it back one stage, in view of the proximity of the predominantly inspiratory population of neurones to the nucleus ambiguus could I ask if in any of your experiments on non-breathing, non-speaking cats you have had electrodes inserted into cells in the nucleus ambiguus. Of course, if you haven't the question is irrelevant; but if you have, do the neurones there describe patterns of activity resembling that of phrenic motoneurones, or are there neurones there with patterns of activity that resemble those of the adjacent inspiratory neurones?

Cohen. I would say the pattern of discharge of neurones in this region is similar to that of the recurrent laryngeal nerve rather than the phrenic nerve. I have done the experiment you mention, and will show some relevant slides later [see General Discussion at p. 468—*Ed.*]. In the region of the nucleus ambiguus, I have recorded from neurones which have patterns similar to the inspiratory discharge pattern in the efferent recurrent laryngeal nerve. This pattern [see Fig. D.28, at p. 469] is trapezoidal in form, with a tendency to decrement as inspiration progresses: in contrast, the phrenic discharge pattern is augmenting. It is likely that the neurones in the region of the nucleus ambiguus which have trapezoidal or decrementing patterns are part of the laryngeal system: and in fact Bianchi (1971)[2] has identified some of these by antidromic stimulation of the vagus (which carries the laryngeal fibres).

von Euler. Merrill (1970)[3] has, of course, done this also and identified nucleus ambiguus neurones by antidromic activation from the laryngeal nerves. He does not regard them as inspiratory neurones in the strict sense. From a respiratory point of view, he has disregarded those that could be activated from the laryngeal nerves but not from the spinal cord.

Campbell. Now that we have finished this discussion, I would like to ask Dr. Cohen to respond sympathetically to Dr. Bishop's urging that he put forward the gospel according to Cohen for us. The story has to be told to medical students by

---

[1] Sears, T. A. (1966) Pathways of supra-spinal origin regulating the activity of respiratory motoneurones, in *Muscular Afferents and Motor Control*, pp. 187–96, ed. Granit, R., New York.

[2] Bianchi, A. L. (1971) Localisation et étude des neurones respiratoires bulbaires. Mise en jeu antidromique par stimulation spinale ou vagale, *J. Physiol. (Paris)*, **63**, 5–40.

[3] Merrill, E. G. (1970) The lateral respiratory neurones of the medulla: their associations with nucleus ambiguus, nucleus retro-ambigualis, the spinal accessory nucleus and the spinal cord, *Brain Res.*, **24**, 11–28.

somebody each year, and I am sure it is your story that all of us would like to have firmly stated—not least so that we can ask the right questions in examinations and get the right answers back!

Cohen. In that case, I should have brought along the sheet that I hand out to the students—but that is always being revised!

Campbell. As long as it is your story, we would like to have the current version.

13 Cerebral Control of Breathing

F. PLUM

Department of Neurology,
New York Hospital-Cornell Medical Center,
New York, U.S.A.

Breathing is a unique 'autonomic' function in that it is completely under the control of the central nervous system and regulated by skeletal muscle, rather than by smooth muscle as are most other autonomic functions (Plum, 1970). This remarkable organization undoubtedly reflects the fact that the act of breathing subserves a major behavioural rôle, as well as its more immediately obvious metabolic-respiratory purposes. As one would expect from such a twin behavioural-autonomic responsibility, the act of breathing is under the influence of both the somatic sensorimotor system and the reticular formation. In the somatic system, one finds influences on respiration involving almost every structure from the cerebral cortex rostrally to the spinal motoneurone pool caudally. In the reticular formation, respiratory-influencing structures extend from the rostral pons to the internuncial pools of the spinal cord, with the main integrating areas lying in the paramedian reticulum of the lower pons and the medulla.

On several previous occasions I have reviewed the deductions made from observations on patients with neurological lesions, and combined these clinical observations with inferences drawn from experimental studies im animals which suggest how each level of the nervous system influences the act of breathing. Today I will briefly recapitulate that story, and add some recent evidence that different functions of the spinal respiratory motoneurones may be served by different descending spinal pathways.

The forebrain exerts a number of obvious influences over the respiratory act that are summarized in Table 13.I. In man, of course, the voluntary control of breathing subserves our main method of communication within our own species. Respiration also is involved in the behavioural acts of laughing and sobbing, both of which are under the control of the limbic system of the forebrain. There is also evidence—in the phenomenon of epileptic respiratory arrest—of vestigial forebrain inhibitory influences on respiration emanating from control mechanisms in the limbic system which may be analogous to the arrest of breathing that occurs in lower forms as part of the diving reflex.

Less obvious, but nevertheless normal, forebrain inhibitory and facilitatory influences affect various stages of the respiratory act and make themselves known by the changes which one observes when disease or dysfunction removes their presence. In this category of disease-induced abnormalities are the phenomena

TABLE 13.I

FOREBRAIN INFLUENCES ON THE RESPIRATORY SYSTEM

FUNCTION	CLINICAL DEFECT IF DAMAGED	NEUROANATOMY
Verbal communication	Dysphasia	Left-hemisphere language area
Voluntary deep breathing or breath-holding	Dyspraxia	Unknown, probably bifrontal
	Pseudobulbar palsy	Interruption of corticobulbar tracts
Emotional behaviour	Forced, 'pseudobulbar' laughing and crying	Limbic system and subcortical connexions
Voluntary or reflex respiratory inhibition	Epileptic respiratory arrest	Medial temporal lobe: limbic system
Wakeful rhythmic breathing	Post-hyperventilation apnoea	Unknown
Damping response to stimulation	Increased responsiveness to carbon dioxide Cheyne-Stokes respiration	Bilateral pyramidal motor system

of post-hyperventilation apnoea and the increased responsiveness to CO_2 stimulation which follow after bilateral injury to the cerebral hemispheres in man. The combination of these two neurophysiological disorders, particularly when they develop acutely or are associated with prolongation of the circulation time (as in hypertensive heart disease) leads to the clinical phenomenon of Cheyne-Stokes respiration (Brown and Plum, 1961). In Cheyne-Stokes breathing one has alternating periods of hyperventilation, during which the ventilatory response to endogenously-produced CO_2 exceeds the rate of CO_2 production, followed by periods of post-hyperventilation apnoea during which chemoreceptor stimuli are in abeyance and the absence of intrinsic cerebral stimuli to breathing results in cessation of the respiratory act. The over-all outcome is a regularly recurring oscillation, whose period length depends on the circulation time required to bring fresh or stale gases from the lungs to the chemoreceptors.

Below the cerebral cortex it is difficult to select a specific influence of subcortical motor centres on the act of breathing, except in so far as the limbic system projects into the diencephalon and hypothalamus to exert the influences that have already been mentioned. Lower animals, of course, employ polypnoea to effect heat loss; and this stimulus, which originates in the hypothalamus, can produce a substantial hypocapnic hyperpnoea that is neurogenically stimulated to the point where it reduces chemical stimuli to breathing far below their threshold value. In overheated, conscious sheep (studied by Gaudio and Abramson (1968), for example) panting continued despite a rise in jugular venous blood pH from 7.38 to 7.77, and a rise in cisternal CSF pH from 7.32 to 7.59. Thermal hyperventilation occurs in man, but is less prominent; and there are no physiological observations on humans with extensive cerebral injury to indicate whether or not any non-chemical stimuli (other than noxious influences from somatic sensory or pulmonary reflex endings) can modify the respiratory act.

Below the anatomic level of the diencephalon, the main integrated influences on the respiratory act appear to arise within nerve centres of the reticular formation. Special kinds of damage to the central reticular core in the posterior part of the midbrain or the rostral pons in man appear, in rare instances, to give a pattern of primary neurogenic hyperventilation. Even more commonly, injury in this area initiates a set of autonomic reflex abnormalities that induce pulmonary oedema (Ducker *et alii,* 1968).

Injury to the tegmentum of the lower pons, both in lower animals and in man, involves that part of the reticular formation which organizes the individual act of breathing. If they come on acutely, such lesions either totally disrupt the respiratory act or, if they selectively involve the lateral portion of the pontine reticulum but spare the central core, they give rise to apneustic or cluster breathing.

Any significant degree of injury to the reticular formation lying between the mid-pons and the cervical spinal cord interferes with the normal ventilatory response to CO_2 stimulation, leading one to suspect that this neurone bed may itself represent the primary central chemosensitive system. Eventually, the CO_2 response declines to next to nothing in the presence of progressive lesions of the dorsal central medulla.

The exact limits of the reticular formation that are required to sustain automatic breathing can hardly be guessed at from studies of human pathology. Extensive lesions, such as infarcts or demyelinating plaques, limited to the pontine reticulum and sparing the medulla altogether will abolish breathing soon after their onset only to have the rhythmic act re-appear subsequently if the patient be kept alive. Thus, border-line oedema and ischaemia may account for some of the acute effects. In patients with poliomyelitis who died after suffering from central respiratory failure the major areas of destruction were in the intermediate lateral reticular formation of the medulla at roughly the level of the nucleus ambiguus. However, the lesions of poliomyelitis are diffuse and it is hardly possible from the available studies to judge the precise location or extent of the critical minimal destruction.

During the past few years one aspect of the neurological control of breathing that has particularly intrigued us is whether there is more than one descending respiratory pathway that projects to the spinal respiratory motoneurones. The speculation that voluntary and involuntary control of the breathing act might descend via different pathways was first suggested by some findings in patients with poliomyelitis.

Sarnoff *et alii* (1951) noted that occasional patients with bulbar poliomyelitis breathed seemingly normally while awake, but hypoventilated or became apnoeic if they fell asleep. Swanson and Plum (1958) subsequently confirmed this finding, and were able to take the observation further in a young woman with poliomyelitis who breathed quite adequately when she was specifically commanded to do so on a breath-by-breath basis, but literally 'forgot' to breathe and became apnoeic whenever the commands were stopped. This unfortunate person died during early convalescence at a time when she had been removed from the mechanical respirator to receive nursing care. At the post-mortem examination, the brain stem contained extensive inflammatory and micronecrotic changes in the medullary reticular formation, but the cervicothoracic respiratory motoneurones were comparatively well preserved. The findings implied that voluntary pathways influencing the act of breathing either were capable of increasing the output of a badly diseased medullary reticulum; or, conceivably, bypassed the medullary centres altogether so as to project directly to the spinal cord—but at the time there was no experimental support for such a speculative concept.

However, a few years later, surgeons performing high cervical cordotomy of the spinothalamic tracts (which run in the anterolateral quadrant of the cord) noted post-operatively that some such patients breathed well when awake and retained excellent voluntary breathing control, but hypoventilated or even became apnoeic during sleep when involuntary respiratory control assumed command (Severinghaus and Mitchell, 1962; Belmusto and Owens, 1963). Since then, experimental observations have established firmly what the clinical studies suggested; for von Euler (Eklund *et alii,* 1964; Corda *et alii,* 1966) and Sears (1971), together with their various co-workers, have discovered and analysed elaborate spinal reflex mechanisms that modulate and integrate respiratory motoneurone activity. Just last year, Aminoff and Sears (1971) described (in the cat) a direct pathway that projects from the cerebral cortex to spinal expiratory motoneurones via what appears to be the corticospinal or rubrospinal tract. Thus, there now seems to be strong evidence that descending spinal pathways conducting voluntary and involuntary respiratory information can be separate from each other in their course to the respiratory motoneurones.

In this regard, Newsom Davis (working in our Laboratory) has found additional evidence in cat and man for both functional and spatial separation in the spinal cord of descending pathways to respiratory motoneurones (Newsom Davis and Plum, 1972). We used mainly decerebrate animals and a few animals anaesthesized with pentobarbitone for our studies, and recorded the direct and

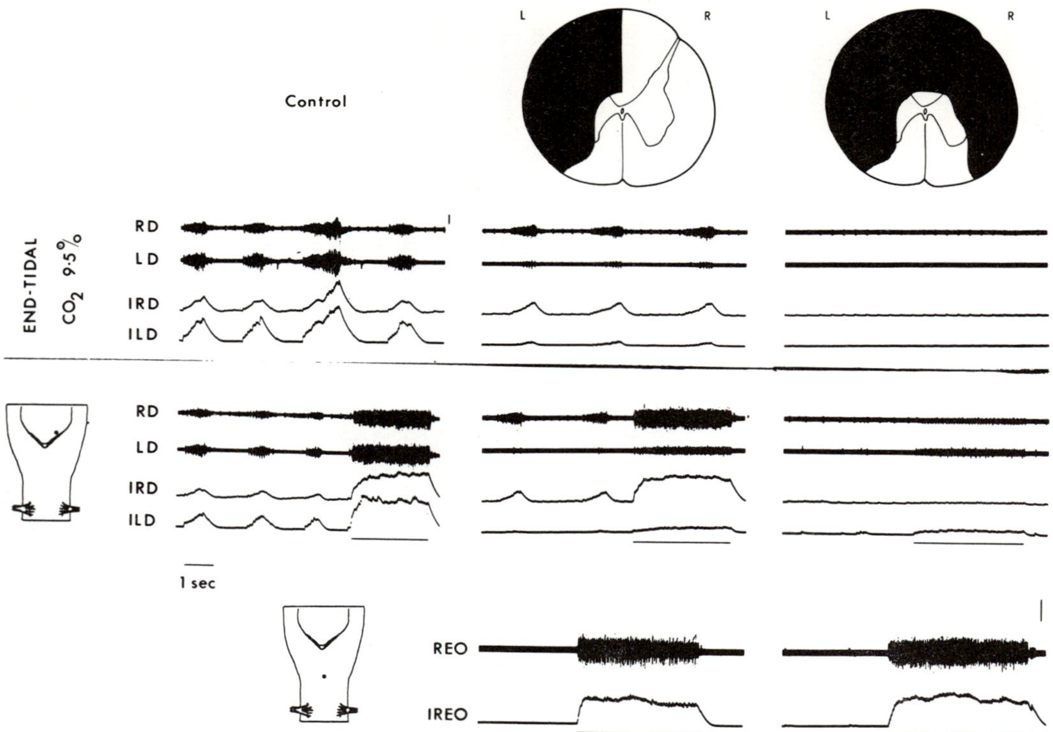

Fig. 13.1. Selective abolition of inspiratory muscle activity by C.2 cervical spinal cord section in the cat. The diagrams illustrate the extent of anatomic section; the points on the brain stem indicate the site of electrical stimulation (5V at 250/sec); the baseline marks the duration of stimulation. Note disappearance of diaphragmatic activity both at rest and during stimulation with preservation of abdominal oblique muscle response to stimulation (RD = right diaphragm; LD = left diaphragm; IRD, ILD = integrated R and LD, respectively; REO = right external oblique).

integrated EMG of both diaphragms (for inspiratory muscle activity) and both abdominal obliques (for expiratory muscle activity). The resting tidal volume was measured with a pneumotachograph, and the end-tidal CO_2 was recorded with an infra-red analyser. The ventilatory and respiratory muscle response was determined, respectively, at rest, to CO_2 stimulation, to inflation of the lungs and to electrical stimulation of the medullary inspiratory apneustic zone, as well as to a standard cough stimulus produced by mechanical stimulation of the larynx. Spinal cord lesions were made at the upper cervical level, either by direct section with a narrow slip of razor blade or by electrocoagulation—using nichrome electrodes inserted through glass capillaries and placed in the cord at appropriate points with a micromanipulator.

In these animals it was possible to eliminate selectively either the inspiratory or the expiratory muscle component of automatic respiration. Fig. 13.1 illustrates the breathing pattern of an animal in which only the ventromedial portion of the spinal cord remains. Note that diaphragmatic activity has been selectively removed and spontaneous ventilation is maintained almost entirely by the abdominal muscles (recorded electromyographically, as illustrated in the lower channels). Thus, in these animals the descending pathway to inspiratory muscles lay in the most ventral portion of the lateral columns, which is in accord with earlier findings by Pitts (1940) in the cat and Nathan (1963) in man. Separation of descending pathways was also obtained by placing lesions ventral and medial to the ventral horn, which selectively reduced and sometimes abolished the abdominal muscle component of quiet breathing (Newsom Davis and Plum, 1972).

In these animals it was possible to demonstrate not only dissociation between descending pathways to inspiratory and expiratory muscles but also dissociation between pathways influencing respiration and those stimulating the same motoneurones when they were involved in certain respiratory reflexes. As Fig. 13.2 illustrates, a large spinal cord lesion—sparing only the ventral funiculus—has markedly attenuated the abdominal muscle contribution to natural breathing, as well as the expiratory response to chest inflation. Nevertheless, the interruption has had no significant effect in reducing the diaphragmatic or abdominal expiratory muscle component of the cough response. Further evidence of functional dissociation in descending spinal pathways to the muscles of breathing is illustrated in Fig. 13.3. In this cat, the participation of the abdominal muscles in breathing as well as in the cough response was blocked by a large, ventrally-placed lesion; nevertheless, the activity of these expiratory muscles was stimulated vigorously as the animal spontaneously defaecated—as recorded at the arrow on the graphic tracing. During defaecation, intense activity developed in the previously quiescent expiratory abdominal muscles, as well as in both diaphragms, in contrast to the absence of any such activity immediately before, during either natural breathing or the cough reflex. One must conclude that whatever the spinal or supraspinal mechanisms required to co-ordinate the respiratory muscles in the act of defaecation, their connexions to the appropriate somatic motoneurones are mediated outside the descending respiratory pathways.

Supporting evidence that descending pathways to the abdominal and phrenic motoneurones mediating the cough response are spatially independent of pathways concerned with the functions of respiration was also obtained in man.

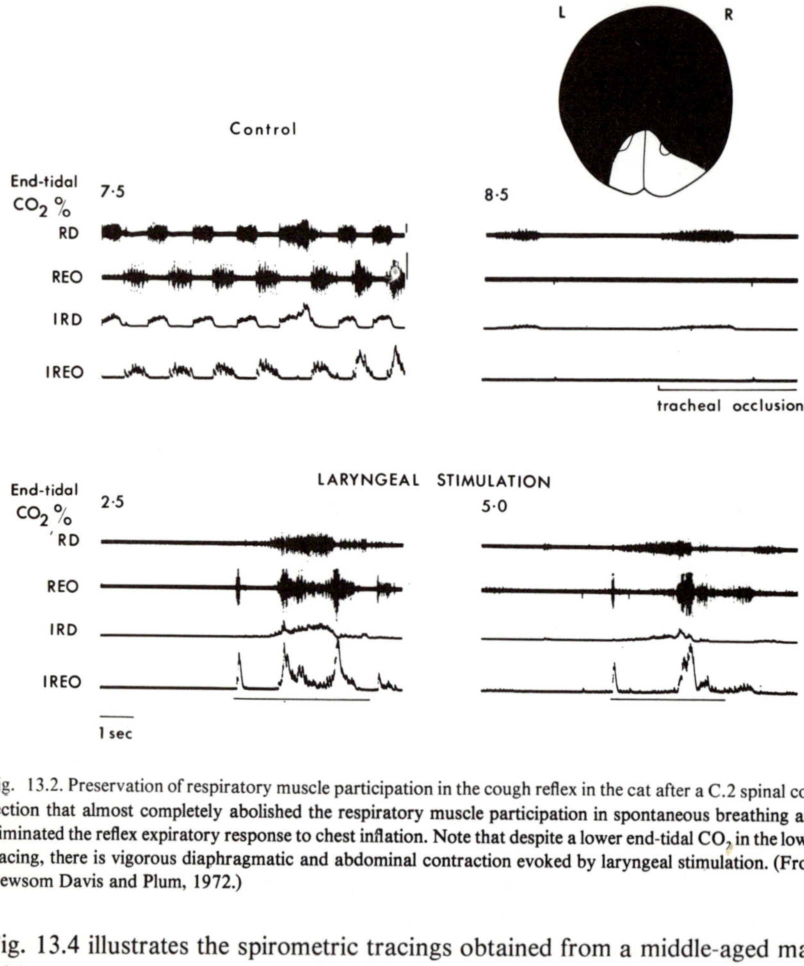

Fig. 13.2. Preservation of respiratory muscle participation in the cough reflex in the cat after a C.2 spinal cord section that almost completely abolished the respiratory muscle participation in spontaneous breathing and eliminated the reflex expiratory response to chest inflation. Note that despite a lower end-tidal CO_2 in the lower tracing, there is vigorous diaphragmatic and abdominal contraction evoked by laryngeal stimulation. (From Newsom Davis and Plum, 1972.)

Fig. 13.4 illustrates the spirometric tracings obtained from a middle-aged man with cancer who, two hours earlier, had undergone the second side of a bilateral, high cervical cordotomy performed for the relief of abdominal pain. Post-operatively his spontaneous breathing had deteriorated progressively to the point of acute respiratory failure, yet he continued to cough vigorously. Despite the stimulation engendered by a P_ACO_2 of 50 mm Hg and a P_AO_2 of 72 mm Hg

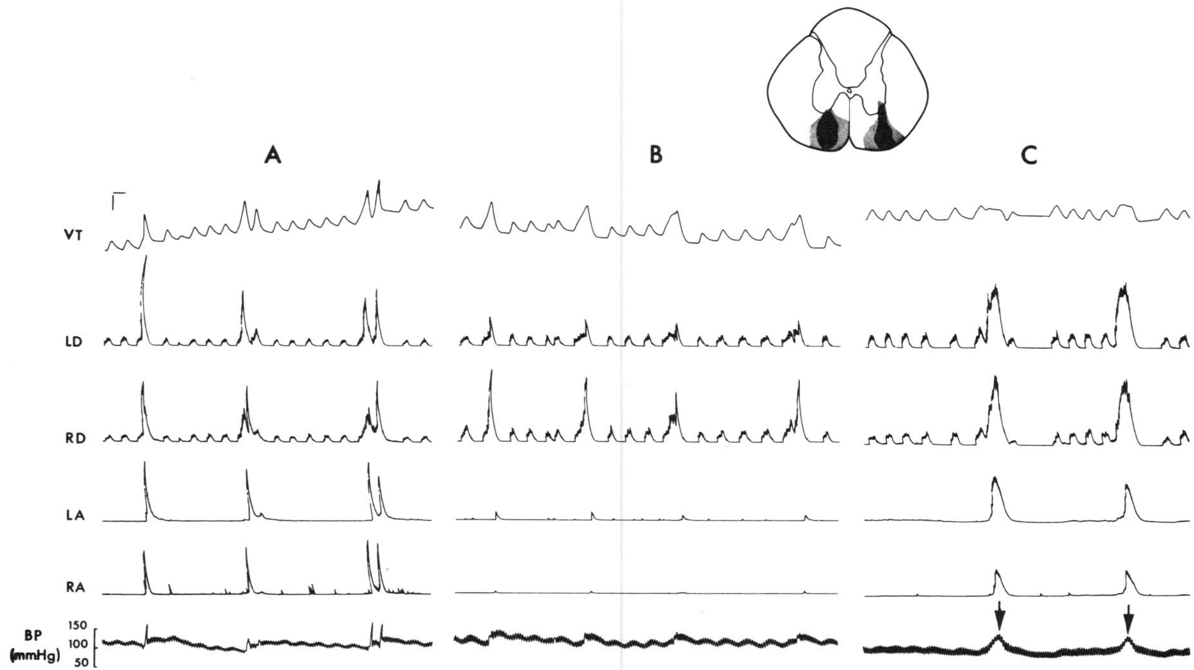

Fig. 13.3. The selective effect of ventral column lesions on abdominal muscle participation in the cough response. [A] control; [B] after bilateral ventral column lesions: laryngeal stimulation to elicit the cough response is indicated by the rise in blood pressure in [B]. In [C], recorded subsequently, the animal spontaneously defaecated (note BP change), producing intense activity in the integrated EMG record from the abdominal muscles and both diaphragms. (From Newsom Davis and Plum, 1972.)

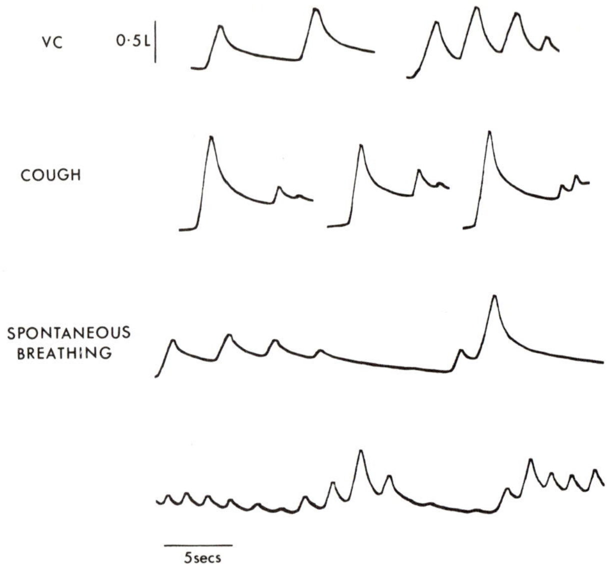

Fig. 13.4. Spirometric tracings illustrating relative sparing of the cough response in man following bilateral C.2 cervical cordotomy that markedly attenuated spontaneous breathing and precipitated respiratory failure (with $P_tCO_2 = 50$ mm Hg). The maximal spontaneous breath or vital capacity is only 720 ml, whilst volumes expired during coughing rise to 1300 ml. (From Newsom Davis and Plum, 1972.)

while breathing room air, his spontaneous breathing through a tracheostomy was shallow (V_t = 50–700 ml), irregular and marked by intermittent periods of brief apnoea. His maximal voluntary vital capacity was 720 ml. Nevertheless, during cough responses evoked by tracheal stimulation (he had a tracheostomy), this man exhaled single volumes of up to 1500 ml. Since coughing is an act integrated no longer than the medulla, it must be concluded that in man, as in the cat, descending pathways mediating the act pass independently of those mediating automatic breathing.

Thus, the several studies of von Euler, Sears and those in man cited here increasingly suggest that we must modify our earlier concept that the medullary reticular formation integrates all the various stimuli to the respiratory muscles, and transmits an immutable programme to the spinal motoneurones. Sears' work increasingly underscores the powerful integrative capacity of the spinal cord in setting the programme of the respiratory act; and our own work in man and animals makes it appear that descending pathways to the respiratory motoneurones of the cord are spatially separated according to the function they serve. It is too early to say whether any part of the pathways mediating the cerebral behavioural control of breathing pass through the descending corticospinal and rubrospinal pathways to bypass the brain stem, but the available experimental work at least suggests this as a possibility.

REFERENCES

Aminoff, M. J., and Sears, T. A. (1971) Spinal integration of segmental, cortical and breathing inputs to thoracic respiratory motoneurones, *J. Physiol. (Lond.),* **215,** 557–75.

Belmusto, L., Brown, E., and Owens, G. (1963) Clinical observations on respiratory and vasomotor disturbance as related to cervical cordotomies, *J. Neurosurg.,* **20,** 225–32.

Brown, H. W., and Plum, F. (1961) The neurologic basis of Cheyne-Stokes respiration, *Amer. J. Med.,* **30,** 849–61.

Corda, M., von Euler, C., and Lennerstrand, G. (1966) Reflex and cerebellar influences on alpha and on 'rhythmic' and 'tonic' gamma activity in the intercostal muscles, *J. Physiol. (Lond.),* **184,** 898–923.

Ducker, T. B., Simmons, R. L., and Anderson, R. W. (1968) Increased intracranial pressure and pulmonary edema, *J. Neurosurg.,* **29,** 475–83.

Eklund, G., von Euler, C., and Rutkowski, S. (1964) Spontaneous and reflex activity of intercostal gamma motoneurones, *J. Physiol. (Lond.),* **171,** 139–63.

Gaudio, R. Jr., and Abramson, N. (1968) Heat-induced hyperventilation, *J. appl. Physiol.,* **25,** 742–6.

Nathan, P. W. (1963) The descending respiratory pathway in man, *J. Neurol. Neurosurg. Psychiat.,* **26,** 487–99.

Newsom Davis, J., and Plum, F. (1972). Separation of descending motor pathways to respiratory motoneurones, *Exp. Neurol.,* **34,** 78–94.

Pitts, R. F. (1940) The respiratory centre and the descending pathways, *J. comp. Neurol.,* **72,** 605–25.

Plum, F. (1970) Neurological integration of behavioural and metabolic control of breathing, in *Breathing: Hering-Breuer Centenary Symposium,* pp. 159–74, ed. Porter, R., London.

Plum, F., and Swanson, A. G. (1958) Abnormalities in central regulation of respiration in acute and convalescent poliomyelitis, *Arch. Neurol. Psychiat. (Chicago),* **80,** 267–85.

Sarnoff, S. J., Whittenberger, J. L., and Affeldt, J. E. (1951) Hypoventilation syndrome in bulbar poliomyelitis, *J. Amer. med Ass.,* **147,** 30–34.

Sears, T. A. (1971) Breathing: A sensori-motor act, *Sci. Basis Med.,* pp. 129–47, London.

Severinghaus, J. W., and Mitchell, R. A. (1962) Ondine's curse—failure of respiratory automaticity while awake, *Clin. Res.,* **10,** 122.

DISCUSSION ON COMMUNICATION [13]

Proctor. I am sure you are aware of this, but I have often observed in patients having thiopentone anaesthesia to the point where they become completely apnoeic that if you lean down and tell them to breathe they take a breath quite promptly. If you then deepen the anaesthesia a little further they won't respond, but if you put a towel clip into the skin they will start breathing again. This seems to be comparable to your poliomyelitis patient who breathes on demand.

Plum. This is really what Dr. Cohen was talking about—it is like hitting a new-born baby on the bottom to make it breathe, isn't it? I think this is the same sort of thing as switching the system by delivering peripheral electrical pulses or lung inflation, as Robson *et alii* (1963)[1] did during barbiturate narcosis. It is a striking thing to watch patients come out of anaesthesia, and see subjects who have 'forgotten' to breathe. It is a nice experimental setting.

Widdicombe. I think the results on the cough reflex are fascinating. Perhaps it is unfair for me to ask if you have done an experiment that I know perfectly well you haven't done. Of these respiratory tract reflexes, you showed that the expiratory efforts bypass all those blocks. The one respiratory tract reflex that was predominantly inspiratory (and the question is whether you have looked at this reflex) is the one from the epipharynx (or the nasopharynx) that produces vigorous contractions of the diaphragm with no alternating expiratory efforts. Have you in your cats seen if the epipharyngeal reflex bypasses your block, or have you in your patients ever passed nasal tubes that tickle the epipharynx and seen if there are any sniff-like or gasp-like efforts?

Plum. That is a good suggestion, but we haven't done it. An interesting thing about the cough, of course, was that the diaphragm was likely to move much more vigorously and simultaneously than it had been during breathing with 8.5 per cent CO_2 stimulation. So whatever the pathway, it is capable of evoking a better than maximal respiratory response in the presence of this kind of spinal cord injury.

Anthony. I don't want to interrupt this, but I am interested in some of the details. How fast can the PCO_2 level change, and how fast can your PCO_2 graphing procedure respond to this change? In one of your slides (on the top line) the PCO_2 changes were shown as a function of time—from this, can you say how fast the changes were actually occurring?

Plum. Unfortunately the infra-red analyser is useless for quantifying the CO_2 response under these circumstances for several reasons—one is because all patients *that* sick have disease of the lungs, possibly for neurogenic reasons; and another is that the patient is by no means emptying his alveoli, particularly during the apnoeic periods. So really the slide was not quantitative. In order to get the data that I told you about, we had indwelling arterial needles and drew repeated blood samples, so the blood gas figures were based on direct measurement and not on end-tidal data.

Beverly Bishop. I feel I should comment on your experiment on the abdominal muscles. I can only say I am pleased to see these results and how beautifully they are put together. I would like to ask something completely different. I found it interesting that these irregular patterns in the human subjects (with a medullary pattern of breathing) had an accentuation of their problems during sleep. I

---

[1] Robson, J. G., Houseley, M.A., and Solis-Quiroga, H. (1963) The mechanism of respiratory arrest with sodium pentobarbital and sodium thiopental, *Ann. N.Y. Acad. Sci.*, **109**, 494–503.

assume you interpret this to mean that the cortical control is important, and that whatever sleep does to cortical functioning, it seems to make things worse. Interestingly enough, during sleep cortical activity becomes more synchronous, and you might expect that this rhythmicity and regularity would also be seen in the lower ventilatory control system.

Plum. The sleep people make a convincing story that all brain activity is not turned off but is merely altered during sleep. It seems to me that one cannot take the increased abnormality during sleep as a simple business of the fact that there is less coming into the system from the cortex. It is more accurate merely to say that wakeful or 'voluntary' behaviour has ceased. A couple of patients with poliomyelitis have told us that they were afraid to go back to sleep, and often they could not say why. One man did actually say, however, 'I can't go to sleep, because if I do I'll stop breathing'. We tested him,[1] and in fact he did stop.

Beverly Bishop. I have heard fellows who have been on Himalayan expeditions say that they experience the same thing at high altitudes.

Plum. I haven't been there myself; but I have been told by many climbers that in the first 24 to 48 hours this is a frightening experience—to wake up apnoeic. I do not know whether that is hypoxic depression. There is this curious business of hypoxia which depresses the brain stem while exciting the chemoreceptors, so that the outcome is the resultant of these two influences—and I do not know the answer to that. The books say there is Cheyne-Stokes breathing in these circumstances; but the tracings I have seen often look clustered—that is, there is a cluster and a pause, a cluster and a pause, without there necessarily being a constant cycle between the clusters.

Wyke. Apropos of this sleep business—which, I agree with you, is of tremendous importance in the types of patient you described—it might be pertinent to record at this point that it is a fact empirically familiar to electroencephalographers doing sleep studies that various changes occur in pulmonary ventilation in association with the different stages of sleep. Thus, it appears that the breathing activity of the REM (rapid eye movement) stage of sleep may be a considerable multiple of the pulmonary ventilation seen in some of the other stages of sleep in the same subject.

Plum. Do you have any blood gas data on that?

Wyke. No I don't. In fact, I was just going on to ask if any of the respiratory experts here had joined their electroencephalographic colleagues in this, and actually got the blood gas data or made reliable measurements of what is going on in terms of ventilatory mechanics at the various stages of sleep—because I don't know of any such data.

---

[1] Plum, F., and Swanson, A. G. (1958) Abnormalities in central regulation of respiration in acute and convalescent poliomyelitis, *Arch. Neurol. Psychiat. (Chicago)*, **80**, 267–85.

Plum. There are three papers, in fact—one by Robin *et alii* (1958)[1], one by Birchfield *et alii* (1958)[2] and one by Pierce *et alii* (1966)[3]. They give slightly different results, but the net outcome is that in normal subjects there was a rise in P_ACO_2 during sleep of about 6 mm Hg. The difficulty is that these studies were all done before the modern neurophysiological studies on sleep began, and therefore do not give the answer to the crucial question of what happens to the PCO_2 during the different stages of sleep. When the people in Kety's laboratory[4] studied cerebral blood flow (using an autoradiographic technique) in animals during REM sleep (during the REM stage of sleep the reticular formation of the brain stem is more active), they found the blood flow to be increased—and they were at pains to say that this was not associated with any rise in PCO_2 in the arterial blood. But whether this latter observation applies to man, like Dr. Wyke I don't know.

Otis. I understood that it has been suggested that the dolphin might have its respiration controlled from the cerebral cortex. This was thought to be because workers with these animals had great difficulty in anaesthetizing them—they just stop breathing—but I believe they have now discovered an anaesthetic that can be used. Do you have any information about this?

Plum. Apparently, Dr. Otis, that is true of all aquatic beasts. The dolphin seems to sleep with one eye open, and I haven't seen other data on this. Lilly's book[5] on porpoises shows some pictures of the animal with one eye open, and he says that the animal is asleep. According to recent information, I have heard that sharks also stop breathing with anaesthesia, and the people who work with sharks have to shoot them with an anaesthetic agent and then intubate them very quickly for artificial ventilation. Apparently, aquatic animals are 'neurobreathers' and not chemosensitive breathers—perhaps this is an evolutionary adaptation to the requirements for 'sounding', at least in the aquatic mammals. I would assume that dolphins are at least partly adapted to much wider fluctuations in PCO_2 than are terrestrial mammals.

Campbell. I would like to urge the need to consider variability between subjects. Much of what we have gleaned over the last century or so has been in an attempt to get the common pattern of *man*; but in the last few years studies of such things as the ventilatory response to CO_2 and the effects of loading, and their interactions, have shown a great variation between *men*—for example, the

---

[1] Robin, E. D., Whaley, R. D., Crump, C. H., and Travis, D. M. (1958) Alveolar gas tensions, pulmonary ventilation and blood pH during physiologic sleep in normal subjects. *J. clin. Invest.*, **37,** 981–94.

[2] Birchfield, R. I., Sieker, H. O., and Heyman, A. (1958) Alterations in blood gases in natural sleep and narcolepsy, *Neurology,* **8,** 107–12.

[3] Pierce, A. K., Jarrett, C. F., Werkle, G., and Miller, W. F. (1966) Respiratory functions during sleep in patients with chronic obstructive lung disease, *J. clin. Invest.,* **45,** 631–36.

[4] Reivich, M., Isaacs, G., Evarts, E., and Kety, S. S. (1968) The effect of slow wave sleep and REM sleep on regional cerebral blood flow in cats, *J. Neurochem.,* **15,** 301–6.

[5] Lilly, J. C. (1961) *Man and Dolphin,* New York.

normal range of the slope of the ventilatory response to CO_2 is sixteen-fold. I feel that in future we may have to be careful when making observations such as what happens to the PCO_2 during sleep in a few specific groups, and extrapolating to what happens in people studied during some other sleep manoeuvre.

Anthony. This something that has puzzled me for a bit. In regard to most of the data given in books on respiration, one would like to know precisely who the subjects were. One suspects, of course, that most of them were medical students—and of course, these are about the last subjects we want to know about in practice.

Godfrey. What about the patients with so-called primary or 'idiopathic' hypoventilation? Do you always find pathology in the brain stem if you look for it? Is there really a consistent pathological picture in people who just don't breathe to the extent that they develop polycythaemia and pulmonary hypertension?

Plum. No, it is not always possible to find pathology in the brain stem. There are at least three published examples with autopsies, and we have studied one, of patients with alveolar hypoventilation who have either no primary pathology, or only non-specific changes, in the brain stem.[1] Thus, the disorder can occur without structural lesions of the brain stem—as after cervical cordotomy, for example.

Godfrey. I asked this, because you said earlier that reflex stimulation may break through in the situation where you have lesions causing apnoea or other ventilatory disturbances.

Plum. In the presence of a depressed central mechanism? Well, the central mechanism can be depressed because of anaesthesia, because of destruction of part of its structure or of its reticulospinal efferents, or because of some intrinsic biochemical lesion the nature of which we don't know. Under any one of these conditions, increased afferent driving would have the effect of raising the excitatory state of the remaining neuronal population.

Godfrey. I recently studied a woman who had massive recurrent pulmonary emboli—which is usually reckoned to be a fairly potent stimulus to breathing, and causes hyperventilation and hypocapnia, as we heard earlier—who was also by all criteria a classical example of the Pickwick syndrome. That is, she was an underventilater with a resting PCO_2 in the region of 65 mm Hg or so in the absence of obstructive lung disease. This, to me, seemed a bit odd, because she must have had enormous drives from her J-receptors or other receptors in her lungs and yet she was quite definitely a primary underventilater.

Plum. Was she fat?

Godfrey. She was moderately fat. She weighed about 75 kg, and was about 150 cm in height.

---

[1] Plum, F. (1966) Breathlessness in neurological disease: the effects of neurological disease on the act of breathing, in *Breathlessness*, pp. 203–22, ed. Howell, J. B. L., and Campbell, E. J. M., Oxford.

Campbell. By implication, Dr. Godfrey, she did *not* increase her ventilation although she *did* have multiple pulmonary emboli?

Godfrey. She did not increase her ventilation though she did have pulmonary emboli—but she could demonstrably increase it on voluntary command. Her maximum voluntary ventilation was 40–50 litres per minute. This seems to be a little odd, if she was getting a colossal drive coming from her lungs.

Plum. It is very odd, although well known—and completely unexplained at present.

Campbell. Could I ask you (for the record) Dr. Plum, how 'high' in the nervous system and how 'low' in the nervous system does your clinical experience allow you to suggest that lesions may lead to indifference to the chemical drives to breathing?

Plum. Many of the patients are like Dr. Godfrey's patient, in whom the effects are secondary to a combination of congestive heart failure, obesity, chronic CO_2 retention and (sometimes) multiple pulmonary infarctions with underperfusion of the lung and ventilation/perfusion defects, leading to a slowly accumulated decompensation: others have neurological lesions of the brain stem or upper cervical spinal cord, as noted in our review in *Breathlessness*[1]. But exactly the same thing can occur with disease of the respiratory muscles, in the absence of any central nervous system defect—the trouble is then all peripheral—with gradual respiratory decompensation and eventually hyposensitivity either to reflex or to chemical stimuli. As Sears[2] has said, breathing is a sensory-motor act—and once the loop is badly broken at this more primitive level, I often cannot separate out the specific cause of failure. Can you?

Campbell. No. This may not be helpful to this Symposium, but I think clinically one would have to agree with Dr. Godfrey that experience with patients indicates that as the capacity to breathe becomes decreased, they reach a point at which the capacity to breathe is so reduced that their PCO_2 goes up. To put it crudely, people are not entitled to have CO_2 retention until their FEV falls below 1.0. However, one sees a number of patients whose PCO_2 rises above 40 mm Hg when they still have a much greater breathing capacity than that, and also as judged by their own ability to breathe when stimulated in other ways (as by exercise) or voluntarily. In the breathing business, one refers to 'won't breathe' and 'can't breathe' situations to separate out these components. I have seen patients with lesions from the cervical spinal cord up to very high levels in the brain stem who may not be able to breathe normally—but who breathe much less than they could.

Otis. I was interested in those patients you mentioned, who came out of anaesthsia and breathed on command. How would they respond if you did some of these loading experiments—adding resistances, and so on?

---

[1] Plum, F. (1966) Breathlessness in neurological disease: the effects of neurological disease on the act of breathing, in *Breathlessness*, pp. 203–22, ed. Howell, J. B. L., and Campbell, E. J. M., Oxford.

[2] Sears, T. A. (1971) Breathing: A sensori-motor act, *Sci. Basis Med.*, pp. 129–47, London.

Plum. We haven't studied that. But in patients with Cheyne-Stokes breathing, the reflex responses to chemical stimuli are quite intact—or even enhanced. Their cough response is certainly enhanced. If one puts something in the mouth (in the posterior pharynx) during the apnoeic period, they will swallow it: if one puts something in the posterior pharynx during the hyperpnoeic period they will swallow it and inhibit the hyperpnoea reflexly even though they can't stop the act of breathing as a voluntary act. So essentially they 'disarticulate' the automatic from the behaviourally-controlled act of breathing.

Otis. Would they find one of these insults, like adding a resistance, uncomfortable?

Plum. I don't know.

Martin. Can I ask you to go a little bit further with some of the dyspraxic patients, the data on which are intriguing. The lesions you described seemed pretty gross to induce this phenomenon, and what you have just said slightly anticipated my question, but did these patients have other forms of dyspraxia?

Plum. Let me separate this into two things, because neurologically I think there is an important distinction here that is not always properly made. By apraxia I don't mean pseudobulbar palsy—I don't mean that the patients have lost all voluntary control over a certain group of neural structures on command. What happens is that apraxic subjects may use those structures for other acts, but they can't voluntarily activate the respiratory muscles for the non-automatic act of breathing (or swallowing). The best example of this is the fact that they can't take a deep breath on command; but yet if you say 'blow out that match' they take a deep breath and go 'puff'. I can't interpret that in terms of anything other than what Lashley would have called 'the forgetting of the engram', with loss of the engrammatic motor act: the neuropathological lesions in such subjects are so diffuse as to be useless for purposes of aetiological explanation. It is clear that these patients understand the command, because they can carry out other simple commands of a comparable degree of complexity with no limitation or resistance. One of the useful things this knowledge provides is to enhance the humanity of the doctor, and keep him from getting angry with his patient for not 'co-operating'.

SECTION III

Physical Aspects of Ventilatory and Phonatory Behaviour

CHAIRMAN

Professor K. N. Stevens

Cambridge, Massachusetts, U.S.A.

14 Computer Models of Gas Exchange Processes in Pulmonary Ventilation

A. I. PACK, B. EMERY, F. MORAN, AND D. J. MURRAY-SMITH

Centre for Respiratory Investigation, Glasgow Royal Infirmary, and the Department of Electrical Engineering, University of Glasgow, Scotland

A satisfactory model of that part of the human respiratory system concerned with gas exchange in the lungs and tissues is a prerequisite for the development of more complex simulations required for the successful study of respiratory control. This communication reviews briefly the development of mathematical models of respiratory gas exchange, and describes some new models designed to aid clinical investigation.

Models of respiratory gas exchange can be conveniently classified according to the mathematical representation of ventilation that they employ. The commonest technique employed is to use the total minute ventilation in the analysis of steady-state gas exchange. The advantage of this approach is that respiratory gas exchange can be described by a system of simple algebraic equations. Much of respiratory physiology is based on this approach. However, mathematical description of non steady-state gas exchange is most simply achieved by means of differential equations.

There are several models that use differential equations. Filley *et alii* (1968) considered ventilation as if it took place instantaneously, so that all gas exchange occurred in that breath-holding period between inspiration and expiration. The description, by means of a differential equation, of gas uptake from a static lung volume is simple enough to allow an analytic solution of the equation, so yielding an explicit relationship between the variables. An alternative approach, first used by Grodins *et alii* (1954) and later by Grodins and other workers (Defàres *et alii*, 1960; Horgan and Lange, 1962, 1963, 1965, 1968; Grodins and James, 1963; Milhorn and Guyton, 1965; Milhorn *et alii*, 1965; Longobardo *et alii*, 1966; Yamamoto and Raub, 1967; Grodins *et alii*, 1967; Milhorn and Brown, 1971) for simulations to study respiratory control, is to consider the lungs as a gas store into which there is a continuous and constant flow of fresh air and from which there is a continuous exhaust to the atmosphere. Ventilation is, however, cyclic; and some models in the literature (Chilton and Stacy, 1952; Dubois *et alii*, 1952; Chilton *et alii*, 1954; Yamamoto, 1960; Murphy, 1966; Flumerfelt and Crandall, 1968), used mainly to study the time course of alveolar events, incorporate a cyclic ventilatory pattern.

STATIC MODEL

We have investigated a number of models for the study of respiratory gas exchange. The first is based on an analysis of steady-state conditions, and its equations are, therefore, algebraic. This model has been constructed to allow quantification of ventilation-perfusion abnormality in the individual subject.

The ratio of ventilation to blood flow ($\dot{V}/\dot{Q}$) in a pulmonary alveolus is of critical importance in determining its gas exchanging capabilities. The most extreme abnormality occurs when either the ventilation or the blood flow to the alveolus is totally blocked; in both cases no gas exchange can occur. An alveolus with no ventilation has a $\dot{V}/\dot{Q}$ of zero. The blood which leaves this alveolus is unaltered, and hence has the same partial pressures of oxygen (PO_2) and carbon dioxide (PCO_2) as mixed venous blood. At the other extreme, an alveolus with no perfusion has a $\dot{V}/\dot{Q}$ which is notionally infinity. Since the gas that leaves such an alveolus is unaltered, a $\dot{V}/\dot{Q}$ of infinity is associated with the partial pressures of

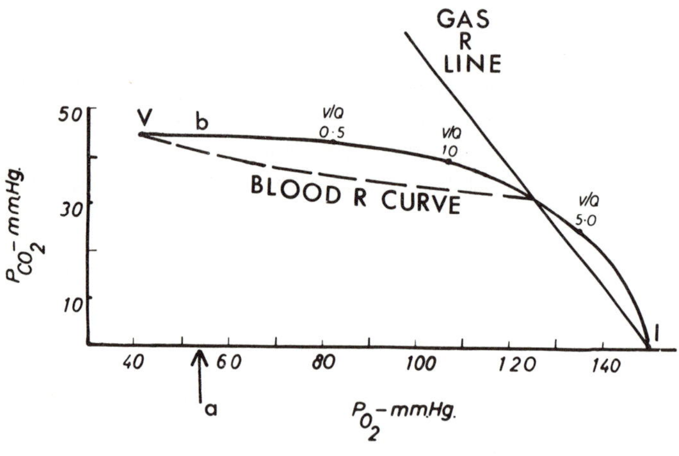

Fig. 14.1 Ventilation-perfusion line. I = inspired air point; V = venous point. Other details in text.

inspired gas. The whole range of $\dot{V}/\dot{Q}$ lies between these two extremes. The partial pressure of oxygen and carbon dioxide in the blood and gas which leave an alveolus are determined by the $\dot{V}/\dot{Q}$ of that alveolus. The relationship is best expressed graphically by means of the ventilation-perfusion line [Fig. 14.1]. The ends of the line are at the venous point ($\dot{V}/\dot{Q}$ = O) and the inspired air point

($\dot{V}/\dot{Q} = \infty$). Each point on the line is associated with a different $\dot{V}/\dot{Q}$ and the partial pressures of oxygen and carbon dioxide that pertain to the blood and gas leaving an alveolus with that $\dot{V}/\dot{Q}$.

Construction of the line (Rahn, 1949; Riley and Cournand, 1949) is based on the recognition that in a steady state the blood *R* equals the gas *R*, where *R* is the respiratory exchange ratio.

$$R = \frac{CO_2 \text{ output per minute}}{O_2 \text{ uptake per minute}} \quad . \; . \; . \; . \; . \; . \; . \; . \quad (1.1)$$

The R_{BLOOD} for an alveolus is calculated from the oxygen and carbon dioxide contents in blood by the following relationship:

$$R_{BLOOD} = \frac{C_{\bar{v}}CO_2 - C_cCO_2}{C_cO_2 - C_{\bar{v}}O_2} \quad . \; . \; . \; . \; . \; . \; . \; . \; . \quad (1.2)$$

whereas the R_{GAS} is calculated from the oxygen and carbon dioxide partial pressures in alveolar gas, by use of the alveolar air equation (Rahn and Fenn, 1955).

$$P_AO_2 = \frac{P_IO_2 \, . \, R + P_ACO_2F_IO_2(1-R) + P_ICO_2 - P_ACO_2}{F_ICO_2(1-R) + R} \quad . \; . \quad (1.3)$$

In manual construction of the line (Rahn and Fenn, 1955; West, 1965) an *R* value is chosen, and the gas *R* line [see Fig. 14.1] which radiates from the inspired air point is constructed. This line describes all possible combinations of PO_2 and PCO_2 that could exist in the gas of an alveolus with that *R* value. The corresponding blood *R* curve is also drawn, and the point of intersection of these is on the $\dot{V}/\dot{Q}$ line. The $\dot{V}/\dot{Q}$ is calculated at this point by an independent formula (Olszowka, 1970):

$$\dot{V}/\dot{Q} = \frac{8.63\left((C_{\bar{v}}CO_2 - C_cCO_2) + (C_cO_2 - C_{\bar{v}}O_2)\dfrac{P_ICO_2}{P_IO_2}\right)}{P_ACO_2 - P_AO_2 \, . \, \dfrac{P_ICO_2}{P_IO_2}} \quad . \quad (1.4)$$

Non-linearity of the blood *R* curve, and the expression for R_{BLOOD} in terms of contents (1.2), make the problem of identifying the point of intersection of the corresponding blood *R* curve and gas *R* line a trial-and-error procedure, each step in the process requiring conversions of partial pressures into gas contents in blood. Manual construction of the line is, therefore, tedious (taking several hours to complete) and inaccurate.

Methods for construction of the line using a digital computer have been described (Kelman, 1968; Olszowka and Fahri, 1969). The first (Kelman, 1968) uses a procedure similar to the manual method, in that points on the line are identified by searching along a gas *R* line for the point of intersection with the corresponding blood *R* curve. The other method (Olszowka and Fahri, 1969) treats the $\dot{V}/\dot{Q}$ as the 'independent variable'. Thus the problem is to find the PO_2 and PCO_2 that are associated with a given $\dot{V}/\dot{Q}$. This is somewhat inefficient, since it incorporates a search procedure in two-dimensional space. The simplest procedure, which we have adopted, is to eliminate the variable *R* and reduce the search problem to finding the root of an equation. Thus given a PO_2—for example, point (a) in Fig. 14.1—the problem of identifying a point (b) on the line reduces to finding a root. For root-finding we have used the method of continuous bisection. In the digital computer programme which we have developed (Pack *et alii,* 1972), sixty points on the $\dot{V}/\dot{Q}$ line are identified in this way. These points correspond to sixty PO_2 values equally spaced between the values of PO_2 in mixed venous blood and inspired gas. Given a $\dot{V}/\dot{Q}$, it is also possible to move in the reverse direction—i.e. to find the associated partial pressures of oxygen and carbon dioxide—by use of the technique of inverse interpolation. We have used second-order Lagrangian interpolation which has given completely satisfactory results.

Development of an efficient digital computer programme for construction of the ventilation-perfusion line has allowed us to investigate a 'parallel static' model with two alveolar compartments. The model has an anatomical dead space which represents the conducting airways, two alveolar compartments with variable blood flows and ventilations, and a true right-to-left shunt. Application of this model to the quantification of ventilation and perfusion abnormality in patients necessitates measurement of the arterial and venous PO_2 and PCO_2, the total minute ventilation and the magnitude of the shunt measured during 100 per cent oxygen breathing. By use of a system of simultaneous algebraic equations and the $\dot{V}/\dot{Q}$ line for each subject, one can partition the blood flow and ventilation between the two compartments in such a way as to produce the measured arterial blood gases. This method of quantification of $\dot{V}/\dot{Q}$ distribution is under study as a diagnostic aid.

The main defect of this approach is the necessity for ensuring that the subject is in a steady-state condition, because the results will be affected by variation in the ventilatory pattern. This has led us to investigate dynamic rather than static lung models.

DYNAMIC MODELS

STRUCTURE AND EQUATIONS

The model [Fig. 14.2] consists of a dead space compartment with constant volume V_D representing the conducting airways, a single homogeneous alveolar compartment (volume V_A) and one tissue compartment (volume V_T). A single tissue compartment was considered adequate, because all the experiments performed were of short duration. Gas transfer is represented as taking place directly between the tissues and the alveoli, i.e. without circulatory time delays. The effect of this assumption is to include the venous blood volume in the tissue compartment, and the arterial blood in the alveolar compartment. In the case of carbon dioxide, there is a metabolic input of gas into the tissue compartment.

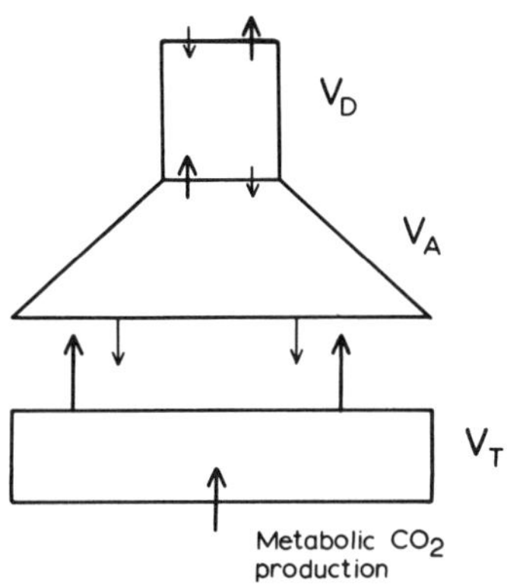

Fig. 14.2. Structure of dynamic model. Symbols are explained in the text.

At the start of inspiration the first gas which enters the alveolar compartment is that which remained in the dead space at the end of the previous breath. As inspiration continues, atmospheric air reaches the alveolar compartment. Within the alveolar compartment, therefore, the respiratory cycle consists of three stages:

I. *Inspiration of dead space gas.*

$$\text{For Stage I} \quad \int_{t_0}^{t} \dot{V}\,dt \leqslant V_D \;\; \& \;\; \dot{V} > 0 \qquad \qquad (2.1)$$

where t_0 is start of inspiration, and t is time from start of inspiration.

II. *Inspiration of atmospheric air.*

$$\text{For Stage II} \quad \int_{t_0}^{t} \dot{V}\,dt > V_D \;\; \& \;\; \dot{V} \geqslant 0 \qquad \qquad (2.2)$$

III. *Expiration.*

$$\text{For expiration} \qquad \dot{V} < 0 \qquad \qquad (2.3)$$

The behaviour of this model can be described by a system of differential equations, one for the alveolar compartment and one for the tissue compartment. The equations are illustrated for carbon dioxide. The symbols are mostly standard, but the entire list of symbols with meanings is given in Appendix I (at p. 239). The equation for the alveolar compartment is

$$V_A \frac{dP_A}{dt} = S\dot{V}(P_I - P_A) + \dot{Q}b\,(\text{B-47})\,(P_T\text{-}P_A) \quad . \; . \; . \; . \; (2.4)$$

The first term on the right-hand side of this equation represents addition of carbon dioxide from the dead space. Since this transfer of gas from the dead space is different in these three stages of the respiratory cycle, this term has a different form in each stage. This can be conveniently represented by the introduction of the symbol S, when

$$S = 1 \text{ in Stage II} \quad . \; . \; . \; . \; . \; . \; . \; . \; . \; . \; . \; (2.5)$$
$$= 0 \text{ in Stages III \& I.}$$

The other equation is for the tissue compartment, i.e.

$$bV_T \frac{dP_T}{dt} = M - \dot{Q}b(P_T - P_A) \quad . \; . \; . \; . \; . \; . \; . \; . \; (2.6)$$

These equations are almost identical to those used by Murphy (1966). In their construction it has been assumed that $P_{\bar{v}} = P_T$, $P_a = P_A$, and that during Stage I $P_T = P_A$.

EXPERIMENTAL VERIFICATION OF MODEL

Many simulations of respiratory gas exchange have been limited by the use of a sinusoidal representation of ventilation. Exploitation of these models, therefore, has been confined to the objective of reproducing previously published experimental results. However, it is possible to subject the model and the system to the same input so that the outputs of model and system can be directly compared. The difference between them can be expressed as an error function. The parameters of the model, i.e. the constants in the model equations which represent the system's characteristics, can then be adjusted in such a way as to minimize the error. The lung is particularly suited to this approach, since the input—inspired gas flow at the mouth—can be measured continuously, as can an output—the gas concentration of the alveolar portion of the expired record. This approach has been used to test our lung model experimentally.

Ventilatory flow is measured at the mouth using pneumotachographs, and gas concentrations at the mouth are measured continuously using a mass spectrometer (MS4 : AEI). Synchronization of the signals is achieved by a specially constructed delay unit. The flow signal and the gas concentration signals are recorded on a magnetic tape loop, the output from which is subsequently fed into the model equations which are patched on a small analogue computer. The model and lung are thus subjected to the same input, and the model output—partial pressure of gas in the alveolar compartment—is compared directly, by means of a visual display, with the alveolar portion of the expired gas concentration record as measured by the mass spectrometer. In the experimental testing of the model, the parameters were adjusted manually on the computer to minimize the output differences between the system and model. To carry out this step, the magnetic tape loop was used to display each experimental record repetitively.

The model's performance has been tested using a number of different test inputs [Fig. 14.3]. When a relatively insoluble gas such as argon is breathed, the main parameter to be adjusted is lung volume. With a suitable value for lung volume, good agreement between the model's performance and that of the system can be achieved both during a period of argon 'wash-in' [Fig. 14.3A] and argon 'wash-out' [Fig. 14.3B]. When a respiratory gas such as carbon dioxide is studied, manual adjustment of the parameters—lung volume, cardiac output, initial tissue partial pressure, tissue volume and metabolic production— is more difficult since the parameters interact. Nevertheless, optimization can be achieved by manual means, as is demonstrated in Fig. 14.4A, B.

During hyperventilation the model not only predicts correctly the alveolar PCO_2, but also the increase in amplitude of the oscillations of the alveolar PCO_2 (model) parallels the increase in the slope of the alveolar portion of the measured expired records. This experimental result has led us to suggest that whereas the mean level of alveolar PCO_2 is determined by the $\dot{V}/\dot{Q}$, the amplitude of variation in alveolar PCO_2 during one respiratory cycle is determined mainly by the ratio between alveolar ventilation and alveolar volume. The variation in alveolar PCO_2 that occurs during one respiratory cycle is due to the fall in PCO_2 which occurs during expiration and in Stage I of inspiration (2.1). In a steady state, the inspiratory fall in PCO_2 must equal the expiratory rise. Since the fall in alveolar PCO_2 which occurs during Stage II of inspiration is due to the diluting effect of incoming fresh air ($PCO_2 = 0$), any increase in the ventilation of the same alveolar volume will produce a larger fall in alveolar PCO_2. The rise in alveolar PCO_2 during expiration and Stage I of inspiration is the result of addition of carbon dioxide to the alveolus from the blood stream. Hence an increase in the blood flow to an alveolus will also result in an increased amplitude of variation of alveolar PCO_2. An increase in ventilation has, however, a greater effect on this amplitude of variation than an increase in perfusion. This is because the difference between inspired gas ($PCO_2 = 0$) and the mean alveolar PCO_2 (at most values of $\dot{V}/\dot{Q}$) is considerably larger than the difference between mixed venous blood and alveolar gas. Thus in alveoli with high ventilation-volume

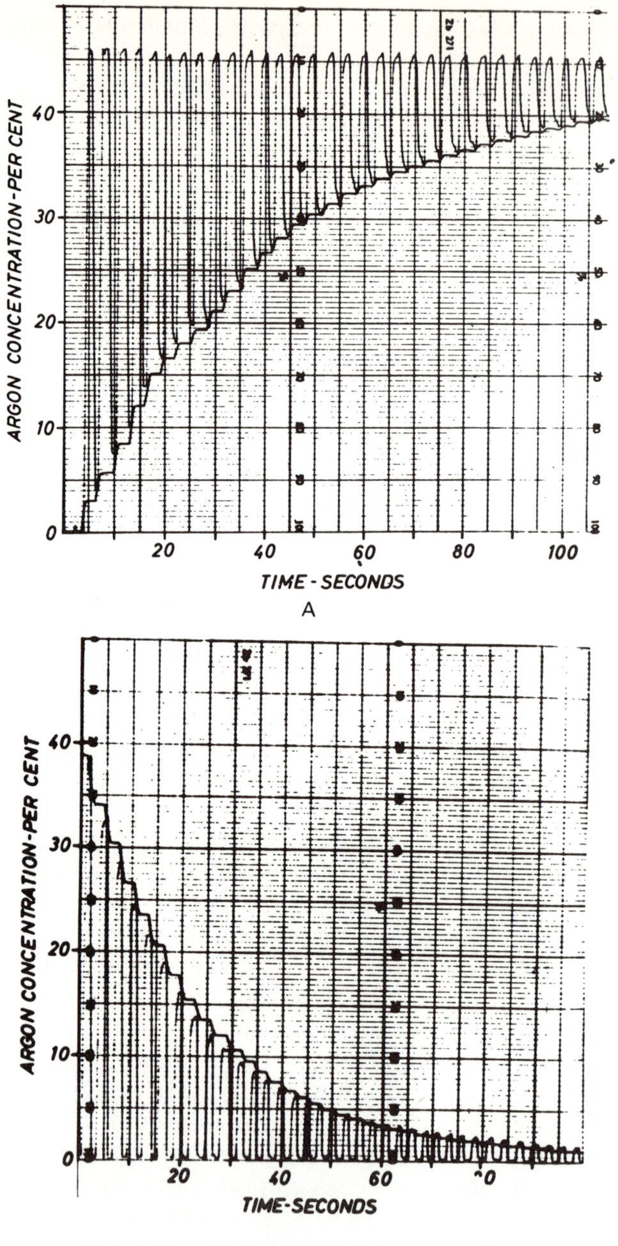

Fig. 14.3. Testing of dynamic model, using argon. [A] Continuously measured record of argon concentration at the mouth during argon 'wash-in' (continuous record that oscillates between the inspired level of 46 per cent and the rising expiration peaks) compared with the model prediction of alveolar argon concentration (heavier stepped line). [B] Continuously measured record of argon concentration at the mouth during argon 'wash-out' (continuous record that oscillates between the inspired level of zero and the falling expiration peaks) compared with the model prediction of alveolar argon concentration (heavier stepped line).

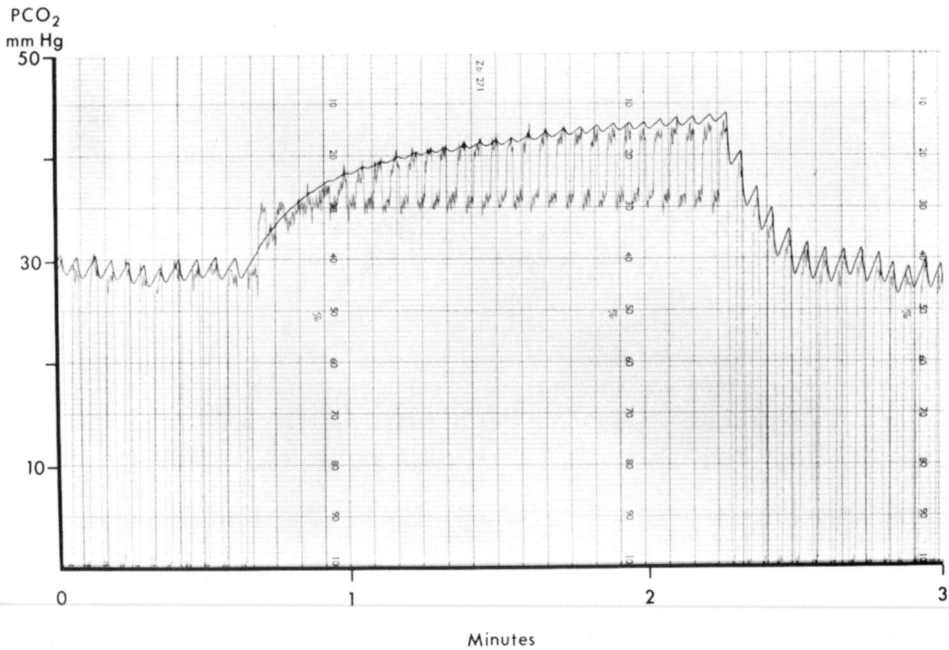

A

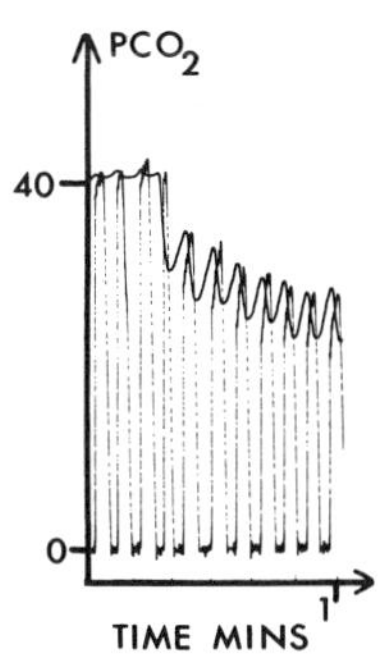

B

Fig. 14.4. Testing of dynamic model, using carbon dioxide. [A] Manual optimization of model parameters during response to carbon dioxide inhalation. The fainter oscillating record is the expired PCO_2 record as measured at the mouth. Note the abrupt change from breathing air ($PCO_2 = 0$) to breathing a gas mixture containing CO_2 ($PCO_2 \times$ mm Hg). The heavier line is the model prediction of alveolar PCO_2. Note the agreement between the model prediction and the expired peaks. [B] Hyperventilation. Note concordance between peaks in expired CO_2 record and model prediction.

ratios there will be large oscillations in alveolar PCO_2 during the respiratory cycle.

We further suggest that large oscillations in alveolar PCO_2 will produce a significant 'alveolar slope'. The commonly proposed explanations for a steep alveolar slope on the measured expired PCO_2 record, which occurs in normal subjects during hyperventilation or exercise and in some abnormal subjects, are on the one hand asynchronous emptying of alveoli with different $\dot{V}/\dot{Q}$ (Otis *et alii,* 1956) and hence containing gas at different PCO_2, and on the other series inhomogeneity (Sikand *et alii,* 1966). Series inhomogeniety is the result of the transition, probably at the level of the respiratory bronchioles, from convective flow to diffusion in the gas phase. It is postulated that the resulting variation in PCO_2 throughout the terminal lung unit will produce an alveolar slope on expiration.

Oscillations in alveolar PCO_2 should also produce a slope. The conducting airways of the lung are a transit delay between the alveoli and the mouth, although they do not represent a pure time delay because of the distorting effect of flow through the airways. Nevertheless, allowing for this distorting effect, an alveolar slope will result when the PCO_2 of the gas transferred from the alveoli to the airways is significantly lower at the start than at the end of expiration.

These hypotheses have been supported by a number of simulation studies of events in a single alveolus and in the summed performance of two alveoli in parallel. In these simulations the equation of the alveolar compartment has been used, since it is applicable not only to a single compartment representing a homogeneous alveolar population but also to a single alveolus. These simulations are also of value for teaching purposes, since they provide a useful insight into the physiology of pulmonary gas exchange—in particular, the ventilation-perfusion concept and the mechanism of production of an arterial-alveolar gradient for carbon dioxide.

ADAPTIVE MODELLING

Although manual optimization of model parameters can be achieved, the use of the model for routine physiological measurements or diagnostic tests requires some form of rapid and fully automatic parameter adjustment. This requirement of self-optimization of the model has led to the study of adaptive modelling procedures.

In a self-adaptive model the error between the measured system output and the model output is used to adjust parameters of the model in such a way that some appropriate functional of the error is minimized. The optimization procedure may be based either on periodic adjustment of the parameters in a series of discrete steps, or on continuous adjustment of the parameters. Both the discrete and continuous methods of parameter adjustment have been used in models of engineering systems, and the general principles of adaptive modelling are well established (Bekey, 1964; Davies, 1970).

The application of adaptive techniques to the lung model has so far been concerned only with continuous parameter adjustment methods. Here the functional of the error is defined as

$$J(t, Q) = \left(Y_s(t) - y_m(t, Q)\right)^2 = e^2(t, Q)$$

where $Y_s(t)$ is the system output
$y_m(t, Q)$ is the model output
$e(t, Q)$ is the error.
and Q is the vector defining the adjustable parameters of the model.

The use of the square of the instantaneous error between the system output and the model output ensures that J (the criterion function) is always positive.

For a single adjustable parameter, q_i, of the model it follows that the rate of change of J with respect to that parameter is given by

$$\frac{\partial J}{\partial q_i} = 2e\frac{\partial e}{\partial q_i} = -2e\frac{\partial y_m}{\partial q_i}$$

The partial derivative $\frac{\partial y_m}{\partial q_i}$ gives an indication of how sensitive the model response is to variation of the parameter. This quantity is appropriately known as a 'sensitivity function', and such functions have an important rôle in self-adaptive models. The sensitivity function may be generated conveniently in the simulation by means of an auxiliary model or cosystem (Tomovic, 1963). This cosystem has a structure that is closely related to the structure of the system model itself, and has a particularly simple form when the differential equations defining the model are linear.

Minimization of the criterion function may be achieved by applying an adjustment algorithm of the form

$$\frac{\partial q_i}{\partial t} = g_i\frac{\delta J}{\partial q_i}$$

where g_i is a constant.

This ensures that the parameter q_i is adjusted at a rate proportional to the gradient $\frac{\partial J}{\partial q_i}$. Near the optimum setting the gradient is small, since the error is small and the parameter value will, therefore, change slowly. Well away from the optimum setting the gradient is larger, and the rate of parameter adjustment will be correspondingly greater. Fig. 14.5 shows the general structure of a continuous adaptive model. This block diagram indicates clearly the way in which the model, the cosystem and the parameter adjustment loop are interconnected.

The performance of an adaptive model with continuous parameter adjustment depends upon the characterization of the model, the choice of initial values of the adjustable parameters and the values of the constants used in the parameter

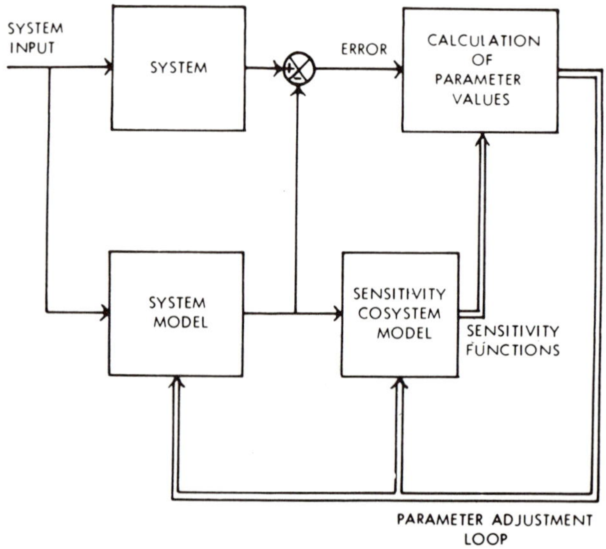

Fig. 14.5. Structure of a continuous adaptive model.

adjustment algorithm. In general there is a limited domain of parameter values for which a model will adapt and it is, therefore, desirable that all relevant information be used in establishing the initial conditions.

The choice of the gain constant g_i in each of the parameter adjustment loops is very important in determining the over-all behaviour of the adaptive model. Larger values of gain constant in these feed-back loops may cause instability, whilst very small values of gain give a sluggish performance.

Figure 14.6 shows automatic parameter adjustment of the lung model for test results obtained using a mixture containing argon. The experimental responses were recorded on magnetic tape and replayed repetitively using a tape loop.

The Figure shows the measured output from the mass spectrometer, the corresponding model output, the error and the value of the parameter being adjusted. In this particular record, the parameter being considered is the inverse of the volume of the alveolar compartment. The record shows clearly the adjustment of the parameter to produce agreement between the model output and the

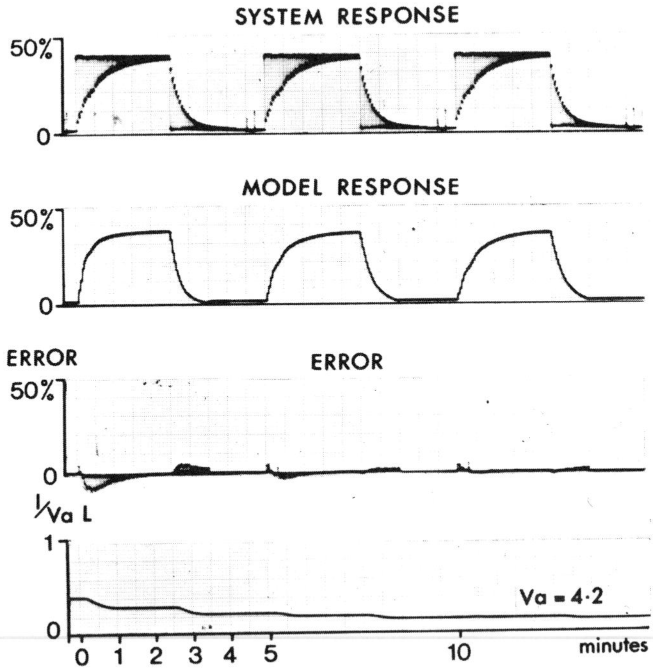

Fig. 14.6. Automatic parameter adjustment, illustrated for argon breathing. The system response record is a repetitive display of the argon 'wash-in' and 'wash-out' records, illustrated separately in Figs. 14.3A, B. Further details are given in the text.

system output. A numerical value for lung volume may be calculated from the final setting of the adjustable parameter in the model.

Similar automatic parameter adjustment has been achieved in an experiment in which partial pressure of carbon dioxide was measured in a normal subject at rest and in hyperventilation. In this case, the parameters do not converge to a unique set of values but rather to a small region in the parameter space. By using larger stimuli, convergence can be made more rapid and the optimum point is better defined.

APPENDIX I

b = Slope of physiological dissociation curve for CO_2, in litres per litre/mm.

B = Barometric pressure, in mm Hg.

C_xCO_2 = Carbon dioxide content of blood, vols. per cent.

C_xO_2 = Oxygen content of blood, vols. per cent.

F_xCO_2 = Fractional concentration of carbon dioxide in gas.

F_xO_2 = Fractional concentration of oxygen in gas.

M = Metabolic production, l/min.
P_xCO_2 = Partial pressure of carbon dioxide, mm Hg.
P_xO_2 = Partial pressure of oxygen, mm Hg.
$\dot{Q}$ = Cardiac output, l/min.
R = Respiratory exchange ratio.
$\dot{V}$ = Ventilatory flow, l/min.
V_A = Volume of alveolar compartment.
V_D = Volume of dead space compartment.
V_T = Volume of tissue compartment.

Subscripts

a Arterial blood.
A Alveolar gas.
c End-capillary blood.
I Inspired gas.
T Tissue compartment.
$\bar{v}$ Mixed venous blood.

REFERENCES

Bekey, G. R., and McGhee, R. B. (1964) Gradient methods for the optimization of dynamic system parameters by hybrid computation, in *Computing Methods in Optimization Problems*, pp. 305–27, ed. Balakrishnan, A. V., and Neustadt, L. W., New York.

Chilton, A. B., Barth, D. S., and Stacy, R. W. (1954) A mathematical analysis of oxygen respiration in man, *Bull. math. Biophys.*, **16**, 1–14.

Chilton, A. B., and Stacy, R. W. (1952) A mathematical analysis of carbon dioxide respiration in man, *Bull. math. Biophys.*, **14**, 1–18.

Davies, W. D. T. (1970) *System Identification for Self-adaptive Control,* New York.

Defares, J. G., Derksen, H. E., and Duyff, J. W. (1960) Cerebral blood flow in the regulation of respiration, *Acta physiol. pharmacol. neerl.*, **9**, 327–60.

DuBois, A. B., Britt, A. G., and Fenn, W. O. (1952) Alveolar CO_2 during the respiratory cycle, *J. appl. Physiol.*, **4**, 535–48.

Filley, G. F., Lacquet, L. M., Olson, D. E., Speck, R. P., and Sodal, I. E. (1968) A mathematical model of pulmonary gas transport, in *Form and Function in the Human Lung*, pp. 199–222, ed. Cumming, G., and Hunt, L. R., Edinburgh.

Flumerfelt, R. W., and Crandall, E. R. (1968) An analysis of external respiration in man, *Math. Biosci.*, **3**, 205–30.

Grodins, F. S., Buell, J., and Bart, A. J. (1967) Mathematical analysis and digital simulation of the respiratory control system, *J. appl. Physiol.*, **22**, 260–76.

Grodins, F. S., Gray, J. S., Schroeder, K. R., Norins, A. L., and Jones, R. W.

(1954) Respiratory responses to CO_2 inhalation: A theoretical study of a non-linear biological regulator, *J. appl. Physiol.*, **7,** 283–308.

Grodins, F. S., and James, G. (1963) Mathematical models of respiratory regulation, *Ann. N.Y. Acad. Sci.*, **109,** 852–68.

Horgan, J. D., and Lange, R. L. (1962) Analog computer studies of periodic breathing, *IRE Trans. Bio-Medical Electronics,* **BME-9,** 221–8.

Horgan, J. D., and Lange, R. L. (1963) Digital computer simulation of the human respiratory system, *IEEE Internat. Conv. Rec.*, pt. 9, 149–57.

Horgan, J. D., and Lange, R. L. (1965) Digital computer simulation of respiratory response to cerebro-spinal fluid PCO_2 in the cat, *Biophys. J.*, **5,** 935–45.

Horgan, J. D., and Lange, R. L. (1968). Chemical control in the respiratory system, *IEEE Trans. Bio-Medical Engineering,* **BME-15,** 119–27.

Kelman, G. R. (1968) Computer program for the production of O_2–CO_2 diagrams, *Resp. Physiol.*, **4,** 260–9

Longobardo, G. S., Cherniak, W. S., and Fishman, A. P. (1966) Cheyne-Stokes breathing produced by a model of the human respiratory system, *J. appl. Physiol.*, **21,** 1839–46.

Milhorn, H. T., Jr., Benton, R., Ross, R., and Guyton, A. C. (1965) A mathematical model of the human respiratory control system, *Biophys. J.*, **5,** 27–46.

Milhorn, H. T., Jr., and Brown, D. R. (1971) Steady state simulation of the human respiratory system, *Comput. biomed. Res.*, **3,** 604–19.

Milhorn, H. T., Jr., and Guyton, A. C. (1965) An analog computer analysis of Cheyne-Stokes breathing, *J. appl. Physiol.*, **20,** 328–33.

Murphy, T. W. (1966) A two compartment model of the lung., *RAND Corp. Memo.*, RM-4833-NIH.

Olszowka, A. J. (1970) Personal communication.

Olszowka, A. J., and Fahri, L. E. (1969) A digital computer program for constructing ventilation-perfusion lines, *J. appl. Physiol.*, **26,** 141–6.

Otis, A. B., McKerrow, C. B., Bartlett, R. M., Mead, J., McIlroy, M. B., Selverstone, N. J., and Radford, E. P. (1956) Mechanical factors in distribution of pulmonary ventilation, *J. appl. Physiol.*, **8,** 427–43.

Pack, A. I., McCusker, R., and Moran, F. (1972) [In preparation.]

Rahn, H. (1949) A concept of mean alveolar air and the ventilation-bloodflow relationships during gas exchange, *Amer. J. Physiol.*, **158,** 21–30.

Rahn, H., and Fenn, W. O., (1955) *A Graphical Analysis of the Respiratory Gas Exchange,* Washington, D. C., American Physiological Society.

Riley, R. L., and Cournand, A. (1949) Ideal alveolar air and analysis of ventilation-perfusion relationships in the lungs, *J. appl. Physiol.*, **1,** 825–47.

Sikand, R., Cerretelli, P., and Fahri, L. E. (1966) Effects of $\dot{V}_I$ and $\dot{V}_A/\dot{Q}$ distribution and of time on the alveolar plateau, *J. appl. Physiol.*, **21,** 1331–7.

Tomovic, R. (1963) *Sensitivity Analysis of Dynamic Systems,* New York.

West, J. B. (1965) *Ventilation/Blood Flow and Gas Exchange,* Oxford.

Yamamoto, W. S. (1960) Mathematical analysis of time course of alveolar CO_2, *J. appl. Physiol.*, **15,** 215–19.

Yamamoto, W. S., and Raub, W. F. (1967) Models of the regulation of external respiration in mammals. Problems and promises, *Comput. biomed. Res.*, **1,** 65–104.

DISCUSSION ON COMMUNICATION [14]

Campbell. Recognizing that you have made it a breath-by-breath model, I am not too surprised that you were able to get good agreement between events, the blood in the lung area and the expired lung air. But I am a little surprised that you managed to get good agreement when you also introduced the tissues. I say this not as any criticism of your model, but from our experience of a few years ago when we were trying to simulate what went on in such conditions as rebreathing, and found it very difficult to get a simple, realistic one-compartment representation of the tissues. We found it necessary to have a multi-compartment tissue model. Also, have you looked at such factors as hydrogen ion, as well as just PCO_2 itself?

Pack. We haven't used rebreathing as a stimulus and we have not looked at pH. With regard to the other point, the tissues consist (as you have said) of a number of compartments with different time constants; but the values of these time constants are large in relation to the time constant of the alveolar compartment. One tissue compartment is therefore adequate for short duration experiments such as we have described.

Campbell. Maybe the answer is a quantitative one; but I notice, for example, that your CO_2 breathing experiment went on for 3 minutes or so.

Pack. Yes, that is so.

Campbell. We found it very difficult, in rebreathing for 4 minutes, to get a fair representation of the behaviour of PCO_2 from one compartment. I noticed (maybe I was looking particularly unfavourably) that there was some discrepancy there. I think in the real life situation there was a higher PCO_2 at the end, whereas earlier on there was a lower PCO_2; and this is the sort of thing that we found difficult to fit with a one-compartment model.

Pack. We are not claiming that the fit is perfect. What we are claiming, however, is that if there is a residual error (and there was obviously a small error in that experiment) it is small, and one can still define a minimum on the parameter surface that will allow one to determine (we hope) the parameters.

Campbell. Would you say that with further development those two models could be combined, or is there some fundamental difference?

Pack. That would be perfectly reasonable. In general terms, however, in constructing models for the identification of parameters the simplest model that works is the best, since the problem of parameter identification does not increase in a linear fashion with the number of parameters but approximately exponentially. Furthermore, the determination of time constants with any degree of accuracy requires an experiment to be carried out for a period longer than those time constants; but for tissue compartments with long time constants, such an experiment is not practical.

Anthony. How far is the time taken to reach the right parameters dependent on the variation from breath to breath. Do you have to train your subjects in how to do this?

Pack. No. In fact, the great advantage of this particular approach is that one is not assuming steady-state conditions. A number of models in the literature, which are often used in standard respiratory physiological techniques, are based on steady-state conditions. This model does not depend on steady-state conditions, and the patient (or subject) is really free to breathe as he wishes.

Otis. Would you care to elaborate on the fluctuations in alveolar PCO_2 with the breathing cycle? There are factors which, I think, tend to minimize these fluctuations that you do not include in your model. I think you assume that the blood flows at a constant rate through the alveoli—and this, I think, is probably not the case. Also, in the case of CO_2 there is, of course, a considerable buffering power in the tissues in the lung—unless you include that as a virtual alveolar volume.

Pack. In the single alveolar compartment model that we have used the volume term of the alveolar compartment is a virtual alveolar volume as you have suggested, and does contain a representation of the buffering capacity of the lung tissues; and I accept the point that the buffering capacity of the lung tissues tends to reduce the oscillations. I don't think the assumption that blood flow is constant will exaggerate the variation in alveolar PCO_2. I think Murphy[1] has shown that the effect of a pulsatile blood flow was not to affect the magnitude of the oscillations (as I remember it) but rather to introduce additional cardiogenic oscillations into the basic respiratory oscillations.

Otis. I was not thinking of the cardiogenic pulsation of blood flow, but of the movement of blood into the pulmonary capillaries at the time of expansion of the alveoli. I don't think this has been very well measured, but I would like to think it is a possibility—perhaps because I look for good things to happen.

Pack. That might be true. You are suggesting that somehow the pulsatile flow is in phase with the ventilatory flow?

Otis. Yes, and it makes sense to me (at least in my imagination) that the blood should be maximized in the capillaries at the time the air rushes into the alveoli. Have you also considered fluctuations in oxygen?

Pack. We have not yet looked at fluctuations in alveolar PO_2. Previous workers[1,2], using simulations similar to our own, have, however, shown fluc-

---

1 Murphy, T. W. (1966) A two-compartment model of the lung, *Rand Corp. Memo.*, RM-4833-NIH.

2 Chilton, A. B., Barth, D. S., and Stacy, R. W. (1954) A mathematical analysis of oxygen respiration in man, *Bull. math. Biophys.*, **16**, 1–14.

tuations in PO_2 similar to those in PCO_2—but they have not particularly considered the effect of large ventilations into small volume compartments. The main point of this part of our Communication is that in lung units with a small volume and relatively large ventilation—such as probably occurs in chronic obstructive airways disease—relatively large oscillations of PCO_2 occur, although such oscillations do not necessarily occur in the combined blood mixture from a number of units (but see Suwa and Bendixen, 1972[1]).

Wyke. You said you were regarding the flow along the respiratory tract system as a straight time delay?

Pack. Yes.

Wyke. I wonder to what extent that assumption would be modified by fluctuations in bronchial diameter and airway resistance, in so far as these might affect mixing rates within the respiratory tract during air flow.

Pack. I said it was a pure time delay to simplify presentation of the hypothesis. The fluid dynamics of the air flow through the respiratory tract are complicated and not fully understood. In the oral cavity and larynx the patterns of flow are complex, and as you go down the trachea the flow becomes turbulent, and then laminar still further down in the bronchial tree. At each bifurcation, however, the parabolic front is bisected and secondary motions occur; and the parabolic front is not re-established for some distance along each daughter branch. Finally, as a result of the large increase in total cross-sectional area of the bronchial tree at the level of the respiratory bronchioles it is probable that the velocity of air flow becomes so slow that gas reaches the alveoli not by a process of bulk transport, but rather by diffusion in the gas phase.

Wyke. It is precisely because there is this enormous array of complexities—about which, as you say, we know very little—that I was wondering about the validity of the straight assumption in the model that there is just a simple time-delay system along the respiratory tract.

Pack. In terms of presenting the hypothesis, the time delay is the most obvious effect of the transport of gas up the bronchial tree. But I agree that there will be an extra distorting effect as the gas flows up the branching bronchial tree.

Campbell. Is it your purpose to restrict this to examination of the lungs, or do you have the longer-term objective of using the model to explore such things as chemical control, and interaction of the lungs with the tissues?

Pack. We certainly do intend to do that. A prerequisite for the study of respiratory control systems is a fully-validated model of alveolar-capillary gas exchange. We believe that the best approach to biological or physiological modelling is to apply the models to individual subjects in the sort of way we have described. A large number of the simulations of respiratory control in the

[1] Suwa, K., and Bendixen, H. H. (1972) Pulmonary gas exchange in a tidally ventilated single alveolus model, *J. appl. Physiol.*, **6**, 834·41.

literature have been done by constructing models on the basis of previously obtained experimental results and validating the model by proving that it will predict other known respiratory phenomena—for example, Cheyne-Stokes respiration.

Anthony. Just one point of information. When you talk about sinusoidal ventilation, I take it that by this is meant the fact that the actual lung volume is changing sinusoidally. It is not the flow pattern, but the actual lung volume that is changing?

Pack. No. It is really the flow pattern that is meant.

Anthony. The flow pattern! I have never seen that sinusoidal, or anything approaching a sinusoidal pattern.

Pack. In our simulations for use in hypothesis construction and education we have used a sinusoidal representation of ventilation as a convenient periodic function approximation. In the adaptive model it is not sinusoidal ventilation that is used, but rather the subject's actual measured ventilatory flow.

Campbell. For the record, because people will be reading this discussion: your point is well taken, but I do feel that if you look at other published papers you will see that these things can vary in individual subjects.

Pack. That, of course, means a large number of simulations.

Campbell. I quite agree with testing the model by studying what an individual does when he hyperventilates to a certain extent, and then comparing the results with his ventilation and steady-states of CO_2, and then rebreathing: in these ways, you can develop the model further.

Otis. Can you tell me a little about the hardware for automatic optimization of values?

Pack. I had better ask my engineering colleague to reply to that.

Emery. I can show you the actual patch diagram if you would care to see it [Fig. D.12]. If we look just at the argon case, for example, the basic structure of the system is simply as shown here. The feed-back contains a multiplier, and we feed the ventilation in here in the multiplier. This takes account of ventilation cycling, and takes account of time variations. The parameter of lung volume is just a potentiometer in there. If you want to make the thing self-adapting, you take the potentiometer out and put in the multiplier, and this gives you the possibility of varying the thing automatically. We now feed in the value of lung volume from the adaptive circuit. This comes from an integrator, and you have to feed into this the rate at which you want that parameter changed. The way you generate the rate of change is rather complicated, actually. I think we will have to discuss it in detail later, but that is the basic technique that is used [see Emery, 1972[1]].

[1] Emery, B. (1972) *An Application of Modelling Techniques to Respiratory Gas Exchange,* Ph.D. Thesis, University of Glasgow.

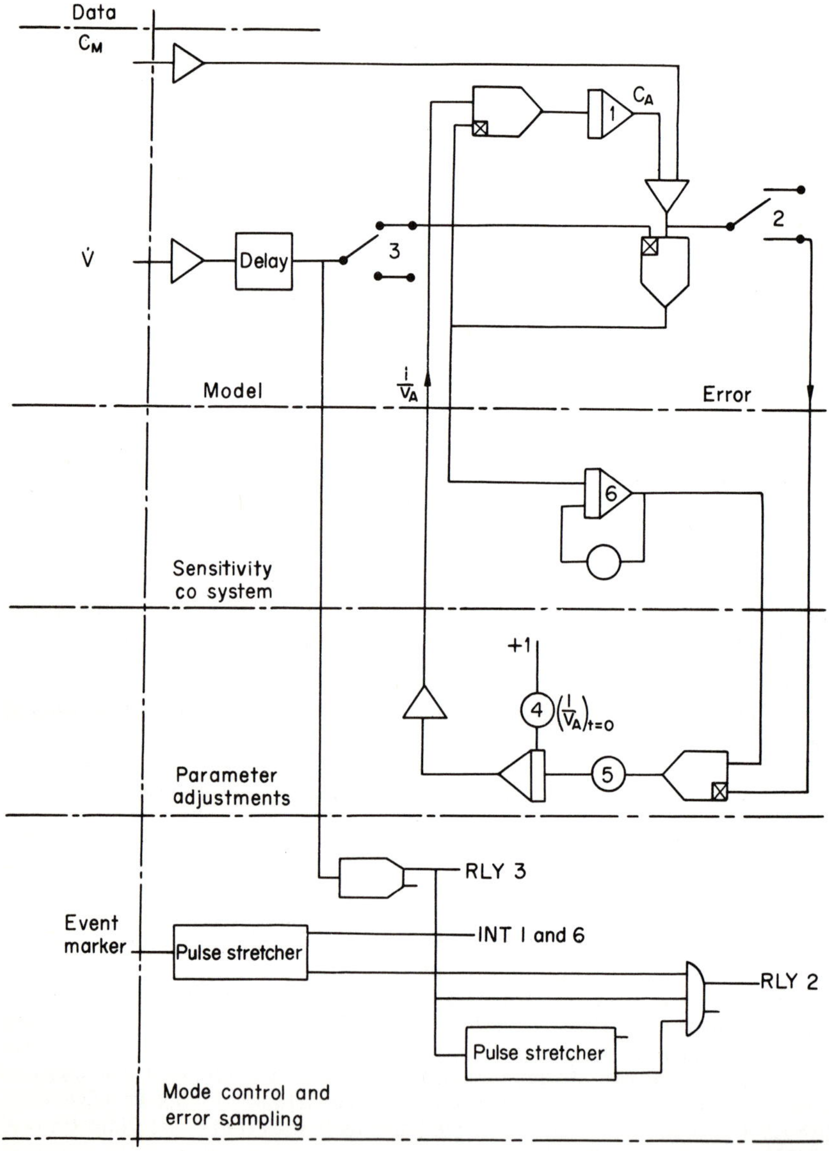

Fig. D.12. Self-adaptive model of argon breathing.

Otis. Do you have several potentiometers involved in this?

Emery. No. Of course, there are lots of multipliers.

Moran. Regarding Dr. Otis's point about possible fluctuations in blood flow in the alveolar walls during the respiratory cycle, one can only agree that obviously this is a pretty complex matter. One recalls the ciné-photographs that Dr. Giles Filley and his colleagues (of Denver, Colorado) made some years ago, in which they made a window in the side of the thorax of an anaesthetized dog, at a point where gross movement of the lung relative to the chest wall would be near minimum; and they were able to demonstrate quite clearly that not only are there cardiogenic pulsations in the blood flow, but that sometimes there is a retrograde flow of blood. Almost any simulation is bound not to be able to take into account events of this kind: I think all one can do is to test each simulation against different kinds of stimuli.

Stevens. Are there other fields in which this kind of adaptive procedure is used on-line to infer something about the parameters, or some other aspect, of system behaviour?

Emery. I don't think it is a common technique, but it is certainly used in the automatic control and adaptive control of high-performance aircraft, and of chemical plants and nuclear power stations, to name just a few instances.

Stevens. I meant application to biomedical problems.

Emery. There is Dr. Murray-Smith's exercise on muscle spindles[1], and on postural mechanisms[2].

Pack. There has also been some work done on albumen metabolism using this sort of approach[3]. But it has not really been very much applied yet in medical work.

---

[1] Murray-Smith, D. J. (1970) *An Application of Modelling Techniques to the Neuromuscular Control System*, Ph.D. Thesis, University of Glasgow.

[2] Roberts, T. D. M., and Murray-Smith, D. J. (1971) An application of self-adaptive modelling techniques in an investigation of the neural mechanisms for postural adjustments, in *Computers for Analysis and Control in Medical and Biological Research*, pp. 55–59, IEE Conference Publication No. 79.

[3] Carson, E. R., and Finkelstein, L. (1971) Identification of metabolic systems, in *Computers for Analysis and Control in Medical and Biological Research*, pp. 1–16, IEE Conference Publication No. 79.

15 Acoustical Cues in the Voice for Detecting Laryngeal Diseases and Individual Behaviour

FRITZ WINCKEL

Technische Universität, Berlin (West), Germany

The aim of this study is to examine the variance and the limits of the fluctuations of the glottic vibrations in order to gain further insight into the function of the larynx and its correlation with phonatory behaviour. With such information, we may hope to improve the diagnostic examination of the abnormal larynx by replacing the laryngeal mirror by an evaluation of the acoustic output of the voice with the aid of magnetic tape recording.

Using speech utterances as test material for this purpose would be inadequate because, from a physical point of view, speech is a stochastic process involving a concatenation of constantly changing transients from which criteria of physiological and psychophysiological functions can hardly be derived. For this reason, modern statistical methods—even with the aid of computer analysis—have provided only scanty data in this regard. Furthermore, intonation of isolated vowels of constant pitch and intensity can be utilized only with reservation in voice tests, because subjects being examined do not utter isolated vowels as naturally as normal speech, and sometimes use more or less breath pressure than normal in such test situations.

The best results are obtained from examination of a singing voice with a large range of intensity and pitch variation; and one should experiment with trained singers in order to exclude non-linear vocal behaviour (such as uncontrolled crying, for example).

Another research method is the electronic synthesis and analysis of voice with the *Vocoder*—a procedure that has been studied for more than 30 years. In such a model, the parameters of the phonation process can be varied electrically and the result is obtained acoustically at the output of the equipment on a loudspeaker or on recorded plots. Today all these processes can be simulated on a computer more precisely, and with quicker evaluation of the data (Flanagan *et alii*, 1970).

From observation of the fluctuations in the periodic vibrations of the glottis, statistical analysis of the frequency distribution of the amplitudes and of the frequency deviations in cycles per sec during a phonation process provides the basis for a classification of the behaviour of the activity of the glottis; and Lieberman (1961, 1962, 1963) provided such data from his own experiments a long time ago. In one of his studies, the fluctuations were determined for every two vibrations in succession; and he found the difference in duration of oscillation to be up to 0.5 msec for a normal voice at a fundamental frequency of 100 Hz (male voice), with a variation in frequency of 5 per cent (or 5 Hz). This may be caused by fluctuations in subglottic pressure, arising from changes in configuration of the vocal tract during articulation. Speakers with longer oscillation durations displayed more marked variations.

In another study, Lieberman found a correlation between the variations in glottic vibration and emotional influences related to the behaviour of the speaker (Lieberman, 1962); but these fluctuations were not variations from period to period, but rather changes spread over a certain sequence of vibrations. It was suggested that nerve impulses originating in controlling centres in the brain might vary the tension in the intrinsic muscles of the larynx (especially the cricothyroid and thyro-arytenoid muscles) that regulate the frequency of the glottic vibrations, such variations being superimposed on the short-time vibrations.

For further investigation, the following experiments were carried out by Lieberman (1962). The voice was picked up by microphone, and split into the fundamental frequency [f_0] and the part of the partial spectrum that represents, with its formants, the influence of the configuration of the vocal tract, the two components being recorded on a two-track tape. (This technique has long been used in the *Vocoder* analysis of speech.) As the next step, the pulses on the f_0-track that deviated from regular periodicity were erased, and replaced by new pulses in cogrect period order. However, the acoustic reproduction of this manipulated two-track tape by means of the synthesizer part of the *Vocoder* prevents the listener from identifying recorded sentences that were selected to convey eight categories of emotion. Nevertheless, the details of the fluctuations determined in this way are not random, but have a high information content that can be analysed independently of the semantic content of the speech.

In essence, we have to differentiate the following features in the course of an analysis of the larynx vibrations: (1) the carrying frequency [f_0]; (2) the irregularly superimposed deviations from periodicity [Δf_0] (comprising perturbations of (a) normal and of (b) pathological voices); (3) the deviations related to the articulation process in order to transmit consciously the semantic content of the voice; (4) the deviations due to emotional state (unconscious). In such terms, the control network of the larynx may therefore be regarded as consisting of (1) a generator, (2) a feed-back circuit (proprioceptive reflex system) with (a) counterbalance activity and (b) control activity in respect of a deviated zero point of balance. The control system has a period tolerance limit of 0.5 sec Δf_0 for normal voices, with another limit of slow deviations over a sequence of periods, the value of which has not been known until recently.

The values of Δf have now been determined in other synthetic speech experiments. If the prominent periodic pulses on the tape are modulated randomly at frequencies of 1–3 Hz (FM), the listener has the impression of hearing a normal human voice. With a rising frequency modulation of 5–8 Hz (the limit of the tolerance) the voice sounds trembling; and any further increase of FM makes the synthetic voice hoarse (Lieberman, 1962). Wendahl (1963), experimenting with an electro-larynx (LADIC), has found that a vocal fry on a carrier frequency of 100 Hz (male voice) starts with deviations of 1 Hz—on 200 Hz (female voice), the tolerance is wider (Wendahl, 1963).

If we consider diseases of the larynx that have a greater effect in distorting the glottic vibrations than inflammation of the vocal folds (such as growths or carcinoma on or near the vocal folds, and paresis or paralysis of muscles in the larynx), we have found that partial involvement of the vocal folds (with an additional unilateral or asymmetrically fixed mass in the bilateral generator, or reduced elasticity in one or both vocal folds) is a significant cause of abnormal variation in vibration parameters. Another cause may be insufficient subglottic pressure due to airway diseases, lung pathology or general muscular weakness; and the acoustic effect of the resulting minor irregularities is an unclear and harsh voice that becomes worse as the pathological state increases in severity.

A good analysis of the characteristic changes produced by the diseases mentioned is provided by recording the acoustic output on sonograms, whilst more detailed data are furnished by the 'voice print' technique—which delivers a quantitative evaluation of vocal intensity in steps of 6 db. Von Leden and his colleagues were the first to present recordings obtained in this way in 1970; and Iwata and von Leden (1970) suggested a classification of the above-mentioned diseases according to the acoustic correlates of the different degrees of hoarseness.

From the data now available, the acoustician can provide some further explanation of the nature of the disturbances in the glottic system in certain disorders. A high-frequency vocal noise band (above 6000 Hz, according to Nessel (1962); or above 3000 Hz (von Leden and Koike, 1970)) is something like a jet noise streaming through the incompletely closing glottis, as through a narrow slit. Noise in the lower range (described as 'increased noise components in the main formant' (von Leden and Koike, 1970)), especially the first formant, can be identified with disturbances of glottic vibration. One should take into account that a bandwidth up 10 Hz cut out from white noise by a filter-set sounds like a sine wave with a central frequency at that of the limited frequency band; therefore oscillations outside a limit of ± 5 Hz will be perceived as noise, depending, of course, on the precise statistical distribution of the noise components. Taking these facts into account, one may develop an improved acoustic classification of the effects of respiratory and laryngeal diseases.

Another feature to be considered is incomplete closing of the vocal folds, which correlates with a reduction of the Bernouilli force. This diminishes the opening explosion-phase in the glottis, and therefore reduces the velocity with which the vocal folds open. The result is a flattening of the triangular pulses of laryngeal vibration, with a reduction in their amplitude. Further, the reduced pulsation impact results in a damping of the spectrum rising with the partials (as can be calculated with Fourier transform analysis of the vocal output)—the author (Winckel, 1953, 1971) has taken the area around 3000 Hz as representative of the spectral damping effect. From this fact we get another criterion for the classification of the vocal effects of diseases.

A third diagnostic criterion has been found in long-duration statistical evaluation of phonation processes. Some preliminary results were obtained from analysis of the phonation of patients with recurrent laryngeal nerve paresis and some other diseases of the larynx. With our computer programme we made recordings of the acoustic vibration output and the respective spectra, and the statistically evaluated envelopes of the spectra, in 12 patients (Winckel, 1971). In Fig. 15.1 the spectral envelopes of a group of healthy voices and a group of

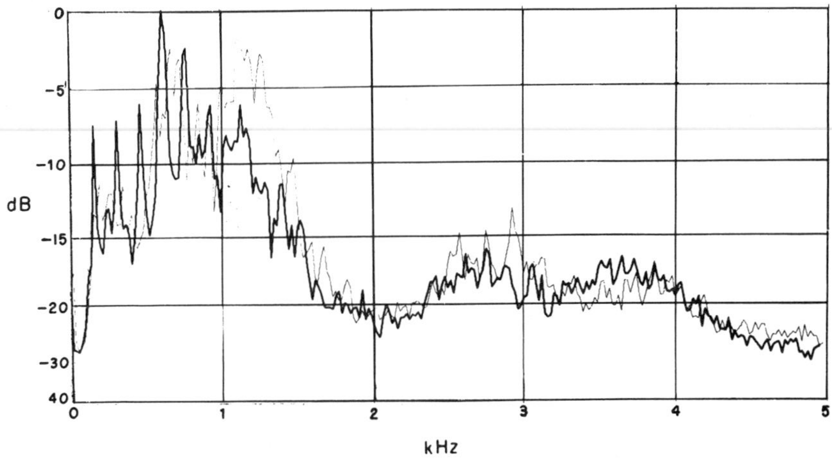

Fig. 15.1.

voices with hypokinetic dysphonia are shown: in the patient group the higher intensity at the critical value of 3000 Hz, together with a second formant maximum in the region of 1200 Hz, seen in the normal subjects is missing. These two maxima are called 'presence' (in German, *Präsenz*) by the sound–control engineer, because of the more lively character of the voice in which they are present. One can amplify these frequency areas selectively on the mixing-console in the studio: and this experience confirms that a healthy voice is characterized by these two additional maxima, as well as by the formant maxima of the different vowels.

In Fig. 15.2 are plotted the acoustic vibrations during the steady state-part of the vowel /*a*/ in a patient with paramedian fixation of the left vocal fold, in whom the voice sounded harsh; and it can be seen that two periods of the glottic vibration pattern are lacking. The diminished elastic forces of the paretic glottic muscles bring the generator temporarily out of balance with the counter component provided by the subglottic pressure (Bernouilli force), providing an extreme

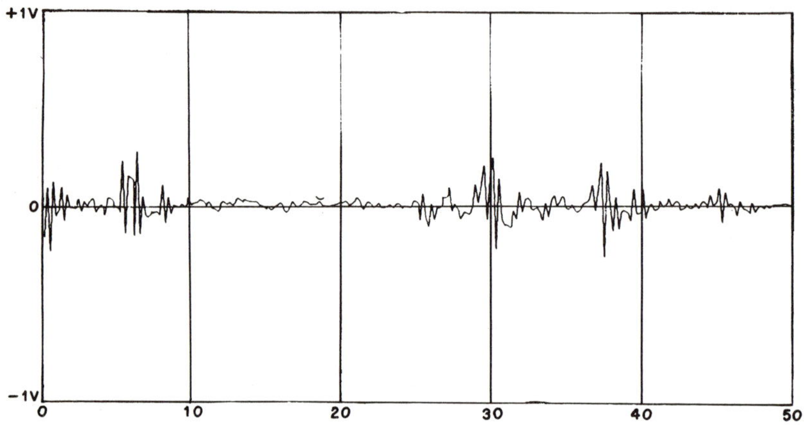

Fig. 15.2.

example of perturbation of vibration. In general, diseases of the larynx that impair the normal kinetics of the muscular and cartilage system require a higher physical force for phonation; whilst a further consequence is restricted adjustment of the muscles controlling vocal pitch, resulting in a more or less monotonous speech (Winckel, 1965) and a diminution of loudness dynamics.

For further research, it will be necessary to develop a more complete model of the processing mechanisms of human voice and speech in order to study particular physiological conditions by simulation on the computer. Such a model should consist of three subsystems: (1) subglottal pressure generation; (2) glottal generator; (3) vocal tract articulation, all with multiple self-control circuits and provision for conscious adjustments. For correlation of the operations of the whole system, a synchronization network is also necessary. A survey, in block diagram form [Fig. 15.3], of the relevant physiological components has been developed by Lieberman (1968); a related electrical circuit has been proposed by Flanagan *et alii* (1970); and the breathing subsystem (Fleming *et alii*, 1969), including the gaseous processes (Bleichert, 1969), the larynx subsystem (Früh and Winckel, 1964) [see Fig. 15.4], and the vocal tract control subsystem (Flanagan, 1970) have all been analysed theoretically.

We have no detailed knowledge of the synchronization of all these subsystems, because we lack essential data—such as the time constants of the members of each subsystem, the timing and patterning of the control pulses emitted from the nervous system to the muscular system, and the respective transit times (especially the different transit times involved in the attack phases of

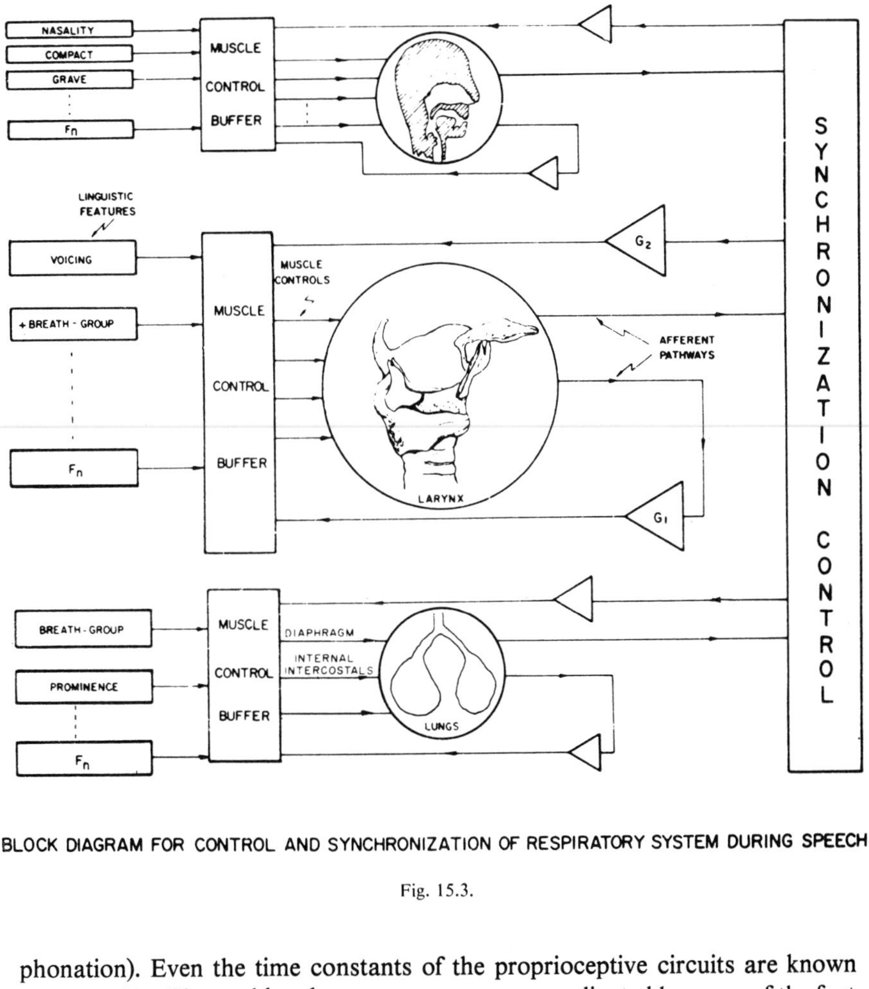

BLOCK DIAGRAM FOR CONTROL AND SYNCHRONIZATION OF RESPIRATORY SYSTEM DURING SPEECH

Fig. 15.3.

phonation). Even the time constants of the proprioceptive circuits are known only roughly. The problem becomes even more complicated because of the fact that the timing pulses or neural bursts are not strict clock frequencies, but are fluctuating discharges—like all parameters governing the operations of biological systems.

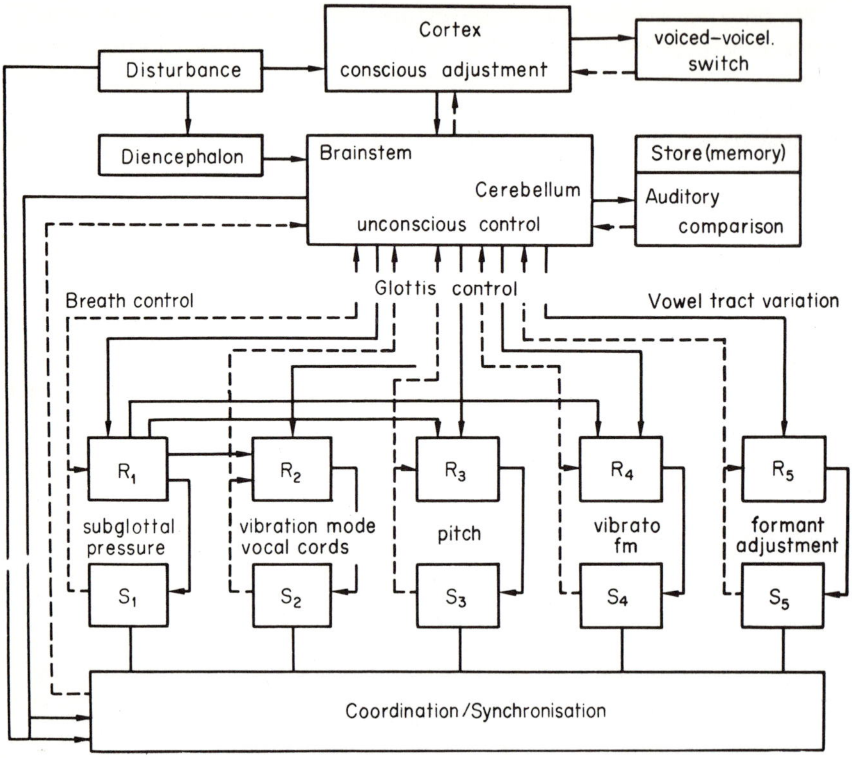

Fig. 15.4.

TABLE 15.I

VIBRATO OF SINGING VOICES

SINGER	*f mod*	*f-hub*
Enrico Caruso	6.3 Hz	7.7 per cent
Enrico Caruso	6.7 Hz	6.8 per cent
Louise Homer	6.3–5.5 Hz	12–14.5 per cent
Frances Alda	7.7 Hz	6.2 per cent
Sigrid Onegin	6.6 Hz	9.0 per cent
Giovanni Martinelli	7.0 Hz	5.7 per cent
Lucrezia Bori	7.5 Hz	7.0 per cent
Lily Pons (quaver)	7.5 Hz	23.5 per cent
Lily Pons	7.5 Hz	8.3 per cent (max)

But one periodicity (of 5–7.5 times/sec) in the phonation process should be mentioned, as it can be studied more easily and compared with other better known physiological processes. This phenomenon can be heard in the singing voice as a *vibrato,* analysis of which has shown it to involve a frequency modulation superimposed on the pitch frequency of the voice. There are individual differences in the frequency and the *hub* (frequency swing) of the voices of individual singers, as indicated in Table 15.I.

In this regard it is pertinent to inquire if the mechanism of muscle tremor resembles that producing *vibrato,* in order to throw light on the physiological origin of *vibrato*—the nature of which has long been controversial [see Potter, 1966].

Tremor may be regarded as an aspect of voluntary muscle activity produced in the servomechanism of the feed-back control system of each motor unit (Lippold, 1971); and Fourier analyses of tremor reveal it to consist essentially of an oscillation of the reflex mechanisms controlling the length and tension of stretched muscles (Halliday and Redfearn, 1956). If one repeatedly prods the extended middle finger of a human subject (each prod lasting 33 milliseconds), some roughly sinusoidal waves are generated mechanically as vibrations of the servomechanism (Lippold, 1971), manifest as a tremor at 10 Hz. (The tremor of Parkinsonism, which has a frequency of about 5 Hz, is to be distinguished from this phenomenon.) Another relevant observation is the resemblance of the α-rhythm of the electroencephalogram to the tremor oscillations, in respect of wave-shape and frequency (8–13 Hz). Lippold (1971) remarks that 'during a voluntary contraction, motor units composed of functionally identical muscle fibres start discharging at about 7 Hz and they accelerate to 30–40 Hz as the strength of the contraction increases'.

Other data hinting at relations between tremor and the *vibrato* of phonation emerge from studies of stuttering patients (for example, see Wyke, 1970, 1971; Fiebiger, 1971); and Lippold (*loc. cit*) describes some electromyographic recordings of the facial articulatory musculature in this condition. The significant results can be briefly enumerated: (1) the main tremor frequency is concentrated in the band covering 7–10 Hz; (2) the tremor bursts occur synchronously in symmetrical muscle pairs; (3) sometimes it happens that during phonation every second burst has a smaller amplitude; and (4) the tremor can be suppressed in some cases.

Finally, it may be noted that Fiebiger (*loc. cit.*) interprets the tremor in the experiment with the prodded finger (*vid. sup.*) as arising from an oscillation of the action potential discharges from muscle spindles: thus, 'It appears to be caused by a "hunting" mechanism in the reflex arc controlling the muscle'. And again: 'Several items of evidence indicated that it was in fact nerve-generated. Its frequency was the same as that of physiological tremor. The waves represent oscillations in a self-sustained feed-back loop (Lippold, 1971).'

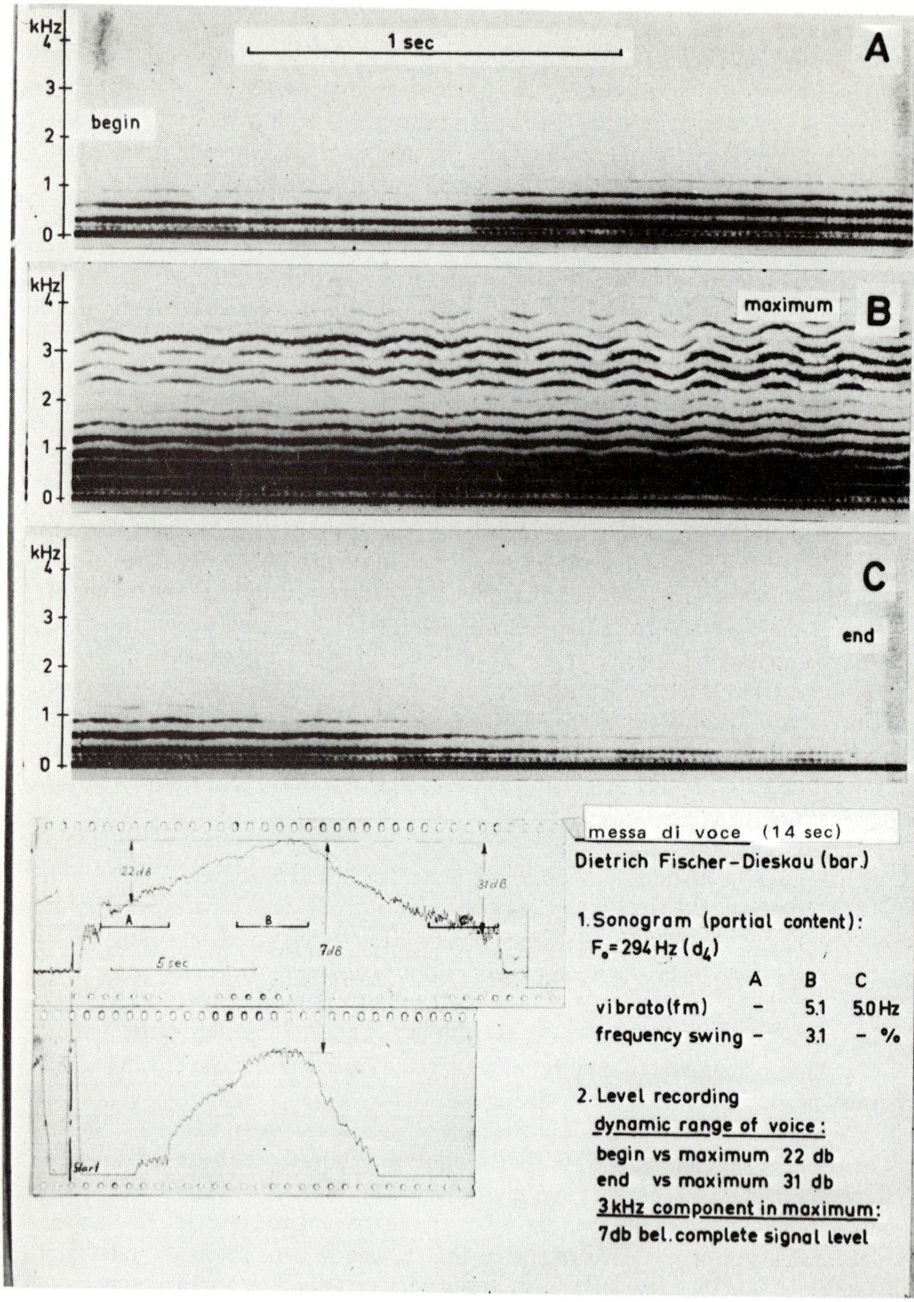
kHz
1 sec
A
begin
maximum
B
C
end
22 dB
31 dB
7 dB
5 sec
Start
messa di voce (14 sec)
Dietrich Fischer-Dieskau (bar.)
1. Sonogram (partial content):
F₀ = 294 Hz (d₄)
A B C
vibrato (fm) - 5.1 5.0 Hz
frequency swing - 3.1 - %
2. Level recording
dynamic range of voice:
begin vs maximum 22 db
end vs maximum 31 db
3 kHz component in maximum:
7 db bel. complete signal level

Fig. 15.5.

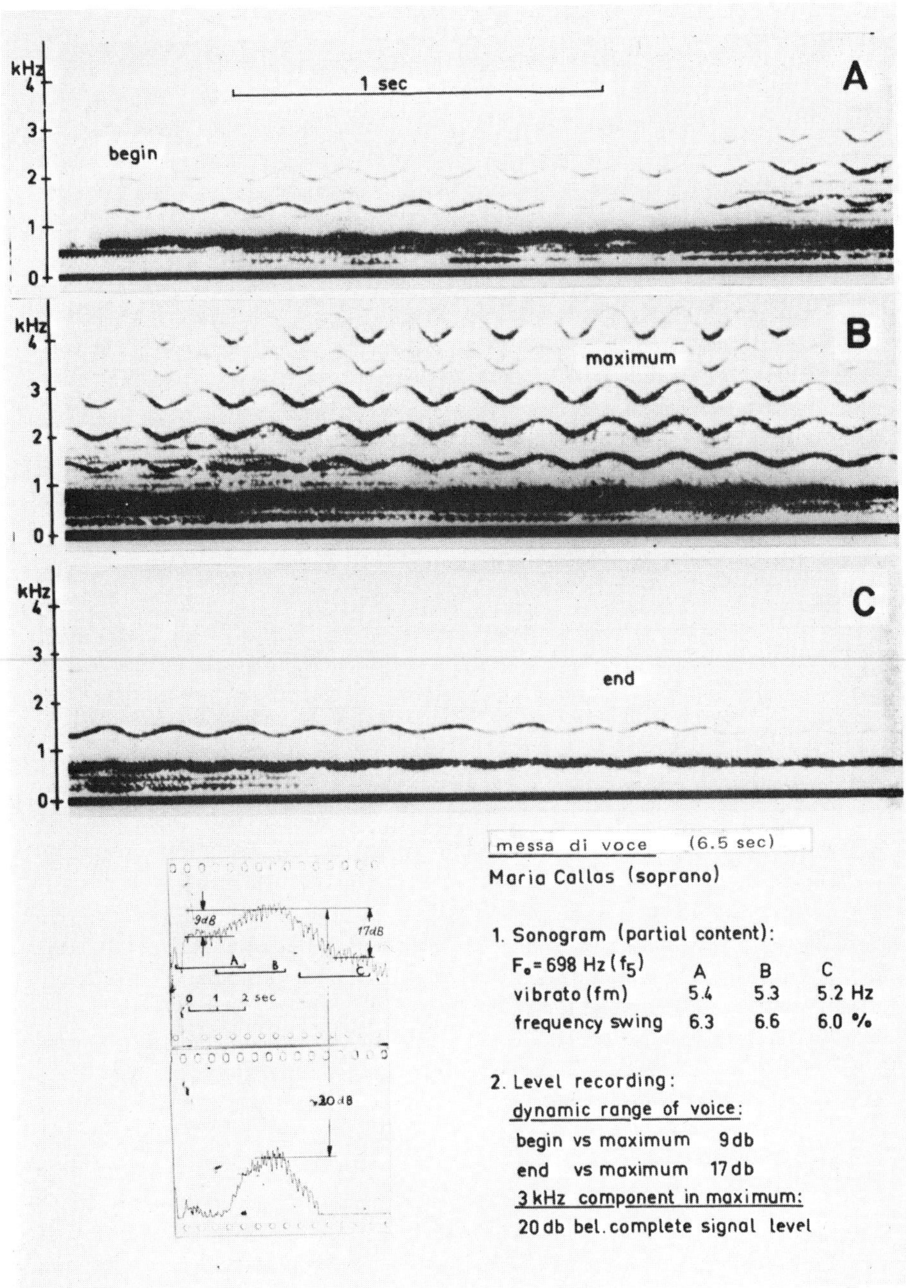
kHz
4
3
2
1
0
1 sec
A
begin
B
maximum
C
end
9dB
17dB
A
B
C
0 1 2 sec
20 dB
messa di voce (6.5 sec)
Maria Callas (soprano)
1. Sonogram (partial content):
F₀ = 698 Hz (f5) A B C
vibrato (fm) 5.4 5.3 5.2 Hz
frequency swing 6.3 6.6 6.0 %
2. Level recording:
dynamic range of voice:
begin vs maximum 9 db
end vs maximum 17 db
3 kHz component in maximum:
20 db bel. complete signal level

Fig. 15.6.

If we take it as now established that there are neuromuscular spindles and other stretch-sensitive receptors in the intrinsic muscles of the human larynx (Wyke, 1967, 1971; Baken, 1971), the observations of Lippold and the EMG studies by Fiebiger on the phonation processes of stutterers may be regarded as indicating that a *vibrato* on an an intoned vowel of intense loudness in a steady state is analogous to tremor oscillation; for although somewhat differing frequency ranges of tremor are given by several authors, the range of the *vibrato* is often 5–8 Hz [see Table 15.I].

In an attempt to demonstrate that *vibrato* is a tremor effect, we have recorded the *messa di voce* intonation of some prominent stage singers. From a level base line, we have measured the frequency and the swing of the vibrato at the beginning (A) and at the end (C) of intonation with low intensity, and in the middle (maximum at (B)) [Figs. 15.5 and 15.6]. One perceives that the swing of subject F.-D. is (in A and C) below the threshold, reaching a maximum deviation of 3.1 per cent in B: the intensity increase A → B is 22 db. In the case of subject M.C. the values for A, B, and C are 6.3, 6.6 and 6.0 per cent respectively, and the intensity increase is 9 db. There are individual differences between about 20 singers that have been tested, but in general the tendency in a correct *messa di voce* is the same.

These observations suggest that in the *messa di voce* process, beginning with a low voice intensity and hence a low subglottic pressure, the stretch applied to the laryngeal muscles is small. With rising intensity (and thus subglottal force) the oscillations of the neurally-engendered bursts of activity in the muscles increase and can then be detected acoustically in voice. Figs. 15.5 and 15.6 also show that the *vibrato* frequency fluctuates, but not in the same way in all singers—although further investigation of this matter is necessary. The process may be further disturbed because the singer constantly corrects the pitch of his output by ear, in order to exert a continuous influence on the muscles adjusting the pitch.

Finally, it should be said that the different sorts of fluctuations in vocal vibration that have been mentioned cannot be replicated simply by a random modulation of the voice carrier in the simulation of speech and voice on the *Vocoder* or the computer—although this has often been tried, in order to avoid the machine-like stiff character of the artificial voice. Our own experiments have shown that even the first two transients of the onset of voicing should be transmitted on an electric line without any distortion in order to avoid an altered sound character.

CONCLUSION

Fluctuations or perturbations of glottic vibration are signs that characterize individual character and emotional status, as well as certain pathological situations affecting the voice, and their analysis may provide the basis for further advances in voice diagnosis and voice pathology, and perhaps in certain aspects of psychiatry.

Consideration of amplitude fluctuations, which are normally correlated with frequency fluctuations, has been excluded from this Communication because this topic has been treated thoroughly by other authors [see von Leden and Koike, 1970].

REFERENCES

Baken, R. J. (1971) Neuromuscular spindles in the intrinsic muscles of a human larynx, *Folia phoniat. (Basel.)*, **23**, 204–10.

Bleichert, A. (1969) Physiologie von Regelvorgängen, *Materia Medica Nordmark*, **21**, 69–78.

Fiebiger, S. (1971) Stuttering explained as a physiological tremor, *Speech Transmission Lab., Roy. Inst. Techn. (KTH), Stockholm*, Oct. 15, pp. 1–23.

Flanagan, J. C., Coker, C. H., Rabiner, L. R., Schafer, R. W., and Umeda, N. (1970) Synthetic voices for computers, *IEEE Spectrum*, **7**, 22–44.

Fleming, D. G., Mesarovic, M. D., and Goodman, L. (1964) Multi-level, multi-goal approach to living organisms, in *Neuere Ergebnisse der Kybernetik*, pp. 269–82, ed. Oldenbourg, R., Munich.

Früh, K. F. (1964) Einfluss von Steuer und Regelvorgängen auf die Stimmgebung (with supplement by F. Winckel), in *Neuere Ergebnisse der Kybernetik*, pp. 210–23, ed. Oldenbourg, R., Munich.

Halliday, A. M., and Redfearn, J. W. (1956) Analysis of the frequencies of finger tremor, *J. Physiol. (Lond.)*, **134**, 600–11.

Iwata, S., and von Leden, H. (1970) Voice prints in laryngeal disease, *Arch. Otolaryng.*, **91**, 346–51.

von Leden, H., and Koike, Y. (1970) Detection of laryngeal disease by computer, *Arch. Otolaryng.*, **91**, 3–10.

Lieberman, P. (1961) Perturbation in vocal pitch, *J. acoust. Soc. Amer.*, **33**, 597–603.

Lieberman, P. (1962) Some aspects of fundamental frequency related to emotional content of speech, *J. acoust. Soc. Amer.*, **34**, 922–7.

Lieberman, P. (1963) Fundamental periodicity of normal and pathological larynges, *J. acoust. Soc. Amer.*, **35**, 344–53.

Lieberman, P. (1968) Vocal cord motion in man, *Ann. N.Y. Acad. Sci.*, **155**,-28–38.

Lippold, O. (1971) Physiological tremor, *Scientific Amer.*, **224**, 65–71.

Nessel, E. (1962) Tonspektrum der pathologisch veränderten Stimme, *Acta otolaryng. (Stockh.)*, **157**, (Suppl.), 1–45.

Potter, R. K., Kopp, G. A., and Kopp, H. G. (1966) *Visible Speech*, New York.

Wendahl, R. W. (1963) Laryngeal analog synthesis of harsh voice quality, *Folia phoniat. (Basel.)*, **15**, 241–50.

Winckel, F. (1953) Physikalische Kriterien fur objektive Stimmbeurteiling, *Folia phoniat. (Basel)*, **5**, 232–52.

Winckel, F. (1965) Ermittlung von Spracheigenschaften aus statistischen Verteilungen von Amplituden und Tonhöhe, *Proc. 5e Congrès Int. d'Acoustique, Liège*, A41.

Winckel, F. (1971) How to measure the effectiveness of stage singers' voices, *Folia phoniat. (Basel),* **23,** 228–33.

Wyke, B. D. (1967) Recent advances in the neurology of phonation, *Brit. J. Disor. Commun.,* **2,** 2–14.

Wyke, B. D. (1970) Neurological mechanisms in stammering: an hypothesis, *Brit. J. Disor. Commun.,* **5,** 6–15.

Wyke, B. D. (1971) The neurology of stammering, *J. psychosom. Res.,* **15,** 423–32.

DISCUSSION ON COMMUNICATION [15]

Malcolm. Can I ask Dr. Winckel whether he has carried out these voice studies on patients suffering from Parkinson's disease, with tremor?

Winckel. I have taken the data from the literature that I mentioned, because I have not myself examined patients with Parkinson's disease. I have used only singers' voices for my studies, and at the beginning of my Communication I explained why. It is not because of my special love or sympathy for the singing voice, but because I found that this is really the best method to examine voices.

Brodnitz. First of all I must confess I heard this beautiful presentation of Dr. Winckel's with a certain amount of sadness, because I realize he is liable to put us laryngologists out of work. As he said in the beginning, we don't need the laryngeal mirror any more: you just put the patient on these machines and you get a beautiful and much more correct impression.

I have just two questions for Dr. Winckel. First, have the fascinating studies you showed on the curves of patients with recurrent nerve paralysis thrown any light on the changes before and after therapy in benign lesions like nodules, polyps and contact ulcers? I would be particularly interested if you have any data on that in relation to practical therapy. It happens very often that at the end of therapy our ear tells us that the patient speaks much better, but the patient says 'I speak "higher" now'. Always I have a feeling that, if my ear is correct, the voice is not higher, but that there is a more even sound production; and I would like to know whether your curves prove that true.

My second question is about that very interesting discussion of yours about the *vibrato.* I was wondering, when I looked at your curves, to what degree is your assumption correct for the good *vibrato* that is characteristic of a fine singer—and what is more, for the *tremolo.* This occured to me particularly when I looked at the last two curves you showed of Fischer-Dieskau's voice. In other words, is there a transition of a *vibrato* into a *tremolo?*

Winckel. That is a very interesting question. You say that you can hear the success of your treatment of a patient, and the patient then tells you that he hears his pitch higher. I have had the same experience in quite another form. I have constructed a little transistorized apparatus to be used for the tuning of orchestras at 440 Hz; and a very experienced conductor who wanted this apparatus told me–when I presented it to him in two different forms with different sound colours—'I prefer the second one, but it has a higher pitch'. But it didn't. Therefore you see how important it is to have a method of recording, especially for the treatment of voices—because these subjective misinterpretations can often happen, even with people of great acoustic experience.

Now your question about the *tremolo*: the *tremolo* occurs, as I told you, as a tremor going up to 30 Hz. What we usually call *tremolo* is at about 11 to 12 Hz, and this occurs in singers with a forced *vibrato*. I have the impression that a *vibrato* at about 7 Hz is unconsciously developed, and the singer does nothing. When he is properly trained, he doesn't know where it comes from. But if a singer is badly trained and the voice doesn't develop freely, then he consciously tries to force it and this results in a higher frequency of *vibrato,* with a *tremolo* effect. I feel that this conforms with my hypothesis of the tremor, because you tighten the intrinsic laryngeal musculature more if you do it consciously and therefore the output frequency is increased.

Fry. I have a couple of questions to ask Dr. Winckel. But before I do, one quick one as I do not want to break the current line of thought. When he had this orchestral tuning device giving a 440 Hz signal, I presume it was giving out a complex tone?

Winckel. Yes. The sound should have the character of the oboe, because an orchestra tunes from the oboe; and the problem is how to get the right tone colour.

Fry. Of course, my point is that if there were a difference in the structure of the tones the conductor might very well hear one as being sharper than the other.

Winckel. That's right.

Fry. Now, about the *vibrato.* I was very interested to see the spectrogram of Fischer-Dieskau's voice, because he is a singer who has a special mannerism of beginning a tone in a quite dead way, without *vibrato.* My question is, simply, is this a matter of his inhibiting the *vibrato* at the beginning? It would be quite interesting, if, in fact, he at first inhibits it and then allows it to take place, because this would rather change one's ideas of the type of mechanism involved.

The second point is that in the same spectrogram one notes, of course, that with the *vibrato* there are very marked changes in intensity through the spectrum. So much so that one sees, as it were in the pattern, one half-cycle of the *vibrato* with high intensity and in the other half-cycle the intensity drops so that there is no trace on the pattern. Could you explain this for me, Dr. Winckel?

Winckel. It is relevant that I had Fischer-Dieskau in an experimental situation, and I asked him not to think about music. His voice diagram is the most symmetrical curve I have from about thirty singers that I have measured up till now. Most of them break off the *messa di voce* after a maximum of 2 or 3 seconds, because then the reserve air is out: most singers produce an asymmetric curve, and are quickly finished. The peculiar processing of the *vibrato* may be explained by the very difficult breath control demanded by a long-duration vocal output.

Stevens. There is also the question about the changing amplitude of the higher harmonics during the *vibrato.*

Winckel. There is an amplitude modulation as well as a frequency modulation, and both are very different in different voices—as you saw in my records of the analyses of Maria Callas and of Fischer-Dieskau. I listed some of the data for a number of other voices in the Table I showed you [see Table 15.I]. You can also hear the differences—and I can play you my tapes of the *messa di voce* analyses of Callas and Fischer-Dieskau to demonstrate this, if you wish. [*Editor's note*: Dr. Winckel played the tapes, at this point].

Ladefoged. I would like to think further about the mechanisms for *vibrato,* and return to Dr. Fry's comment about the variation in amplitude intensity in the spectrum. My suggestion, on which I would like to hear your comments, is that the way in which the pitch is varied is *not* by altering the tension of the cricothyroid muscle (which is probably the strongest and the most usual way of varying pitch), because the *vibrato* is not that type of pitch adjustment at all. Instead, it involves some adjustment of the position of the arytenoid cartilages; and as a result of this adjustment, the vocal folds are not in a position to vibrate with the same intensity—that is to say, their spacing (their distance apart) is varied as well as their tension. The tension variation causes the changes in pitch: whilst the fact that when the vocal folds come together they do so with more or less force causes the variation in intensity.

Now, let us assume that this is a deliberate movement that is causing the *vibrato*; and leading out of that, I would like to consider what is the maximum rate at which one can make deliberate movements. As far as I am aware, it is of the order of 8 or 9 per second; and so the *vibrato* might go up to that frequency. I can tap my tongue (and I will be talking a little more about this tomorrow [see Communication 18] in relation to rates of speech and variations in its timing) about 8 or 9 times per second. I can't go much above that; and I certainly can't go up to 12 or 13 per second. So that anything that is made up of frequencies at 12 or 13 per second would have to be a different mechanism, perhaps not based on separately controlled movements. If there is something at 13 per second, it must be a mechanical effect—but that's pretty slow for a mechanical effect. After all, a tongue-tip trill [Dr. Ladefoged demonstrates—*Ed.*] is of the order of 25 to 30 per second—which is very much higher; so what on earth can trill or vary at 13 per second? I have no answer as to how one makes a movement at a frequency of 13 per second.

Winckel. I made a comparison with stuttering, in relation to the data in the paper by Fiebiger (1971)[1]. This encouraged me to present my hypothesis, because he reviewed physiological tremors in general and gave the maximum frequency, as you say, as 30 per second.

In relation to the amplitude variation, I would point out that frequency modulation is combined with amplitude modulation and amplitude modulation with frequency modulation. I have concentrated in my studies on the frequency modulation for the present, because it is easier to measure: it is very difficult to measure amplitude modulation accurately in the situations I am studying. I am just at the beginning; but I will try to find if I can differentiate the frequency and amplitude modulations, but I don't know at the moment.

---

[1] Fiebiger, S. (1971) Stuttering explained as a physiological tremor. *Pub. Dept. Speech Communic., Royal Inst. Technol., Stockholm*, STL–QPSR, No. 2–3, pp. 1–23.

Klatt. I would like to return to the question of detecting vocal pathologies by using the acoustic wave form. You described two methods that at least show some statistical promise, one being based on detecting perturbations in the fundamental frequency of the vibration of the vocal folds and the second involving looking in detail at the spectrum of a vowel and asking what the amplitude relationships are in the higher frequency region. As to the first method it is questionable whether this is really going to lead to the ability to detect pathology reliably enough to be practically useful. One could conceive of using this simply as a screening device, and I believe this has been tried in the United States[1]—but the results there have been somewhat disappointing. The second technique, the spectrum one, seems even less promising because the amount of difference that you are detecting between pathological voices and normal voices is extremely small (just a few db), and is really smaller than the variations that one sees in a normal voice. Would you care to comment on the potential value of this method?

Winckel. The data I presented are the results obtained only in our Laboratory—in which, of course, we do not have many patients—so my experience with pathological conditions is limited. Now to the other question of amplitude detection—which, in part, I did not follow. In the paper of von Leden and Koike (1970)[2] there is a very thorough evaluation of many sonograms (or so-called 'voice prints')—but as this is the only detailed study I know of I am as doubtful as you, and do not know if this method will prove to be successful. For this reason, I avoided speaking about amplitude modulation and its analysis in relation to vocal behaviour in my Communication today.

Stevens. I believe that paper was referring to a paper of Yanagihara (1967)[3], who looked at spectrograms of patients with various degrees of hoarseness and found different amounts of noise in the high frequencies.

Fourcin. With regard to the long-term spectral analysis of unilateral paralysis (I think that was the pathological voice condition that was basic to your slide) one can, very easily, with that type of laryngeal condition get an excellent voice quality and then, from exactly the same sample of speech but just a second later, get an example of the erratic closure that is so typical of unilateral paralysis. In consequence, I don't think long-term spectra (of the type you showed) are of any use at all to show the effects of any form of therapy—simply because one can get (in a grossly abnormal case) good long-term spectra and bad in the same song. You can only show that there is something wrong by using a timed waveform of the sort you showed—but preferably, I think, one that does not have all the complexities of supraglottic activity associated with it. You have got to do some sort of processing; but even when you have this final record you are still not there, I think, because it is so very difficult to distinguish between one type of pathological condition and another. For example, the erratic speech waveform

[1] Crystal, T. H., Montgomery, W. W., Jackson, C. L., and Johnson, N. (1970) Methodology and results in laryngeal disorders: detection through speech analysis, *Final Rpt., U.S. Public Health Service Contract* No. PH86-68–192, Signatron Inc., Lexington, Mass., U.S.A.

[2] von Leden, H., and Koike, Y. (1970) Detection of laryngeal disease by computer, *Archiv. Otolaryngol.*, **91**, 3–10.

[3] Yanagihara, N. (1967) Hoarseness: investigations of the physiological mechanisms, *Ann. Otol. (St. Louis)*, **76**, 472–89.

you showed could have been found equally well in a case of sore throat. You couldn't say that it was unilateral paralysis just by looking at the record.

Winckel. I have learned singing myself for scientific reasons—and my teacher once told me that he had a patient with unilateral paralysis whom he could present with very good voice quality as a result of training. I therefore think that in such cases training may have significant effects.

Fourcin. Did he ever get to *speak* more normally. You certainly can get such people to sing, and to produce some intonations that are excellent—but can their speech ever become uniform?

Winckel. I am not able to say, as I have been concerned in my studies only with the singing voice and not with the effects of rehabilitation of patients' voices.

16 Resonant Cavities, Dipole Resonance and Phonation: A Description of the Upper Respiratory Tract in Engineering Terms

J. E. MALCOLM

Royal Air Force
Central Medical Establishment,
London, England

INTRODUCTION

In the course of haemodynamic research (Malcolm, 1965), an apparatus was built to study oscillatory flow in a fluid-filled, elastic system at frequencies of the orders 1 and 10 Hz. It was discovered that the device (the oscillator) would oscillate in stationary wave motion when a stream of air was delivered through it at a pressure of about 0.7 kg/cm^2 (the oscillator being adjusted exactly as for water) in two systems of waves of differing length, namely 1–2 cm and 75 cm approximately. Using air, it could be made to produce a great variety of noises, sounds and musical notes similar to those of the human voice, as well as warbling notes resembling those made by the syrinx in birds. Tuning was readily controlled at the site of a pseudo-stationary node, the waves had transverse as well as longitudinal components, and the half-wave length of the short wave matched the diameter of the elastic tube.

In rheology, mathematical models may be employed, derived from mechanical analogues such as the Kelvin Body described by Reiner (1956). Use of such models for the study of flow of blood led to the realization that the physics of wave guides might be applied to the behaviour of the sound-emitting apparatus.

Study of the behaviour of the oscillator, when driven by compressed air, suggested that the larynx might be regarded as a dipole emitter (in this case, of plane polarized waves), the effect being enhanced by the water vapour in the expired air and by para-crystalline compounds in the saliva.

ANALOGOUS STRUCTURE AND FUNCTION

The pharynx is a resonating cavity wherein the soft palate acts as a separator or attenuator. The uvula, however, is a central conductor that shortens the 'electric' lines, making the pharyngeal cavity appear larger and reducing the resonant frequency. The epiglottis is a conductor close to the wall of the pharynx that shortens the 'magnetic' lines, so raising the resonant frequency of the pharynx. The pharynx also acts as a curved horn to match the laryngeal cavity to the mouth and nasal cavities.

The mouth is a sectoral acoustic horn and lens, comprising two shaped flexible plates—the soft palate and the tongue. The system is an example of configuration focusing, the ray path being the geodesic of the mean distance between the tongue and the palate. The lips act as a radiating slot, and also as a filter.

The teeth constitute a dipole array, each tooth corresponding to one or more rods, and they are arranged to act as a reflecting/directing system (i.e. a phased aerial). When the teeth vibrate, as a result of resonant vibration in response to the oscillator (i.e. the larynx), they drive the nasal horn. The teeth stabilize the mode and act as quarter wave plates, to effect or correct elliptical polarization and transfer energy from the mouth to the nasal cavities by bone conduction, or vice versa. The articulation of the mandible with the maxilla implies that the dipole array is 'folded'.

Waves generated in the mouth are 'transverse magnetic'. The 'magnetic' force lines encircle the tongue, the 'electric' lines of force run between the upper and lower teeth, and also from there to the exterior via the mouth. The parotid and submandibular glands supply chemical substances that 'enable' magnetic and electric force lines.

The nasal conchae correspond to an acoustic lens, being a 'stack of plates' mounted inside a short acoustic horn: the lens is dispersive, focusing being affected by the larynx. The 'near zone' waves emitted by the larynx into the laryngeal sinus are standing; the 'far zone' waves, leaving the mouth, are travelling. The intermediate zone waves, already plane polarized, strike the edge of the plates (conchae) at an angle and set up two modes of oscillation of the plates, 90 degrees out of phase and hence circularly polarized. When these waves are emitted by the nostrils, they are in opposite sense and are modulated by movement of the external nares.

The tonsils act as damping devices to absorb unwanted acoustic reflections from the mouth, otherwise such reflected waves would be directed back to the oscillator to interfere with its function. Similarly, the adenoids absorb reflected acoustic waves from the nasal lens. A plate stack lens is particularly liable to set up reflected waves that will choke an oscillator, but this can be prevented by placing the oscillator off the optical axis. This measure, however, produces spherical aberration and astigmatism if the object-source distance is extended. An alternative measure is to divide the lens into two parts and locate one of these a quarter of a wave-length further from the source than the other. The lenses represented by the nasal cavity and the mouth may be located in this manner with respect to the larynx.

DIPOLE RESONANCE IN THE NASAL CAVITIES

In phonation, there is a steady flow of air expelled from the lungs via the space between the arytenoids into the mouth or nasal cavities, there producing a static pressure. Superimposed upon this is an alternating or pulsed pressure, derived from vibration of the vocal folds, that excites dipole resonance in a manner analogous to the phenomenon of dipole resonance in water molecules excited by infra-red waves.

The conchae may be regarded as a variable capacitor that links an input signal directly from the oscillator, an output signal via the nostrils and a driven dipole non-linear system with several modes of resonance, related to the frontal sinus and the ethmoidal and sphenoidal cavities. The dipole system is driven by sound waves originating in the mouth, conveyed by the premolars and first and second molar teeth to the maxillary sinus from which, arranged at right angles to the input signal, 'electric' force lines streaming out through the aditus ad antrum impinge upon the middle nasal concha and set it, together with the others, into vibration. This constitutes a 'pump' corresponding to the analogous device found in a parametric amplifier. In addition to parametric amplification, the entire assembly acts as a frequency up-converter; and being non-linear, tends to produce harmonics. In this case, the nasal resonant cavities are 'centrally driven' by the teeth in a 'transverse electric' mode.

When the acoustic energy arises in the incisors and canines, it is transmitted to the frontal sinus by bone conduction. Similarly, when it arises from the third molar it is transmitted to the sphenoidal sinus (the teeth of the upper and lower jaw vibrating in sympathy) whilst the mandible and the maxilla, together with the teeth, act as a folded dipole array. In each of these cases the system is 'end driven'.

A MATHEMATICAL MODEL

It is possible to explain the interaction of the resonant cavities of the nose and mouth with the aid of a mathematical model that was devised originally for the hąemodynamic investigations mentioned earlier, the details of which will be published elsewhere.

Briefly, the theory depends on the fact that all natural things (animal, vegetable and mineral) are crystalline in nature—with the exception of volcanic glass and liquids. Intermediate in properties between liquid and crystal are 'liquid crystals'; and the smectic phase of these forms structures that display the geometry of confocal conic sections [Fig. 16.1], and the cholesteric phase, that of the helix. When biological crystals form and aggregate into the anatomical shapes of the body they pass through the paracrystalline state, which may leave

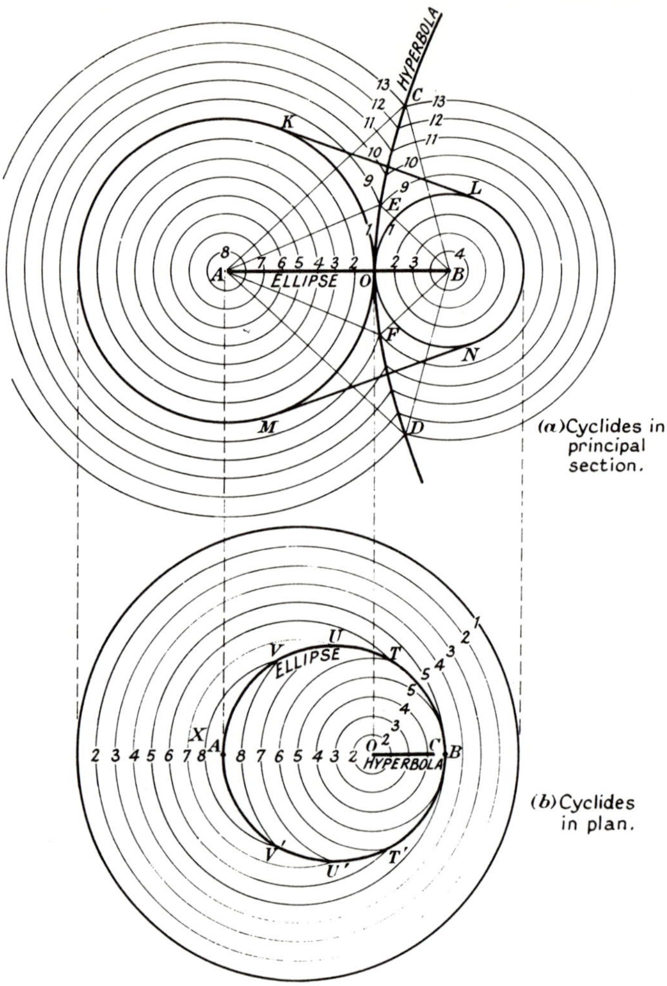

Fig. 16.1. Confocal conic section: the cyclides of Dupin. (Reproduced from Harteshorne, U.H., and Stuart, A. (1970) *Crystals and the Polarising Microscope*, London.

permanent trace of the cyclide figures (although these are probably pseudomorphs). This applies to the nasal cavities; but it must be appreciated that the structure is skeletal in the crystallographic sense, and the system is one of air-filled cavities—as opposed to being fluid-filled.

The sphenoidal, ethmoidal and frontal cavities are parts of a hollow spiral structure, the sphenoidal cavity being the blunt end and the frontal region the sharp end. They are resonating cavities, and act as reflectors. The teeth are mechanical dipoles, and have apparently passed through a developmental stage in which they are liquid crystals in the smectic mesophase (of which their anatomy bears evidence). They are readily identified as examples of polygonal texture—being cones of revolution with an apex (the root) and an ellipsoidal base (the cusps): the geometry is that of confocal conic sections, and subparallel twinning, in calcite aggregates, has occurred. The nasal and oral cavities can then be described by the concentric circles of the cyclides, but taken in sectors. The modes of resonance of mouth and nose resemble those found in ellipsoidal cavities, the conjugates being derived from the confocal conic sections and modified by the helix, which is probably logarithmic. The relevant simultaneous equations depend for their solution upon the even and odd solutions of Mathieu's Equation, of which an account is given by McLachlan (1947).

The resonant nasal cavities are, however, of differing size and therefore of differing natural frequencies; and there are two interacting sources of energy that set them into resonant vibration, as explained above. The behaviour of such systems may be analysed by means of a mathematical approach utilizing the Manley-Rowe relations (Manley and Rowe, 1956).

The principles of the Manley-Rowe relations are exemplified by the model shown in Fig. 16.2. A variable reactance (which may be inductor or capacitor,

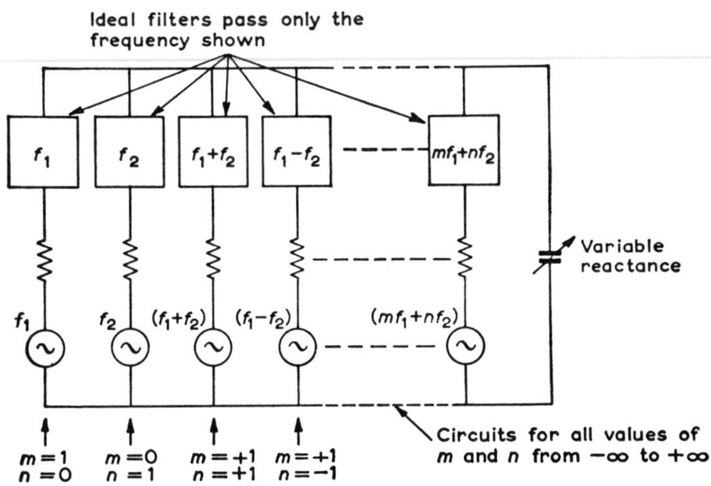

Fig 16.2. Analogous electrical circuit of the Manley-Rowe relations. (Reproduced from Sims, G. D., and Stephenson, I. M. (1963) *Microwave Tubes and Semiconductor Devices*, London and Glasgow.

and of which the reactance is an instantaneous function of the applied voltage or current) is connected to a number of circuits, each of which comprises in series a lossless filter, a load and a signal source that may or may not be active. Each filter is tuned to one particular frequency and presents a high impedance to all other frequencies.

Circuits of this nature are employed in the parametric amplifier and similar devices, where the source of power is alternating current and not the more customary direct current. One of the circuits supplies energy to 'pump' the reactor in a parametric manner. A signal is introduced by a second circuit; and generally there will be a third circuit, known as the 'idler', from which the output may be taken (as in the parametric frequency converter). Alternatively, the power outlet may be taken from the input circuit with some device to prevent this energy being lost at the input, as in the parametric amplifier.

Each circuit responds to or generates one of the frequencies $mf_1 + nf_2$—where m and n are integers that may be positive, negative or zero. The relations are given by two independent equations:

$$\sum_{m=0}^{\infty} \sum_{n=-\infty}^{\infty} \frac{mW_{m,n}}{mf_1 + nf_0} = 0$$

$$\sum_{m=-\infty}^{\infty} \sum_{n=0}^{\infty} \frac{nW_{m,n}}{mf_1 + nf_0} = 0$$

where W_{mn} is the power of the circuit $(mf_1 + nf_2)$, which is positive if it supplies power to the variable reactance and negative if the circuit receives power coming out of the reactance. As a rule the system will contain only three frequencies; namely, f_1, f_2 and a frequency that is one of all their possible sums and differences. The relations are a general statement of the Law of Conservation of Energy and also cover the case of linear reactors when $W_{mn} = 0$ for all values of m and n.

CONCLUSION

The theory of wave guides may be applied by analogy to explain the form and behaviour of some of the anatomical structures involved in phonation. Thus, the larynx may be regarded as an oscillator, the tongue as a path-length lens, the teeth as a dipole array, the nasal conchae as a plate stack lens and the mouth as a sectoral horn. The theory of parametric amplifiers applies to the interaction between the teeth set into vibration and dipole resonance in the nasal cavities excited by the larynx acting as an oscillator. Mathematical formulae are proposed for a 'model' of the system.

The Manley-Rowe relations appear to be applicable to explain the origin of the formants of speech, of which a classic description has been given by Fletcher (1953). The concept of 'end drive' of the nasal cavities by the teeth introduces the further possibility that the system may be acting as a kind of maser, effecting microwave amplification by stimulated emission of radiation, which would apply particularly to the sounds of speech of which the quality is impulsive.

REFERENCES

Fletcher, H. (1953) *Speech and Hearing in Communication,* pp. 25–67, Princeton, New Jersey.

McLachlan, N. W. (1947) *Theory and Application of Mathieu Functions,* London.

Malcolm, J. E. (1965) *Blood Pressure Sounds and their Meanings,* Part 4, pp. 20–22, London.

Manley, I. M., and Rowe, H. E. (1956) Some general properties of nonlinear elements. Part 1. General energy relations, *Proc. I.R.E.*, **44,** 904–13.

Reiner, M. (1956) *Rheology*, Vol. 1. Chapter 2, pp. 45–8, ed. Eirich F. R., New York.

DISCUSSION ON COMMUNICATION [16]

Klatt. Some very basic physical principles lead me to suspect that all this is very strange, if not incorrect. So maybe I can ask a simple question that would clarify, in my mind, what the real analogies are here. You have talked about electromagnetic field theory—you talked about electrical and magnetic fields, which are both vector fields—and then you tried to draw an analogy with acoustics in which we deal with a scalar pressure and a separate philosophy. How can you possibly do this? The equations are completely different: the physics is different!

Malcolm. I have little doubt that things work in this way. To start with, one is not dealing with pure air (this is the crux of the matter) because the expired air includes saliva and other materials. I strongly suspect they are paracrystalline and so have magnetic and electric characteristics, and that it is probably a true analogue.

Klatt. I would just like to interject that you haven't convinced me. Of course, the output of your model must be analogous to acoustic sound pressure and somehow, therefore, you have to get to a point where you draw an analogy to a pressure. What is pressure in your model and what is velocity? You never said that, explicitly.

Malcolm. Your point is well taken.

Stevens. If I may pursue it just a little further I think there is, indeed, one place in the speech production system where air flow and acoustics interact with vibrating bodies, and that is in the larynx. There certainly may be something, I think, in what you say where you have an air flow through the constriction formed by the vocal folds, and the air flow causes some sort of wave-like motion in the vocal folds. This has been looked at theoretically by Ishizaka and Matsudaira (1968)[1,2] recently. But when one comes to the upper vocal tract above

---

[1] Ishizaka, K., and Matsudaira, M. (1968) What makes vocal cords vibrate?, *Rpt. 6th Inter. Congr. Acoustics,* IIB, B9-12.

[2] Ishizaka, K., and Matsudaira, M. (1968) Analysis of the vibration of the vocal cords, *J. acoust. Soc. Japan,* **24,** 311–12.

the larynx, theoretical analysis has shown that the walls of the vocal tract really play a very minor rôle in sound production. They can, to all intents and purposes, be treated as rigid bodies—with some minor effects of damping and so on. One has been able to to make measurements of the vocal tract, and from these measurements to apply the acoustic theory of a pipe with varying cross-sectional area and predict the output and compare it with actual measurements. I am not certain that the vibrating epiglottis or the uvula play that strong a rôle.

Malcolm. The motion that I would suggest occurs, does so in the turbinates. I strongly suspect that the conchal vibrations correspond to the reactor of a parametric amplifier, and that the driving mechanism of the pump is the maxillary sinus connected to a dipole which would be, say, the first molar tooth and the connexion through the antral orifice. In other words, we have a system whereby the mouth and nasal cavities are linked, and in which focusing can occur. This might be the explanation of how a singer can 'project' his voice—in fact, he is not projecting it, he is focusing it.

Stevens. At the frequencies that are of interest in speech, it is hard to see how these effects can play a significant rôle.

Gould. Some of this also disturbs me, as a clinical laryngologist. I have seen patients with large cysts in their antra in whom one could not really tell that there was any vocal disturbance at all. Certainly, if you have nasal congestion there is a change in voice quality; or if you have a nasal cavity that is much too large, you have extensive voice changes. But I find it difficult to understand how your model of the resonating chambers would fit with these clinical observations.

Malcolm. I have a feeling that problems such as those could be only resolved by examining the formants.

Gould. The singer who has had the epiglottis removed still has a relatively normal sonogram.

Malcolm. There is a lot of scope for variation in voice power—and the uvula will, of course, deal with the size of the cavity and the frequency. There is a great deal of scope for adaptation to unusual situations such as those you describe.

Brodnitz. I want to second Dr. Gould's remarks, and add something further. A change of the voice after tonsillectomy may certainly be produced by surgical damage to the soft palate, or to the muscles that move it. But if you do a careful tonsillectomy in a singer, it is remarkable how little the quality of the voice is changed even though the vocal tract configuration is considerably altered. I have no sonograms of these patients, but can just tell you that by ear there is amazingly little change in voice quality and that the singers themselves feel the same.

Malcolm. I think that if the tissues remain soft and vascular, they could quite well absorb the waves. If they become scarred—as would happen if you damaged the muscles—I suspect you might be in trouble.

Wyke. I think that what is really needed in a situation like this, where we have a complex series of analogies piled one on top of the other, are some hard measurements. I would feel that whilst this approach might be defended on purely hypothetical grounds, I don't really think it will stand up until we have some actual measurements of what goes on in the various parts of the model system and can match them with one another and see what the output characteristics are.

The only other comment I would like to make is that the majority of people (except, perhaps, Australians) don't take their vocal output through their noses unless there is something seriously wrong with them. But, in general, I am not sure of the relevance of all this analogue theory to the real hard-core problems we have in trying to decipher what is actually going on during the process of emission of sound.

Malcolm. Could I answer the question backwards, as there are two questions. First of all, I suggest that if Dr. Wyke were to put a peg on his nose it would alter the quality of his speaking voice. Second, as to the mathematics, the kind of physics I am talking about is a general statement of the Law of Conservation of Energy, and is a very elegant method of adding up and subtracting the energies in vibrating systems.

Stevens. Is this still in response to Dr. Wyke's comment?

Malcolm. Yes. This (writing equation on blackboard) is the kind of mathematics I am suggesting. The point is that a parametric amplifier has a lot of capabilities—one of them is as an amplifier, another one is as an oscillator, another one is as a frequency down-converter and yet another is as a frequency up-converter—and being non-linear, it is particularly liable to produce harmonics. I suggest that this is the way you get the formants of speech, that this is the basic physics and mathematics of the formants of speech; and these figures being known, I think there will be wide scope for the mathematicians. Does that answer the question?

Ladefoged. I don't think it does answer the question, because the way in which we get the formants of speech (such as you have got up there on the blackboard) is very well known. The mathematics have been well determined already by people like our present Chairman (Dr. Stevens), and by Dr. Klatt and Dr. Fry. We know which resonating cavities produce these kinds of things—we know what the sources are. I am frightfully sorry, but they are not the things you suggest. They are, as our Chairman suggested, the bodies of air that are involved and the sizes and shapes of the bodies of air in the vocal tract. This is well known already.

Malcolm. I don't think I excluded those factors, because I am talking about tuned cavities. If you take a Helmholtz system, the tuning of cavities depends on their size and the mass of the air in them.

Wyke. May I interject that what may happen if I were to put a peg on my nose is not relevant to consideration of the physical problems of sound emission in normal speech.

Stevens. I think what you are invoking are principles of physics like those of waves in tubes and tuned cavities, and those of interaction of air flow with soft structures like the larynx. If you are saying that such principles should be applied to better understanding of speech mechanisms, then I think we would all agree. But I think there are certain aspects of what you have said that are rather speculative and, in fact (I think many of us would also agree), not correct—things relating to the vibrations of solid structures above the larynx, the teeth, aspects of the nose and so on. I think it has already been worked out that the acoustic aspects of the cavities formed by these structures play by far the largest rôle, and the relevant mathematics are well known. It is the air in these resonators that is the principal influence.

Malcolm. I wonder if the mathematicians have taken into account the Manley-Rowe relations? If all these cavities are resonating and interacting, then somehow they will have to exchange energy with each other; and if they are exchanging energy, then I suggest the most fundamental way of tackling the problem is in terms of Manley-Rowe relations.

Stevens. I agree that perhaps more elegant mathematical procedures might be followed—at least, procedures that are more elegant than the ones that have been used up to now. But the models on which these procedures are based are, in fact, fairly well known.

Freedman. What would happen either to your actual model or to your analysis if you changed the nature of the gas or the fluid going through it?

Malcolm. I have only done water, blood and air. I don't know the answer to the question.

Freedman. What would you expect to happen if you changed from air to a different gas?

Malcolm. I don't know, at present.

Stevens. The effect is well known now in respect of helium. With divers breathing helium there are changes in the velocity of sound propagation. One can predict therefore what would happen—and indeed it does happen. The frequencies of the speech formants shift upwards, because the velocity of sound propagation in a helium-oxygen mixture is greater than in air.

Malcolm. We might be able to provide some information about that in the decompression chamber.

Freedman. You need not go that far. It is a well known comical event in respiratory laboratories that when you breathe odd gas mixtures then your voice goes higher or deeper, depending on the nature of the gases.

Klatt. That brings up another point. That is, there is a difference between the physical device we see here* and the theoretical models you have been

* *Editor's Note:* Dr. Klatt is referring to a mechanical sound-producing device demonstrated by Dr. Malcolm during his Communication.

describing. No one would question that this physical device is producing sound—and perhaps, in a way, similar to the manner in which the larynx produces sound. The question is whether your description, using your analogies, has any relevance at all to the larynx—or even to this device itself.

SECTION IV

Mechanisms Influencing Phonatory Behaviour

CHAIRMAN

Professor D. B. Fry

London, England

17 Current Models of Sound Sources for Speech

K. N. STEVENS AND D. H. KLATT

Department of Electrical Engineering and Research Laboratory of Electronics, Massachusetts Institute of Technology, Cambridge, Massachusetts, U.S.A.

INTRODUCTION

Speech is produced by generating acoustic energy at the glottis and/or in the vocal tract. The acoustic characteristics of these sources of acoustic energy are modified by the resonant properties of the vocal tract, which can assume many different configurations and thus can filter the sources to produce a variety of sounds. This paper will focus on the nature of the sound sources employed to generate spoken language and will not consider the acoustic properties of the vocal tract, which have been treated elsewhere (Fant, 1960). However, it should be noted that it is the dynamic control of the articulators (tongue, lips, velum, etc.) that is primarily responsible for the large inventory of different speech sounds in language, and differentiates the speech of man from the verbal communication of other primates and animals (Lieberman, Crelin and Klatt, 1971).

All of the speech sounds that occur in language are produced by modulating the air flow through a constricted region in the respiratory pathways between the trachea and the lips. For some sounds, a constriction is placed in the airway so as to induce instabilities in the air flow, giving rise to turbulence. Turbulence noise generated at constrictions in the vocal tract above the vocal folds is called frication noise and is present in fricative consonants and immediately following the release of plosive consonants. Turbulence noise generated at the glottis (the space between the two vocal folds) is known as aspiration, and is present for a time following the release of voiceless aspirated plosives such as [*p, t, k*] in English and in the English [*h*] sound.

For other classes of sounds there is a quasiperiodic modulation of the air flow induced by vibrating structures such as the vocal folds. The vocal folds vibrate to produce sound during the utterance of normal vowels and voiced consonants, but are spread apart and remain inactive during voiceless consonants and whispered vowels. In unusual circumstances, the velum, tongue tip or lips may be set into vibration during the production of uvular, dental or labial trills and fricative consonants.

The aim of this paper is to attempt to tie together more closely the articulatory, aerodynamic and acoustic phenomena that occur in utterances when turbulence noise is generated and when the vocal folds are set into vibration. The present study does not report any new data, but attempts to correlate data and theories reported by various investigators (including the authors).

PRESSURE AND AIR FLOW AT A CONSTRICTION IN THE VOCAL TRACT

During speech production, the air passages between the trachea and the lips consist of a relatively wide pathway that contains one or more constrictions along its length. A schematized view of the vocal tract in which there is a glottal constriction and a supraglottal constriction is shown in Fig. 17.1a. The pressure drop across a constriction (ΔP) is a function of the volume velocity of the air flowing in the tube (U), the cross-sectional area of the constriction (A), and the shape of the constriction.

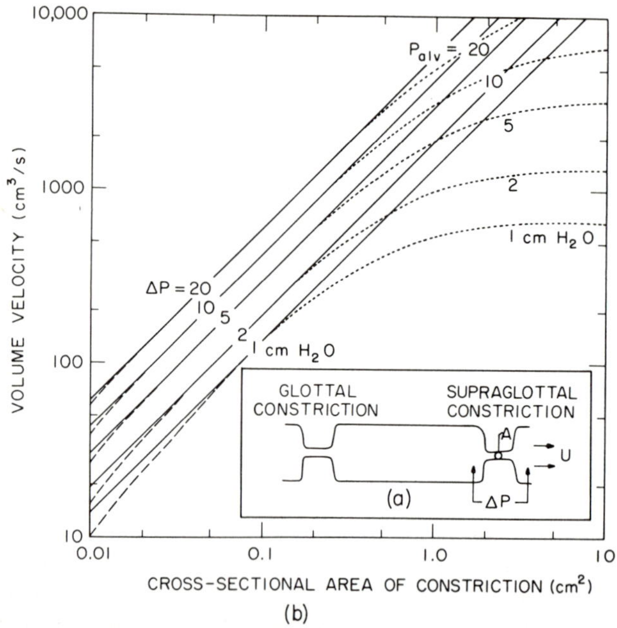

Fig. 17.1. A schematic representation of a vocal-tract configuration having both a glottal and supraglottal constriction is shown in inset (a). The solid lines in part (b) represent contours of constant pressure drop ΔP across a single constriction for air flow with various volume velocities and through constrictions of different sizes. The dashed lines at the lower left take into account the laminar resistance term in Eq. [4] [see text]: the dotted lines at the upper right represent air flows and constriction sizes for constant alveolar pressure, and reflect the pressure drop in the subglottal airways. (Adapted from Stevens, 1971.)

The major contribution to the pressure drop at a constriction is usually caused by losses near the entrance to and exit from the constriction, rather than within the constricted area itself (except when the constriction is extremely narrow). The losses arise because eddies form in the flow in the vicinity of the contraction and expansion (Streeter, 1962). The eddies can be regarded as volume elements of air that perform rotations. Because of the viscosity of the air, the energy loss is manifest as an additional pressure drop in or resistance to the flow.

It is found experimentally that the pressure drop due to eddy losses at the contraction and expansion is given by

$$\Delta P_E = k_2\,(\rho U^2/2A^2) \qquad [1]$$

The value of the constant k_2 depends on whether the entrance to and exit from the constriction are abrupt or gradual. A value of $k_2 = 0.9$ is probably a reasonable average for the kinds of constrictions found in the vocal tract (Heinz, 1956; Van den Berg *et alii,* 1957).

In addition to the losses at the contraction and expansion, there is a contribution to the pressure drop from the laminar resistance, k_1, of the constriction. In the case of a rectangular constriction of dimensions $w \times b$ ($w << b$) and length l, simulating the glottic opening or a constriction formed by the tongue blade, the laminar resistance is given by

$$k_1 = 12\mu l/(bw^3) \qquad [2]$$

where μ is the viscosity of air. For a constriction with a circular cross-section of diameter d, simulating an orifice formed by lip rounding or pharyngeal constriction, the laminar resistance is equivalent to

$$k_1 = 128\mu l/(\pi d^4) \qquad [3]$$

The total pressure drop at a constriction can be written, therefore, as the sum of two terms:

$$\Delta P = k_1 U + k_2(\rho U^2/2A^2) \qquad [4]$$

Calculations show that the contribution of the first term is normally negligible for the constriction shapes and pressure drops of interest in speech, except when the cross-sectional area of the constriction is about 0.05 cm^2 or less, or when the pressure drop is of the order of 1 cm H_2O or less (Van den Berg *et alii,* 1957).

A chart giving an approximation to the volume flow of air for various constriction sizes and values of pressure drop ΔP across the constriction is shown in Fig. 17.1b. The straight lines are representations of the pressure drop (in cm H_2O) due to eddy losses alone, with $k_2 = 0.9$. This chart can be used to estimate the relation between pressure drop, air flow and constriction size for a constriction either at the glottis or above the glottis (i.e. for either constriction in Fig. 17.1a), assuming that the vocal folds are not vibrating. The contribution of the laminar resistance component is suggested by the dashed lines at the lower left corner of Fig. 17.1b. For a large airflow, the pressure drop across a single constriction in the vocal tract is less than the alveolar pressure, since there is a significant pressure drop due to the resistance of the subglottal airways. The dotted lines approximate plots of constant alveolar pressure assuming a fixed airway resistance of 1.5 cm H_2O/litre/sec (Briscoe and Dubois, 1958).

TURBULENCE NOISE GENERATION AT A CONSTRICTION

When there is rapid flow of air through a constriction or past an obstruction in a tube, turbulent eddies are created in the flow. These random fluctuations in the flow act as a source of sound, and the sound generated in this manner is called turbulence noise.

The characteristics of this type of sound source have been examined experimentally and theoretically by Gordon (1968, 1969), by Heller and Widnall (1970) and by many other investigators. The noise source can be modelled by a sound-pressure source in series with two acoustic tubes that model the characteristics of the vocal tract in front of and behind the constriction respectively. The sound pressure, p_s, of the noise source is given by

$$p_s = k_3 \Delta P \qquad [5]$$

where k_3 is a constant and ΔP is the pressure drop across the constriction. The constant k_3 is more or less independent of the size of the constriction and its position in the tube, except that it must be reduced to about 0.3 of its value when the turbulence noise source is near the mouth opening, as it is for sounds like [*f*] and [*θ*] (as in *th*in). Thus the solid lines plotted in Fig. 17.2 also represent lines of constant source noise pressure. Lines which differ by a factor of 2 in ΔP correspond to a difference in source sound pressure level of 6 db.

The spectrum of the noise source is relatively flat over a frequency range of two or three octaves centred on a frequency of about 0.2 $U/A^{3/2}$, and falls off gradually outside this region. For the volume velocities and constriction sizes encountered in turbulence noise generation in speech, this centre frequency is usually in the range of 500–3000 Hz, the lower end of the range being for aspiration noise or noise produced with a less constricted vocal tract and a lower velocity air flow, and the upper end being for most types of frication noise. Available spectral data on fricatives obtained from measurements on natural speech usually indicate that the source spectrum is essentially flat over a few octaves (Heinz, 1957; Fant, 1960; Heinz and Stevens, 1961).

The sound pressure p_r radiated from the lips of a speaker is not directly proportional to the turbulence noise pressure p_s. The properties of the vocal tract superimpose a resonant structure on the noise spectrum (Fant, 1960), and radiation from the lips introduces a frequency-squared dependence into the spectrum. The radiated sound pressure for a given configuration can be shown to be

$$p_r \sim \Delta P^{3/2} A^{1/2} \quad \ldots\ldots\ldots\ldots\ldots \quad [6]$$

due to the effects of the radiation characteristic. The available experimental data relating sound pressure level and subglottal pressure (Meyer-Eppler, 1953; Hixon, Minifie and Tait, 1967) conform to the relationship predicted in equation [6].

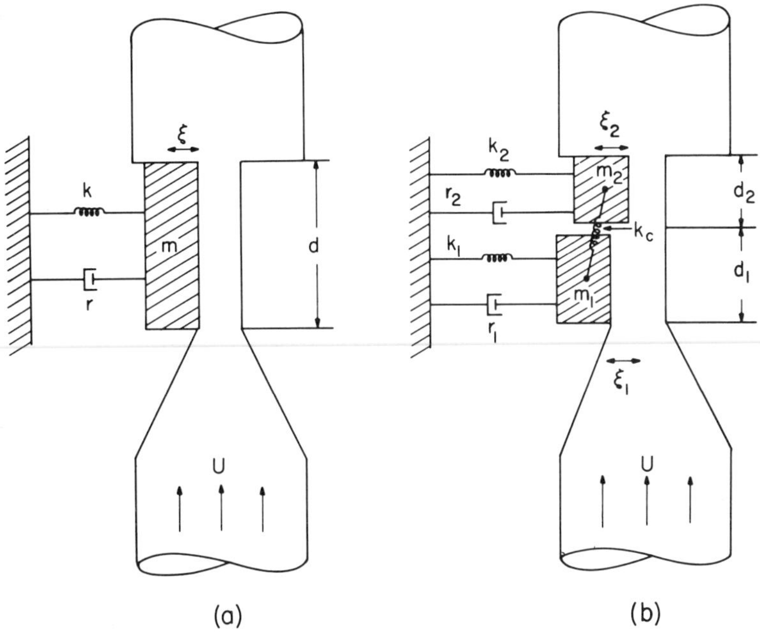

Fig. 17.2 (a) One-mass model of the vocal folds; (b) two-mass model of the vocal folds. Symmetrical vibrations are assumed, and hence it is sufficient to represent only one of the vocal folds in the model. (From Ishizaka and Matsudaira, 1968.)

SOUND GENERATION DUE TO VOCAL FOLD VIBRATION

A number of speech sounds are generated by positioning the vocal folds such that they are set into vibration by the application of a transglottal pressure drop. The mechanism whereby an applied DC pressure induces vibratory activity in the vocal folds has been clarified recently, in a paper by Ishizaka and Matsudaira (1968). They describe in mathematical form the aerodynamic and mechanical events taking place in a model that considers the upper and lower edges of the vocal folds as separate masses that are coupled together by means of a mechanical compliance.

This Ishizaka-Matsudaira model is shown in Fig. 17.2b. (Also illustrated, in Fig. 17.2a, is a simpler, one-mass model, shown by Ishizaka and Matsudaira to provide an inadequate representation of observed larynx behaviour.) The equations governing the displacements ξ_1 and ξ_2 of the lower and upper masses relative to some rest position are of the form

$$m_1\ddot{\xi}_1 + r_1\dot{\xi}_1 + k_1\xi_1 + k_c(\xi_1 - \xi_2) = F_1(\xi_1, \xi_2) \quad \text{[7]}$$

and

$$m_2\ddot{\xi}_2 + r_2\dot{\xi}_2 + k_2\xi_2 + k_c(\xi_2 - \xi_1) = F_2(\xi_1, \xi_2) \quad \text{[8]}$$

where the coefficients m_1, m_2, r_1, r_2, k_1, k_2 and k_c are defined in Fig. 17.2, and F_1 and F_2 are the forces on masses m_1 and m_2 due to the subglottal pressure P_s and the airflow U. These forces are functions of the positions of m_1 and m_2. For small displacement values of ξ_1 and ξ_2, it can be shown that $F_1 \cong k_a(\xi_1 - \xi_2)$ and $F_2 \cong \frac{1}{4}k_a(\xi_2 - \xi_1)$, where the constant k_a is dependent on the subglottal pressure P_s and on the static glottal opening with reference to which the displacements ξ_1 and ξ_2 are specified (Stevens, in preparation). In fact, k_a increases with P_s, and has a maximum value for a glottal opening somewhat less than 1 mm (if appropriate dimensions for the model are selected).

Thus the effect of the aerodynamic forces F_1 and F_2 is to introduce a negative coupling compliance (as can be seen if the terms on the right-hand side of equations [7] and [8] are shifted to the left-hand side). If k_a is sufficiently large (greater than k_c), the solution of equations [7] and [8] indicates instability, i.e. oscillations of ξ_1 and ξ_2 will be initiated. Analysis shows that these oscillations occur with ξ_1 out of phase with ξ_2, ξ_1 leading in phase by about 90 degrees. For $k_a < k_c$, the masses m_1 and m_2 remain stationary in their rest position. Thus there is a range of static glottal openings and subglottal pressures yielding sufficiently large values of k_a such that oscillation will occur. For static glottal openings that are too large or too small, or for subglottal pressures that are too low, oscillations will not be initiated since the constant k_a is too small. Furthermore, oscillations will occur for a wider range of conditions if k_c is small (the coupling between the two masses is slack) than if k_c is large (the vocal folds are stiffened).

While this kind of analysis, based on small perturbations of ξ_1 and ξ_2, indicates the conditions under which vocal fold vibration may be initiated, it does not necessarily provide insight into the mechanism of vibration when the amplitudes of motion of the masses become large. The process whereby the subglottal pressure and the air flow through the glottis create alternating forces on the vocal folds for the normal mode in which the amplitudes are large is illustrated schematically in Fig. 17.3. Part (a) of this Figure shows sketches of sections through the glottis at various points in a cycle of vibration, and indicates by arrows the net external force acting on the glottic walls due to aerodynamic factors. In part (b) of the Figure is a sketch of the waveform of the glottal airflow, labelled with points corresponding to the instants of time represented in the sketches in part (b). During the closed phase (sketches 1 and 2), the subglottal pressure creates an abducting force in the lower edges of the vocal folds, and these edges begin to separate. At the time corresponding to sketch 3, the lower portions of the vocal folds are abducted still further, and because the tissue connecting the lower edges to the upper edges is stretched, this outward displacement also causes an outward force on the upper edges, resulting in an abducting motion of these edges. The airflow through the slit between the vocal folds causes a small inward-pulling Bernouilli force at the narrow opening at the upper edge. The coupling stiffness between upper and lower edges continues to abduct the upper edges, and the same force begins to draw the lower edges together until the situation shown in sketch 4 is reached, where the glottal width is more or less

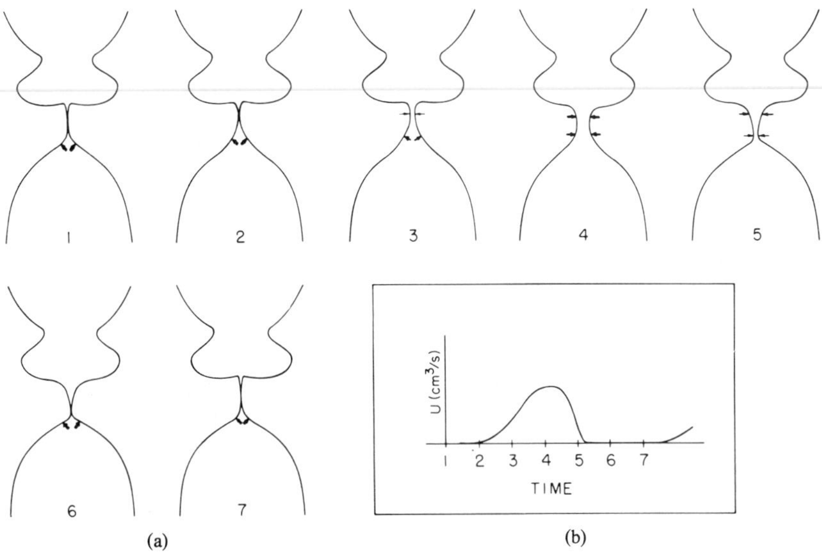

Fig. 17.3. The sketches numbered 1 to 7 in part (a) represent schematized sections through the larynx at various instants of time during a cycle of vocal fold vibration. The small arrows indicate the direction of forces acting on the upper and lower edges of the vocal folds due to the air flow and air pressure, and the width of an arrow is a measure of the magnitude of the force. A sketch of the waveform of the volume velocity through the glottis is given in (b). The points indicated on the sketch correspond to the various sections in part (a) of the Figure. These sketches are not based on actual measurements, but are derived from reports of stroboscopic tomography of the larynx (Hollien *et alii,* 1968), observations of high-speed and stroboscopic pictures of the vibrating larynx, and the computed behaviour of a two-mass model of the vocal folds (Ishizaka and Matsudaira, 1968.)

uniform. The outward motion of the upper edges can be considered as a sort of whipping action that is initiated by the abducted, more massive, lower portions of the vocal folds. In sketch 4, there is an adducting Bernouilli force along the entire thickness of the vocal folds. In sketch 6 the lower edges have come together, whereas the upper edges have not yet completed their adducting movement. At the time corresponding to sketch 7, the vocal folds have returned to their closed position, and the cycle begins again. During the closed position, the energy in the inward-moving vocal folds is dissipated in the tissues.

This analysis shows, then, that vocal fold vibration is maintained because the force acting on the more massive lower section of the vocal folds is outward when the vocal folds are moving outward and is inward when there is an adducting motion. Thus energy is added to the oscillatory movement from aerodynamic forces during each cycle of vibration. The out-of-phase valving action of the upper edges is basically responsible for this mode of energy transfer. A model based on a vibratory movement of the entire vocal fold as a single rigid mass (Flanagan and Landgraf, 1968), such as that shown in Fig. 17.2a, would not produce this kind of energy transfer.

An indication of the range of transglottal pressures, vocal fold stiffnesses and glottal separations within which vocal fold vibration can or cannot occur is given in Fig. 17.4. This chart is similar to that shown in Fig. 17.1, and gives the airflow through the glottis as a function of the glottal opening and the pressure across the glottis. The two heavy lines indicate the boundary between a state of the glottis where vocal fold vibration is not initiated (to the right and below the lines), and one where vocal fold vibration occurs (to the left and above the lines) for the conditions of slack and stiff vocal folds. The positions of these lines are determined from the analysis of the conditions of vibration for the Ishizaka-Matsudaira model. At point A, for example, where the glottal width is 3 mm and the pressure across the glottis is 3 cm H_2O, no vocal fold vibration occurs—irrespective of whether the vocal folds are stiff or slack. At point B, where the static glottal opening is about 1 mm and the pressure drop across the glottis is 7 cm H_2O, vibration is initiated if the vocal folds are slack; but the vocal folds remain stationary if they are stiff. At point C, vocal fold vibration begins for both stiff and slack conditions: that is to say, the state of the vocal folds does not remain stationary when they are placed in a configuration corresponding to point C.

The sound source resulting from vocal fold vibration can be modelled as a volume-velocity source, with a spectrum having the greatest energy at low frequencies and decreasing at about -12 db/octave at higher frequencies. A sketch of the spectrum envelope of the glottal volume-velocity waveform is given in Fig. 17.5. Experimental data show that the magnitude of this source increases as a function of $\Delta P^{3/2}$, where ΔP is the pressure across the glottis. This is the same dependence on ΔP as the sound pressure produced by a turbulence noise source. Thus both types of source increase and decrease together when there is

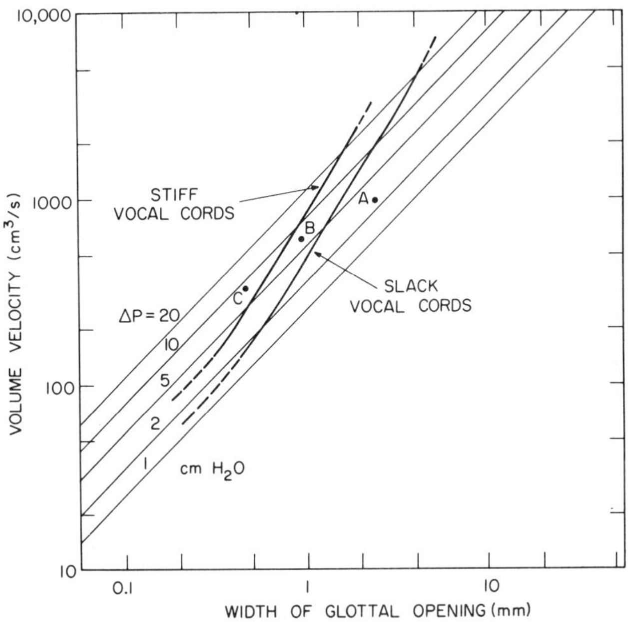

Fig. 17.4. The light lines labelled with different values of P_s represent the approximate airflow through a glottal constriction as a function of the width of the constriction. A glottis length of 1.8 cm is assumed. The curves labelled 'slack' and 'stiff' indicate the limits of oscillation of the vocal folds for these conditions of stiffness, based on the theory outlined in the text.

a change in subglottal pressure. Also shown in Fig. 17.5, for comparison, is the approximate power spectrum of the effective volume velocity of a turbulence noise source that could exist at a constriction in the vocal tract. This turbulence noise spectrum is included simply to indicate the different spectral shapes of the glottal source and the turbulence noise source, and the considerably lower overall level of the noise source. At high frequencies, however, the levels of the two sources (in a given frequency band) are comparable.

Calculations based on the Ishizaka-Matsudaira model—together with experimental evidence drawn from EMG recordings made from selected laryngeal muscles, motion pictures taken of the vibrating larynx through a fibre optic bundle or a laryngeal mirror, air flow studies and acoustical studies, and observations of excised larynges—yield the following observations concerning the behaviour of the larynx: (1) when the DC pressure across the glottis drops to below about 1–2 cm H_2O vocal fold vibration ceases, since not enough energy can be fed from the aerodynamic system to overcome energy losses in the vocal folds; (2) increasing the stiffness of the vocal folds causes an increase in the frequency of vibrations; (3) increasing the stiffness of the vocal folds reduces the range of glottal openings and transglottal pressures within which vocal fold vibration occurs; (4) if the vocal folds are sufficiently spread or are tightly adducted, glottal vibration is inhibited—particularly if the folds are stiffened; (5) when the glottis is spread or is adducted, and the vocal fold vibration is maintained, the waveform of the air flow through the glottis may differ markedly from the waveform with the folds in a normal position. These conditions correspond to 'breathy voice' and 'creaky voice' respectively (Halle and Stevens, 1971).

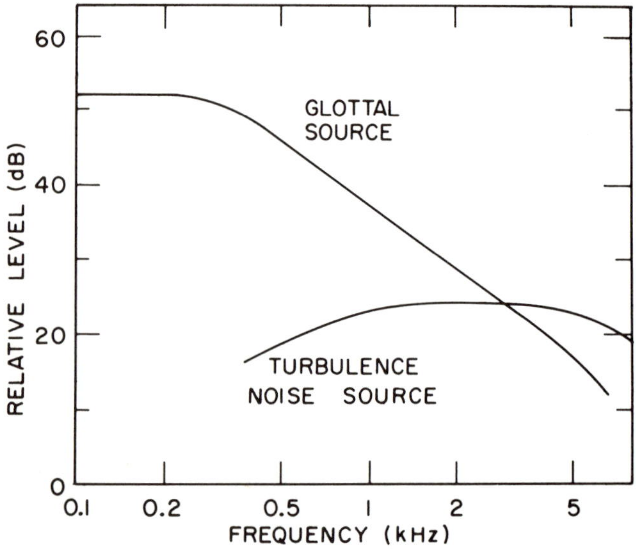

Fig. 17.5. Sketches of the spectrum envelope for a typical periodic volume-velocity flow through the vibrating glottis, and the power spectrum of the sound-pressure source due to turbulence noise at a constriction in the vocal tract. The relative levels of the two sources provide an indication of the relative sound-pressure levels of sounds produced with turbulence noise and sounds produced with vocal fold vibration.

The principal manipulations that lead to distinctively different sound outputs from the larynx are: (1) adjustment of the stiffness or slackness of the vocal folds; (2) manipulation of the static opening to the glottis (corresponding to adjustment of the separation of the arytenoid cartilages); and (3) control of external variables such as pressure across the glottis—either by changing the subglottal pressure, or by increasing the pressure above the glottis by creating a supraglottal constriction. These variables determine whether aspiration will be present and/or vocal fold vibrations will begin. They also determine the fundamental frequency and spectral characteristics of the vibration pattern.

EXAMPLES OF SOURCE ACTIVITY IN NATURAL SPEECH

Figure 17.6 shows a spectrogram of the sentence 'Ten grey shoes are here' as recorded by a normal adult male speaker. Such a spectrogram is a 3-dimensional plot of energy *v*. frequency *v*. time that is used to display certain acoustic features of speech waveforms. The blackness of any point on a spectrogram is monotonically related to the energy contained in a 300 Hz region about the point, as averaged over a time period of a few msec.

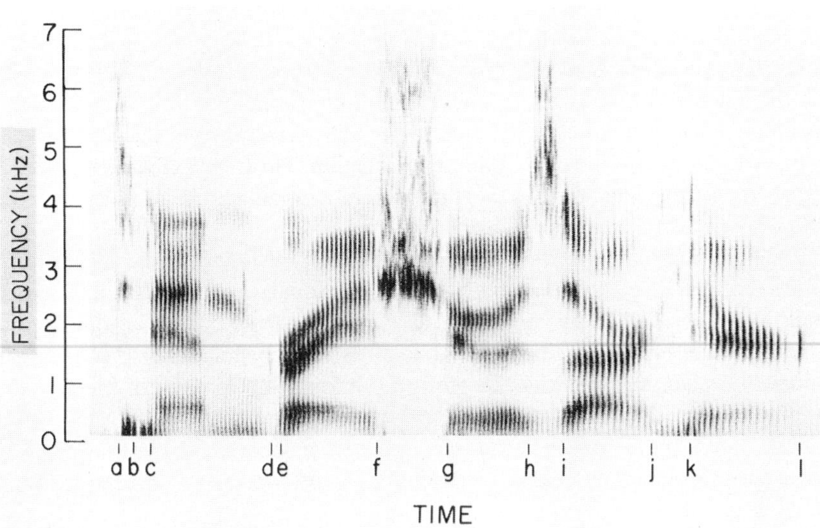

Fig. 17.6. Broadband spectrogram of the utterance 'Ten grey shoes are here'. This spectrogram indicates the energy distribution in a frequency range from 0 to 7 kHz over the 1.5 sec time course of the speech sample. The times of transition between different sound sources are indicated by the letters 'a' to 'k' along the horizontal time axis.

The vertical striations in Fig. 17.6 indicate the presence of vocal fold vibrations, each vertical line corresponding to a single period of the vibration pattern. The horizontal dark bands indicate the resonant frequencies of the vocal tract. The resonant frequencies move about in time as a function of the articulatory configuration of the vocal tract.

A brief interval of high-frequency frication noise can be seen in Fig. 17.6 at the release of the plosive consonant [*t*] (interval *a-b* on the spectrogram). As the oral closure expands, the major constriction shifts to the glottis and the frication turns into aspiration (*b-c*). In this region, the vocal folds are sufficiently abducted so that vibration is not initiated (i.e. the constant k_a in the analysis of the vocal fold model discussed on pages 283–4 is not sufficiently large). The vocal folds are then brought together in such a way as to permit the onset of vocal fold vibrations for the vowel and following nasal consonant (*c—d*).

The voiced plosive [*g*] at the beginning of the word 'gray' contains another example of a brief interval of frication (*d-e*); but there is no aspiration, because the vocal folds are already in a position appropriate for voicing. At the onset of the word 'shoes', a constriction is formed in the vocal tract behind the alveolar ridge and the pressure in the mouth builds up, resulting in a drop in pressure across the glottis and hence a cessation of vibration, particularly if the vocal folds are stiffened. These activities result in the production of the voiceless fricative [*š*] (*f–g*). The glottal and supralaryngeal configurations appropriate for a vowel are then resumed (*g-h*).

During the voiced fricative [*z*] (*h-i*), the vocal folds are spread slightly to permit the increased air flow required for frication production, the vocal folds are probably slackened, and the pressure in the mouth increases—but the pressure across the glottis is sufficient to maintain vibration. The frication noise seen in the upper frequency range for the [*z*] is periodically modulated in intensity to some extent by the vibrating vocal folds.

The aspirated [*h*] at the onset of the word 'here' (*j-k*) is another example of two sound sources being active at one time. In intervocalic positions, the [*h*] is normally partially voiced: the vocal folds are spread apart to produce a glottal constriction appropriate for the generation of aspiration noise, but the folds are somewhat slack and continue to vibrate (although they may not meet at the midline during the period). In the final word in the sentence, the increased spacing between the vertical striations indicates a decrease in the frequency of vocal fold vibration, due either to a fall in subglottal pressure or a reduction in vocal fold stiffness, or both.

CONCLUSION

The models discussed in this paper predict the existence of sound sources that fall naturally into gross categories that are employed in different combinations in various languages (Stevens, in press). The models are not adequate for predicting the fine differences in source characteristics that may occur from one individual to another, or that may be a consequence of some anomalous property of an individual vocal tract, larynx, or respiratory system. A future objective of work of this kind is to account for such individual differences or anomalous behaviour.

ACKNOWLEDGEMENT

Preparation of this paper was supported in part by the National Institutes of Health and the United States Air Force Cambridge Research Laboratories.

REFERENCES

Briscoe, W. A., and Dubois, A. B. (1958) The relationship between airway conductance and lung volume in subjects of different age and body size, *J. clin. Invest.*, **37,** 1279–85.

Fant, C. G. M. (1960) *Acoustic Theory of Speech Production,* 's-Gravenhage.

Flanagan, J. L., and Landgraf, L. L. (1968) Self-oscillating source for vocal-tract synthesizers, *IEEE Trans. Audio. Electroacoust.*, **AU-16;** 57–64.

Gordon, C. G. (1968) Spoiler-generated flow noise. I. The experiment, *J. acoust. Soc. Amer.*, **43,** 1041–8.

Gordon, C. G. (1969) Spoiler-generated flow noise. II. Results, *J. acoust. Soc. Amer.*, **45,** 214–23.

Halle, M., and Stevens, K. N. (1971) Some comments on phonetic features related to glottal configuration and vocal-cord stiffness. *Quarterly Progress Report No. 101, Massachusetts Institute of Technology Research Laboratory of Electronics, Cambridge, Massachusetts,* pp. 198–213.

Heinz, J. M. (1956) Fricative consonants, *Acoustics Laboratory Quarterly Report, Massachusetts Institute of Technology, Cambridge, Massachusetts,* pp. 5–7.

Heinz, J. M. (1957) Fricative consonants, *Acoustics Laboratory Quarterly Report, Massachusetts Institute of Technology, Cambridge, Massachusetts,* p. 1.

Heinz, J. M., and Stevens, K. N. (1961) On the properties of voiceless fricative consonants, *J. acoust. Soc. Amer.*, **33,** 589–96.

Heller, H. H., and Widnall, S. E. (1970) Sound radiation from rigid flow spoilers correlated with fluctuating forces, *J. acoust. Soc. Amer.*, **47,** 924–36.

Hixon, T. J., Minifie, F. D., and Tait, C. A. (1968) Correlates of turbulence noise production for speech, *J. Speech Res.* **10,** 133–40.

Hollien, H., Coleman, R., and Moore, P. (1968) Stroboscopic laminagraphy of the larynx during phonation, *Acta oto-laryng. (Stockh.),* **65,** 209–15.

Ishizaka, K., and Matsudaira, M. (1968) What makes the vocal cords vibrate? *Reports of the 6th International Congress on Acoustics, IIB,* B-9–12; also (1968) Analysis of the vibration of the vocal cords, *J. acoust. Soc. Japan,* **24,** 311–12.

Lieberman, P., Klatt, D. H., and Crelin, C. S. (1971) Anatomial constraints on speech in chimpanzee, newborn *homo sapiens* and Neanderthal man. Presented at 81st Meeting of the Acoustical Society of America, Washington, D.C., 19–23 April, 1971. (*J. acoust. Soc. Amer.*, **50,** (A) 139–140.)

Meyer-Eppler, W. (1953) Zum Erzeugsmechanismus der Geräuschlaute, *Zeit. Phonetik,* **7,** 196–212.

Stevens, K. N. (1971) Airflow and turbulence noise for fricative and stop consonants, *J. acoust. Soc. Amer.*, **50,** 1180–92.

Stevens, K. N. [in press]. The quantal natural of speech: Evidence from articulatory-acoustic data, in *Human Communication, A Unified View,* ed. David, E. E. Jr., and Denes, P. B., New York.

Stevens, K. N. [in preparation] An analysis of glottal activity during speech sound production.

Streeter, V. L. (1962) *Fluid Mechanics*, New York.

Van den Berg, J. W., Zantema, J. T., and Doornenbal, P. (1957) On the air resistance and the Bernouilli effect of the human larynx, *J. acoust. Soc. Amer.*, **29**, 626–31.

DISCUSSION ON COMMUNICATION [17]

Kirchner. I have one question, just to be sure that I have understood your explanation of what the subject can do with his larynx in order to produce a particular type of sound. You mentioned that he can adjust the stiffness of the vocal folds, and that he can adjust the position of the vocal folds in relation one to another. If we look at frontal laminograms, the area of contact of the vocal folds at high frequencies is very thin, as you know, whereas at lower frequencies there is a relatively broad contact surface. Cannot the subject also vary the mass of the vocal folds?

Klatt. Yes. In fact, I don't think one can really vary the stiffness without varying the effective mass of the vibrating vocal folds. The stiffness is varied principally by stretching the vocal folds.

Stevens. The mass would change with the stiffness. I am not certain whether the mass can be manipulated independently of the stiffness—but I think not.

Fourcin. In your mechanical models of the vocal folds you showed linear damping which is going to be operative in a uniform manner, as I understood it, throughout the course of vibration. But in practice, the greatest energy loss is going to occur during the closure.

Klatt. That's right. The model I showed was only used, particularly in this Communication, to determine whether vocal fold vibration would occur, or not, from an initial configuration. It is not a model that describes what happens *when* the vocal folds come together. One really needs to change suddenly to a different model at that instant in time.

Martin. You have given us a very useful model of what happens in the vocal tract in engineering terms. You haven't given us very much on the control side, however, or said how this is achieved. It seems to me, from your description, that there are two important factors—one is the cross-sectional area of the constriction, and the other is the change of pressure [ΔP].

Could I switch to the biological aspect then, and say it seems to me that we might control this in two ways—either by a variety of afferent inputs coming from the muscles and joints, or also acoustically through our ears. We can probably get a lot of information about the cross-sectional area in the biological situation, but I don't think we can get very much information about ΔP. We know what happens within the lung, on the subglottic side; and is it necessary to postulate that there are pressure receptors above this level to enable this model of yours to operate?

I am afraid the next question is very much more speculative. How do you see the interrelationship of these two different levels of afferent activity—engineering relations, so to speak—between the processing of afferent information from the muscles and joints and the acoustical control system?

Klatt. I should say that we are really here to learn, and you probably know more about the physiology than I do. With respect to control, there have been some introductory experiments in which various structures in the vocal tract have been anaesthetized in an attempt to see whether speech alters. Similar experiments have attempted to mask auditory feed-back during speech production. But before I tell you the results of these experiments there is another issue here; namely, how precise does the control really have to be?

The models we have talked about, particularly for frication generation, indicate that there is a rather wide range of areas within which one gets approximately the same intensity of frication noise. You either get frication noise or you don't, and if you do it has roughly the same intensity: so for control purposes, it may not be necessary to be quite this precise in our laryngeal adjustments, or with one's tongue tip or one's lips, as you might think. The masking studies I mentioned generally show that there is no immediate apparent degradation in the control of speech under the masking noise condition—so the influence of ongoing auditory feed-back is probably not very important once language has been acquired. In respect of anaesthetizing some of the afferent systems in the mouth, there have been informal reports that this does disturb speech production somewhat, and (perhaps) affects fricatives more than anything else—but I am not very familiar with this work. Perhaps someone else in the audience can address himself to this issue?

Proctor. Some years ago I was having studies done on myself, measuring subglottic pressures with a trans-cricothyroid needle. The doctor who passed the needle injected a little local anaesthetic into the skin, but by mistake a drop came through into my trachea; and the result was that I was able to talk perfectly well,* but I couldn't sing worth a damn! So it would seem that just superficial topical anaesthesia, perhaps affecting only a portion of the vocal folds themselves, is enough to interrupt the ability to control the singing voice—so much so, that we had to quit that experiment and try again another day even though I was able to talk perfectly intelligibly.

Wyke. Could I follow up the question that Dr. Fourcin asked you, to clarify my understanding of the model? As I understood it, the relationships you described are predicated on parameters that have linear dimensions exclusively. To what extent would the model, and the theory deriving from it, be modified by the introduction of a rotatory component in the movements of the vocal folds? It is now well known from cinéradiographic and tomographic studies[1,2] in living subjects that there is, in fact, a rotatory movement of the vocal folds up and

---

* *Editor's Note:* see Dr. Proctor's later qualification of this statement at page 295.

[1] Ardran, M., and Kemp, F. H. (1966) Mechanism of the larynx. I. The movements of the arytenoid and cricoid cartilages, *Brit. J. Radiol.*, **39**; 641–54.

[2] Damsté, P. H., Hollien, H., Moore, P., and Murray, Th. (1968) An X-ray study of vocal fold length, *Folia phoniat. (Basel)*, **20**, 349–59.

down as they open and close—and we saw a little of this in Dr. Cavagna's splendid film on the first day (see Communication 6). To what extent is it necessary to introduce such a component in order to make the theory work in practice?

Klatt. My impression of the two-mass model is that it really incorporates the essential aspect of this rotatory motion, whereby one part of the upper portion of the vocal fold is subjected to forces somewhat different from other portions of the fold, and thus moves in and out in phases different from the lower surface. Maybe Dr. Stevens ought to be given the chance to comment on this, too?

Stevens. It is quite interesting to us that, in fact, no one as yet understands quantitatively how and why the vocal folds vibrate—no one gets down to the details. It is only recently that this two-mass model has been worked out; and, as you point out, it assumes only an in-and-out motion. But there certainly is, in fact, a rotatory component also—and future models must take that into account. One of the problems is that we do not know the physical constants of this system, and we would be most happy to receive advice on how one might measure those.

Widdicombe. I want to go back to Dr. Proctor's observation—it is fascinating that you could not sing but you could speak. Of course, one possibility is that the local anaesthetic agent had only blocked the mucosal receptors and had not got down to the laryngeal muscle spindles and joint receptors.

Proctor. I am sure that was true. The anaesthesia was only superficial.

Widdicombe. It is difficult to see how you could retest this in man, but if you took cats and cut the internal branch of the superior laryngeal nerve (which is largely afferent), leaving the motor branch to the cricothyroid muscle, you could then virtually de-afferent the larynx. Could the cats then say 'miaow'?

Wyke. We have done this sort of experiment in the course of the electromyographic studies of laryngeal reflexes that I am going to talk about tomorrow [see Communication 24]—but I can't tell if the animals could say 'miaow' or not, as they were anaesthetized.

Ruth Bowden. When you mentioned anaesthetization of the mouth, could you tell us the route by which this was done—was it by local application?

Klatt. Yes—it was topical anaesthesia.

Ladefoged. Actually, I did some of these experiments (Landefoged, 1959[1], 1967[2]). I have also heard tape recordings of other experiments that were more carefully controlled (Ringel and Steer, 1963[3]; Scott and Ringel, 1971[4]). Both my

---

[1] Ladefoged, P. (1959) The perception of speech, in *The Mechanisation of Thought Processes,* Vol I., pp. 399–409, London, H.M. Stationery Office.

[2] Ladefoged, P. (1967) *Three Areas of Experimental Phonetics,* London.

[3] Ringel, R. L., and Steer, M. D. (1963) Some effects of tactile and auditory alterations on speech output, *J. Speech Dis.,* **6,** 369–78.

[4] Scott, Cheryl M., and Ringel, R. L. (1971) Articulation without oral sensory control, *J. Speech Res.,* **14,** 804–18.

own and the other experiments were done in various ways, including topical anaesthesia and including double mandibular nerve blocks—and really, the surprising thing is how *well* one can speak when you have removed quite a lot of sensory input in this way.

Now, getting back to the point about your singing, Dr. Proctor, I don't want to be rude but I just don't believe the fact that you could talk normally but could not sing with a local anaesthetic agent in your larynx. As somebody who can't normally sing anyway, I will maintain that talking is just as skilled an operation as singing; and if you could talk correctly, that means that you still had all the necessary (and very subtle) pitch inflections. Do you really mean to say that you still had your usual gamut of pitch inflections?

Proctor. I should correct my initial statement. Certainly my talking voice was not normal, I am sure. What I should have said was that the very superficial topical anaesthesia eliminated the finer control that yields what one feels is a satisfactory singing voice, while one is still able to continue to communicate in such a way that people understand one. You can see that to a greater degree when you do topical anaesthesia for endoscopy. As I am sure Dr. Kirchner has seen frequently, even after thorough anaesthesia of the entire larynx the patient is able to communicate with you fairly satisfactorily, although certainly not in a normal voice, I am only trying to say that some of the things that are required for satisfactory singing are a little different from those needed for *intelligible* speaking.

Ruth Bowden. Could I ask, did you ever try anaesthesia of the semilunar ganglion? I think this important, because if you block only the mandibular nerve you may have only partially de-afferented the muscles—in my opinion.

Ladefoged. I am quite sure that you are correct in saying that one has only partially de-afferented the muscles. Is it possible to do so in man? I would rather do it on your mouth than mine!

Wyke. Neurosurgeons and neurologists have often injected the semilunar ganglion with local anaesthetic agents, and with alcohol, in patients with trigeminal neuralgia; and my clinical impression is that this makes little difference to their voice quality. But, of course, on these patients the procedure is performed only unilaterally; and I am not sure how this observation would relate to the findings of Leanderson and Persson (1972)[1] mentioned on the first day by Dr. Lindquist [see p. 72], or to the disturbances of speech in patients with abnormalities of oral sensation described by Worster-Drought (1956)[2] [see also Bosma, 1967[3]].

Malcolm. I would like to congratulate Dr. Klatt; and could I add that my theory of phonation does require the larynx to oscillate in two modes as a radiating slot transmitter. In one of them the air streams through the vocal folds in a vortex; and the other stream comes, as it were, off the tips of the arytenoids.

---

[1] Leanderson, R., and Persson, A. (1972) The effect of trigeminal nerve block on the articulatory EMG activity of facial muscles, *Acta oto-laryng. (Stockh.)*, **74**, 271–8.

[2] Worster-Drought, C. (1956) Congenital suprabulbar paresis, *J. Laryng.*, **70**, 453–63.

[3] Bosma, J. F. [Ed.] (1967) *Symposium on Oral Sensation and Perception*, Springfield, Illinois.

Otis. Do you attempt to include any influence of surface tension forces in the larynx, in your model?

Klatt. No. The model we presented does not account for the behaviour of the vocal folds during the time interval when they are approximated to one another.

Stevens. I think it is an interesting speculation that surface tension does play a rôle as far as the release of the vocal folds is concerned, but I don't know whether it has gone beyond speculation. I would be interested to know whether such a force, which is a kind of force quite different from that due to pressure and air-flow effects, should be included in any model.

Plum. A question, please. Before one systematically plots out the innervation and reciprocal relationships, could you tell us whether the morphology of the larynx in man—with his rather greater vocal range and flexibility—is the same as it is in the lower animals? Do we have any comparative data on the structure and innervation?

Ruth Bowden. There is a large variation in the morphology and nerve supply of larynges in different species, and I can bring in some pictures of different species tomorrow. For example, the gross connexions between the superior and the recurrent laryngeal nerves vary quite considerably in different forms.

Plum. Fibre numbers in the laryngeal nerves?

Ruth Bowden. I am afraid we have data on fibre counts only in cats and in man.[1, 2] I can add that the numbers of fibres are just about equal on the two sides—that is, on the left and right sides the numbers of fibres are about the same. I will include these data in my demonstration tomorrow.

[*Editor's Note:* Professor Bowden arranged a demonstration of material from her Department* illustrating the comparative features of the morphology and nerve supply of the larynx in a variety of animal species, for the benefit of the participants in the Symposium.]

Cavagna. I was interested in the slide in which you showed how the air moves past the vocal folds, and I would like to ask you two things. One is about the volume-time relationship in each cycle, which has a rather long rise and then an abrupt fall. I want to ask if this is always the same at all intensity levels—that is, at all different pressures pushing the vocal folds apart, or not. The other problem is why the lower parts of the vocal folds are the first to close and the upper ones the last. Do you have any simple explanation of why the opening and the closing behave in this way?

Klatt. Let me answer the second question first. The idea is (you are quite right) that the superior part does receive the Bernouilli force first, but the total forces

---

[1] Scheuer, J. Louise (1964) Fibre size frequency distribution in normal human laryngeal nerves, *J. Anat. (Lond.)*, **98,** 99–104.

[2] Abo-El-Enein, M. A. (1967) *Functional Anatomy of the Larynx,* Ph.D. Thesis, University of London.

* *Editor's Note:* Dissections of the comparative anatomy of the larynx were the work of Dr. Louise Scheuer: laryngeal nerve histograms were prepared by Dr. Scheuer and Dr. M. A. Abo-El-Enein.

acting on that part still tend to pull it apart. It has momentum moving in that direction. The other masses are pulling it apart, and it just continues to move that way while the other mass starts pulling.

With regard to your first question, the drawing you saw was intended just to give an impression of the nature of the volume-velocity waveform. It is rather difficult to really get precise measurements of the volume-velocity waveform. I think it is generally agreed that the vocal folds close somewhat more rapidly than they open under most conditions. With techniques such as inverse filtering, or estimation of the volume velocity from calculations of the area between the vocal folds seen in high-speed motion pictures of the vibrating larynx, you see a large variety of different (more or less triangular) shapes that may approach equal opening and closing speeds.

Anthony. In essence, of course, this is a steady-state model that you have, though you did say this only applied to part of the opening phase. But in natural speech there is a gradual coming together of the vocal folds while the lung pressure is built up: the two things interact together, so that there is some interaction between the vocal fold closure and the increase of pressure that starts the folds vibrating.

Klatt. I believe that it is true of the models we have shown that the first motion from any initial state of static opening and time course of pressure build-up is outward. Do you recall whether that is true, Dr. Stevens?

Stevens. If the vocal folds happen initially to be positioned slightly apart, the first motion can be either in or out.

Ruth Bowden. Has anyone studied the effect of the shape of the joint between the cricoid and the arytenoid cartilages, because this has a very subtle and sinuous curvature?

Another point is that the fibres in the vocalis muscle are of rather different histological appearance from those in the more lateral parts of the thyroarytenoid musculature—there are many more pale fibres in the vocalis portion than in the other parts. These two different muscular components might effect differential movements.

Klatt. This is, again, a subject that we hope to learn something about tomorrow [see Communications 22 and 23—*Ed.*]. I think that there are other people in the audience who could answer both questions better than I could.

Wyke. Could I refer to one point regarding this matter of the curvature of the crico-arytenoid articular surfaces raised by Dr. Bowden? It has been suggested[1,2,3,4,5]—I think, more often on theoretical than observational

[1] Sonesson, B. (1959) Die funktionelle Anatomie des Cricoarytaenoidgelenkes, *Z. Anat. Entwickl.-Gesch.*, **121,** 292–303.

[2] von Leden, H., and Moore, P. (1961) The mechanics of the cricoarytenoid joint, *Archiv. Otolaryng.*, **73,** 541–50.

[3] Frable, Mary A. (1961) Computation of motion at the cricoarytenoid joint, *Archiv. Otolaryng.*, **73,** 551–6.

[4] Ardran, G. M., and Kemp, F. H. (1966) The mechanism of the larynx. I. The movements of the arytenoid and cricoid cartilages, *Brit. J. Radiol.*, **39,** 641–54.

[5] Kotby, M. N., and Haugen, L. K. (1970) The mechanics of laryngeal function. *Acta oto-laryng. (Stockh.)*, **70,** 203–11.

grounds—that it is this curvature of the surfaces of the crico-arytenoid joint that actually throws in the rotatory component of the upward and downward displacement of the vocal folds that I mentioned earlier [see p. 293]. But whether this is actually so or not, I don't know.

18 Respiration, Laryngeal Activity and Linguistics

P. LADEFOGED

Department of Linguistics,
University of California, Los Angeles,
California, U.S.A.

The sentences of a language may differ from one another in terms of the individual speech sounds, or in terms of larger units. Linguists often refer to the former as *segmental* and to the latter as *suprasegmental* differences. The respiratory and laryngeal systems are involved in producing both kinds of distinction. We will consider first the ways in which suprasegmental information is produced.

SUPRASEGMENTAL MECHANISMS

There are three major kinds of suprasegmental information. In the first place, information may be conveyed by varying the degree of stress. An example from English is the difference between an accented and an unaccented syllable, as in *That's a hot-dog* [a frankfurter on a bun] and *That's a hot dog* [an overheated animal]. Secondly there may be variations in pitch. Thus English differentiates between a rising and a falling intonation, as in *Yes* [I agree with you] and *Yes?* [did you say 'yes'?]. Thirdly, languages use variations in length and timing of the kind that give rise to differences in rhythm. English is said to be a 'stress-timed' language, as opposed to French where there is a greater tendency for the individual syllables to recur at regular intervals of time.

Stress

Definitions of what is meant by stress are often vague. But the data reported in Ladefoged (1967) clearly show that an accented syllable in English is accompanied by an increase in subglottal pressure, which in normal conversational utterances is produced by the action of the internal intercostal muscles. Recent work by Netsell (1970) has shown that in many, if not all, accented syllables there is also additional activity of at least one laryngeal muscle; namely, the cricothyroid. So the production of accented syllables requires adjustments of both the laryngeal and the respiratory mechanisms.

Intonation

Differences in intonation are also dependent on the action of both mechanisms, but there is some disagreement over the extent to which each is involved. Lieberman (1967) has suggested that in an English simple declarative sentence the tension of the vocal folds is relatively constant, and the pitch changes are due to variations in subglottal pressure. He has claimed that a 1 cm H_2O increase in the subglottal pressure will cause a 16–22 Hz increase in the rate of vibration of the vocal folds. If this figure were correct, then it might be true that English intonation was controlled mainly by respiratory mechanisms. But it is now generally agreed that this figure is far too high. In a review of the literature, Ohala and Ladefoged (1970) point out that most investigators find an increase of between 3 and 5 Hz for a 1 cm H_2O rise in subglottal pressure in normal conversational speech. These authors claim that 'at least 90 per cent of the linguistically significant pitch patterns in English sentences are effected by means of controlled changes in the laryngeal muscles'.

Rhythm

Rhythmic differences are more difficult to discuss. There are some indications that the organization of temporal units is in terms of the occurrence of stresses. As has been shown by Allen (1971), syllables and stresses are the only units involved in the rhythm of English phrases. Similar observations have been made by Kozhevnikov and Chistovich (1965) for Russian.

SEGMENTAL MECHANISMS

We may now discuss the ways in which the laryngeal and respiratory systems can produce differences between segments.

Respiratory features

Chomsky and Halle (1968) have claimed that there is an increase in subglottal pressure in sounds such as those at the beginning of the English words *pie* and *tie*, and in their view this increase is one of the features that distinguishes these sounds from those at the beginning of words such as *buy* and *die*. But the data reported in Ladefoged (1967), Lieberman (1967) and Netsell (1969) do not show any systematic increase in subglottal pressure during these stop closures, and the unsupported Chomsky-Halle hypothesis should probably be abandoned. There is no doubt that after the opening of the stop closure (the opening of the lips in a word such as *pie*) there is a release of a great deal of respiratory air before the vocal folds start vibrating for the vowel. During this period the subglottal pressure drops considerably, so that it is possible to claim that more respiratory energy is used up in these sounds. Furthermore, several years ago Ladefoged,

Draper and Whitteridge (1958) reported that there were increases in intercostal activity associated with consonants which were accompanied by a high rate of flow of air. At that time it was not possible to say whether the increase in intercostal activity occurred before the increase in the airflow (so that there could be said to be a systematic increase in respiratory activity associated with these sounds), or whether it occurred a short time after the increased airflow and the consequent loss of subglottal pressure (so that it could be said to be due to a feedback mechanism such that the respiratory system operated to produce the subglottal pressure required for speech). These experiments have still not been repeated so that this question might be resolved.

Data have been reported which indicate that the respiratory system might be ancillary in producing differences between individual speech sounds in some languages. Thus Lee and Smith (1971) report a systematic difference in subglottal activity in one of the series of stop consonants that occur in Korean, and Ladefoged (1971) reports differences in Kachin-Jingpho (spoken in Burma). But all these distinctions involve concomitant changes in the state of the glottis (which will be discussed in the next section of this paper), and it is doubtful if the respiratory system ever plays more than a minor rôle in distinguishing between individual speech sounds.

Glottal features

In order to understand the actions of the larynx in producing differences in segments it is necessary to consider the possible glottal states. The glottis viewed from above is like a triangular slit. The two principal forces acting on it are shown very schematically in Fig. 18.1. One force causes it to be stretched in the anterior-posterior dimension (in the direction of the line a-a^1), mainly by varying the position of the thyroid cartilage. The other force results in movement of the arytenoid cartilages along the lines shown as b-b^1, thus decreasing or increasing the length of the base of the triangle. There are, of course, numerous other forces acting on the glottis. For example, there are movements in the vertical dimension, which cannot be shown in Fig. 18.1. To the best of my knowledge, none of the other possible movements are used linguistically independently of the two forces we have discussed. In fact, even these two forces interact, in that movements of the arytenoids obviously also affect the length of the glottis.

There is a great deal of speculation on the way in which languages use different laryngeal gestures to produce different segments. What little evidence there is seems to suggest that variations in phonation type are principally dependent on movements along the lines b-b^1 in Fig. 18.1, just as variations in pitch are principally dependent on movements in the direction of the line a-a^1.

In support of this contention, consider the photographs shown in Fig. 18.2. These photographs, which were taken by John Ohala and Ralph Vanderslice, show the state of the glottis in: (a) voiceless sounds, such as the voiceless vowel symbolized by /*h*/ at the beginning of the word *hot*; (b) murmured or breathy voiced sounds of the kind that occur in Hindi and other Indo-Aryan languages; (c) voiced sounds such as the vowel /*a*/ in *hot*; and (d) laryngealized or creaky voiced sounds of the kind that occur in Indo-Chadic languages. Immediately below each photograph is a schematic representation of the glottis, with the vibrating parts of the vocal folds indicated by dashed lines and the centres of the arytenoid cartilages shown by solid circles. The distance between the arytenoids is arbitrarily taken to be 100 per cent in voiceless sounds, as in (a). In sounds of this kind the arytenoids are far apart, and the vocal folds cannot be set into vibration. In breathy voiced sounds, as in (b), the distance between the arytenoids is

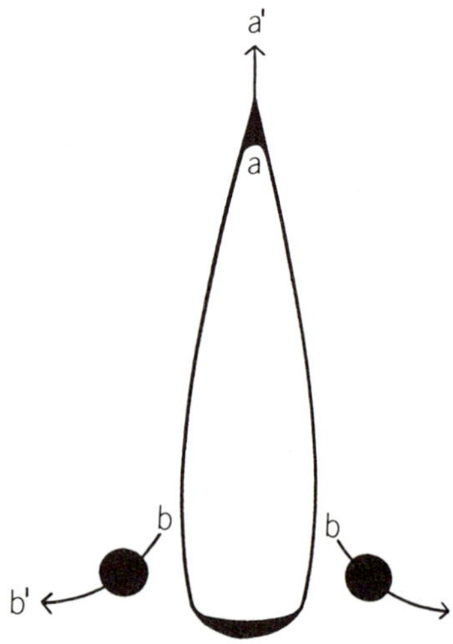

Fig. 18.1. The two major forces acting on the glottis. The force along a-a' (mainly due to the movement of the thyroid cartilage) affects the PITCH. The forces along b-b' (the movements of the arytenoid cartilages) affect the GLOTTAL STRICTURE.

only 90 per cent of that in (a). Because of the basically triangular shape of the glottis, the anterior portions of the vocal folds are now close enough together to be set into vibration. In ordinary voiced sounds, as in (c), the arytenoids are still closer together (about 75 per cent of the distance in voiceless sounds) so that the whole glottis is able to vibrate. Finally, in creaky voiced sounds, as in (d), the arytenoids have come even closer (to 67 per cent of the distance in voiceless sounds) so that the posterior portion of the glottis is held together and unable to move, leaving only a small part of the ligamental vocal folds vibrating. Note that, because the arytenoid cartilages move forward as they come together, the vocal folds tend to be less stretched in creaky voiced sounds; they are therefore likely to vibrate at a lower frequency. But the coming together of the arytenoids and the movements of the thyroid cartilage that stretch the vocal folds are independent laryngeal gestures, so that it is quite possible for creaky voiced sounds to occur on any pitch. They are, in fact, used with contrastive pitches in tone languages such as Kuman.

The four states of the glottis shown in Fig. 18.2 represent points in a continuum of possible arytenoid positions. Other photographs of a similar kind have shown quite clearly that the arytenoids are further apart than in Fig. 18.2(a) when breathing in, and perhaps also in strongly aspirated sounds. They are also closer together than in Fig. 18.2(d) when making a glottal stop, which is a laryngeal gesture used by many languages (e.g. Tagalog and other Malayo-Polynesian languages; or for most Americans as the form of /*t*/ that occurs in a word such as *button*). Furthermore, several stages intermediate between those shown in Fig. 18.2 are possible. Thus, there is a noticeable difference in the degree of breathiness that occurs in the murmured stops in the languages spoken in different parts of India; and the murmured sounds that occur in Southern Bantu languages are different from any that occur in India in that they are much more like sounds with regular voicing. Differences in the degree of laryngealization (creaky voicing) also occur. For example, a very slight degree of laryngealization may be heard in the vowel after so-called 'fortis' unaspirated stops in Korean. Kagaya (1971) has shown, by a motion picture film of the larynx taken through a fibrescope, that the 'fortis' Korean stops have a much more constricted glottis than the corresponding 'lenis' stops. But the slightly laryngealized Korean stops are very different from the strongly laryngealized stops that occur in Hausa. Furthermore, the difference between the two sets of vowels that occur in a Nilotic language such as Luo is that one set is slightly laryngealized, whereas the other set has regular voicing or a tendency towards murmur. But the degree of laryngealization that occurs in Luo is nothing like so extensive as that found in the creaky voiced semivowels in Chadic languages such as Margi.

The multivalued feature underlying all these differences in phonation type may be referred to as glottal stricture. A set of terms for possibilities within this feature might include: (1) glottal stop; (2) creak; (3) creaky voice; (4) stiff voice; (5) voice; (6) slack voice; (7) murmur; (8) breathy voice; (9) voiceless. The terms 'stiff voice' (for a moderate degree of laryngealization) and 'slack voice' (for a tendency towards murmur) are taken from Halle and Stevens (1971), where they are defined slightly differently. It should be emphasized that each member of the set of terms corresponds to an arbitrary area in the range of possibilities. It would have been quite possible to have added other terms. Alternatively, we could have collapsed some adjacent pairs of terms. As far as I am aware, no language contrasts more than three possibilities within this feature. There is, for example, no language that differentiates between a group of sounds solely by making one with creaky voice, another with regular voice, a third with breathy voice, and a fourth which is voiceless.

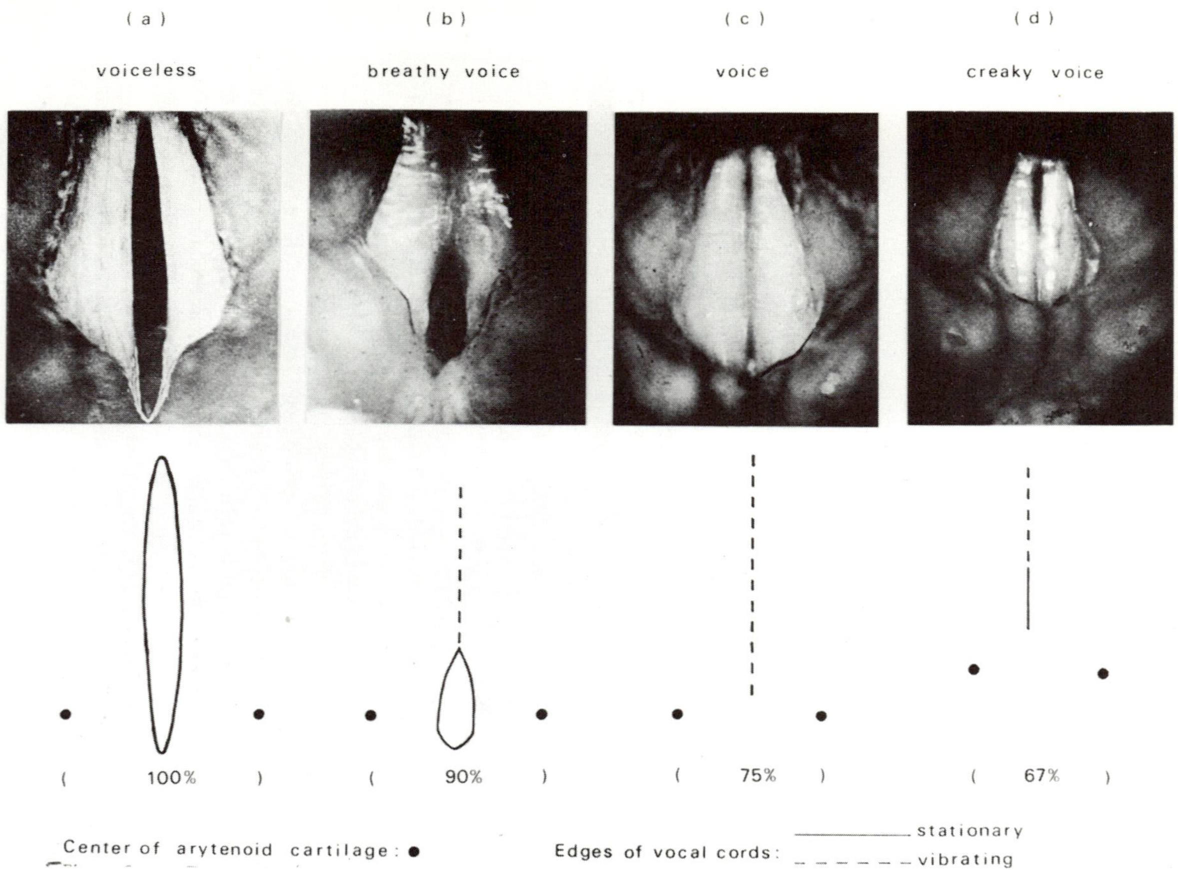

Fig. 18.2. Photographs (taken by John Ohala and Ralph Vanderslice) and schematic representations of four states of the glottis. The distance between the centres of the arytenoid cartilages in the voiceless position is arbitrarily taken to be 100 per cent.

Another way in which languages use laryngeal features is by varying the timing of laryngeal events, with respect to the articulatory gestures. Thus in stops and fricatives the vocal folds may be vibrating throughout the articulation (as in English *zoo* or Burmese /*zan*/ 'levitation'); or they may start vibrating just after the articulation (as in English *Sue* or Burmese /*san*/ 'example'); or there may be a short period of aspiration before regular voicing starts (as in Burmese /*shan*/ 'rice'). No language uses this feature to distinguish more than three possibilities. Linguistic differences in voice onset time have been well described by Lisker and Abramson (1971, and references therein) and need not be elaborated further here.

There are, therefore, from the linguistic point of view, three glottal features: speech sounds have to be specified in terms of (1) pitch, (2) glottal stricture, and (3) voice onset time. But there is no reason to believe that these three glottal features are independent. It is clear that there is an interaction between the value of the voice onset feature and the glottal stricture feature, in that only a limited range of phonation types can be potentially aspirated or unaspirated. Voiceless sounds and sounds accompanied by a glottal closure may, or may not, be followed by a period of voicelessness (aspiration); but laryngealized sounds, and sounds with regular voicing are, to the best of my knowledge, never followed by a period of voicelessness before regular voicing starts. However, the voice onset and the glottal stricture features cannot be collapsed into one feature because sounds with several different phonation types (e.g. glottal stops, murmured stops, and voiceless stops) can all be aspirated.

There is also an interaction between the value on the pitch feature and that on the glottal stricture feature. We have already noted that laryngealized sounds tend to have a lower pitch, because they are the result of a glottal state in which the arytenoid cartilages are not only held together but also moved forward, so that the vocal folds become less stretched. Conversely, after a voiceless sound in which there is a high rate of flow, the pitch tends to be higher because of the increased Bernouilli effect. We must emphasize, however, that there is only a tendency towards a correlation between the rate of vibration of the vocal folds and the degree of glottal stricture. The glottal stricture and the pitch of a sound can be varied independently in many Nilotic languages, so at least in these languages (and therefore within the theory of general phonetics) these two features cannot be collapsed into one.

LINGUISTIC IMPLICATIONS

The foregoing discussion seems to indicate quite conclusively that we need at least three laryngeal features in linguistic description; and I would not be at all surprised to find that we need even more. There are some states of the glottis (such as whisper) which cannot be characterized in terms of a single variable such as distance between the arytenoids; and many authors (cf. Fromkin, 1971) have shown that a single feature, pitch, is not sufficient to account for the contrasts that occur in tone languages.

Halle and Stevens (1971) have suggested that all the phenomena we have been discussing can be described in terms of only two variables—the degree of stiffness of the vocal folds, and the degree of constriction of the glottis. (They in fact characterize each of these variables in terms of two binary oppositions: [± stiff], [± slack]; and [± spread], [± constricted], with a convention that neither [+ stiff, + slack] nor [+ spread, + constricted] can co-occur. They are thus expressing two ternary oppositions in terms of four binary oppositions; but this idiosyncrasy is irrelevant to the present discussion). Halle and Stevens use their features to capture a number of interesting linguistic generalizations. For example, they claim that they are able to formalize the reason for the occurrence of alternations in which a vowel occurs with a high tone after voiceless stops but with a lower tone after voiced stops. In their view, this is just an assimilation in which the vowel is taking on the stiffness value of the consonant. High tone vowels have stiff vocal folds, and so (they claim) do voiceless consonants; but low tone vowels have slack vocal folds, and (in their view) so do voiced consonants.

Halle and Stevens do not produce any evidence for their claim that the pitch tends to rise after voiceless consonants because voiceless consonants have stiff vocal folds. Personally, I would rather assume there is no special stiffening in voiceless consonants; it seems to be an unmotivated and, indeed, unlikely additional complication. There are two other reasons why there may be an increase in pitch after voiceless consonants. First, voiceless consonants may have a higher larynx position than the corresponding voiced consonant. As we noted earlier, this may result in stretching of the vocal folds; but it seems curious to call this stretching a stiffening of the vocal folds. Secondly, the increase in pitch may be due to the increase in the air-flow rate that is associated with voiceless sounds. Probably the best way of dealing with this phenomenon in phonological theory is by means of marking conventions, rather than by combining the feature specifying the degree of glottal stricture.

Furthermore, as many of the examples discussed above have shown, a correct feature system must make it possible to separate out differences in phonation type, which are due to differences in glottal stricture, from differences in pitch, which depend on another mechanism. It must be possible to show that voiced sounds, breathy voiced sounds, and creaky voiced sounds can all occur over a range of different pitches. In addition, I cannot see any reason for dispensing with any simple way of specifying voiced and voiceless sounds (which, under the Halle-Stevens proposal, are specified in terms of differences in stiffness-slackness, and are inextricably confused with pitch variations). As has been shown by Fromkin (1971), the use of a feature system of this kind results in some ridiculously unnatural rules. It seems clear that these must be at least three multivalued glottal features, not just two ternary oppositions.

REFERENCES

Allen, G. (1971) The location of rhythmic stress beats in English: An experimental study, *Language and Speech.* [In press.]

Chomsky, N., and Halle, M. (1968) *The Sound Pattern of English,* New York.

Fromkin, Victoria (1971) Tonal features and rules and generative phonology, *Working Papers in Phonetics, 21.* [In press.]

Halle, M., and Stevens, K. N. (1971) A note on laryngeal features, *Quarterly Progress Report of the Research Laboratory of Electronics, M.I.T.,* **101,** 198–213.

Kagaya, R. (1971) Laryngeal gestures in Korean stop consonants, *Bulletin of the Research Institute of Logopedics and Phoniatrics, University of Tokyo,* **5,** 15–24.

Kozhevnikov, V. A., and Chistovich, L. A. (1965) *Speech: Articulation and Perception,* U.S. Dept. of Commerce translation, JPRS, **30,** 543.

Ladefoged, P. (1967) *Three Areas of Experimental Phonetics,* London.

Ladefoged, P. (1971) *Preliminaries to Linguistic Phonetics,* Chicago.

Ladefoged, P., Draper, M. H., and Whitteridge, D. (1958) Syllables and stress, *Miscellanea Phonetica,* **3,** 1–14.

Lee, C., and Smith, T. (1971) Oral and direct subglottal pressure in Korean stops, *J. acoust. Soc. Amer.* [In press.]

Lieberman, P. (1967) *Intonation, Perception and Language,* Cambridge, Mass.

Lisker, L., and Abramson, A. (1971). Distinctive features and laryngeal control, *Language,* **47,** 767–85.

Netsell, R. (1969) Subglottal and intraoral air pressures during the intervocalic contrast of /t/ and /d/, *Phonetica (Basel.),* **20,** 68–73.

Netsell, R. (1970) Underlying physiological mechanisms of syllable stress, *J. acoust. Soc. Amer.,* **470,** 103–4.

Ohala, J., and Ladefoged, P. (1970) Further investigation of pitch regulation in speech, *Working Papers in Phonetics,* **14,** 12–24.

DISCUSSION ON COMMUNICATION [18]

Klatt. You suggested that stress would have the physiological correlate of increased subglottic pressure on stressed syllables. For example, in one of the sentence examples you gave, the subglottic pressure in '*dead*' might be greater than it is in '*head*'. Do you have any data that would demonstrate this type of relationship?

Ladefoged. This (showing slide—Fig. D.13) is an early picture of the activity recorded while saying '*the old man doddered along the road*', where we tried to quantify the impulse rate. We were fortunate enough to get a single motor unit in the internal intercostals (recorded in trace [2]); and line [4] is the impulse rate (or the reciprocal of the time difference between motor unit potentials). It does seem here as if there is a burst of activity corresponding to the stresses, at least as far as the electrical activity in the muscles is concerned. As far as the sŭbglottic pressure is concerned I haven't got a specific slide here; but I found (absolutely consistently) that on every single occasion when one would say that there is a

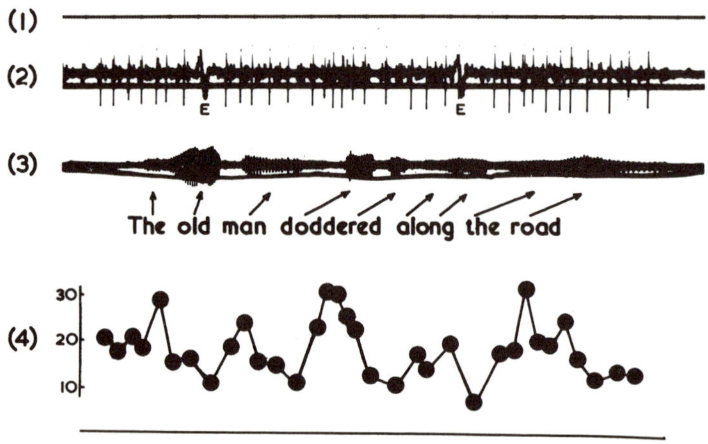

Fig. D.13. Description in text.

primary stressed syllable or a secondary stressed syllable (or what I would call an accented syllable) there is an increase (beyond the mean of what one would expect) in the subglottic pressure. We have quite a lot of published records showing increases in subglottic pressure for all stressed syllables, so I would say there is a very strong physiological correlate to accent in English; namely, subglottic pressure.

Klatt. Just to clarify it, how many stressed syllables would there be in a sentence like that?

Ladefoged. It depends how you say it. You could say it slowly (demonstrates by tapping), or you could say it more quickly (demonstrates by tapping). I think, roughly speaking, one would get an accent for each tap.

Winckel. About the timing process. You mentioned piano-playing of notes at 8 per second, and then you said that in your sentence 8 per second was the maximum rate of utterance, did you not?

Ladefoged. 8 to 9 per second, yes.

Winckel. But I know that one can play the piano, or typewrite, up to a frequency of 12 per second. Do you perhaps think that when we do this sort of fingering it is programmed by a pattern-like processing control? Do you think that with continuous conscious control one can get up only to 8 per second rather than 12 per second—because I know it is possible, as we have measured it with typewriting.

Ladefoged. With a single finger? You can type twelve times per second with a single finger? Of course, if you are using two fingers the situation is very different!

Winckel. I agree. I think there is a difference, using two fingers.

Ladefoged. I think Dr. Wyke knows something about this.

Wyke. To get the blame where it properly belongs in reference to the statement that the maximum rate of piano playing is 8 per second, this was, in fact, a personal communication to Dr. Ladefoged from me, so I am responsible for it. The figure is based on the only (unpublished) study I know of that has actually measured the rate of change in motor unit activity in the muscles controlling the fingers in these circumstances; but of course it refers to the maximum rate at which the activity can be changed in a single muscle by an experienced pianist while operating with just a single finger. Can I emphasize that what Dr. Ladefoged said was not that one can play piano notes 8 times per second, but that one could move a finger up and down 8 times per second on a single piano key—which is about the fastest touch rate at which a skilled pianist can go, as far as I know. I want to ask you, Dr. Winckel, do you have any data that would push that figure up to 12 times per second, because I am not aware of the fact?

Winckel. That is what I wanted to say before. There is a fixed programme (as in a processing control system); and this fixed programme operates faster—as in the situation you are describing now—with one muscle, and can really go up to 12 per second. We have a lot of examples indicating that this is the real limit of what is possible. But returning to Dr. Ladefoged's sentence, is this an analogue of what we have been discussing, in that you have here the same muscles operating in this phonetically constructed sentence, and the situation would be different for other sentences with other phonemes?

Ladefoged. That is correct. I think that the fastest rates that we have been able to observe have been with tip-of-the-tongue consonants, such as /*d*/. There are additional data in the literature that indicate that back-of-the-tongue consonants, or lip consonants, are slightly slower—so I guess it does depend on what specific sounds you choose to use.

Winckel. I think you have made more of these experiments with other sentences that are constructed with other combinations of phonemes—have you the timing for these other sentences? This would be very interesting, because it seems that 8 to 9 per second is the maximum for this particular sentence, and other sentences and other phonemic combinations are perhaps less slow—but I don't know.

Ladefoged. We are working on this. We are measuring the degree of difficulty of articulation by measuring the length of time it takes to say things. Let me quote one of the sentences; namely—'*She sells sea shells on the sea shore*'. This is also a tongue-tip action, and the maximum frequency is also of the same order of magnitude as in the example I gave earlier.

Godfrey. First of all, I want to say how fascinating I found your talk. Now may I just inject a slightly perverse note? In the electromyograms that you related to the speech sounds, have you any evidence of which comes first? Could not the electromyographic change be a reflex response to the interruption of air flow—in a manner similar to what happens to the muscles when you put positive end-expired pressures on, and that sort of thing? Is there any evidence that the muscle changes occur before the speech changes, or is it the other way round?

Ladefoged. People will keep asking just the right questions! The direct answer is no, I haven't any such evidence—but let me word the problem in a slightly different way. Here (drawing on the blackboard) is a very schematic representation of somebody saying '*Say bat now*' and '*Say hat now*', and obviously the only difference is between the '*b*' and the '*h*' in the two sentences. When we recorded the electromyographic activity, we observed that you got a certain amount of activity on the stressed syllable in 'Say *b*at now' but that you got a lot more for the '*h*' in 'Say *h*at now'—where the lungs, of course, collapse at a great rate because the air comes out.

Now the question to ask, if I understand you correctly (and I want to know the answer too) is—to put it anthropomorphically—does the brain say 'I've got an '*h*' coming, I must get ready for it by pulling out a lot of internal intercostal activity to put the pressure up ahead of time'; or is it instead, 'My God, I've lost a lot of air, I must do something about that', and it's a reflex where the internal intercostals come in afterwards. We now have sufficient programming and computer technology to process our electromyographic records to answer that question—that is, does it occur just before the '*h*' or just after the '*h*'? The only thing I haven't got now is the electromyographic set-up (having left Professors Whitteridge and Draper) to do work on the internal intercostals; so if anybody is currently working on the electromyographic activity of the internal intercostals, this is an experiment I would very much like to get involved in.

Stevens. You mentioned that the control of fundamental frequency was primarily a stretching of the a-a^1 axis. Isn't it also possible that one could control the fundamental frequency by contracting the vocalis muscle, and thereby change the coupling stiffness in the diagram (it shows coupling between the upper and lower edges)? Contraction of that muscle would tend, in fact, to shorten the vocal folds. Is that a possible additional mechanism, or an independent one?

Ladefoged. I agree with you. We must also do more work to find out the relation between the two. The kinds of pictures that I have shown are very typical of people going up in pitch. When they go up in pitch, generally they do so by stretching in the a-a^1 direction. Their vocal folds, of course, also become thinner because when you stretch them out in that direction the mass gets thinner—and in that sense it may affect the stiffness. I just don't know the relative importance of these changes. I think I am biased because I happen to have looked at lots of pictures of the glottis taken from above: but from looking at pictures in the opposite direction, or from X-ray studies, one might get a different view of the situation.

Wyke. I want to respond briefly to the point that was made about the relative time relations between the electromyographic activity and what one might call the pneumatic events associated with speech. You will doubtless recall that there are a number of scattered observations in the literature[1,2,3,4,5,6] indicating that as far as the intrinsic laryngeal muscles are concerned (I am not so sure about the data on the intercostals[4,7,8,9]) there is reasonably good evidence, with which my own observations would agree, that during phonation there is an increase in motor unit activity (in the vocal fold adductors, for example) that precedes the actual emission of sound—and which is correlated with the initial oscillation of subglottic pressure that I think you showed some time ago[7,9]. So while we may not yet know exactly what is happening in the intercostals in regard to this relationship, I think[10] the phonatory behaviour of the intrinsic laryngeal muscles would make it probable that in speech muscle activity in general antedates the pneumatic events, rather than the other way round. In other words, I feel that what you said first is right: namely, that the brain decides that there is an '*h*' coming and sets the muscular programme for that, and then acoustically monitors the uttered end-result to see whether it had got it or not[10].

Ladefoged. That's my guess too—that the activity is pre-programmed, rather than the result of feed-back.

Malcolm. These slow vibrations are very interesting. I wonder if it is possible that the inter-arytenoid distance might be varied periodically, say at about 7 Hz. Second, I wonder whether it is possible that the arytenoids themselves might be slightly out of phase at that period in view of the different lengths of the two recurrent nerves—which might introduce an asymmetrical delay in the production of their movements.

Ladefoged. I think, in fact, that I suggested yesterday [see p. 262] that the mechanism for *vibrato* is possibly a variation in the inter-arytenoid distance—and it is, as we were saying yesterday, of the order of 7 Hz. So you may be right. I don't know about the phase possibilities.

Cavagna. About Dr. Stevens' question. It may be that the relationship

---

[1] Faaborg-Andersen, K., and Buchthal, F. (1956) Action potentials from internal laryngeal muscles during phonation, *Nature (Lond.)*, **177**, 340–1.

[2] Faaborg-Andersen, K. (1957) Electromyographic investigation of intrinsic laryngeal muscles in humans, *Acta physiol. scand.*, **41** (Suppl. 140), 1–149.

[3] Buchthal, F. (1959) Electromyography of intrinsic laryngeal muscles, *Quart. J. exp. Physiol.*, **44**, 137–48.

[4] Buchthal, F., and Faaborg-Andersen, K. (1964) Electromyography of laryngeal and respiratory muscles: correlation with phonation and respiration, *Ann. Otol. (St. Louis)*, **73**, 118–23.

[5] Faaborg-Andersen, K. (1965) *Electromyography of Laryngeal Muscles in Humans: Technics and Results*, New York.

[6] Sutton, D. Larson, C. R., and Farrell, D. M. (1972) Cricothyroid motor units, *Acta oto-laryng. (Stockh.)*, **74**, 145–151.

[7] Ladefoged, P. (1960) The regulation of subglottal pressure, *Folia phoniat. (Basel)*, **12**, 169–75.

[8] Sears, T. A., and Newsom Davis, J. (1968) The control of respiratory muscles during voluntary breathing, *Ann. N.Y. Acad. Sci.*, **155**, 183–90.

[9] Ladefoged, P. (1968) Linguistic aspects of respiratory phenomena, *Ann. N.Y. Acad. Sci.*, **155**, 141–50.

[10] Wyke, B. D. (1967) Recent advances in the neurology of phonation: phonatory reflex mechanisms in the larynx, *Brit. J. Disor. Commun.*, **2**, 2–14.

between the lengths of the vocal folds and their tension is influenced simply by the isometric length/tension relationship of the vocalis muscle. Maybe the greater the length, the greater is the possibility for the vocal fold musculature to develop a greater tension.

I also want to ask you a question about the 'creaky' voice. I would think that when we press the vocal folds together in order to have minimal flow the problem arises of pushing them apart, particularly at low intensity, with a low subglottic pressure; but now you tell me that there is a possibility of pulling the vocal folds still more tightly against each other, and that that produces a creaky voice. I want to ask you, how free are you to change frequency with this creaky voice, and to go down to a low intensity level?

Ladefoged. I will show you that it is possible to go up and down over a fair range. [*Editor's note:* Dr. Ladefoged demonstrated frequency shifts with a creaky voice at this point.] I don't know what you are going to do about variation in intensity, I must admit—I just don't know how much I can vary that.

Cavagna. There was a big glide there. Do you know the glottal resistance and the subglottic pressure in these conditions you just demonstrated?

Ladefoged. I think there are measurements. Perhaps Dr. Stevens can tell us better about them. The pressure is substantial, as I think Dr. Klatt was saying earlier on. This pressure (apart from the extra pulses related to stressed syllables) is kept fairly constant.

Kirchner. In response to several of the questions that have come up about the lengths of the vocal folds—the a-a^1 business—isn't it true that in cadaver larynx specimens removed from the human, when air was passed through subglottically, the pitch could be varied directly as a function of the lengths of the folds simply by rocking the cricoid on the thyroid cartilage to stretch the folds? This suggests that the frequency of the emitted sound may vary according to the lengths of the vocal folds, even though there is no muscle activity.

Ladefoged. Yes—but normally there is muscle activity that rocks the cartilages on one another. I agree that that is a kind of answer to Dr. Stevens' question, and suggests that maybe the action of the actual intrinsic muscles is not that important—but I don't want to extrapolate too much from dead bodies to live ones. I know Mr. Anthony has worked on cadaver larynges more than I have, and can probably comment on this point.

Anthony. The problem is that the vocalis muscle of the cadaver larynx has a constant mass and its own constant tension. If you just treat it as a physical device and pull on the thyroid cartilage to stretch the vocal folds, then the pitch goes up very definitely with the pull. But, of course, it also goes up with an increase in pressure.

Proctor. I think it should be pointed out that you can't lengthen the vocal folds in the way you are talking about without changing their tension and mass. You are bound to thin them and increase their tension at the same time as you lengthen them.

Stevens. Just a comment. These cadaver experiments do not necessarily show that the vocalis muscle does not play a significant rôle in the live larynx.

Ladefoged. I agree entirely. I said I did not want to generalize from them.

Stevens. You are suggesting here that one of the variables over which one has control is the time from the release of certain consonants to the onset of voicing, as though one is saying to oneself 'I will release the supralaryngeal constriction, and then I will wait a few tens of milliseconds before I turn the vocal fold muscles on'. I just wonder could that not be an automatic consequence of the fact that the vocal folds begin in an abducted position—as, let's say, with the */p/* sound. At the release of the consonant the vocal folds are abducted, and it just takes a certain amount of time for them to close after the release. Now this, of course, would mean that there has to be some sort of reflex action: in other words, when the mouth is released and air flow starts through the glottis a reflex would have to be initiated at that instant to begin the closure. My question to you, or to others, is whether such a reflex is possible?

Ladefoged. You have also to take into account the fact that consonants can be 'breathy voiced', so to speak. You get the '*bha*' type of thing, with a certain length of time of breath-voiced aspiration before the regular voicing starts; and any mechanism we suggest to account for the length of time between the opening of the lips and the start of regular voicing has to apply to that kind of thing, as well as to the more usual kind.

Wyke. I agree with Dr. Stevens in thinking that there is probably a form of significant reflex relationship here. But I think there are other possibilities that we have to keep in mind—one of which is that the conduction times from the command centre in the brain (wherever it happens to be) to the motoneurone pools controlling the muscles of the lips and to those controlling the muscles of the larynx must be significantly different, simply in terms of the differences in the anatomical geometry of the two pathways. Furthermore, if one were to look at the fibre diameter spectra in these two pathways I think it would be quite possible, with a little arithmetic, to calculate whether the timing difference might also in some degree reflect relative differences in the conduction velocities. But I don't think we have the data to decide between these various possibilities yet*—although I must confess that I incline intuitively to a more specific form of reflex adjustment.

Stevens. The reflex could be triggered by the release of the lips; or it could be triggered secondarily by the sudden increase in air flow occasioned by the release of the lips—that is, by the air flow through the glottal opening. In breathy-voiced and voiceless consonants you get rapid air flow—and this might stimulate receptors somewhere in the airway.

---

* *Editor's Note:* In a paper that appeared after the Symposium (Leanderson, R., and Lindblom, B. E. F. (1972) Muscle activation for labial speech gestures: an electromyographic and acoustic study of normal speakers, *Acta oto-laryng. (Stockh.)*, **73,** 362–73), prephonatory activity was identified in the lip musculature in association with the utterance of certain sounds.

Ladefoged. The problem is that perhaps I spoke as if one couldn't vary that length of time—but one can; and it can't just be a simple reflex, because one can make a fair number of steps in between those sounds. One can learn to do that—and I think Dr. Stevens' earlier comment may be right. Chin Kim[1] suggested that perhaps one does that trick by setting the vocal folds at a certain degree of apartness during the time the lips are closed, so that when the lips open they take longer to come into vibration than if they have been set at a further distance apart. Maybe that is what one is doing; but certainly one is doing some deliberate trick of muscular adjustment, rather than something to be explained simply in terms of reflexes.

Anthony. I have been doing some work with the deaf, in which we give them a visual presentation of what is happening—in this case, information just that the larynx is vibrating or not vibrating—and the two subjects that we have tested so far seem to have reasonable control over their voicing quotient time. The great difficulty is in keeping the voicing going right through, unfortunately.

I wonder, when you were doing the full vital capacity business, when these muscles came into action. When you got down to FRC, then the internal intercostals came in; but at what point in the vital manoeuvre did the next set of muscles come in?

Ladefoged. I can't recall very clearly, and I am not absolutely sure of the measurements that we did—they were pretty sloppy. I wouldn't like to put an awful lot of faith in their accuracy.

Winckel. Could I come back again to the time process. There is the other example of the 'trill', as well as the *vibrato.* I suggested yesterday (but I am not sure that it is right) that the trill is an enforced vibration—and this could be the same as with the finger. This, I know, is 11 per second.

Ladefoged. If you are talking about a '*rrrrrrr*', that is about 25 per second and, of course, is simply an aerodynamic effect. If I get my tongue in a certain position, I can set it trilling with the airstream going over it.

Winckel. But a trill is not just an aerodynamic effect. I think it is a combination of two different effects—one aerodynamic, and the other involving conscious muscular control.

Freedman. Just a point of clarification. Where do we think the rate-limiting event is in the inability to do fast alternating movements—in conscious control, neural conduction, change of muscle membrane potential, transduction of this into muscle movement, or in the mechanical linkages of the parts we are moving. I would put my bet on the last one, but I wonder what other people feel.

Ladefoged. As Dr. Klatt said at the end of the last session, that is what we are here to learn!

Fry. I think that that is as good a note as any on which to end this particular discussion.

---

[1] Kim, C. (1970) A theory of aspiration, *Phonetica (Basel),* **21,** 107–16.

19 Laryngographic Examination of Vocal Fold Vibration

A. J. FOURCIN

Department of Phonetics and Linguistics,
University College London, England.

The work described in this Communication involves the use of a new development (Fourcin and Abberton, 1971) of an electrical impedance technique [see Fabre, 1957] for the direct examination of vocal fold closure, which does not interfere with phonation. Our original aim, however, was to provide a simpler, and yet more accurate and instantaneous, basis for the determination of voicing frequency than is available when the acoustic waveform of speech itself is used. The difficulty of disentangling the fundamental excitation frequency from the vocal tract response components of the normal speech waveform has been largely resolved by the use of cepstral techniques (Noll, 1967); but the processing involved is complex and necessarily slow, response is impossible within the time of a single larynx vibration, and only an average value for the period of a complete vocal fold vibration can be obtained. With information derived directly from the motion of the vocal folds themselves, the computation is simpler and can be accomplished in real time. The particular feature of the electrical impendance approach that is discussed here results, however, from the observation that its output contains information principally about the form of contact between the vibrating folds and gives almost no direct indication of glottal area variation.

Direct and indirect laryngoscopy both provide information which is immediately related to vocal fold movement, and although particularly relevant to the description of the supraglottal view of their separation, they also make it possible to infer some aspects of the nature of the contact during the closure phase. However, only ciné-strobo-laminagraphy (Hollien, 1962) provides an established method of examining the nature of the contact phase; and even this technique is only suitable for sustained phonation, rather than normal speech.

There are three other methods of essentially instantaneous examination of vocal fold movement—subglottal illumination (Sonesson, 1960), together with supraglottal illumination (Lisker *et alii*, 1969); inverse filtering of the acoustic speech waveform (Miller, 1959); and ultrasonic detection (Asano, 1968). They all, however, give information primarily about the separation of the vocal folds and very little indication of the nature of their contact.

Fabre devised his glottograph on the basis of his earlier work on the recording of arterial blood flow (Fabre, 1940), where he detected the dimensional variation that accompanies the pulsatile dilatation of an artery by measuring the accompanying electrical impedance variation. The electrical measuring circuit applied to the larynx involves the application of a constant current (150 KHz) source via two electrodes which are positioned superficially on the wings of the thyroid cartilage. The output is derived by the demodulation of the amplitude-modulated voltage variations that appear between the electrodes as the electrical impedance of the circuit is altered by the changing configuration of the vocal folds. Fabre (1957) described the waveform of this output voltage as being a function of the spacing of the vocal folds.

Although this apparatus arrangement is straightforward, its practical employment has not been widespread because the output appears to be unstable—being variable not only from subject to subject, but also variable with time for a particular speaker. These difficulties also occur with other versions of Fabre's device: even though matching to the individual speaker is employed, it is not satisfactory for all speaking conditions. An arrangement is required that reduces the variability of the output for a given speaker and allows for automatic compensation for interspeaker variation.

Our own work (Fourcin and West, 1972) on this approach involved first the study of the electrical impedance variations which were likely to be encountered in practical situations, and the subsequent development of an arrangement that was electrically capable of compensating at least in part for the large impedance variations which can occur between one subject and another and for the slow impedance changes that occur in a particular subject during the course of speaking, and which result primarily from the continuous laryngeal adjustment that takes place as the speaker monitors and controls his voice output. The intersubject variation has been reduced by the use of a constant voltage source rather than a constant current or bridge output as the input to the subject's neck, and by detecting the current output from the neck in a low impedance circuit. Both input and output are now at a low impedance level and the electrodes can be given guard rings, at earth potential, to reduce surface conduction across the skin. The slow impedance variations that occur during normal speech can, in principle, be removed simply by filtering—but they are very much larger in their effect on the output current amplitude than the variations due to vocal fold vibration, and this can make impossible demands on the dynamic range of the detecting circuits. This difficulty has been reduced by the use of a feed-back circuit which operates on the carrier level itself rather than on its modulating envelope. An example of the type of electrode used is shown in Fig. 19.1: the guard rings are connected together to provide a three pole connexion to the neck.

This electrical arrangement proved to be quite satisfactory, and could be employed for new-born babies and fat or loose-necked subjects as well as normal adults, as a basis for larynx frequency measurement. The relation between its output and the corresponding vocal fold movement, however, was largely unknown. Informal observation confirmed the oscillographic evidence (from Fant *et alii,* 1966) that there was an antiphase relation between the output of the electrical impedance device on the one hand and those provided from a photoelectric glottograph and inverse filtering on the other. The beginning of improved

Fig. 19.1. Laryngograph electrodes. These are held in place by an elastic band around the speaker's throat, on either side of his larynx. The two outer guard rings are directly electrically connected. The two inner electrodes provide a low impedance voltage source output (3 V rms., 10^6 Hz), and the input to a low impedance current detector. Smaller electrodes of less elongated form are used for babies; and larger, more elongated electrodes are used for large necks where subcutaneous fat provides a short-circuiting path and makes precise positioning difficult.

understanding of the relation, however, seemed best obtainable by the direct visual comparison of vocal fold movement with the output waveform from the electrical impedance arrangement.

A separate study has been made of this, and one direct comparison is shown in Fig. 19.2.* The antiphase relation between the curves is clearly apparent. The results of a similar experiment performed elsewhere have been published (Michel

*I am grateful to my co-workers, Rupert Donovan and Peter Roach, for permission to publish this Figure.

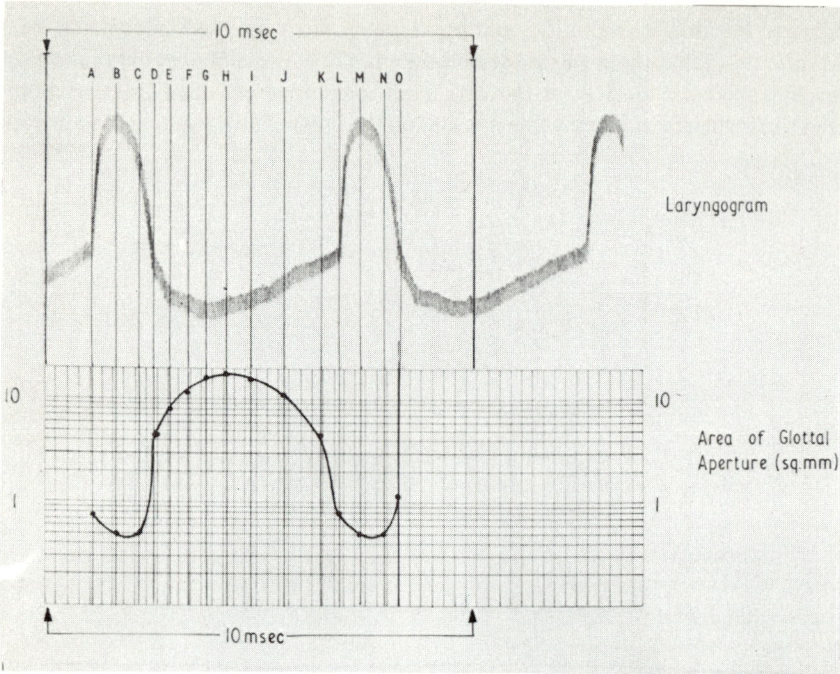

Fig. 19.2. Comparison of the laryngograph output with the glottal area curve. A recording was made of the laryngograph output (Lx) while the subject's vocal folds were photographed stroboscopically [see opposite page]. The stroboscope light source was synchronized with the laryngograph output and the ordinary ciné-camera such that the photographs were made, in the order displayed on page 319, on successive film frames. In this particular example the subject sustained a breathy vowel [*a*]. The anterior ends of the folds are slightly obscured by the epiglottis, but the antiphase relation between Lx and the glottal area is quite clear, and typical of all our results.

et alii, 1970), and show an essentially in-phase relation between the output of the glottograph and the glottal aperture area. It is quite clear, however, from Fig. 19.2 and from our other stroboscopic observations on the same subject and on three others, that the electrical output is only really significant during the period of vocal fold closure—and indeed, this result springs from the basis of its operation. In consequence, it is better not to call the instrument a 'glottograph', since glottal area variations do not contribute substantially to its output. For this reason the apparatus which we have developed is called a 'laryngograph'; and we refer to its output waveform, positive-going for increasing glottal closure, as 'Lx'.

Typical Lx waveforms for a normal adult male are shown in Fig. 19.3, accompanied by their related speech pressure waveforms. In all instances, the beginning of closure in the Lx waveform is followed by the onset of maximum acoustic response from the vocal tract—which occurs during closure. This relation

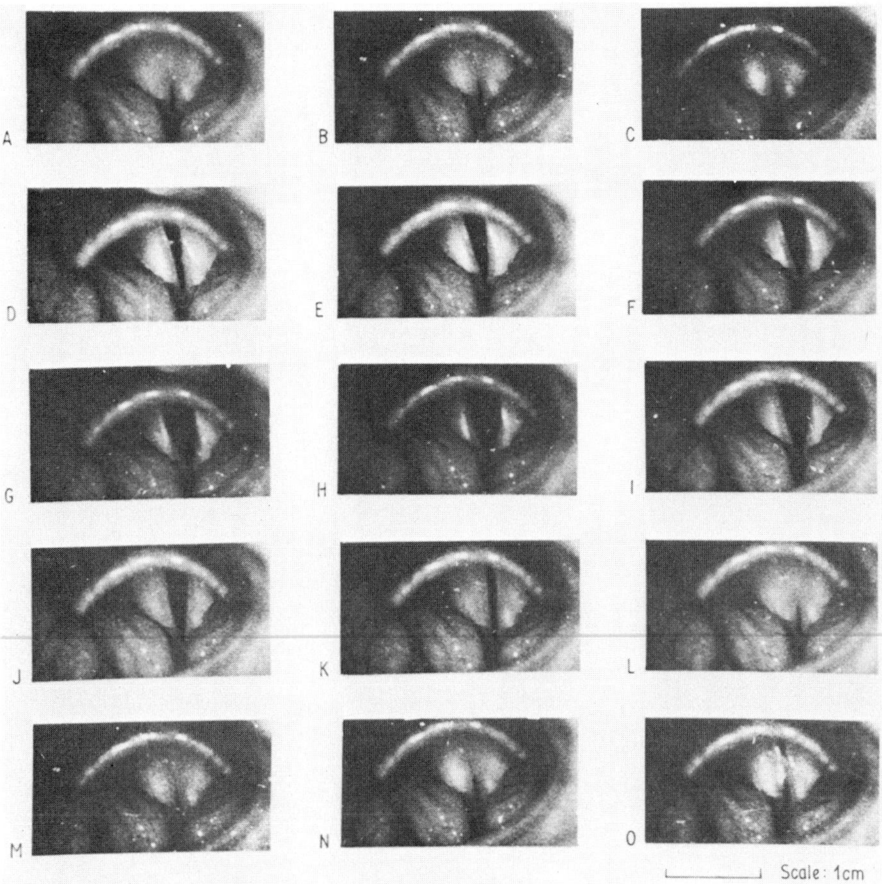

Fig. 19.2. (*Contd.*)

between glottal closure and the peak of acoustic response is also obtained from comparison of the volume velocity and speech pressure waveforms (Holmes, 1962). The opening phase in Lx for chest voice occurs more gradually than the closing phase and, although the acoustic output can show the increased decrement resulting from the coupling of the acoustic loss from the subglottal system, the slow change often obscures its introduction. Chest voice waveforms, from the same speaker as in Fig. 19.3, for two different sustained vowels are shown in Fig. 19.4(a) and (b), but with an enlarged time scale. The speech waveforms were recorded with an extremely small microphone which could be put at the speaker's lips, and the time between Lx closure onset and the onset of maximum acoustic output is a function only of the speaker's vocal tract length configuration (this relation allows the microphone distance from the speaker's lips that was employed by Michel *et alii* (1970) to be calculated as 30 cm).

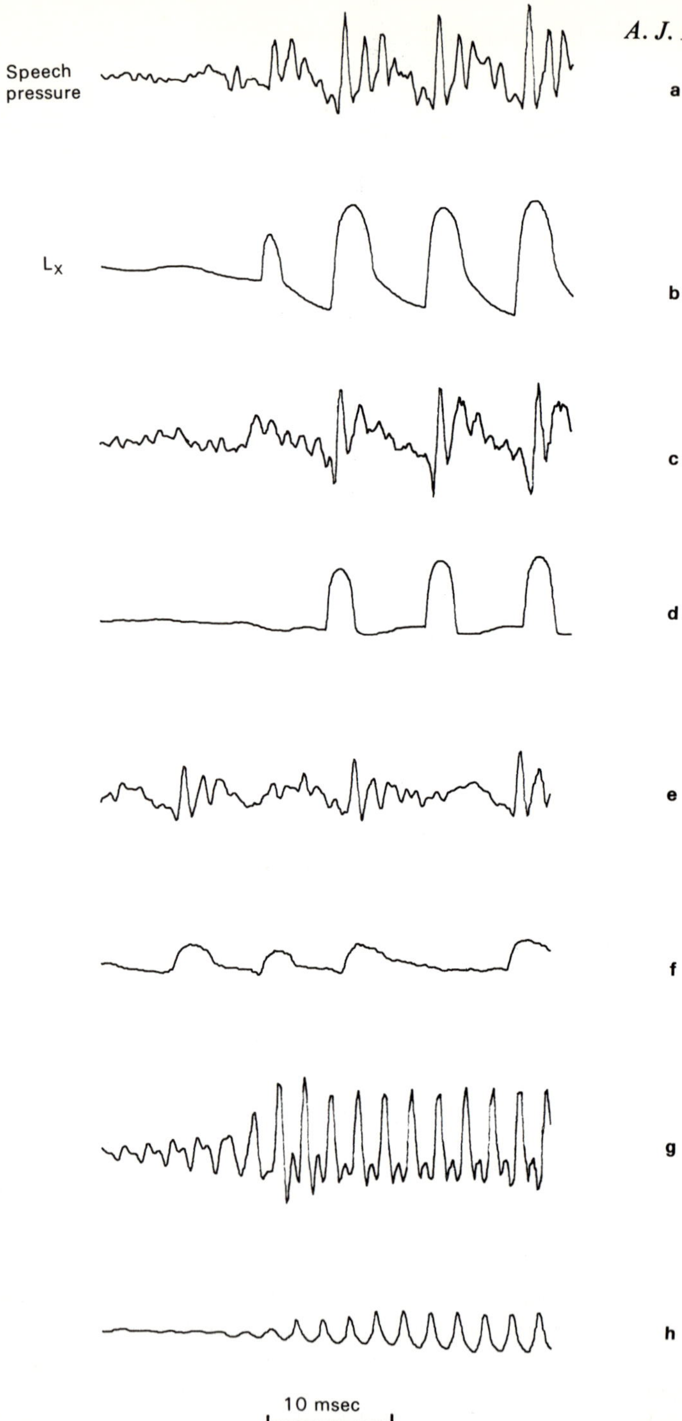

Fig. 19.3.

FORM OF Gx

The large changes that can take place in the output of the laryngograph when no feed-back is employed (this output is called Gx) are shown in Fig. 19.4(e). The output with feed-back applied to counter the slow variations is shown in Fig. 19.4(f). The Gx output is primarily a function of laryngeal adjustment prior to, during and after actual phonation; but other activity in the region of the larynx can contribute to it—including swallowing, of course, and movements of the neck and head. Initial vocal fold adduction is associated with a small, sharp, precursive positive peak in Gx.

FORM OF Lx

The main contribution to the detailed form of Lx comes from the area and nature of the electrically conducting surfaces that are in contact during vocal fold adduction. In consequence, if there is a change in area, or in the nature of a contacting area, from one cycle of vibration to another this will be shown by a corresponding change in Lx amplitude. If, in an extreme case, there is failure to close the folds together, there will be no corresponding peak in Lx.

Figure 19.5(a) shows a particular example of Lx for a new-born baby's cry (recorded with Mrs. Ann Parker of University College Hospital) which, in our experience, is typical of neonatal laryngeal activity. It seems possible that the triplet sequence of closures indicates a progressively reducing area of contact, and this may be due to asymmetric tensioning of the child's vocal folds. A similar type of Lx activity is shown in Fig. 19.5(b), where the Lx recordings were obtained from a 72-year-old man with unilateral vocal fold paralysis. In both of

---

Fig. 19.3. Four pairs of waveforms are shown. The top member of each pair is a speech pressure waveform; the lower is the corresponding, simultaneously recorded, Lx. Positive-going Lx corresponds to increasing vocal fold closure. All pairs come from the same normal adult male producing the vowel [*a*], and the waveforms are taken from the beginning of each utterance. The beginning of vocal fold closure always corresponds to the initiation of the peak of acoustic response from the vocal tract. An Lx output occurs only when the folds are actually in contact, so vocal response always starts before appreciable Lx activity can be seen.

(a) and (b) Chest voice. The closing and opening of the vocal folds occurs consistently, and the peak amplitude of Lx is soon stabilized. There is a relatively large closed-to-open phase ratio once phonation has begun, and the opening of the folds is not as well defined as the onset of closure.

(c) and (d) Breathy voice. Both opening and closing of the vocal folds are well defined, and the closed-to-open phase ratio is relatively small. The form of fold closure appears to be simpler than for chest voice: the vibration, once initiated, is as consistent.

(e) and (f) Creaky voice. The basic pattern of Lx for this voice quality indicates a long closure duration only moderately regularly repeated—and apparently with a small degree of closure—superimposed on which is a shorter period vibration of variable occurrence. The onset of closure is always sharp and opening always slow.

(g) and (h) *Falsetto* voice. The degree of closure is small, and closing and opening occur with equal rapidity. The shape of the Lx waveform is consistent with the time intervals for closing and opening being comparable with the total time of close contact.

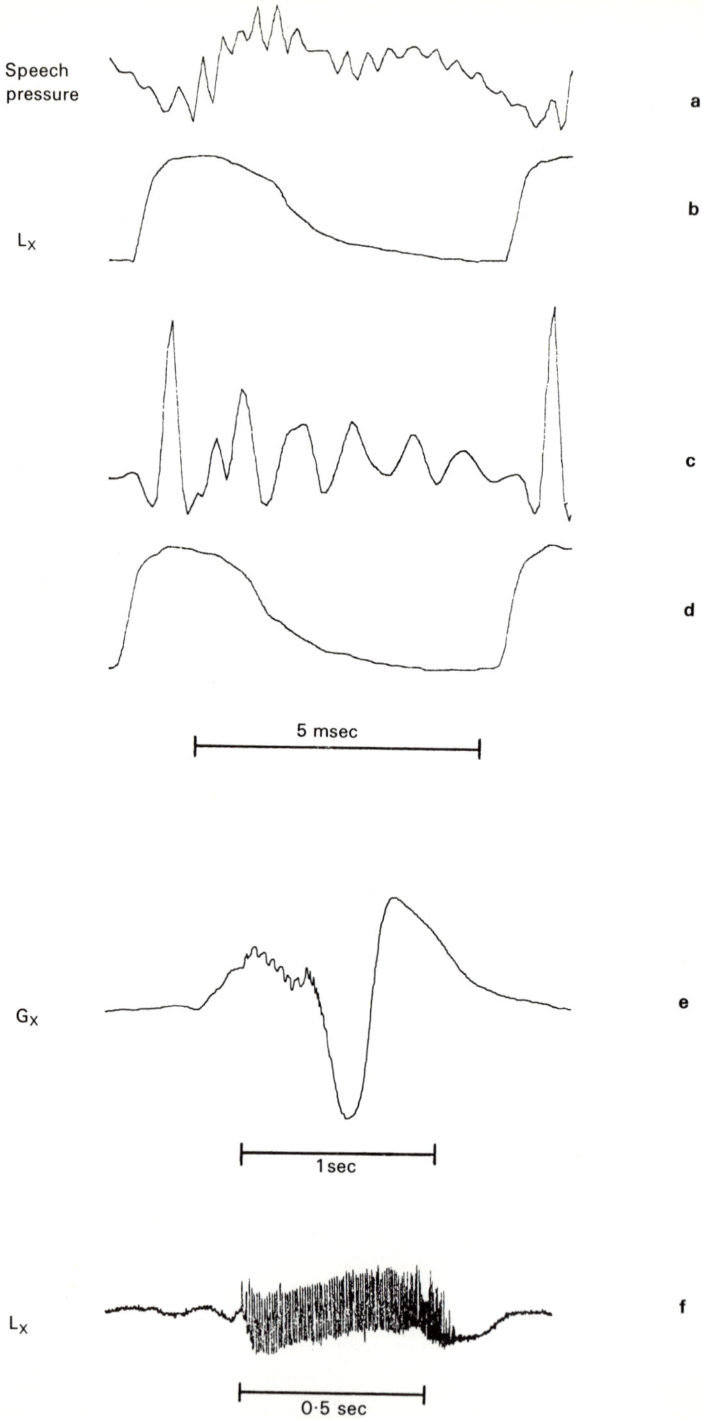

Fig. 19.4.

these examples, asymmetric fold tensioning could result in the two folds of each individual having different natural frequencies. Synchronization takes place every third cycle when relatively violent closure occurs, and the consequent energy loss resets the two vibrating fold systems.

As different intonation patterns are attempted with unilateral paralysis, quite different Lx waveforms can be obtained, however [Fig. 19.5(c)]. The roughness of voice quality found with unilateral paralysis can also occur with other pathological conditions of the vocal folds, but their influence on the form of Lx can be quite different from one another. Two particular examples of this from two other speakers are shown in Fig. 19.5(d) and (e). Since the apparatus is simple to employ, requires no special adjustment from one individual to another, and the electrodes can be applied more easily than a laryngoscopic mirror, it may be that the different wave shapes will prove to be of some clinical diagnostic value.

FORM OF Fx

The most important present application of the laryngograph is in producing a visual correlate of intonation. This is done by displaying the instantaneously derived fundamental frequency, Fx, on a logarithmic scale against time on the face of a storage display tube. We are using this arrangement both for teaching foreign intonation to normally-hearing students, and normal intonation to deaf speakers (Fourcin and Abberton, 1971). Figure 19.6 gives two examples of the normal form of Fx and two examples of abnormal intonation. In some respects, the Fx display can provide more information about abnormal vocal fold vibration than is available from a detailed examination of Lx: a restricted range of Fx variation is immediately visible, and the abrupt changes in mode of vibration that result from a new larynx adjustment can be easily seen in Fx whilst they have to be sought in the long sequences of Lx. Fx also provides a first approximation to a statistical examination of the Lx data, and period-to-period variations can be detected that are not discernible in Lx.

Fig. 19.4.(a) and (b), (c) and (d). The same adult male speaker as in Fig. 19.3 has been asked to sustain /*i*/ for the speech pressure waveform of (a) and /*a*/ for the speech pressure waveform of (c). The corresponding, simultaneously recorded, Lx outputs (closure positive) are shown in (b) and (d). The expanded time scale makes it easy to see the time difference between the onset of closure, as shown by Lx, and the beginning of maximum acoustic output at the speaker's lips. This interval is a function of the speaker's vocal tract length, since the acoustic recording was made with a small microphone placed immediately to the side of his lips. Reversed-time tape replay phase correction is used here, and for Figs. 19.3 and 19.5.

(e) When the output from the laryngograph has not been produced with an internal feed-back operating to counter the slow variations in neck impedance due to laryngeal adjustment, large output variations occur due to the setting and control of the vocal folds. The waveform, Gx, illustrates this output for the utterance '*Do you*?' spoken by a woman with rising intonation. See also Fig. 19.6 (a).

(f) The same utterance repeated by the same speaker as in (e) is shown here with the stabilizing feed-back restored, and with larger amplitude and time scales: only the changing portion of the output has been selected for illustration.

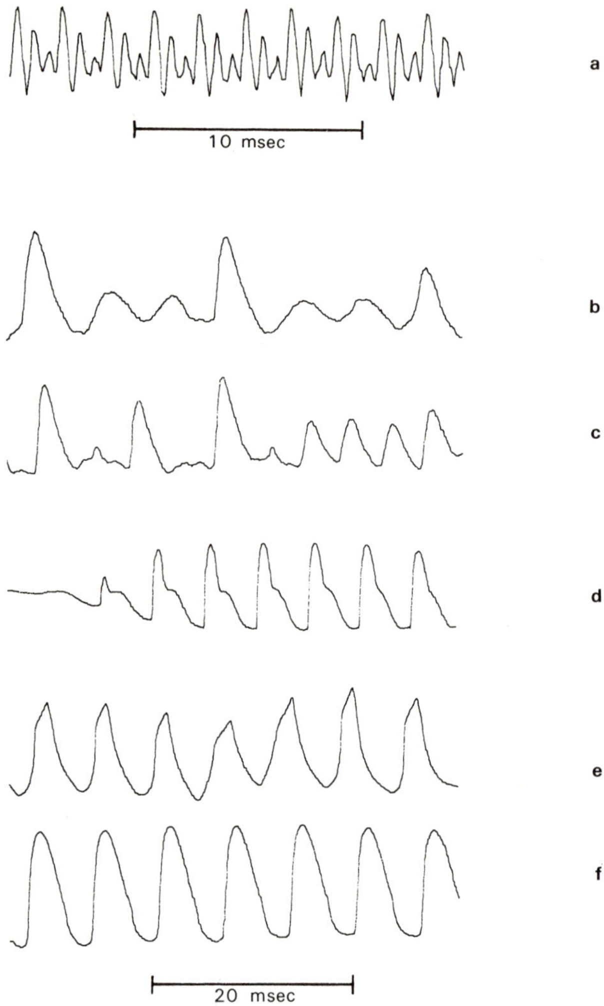

Fig. 19.5. Examples of Lx waveforms which are outside the normal adult range. The waveform in (a) was recorded while a 4-day-old baby was crying vigorously. The sequence of Lx peaks descending in amplitude appears to be quite normal, although it probably results from asymmetric fold tensioning. Lx in (b) was obtained during the conversational speech of a 72-year-old man with unilateral vocal fold paralysis, and shows the same amplitude characteristic as for the child. (c) starts only two cycles later on in this Lx sequence, and shows how a change in larynx adjustment can return the vibration to a more normal mode. The waveform of (d) was made at the onset of phonation from a speaker with laryngitis, and the sharp initial Lx closure peak in the first four cycles may result from non-uniform mucosal irrigation producing a variation in electrical conductivity over the contacting surfaces of the vocal folds. The speaker in (e) had a pedunculate polyp which appears to have been impeding—in rather random fashion—the normal closing of his vocal folds. The same utterance from a corresponding passage of continuous speech was used to provide the basis for waveform (f), where closure appears to be normal: (f) is characteristic of all his Lx recordings in the post-operative state.

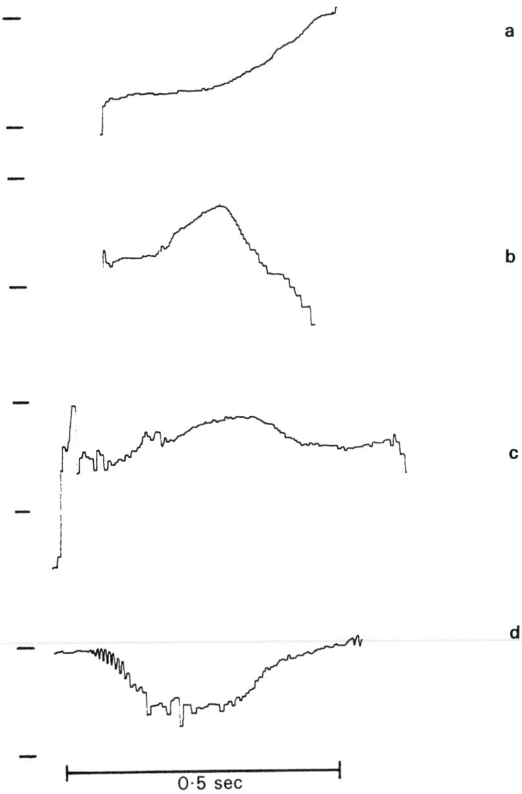

Fig. 19.6. Examples of normal and pathological Fx. Waveforms (a) and (b) show rising and rise-fall intonations from a normal adult female saying '*Do you*': (a) corresponds to the same utterance shown in Fig. 19.4 (e) and (f). The ability of the apparatus to follow individual period variations is especially evident towards the end of the fall in (b). The rise-fall of (c) was obtained from the same speaker when she had laryngitis; the restricted range, irregular voicing onset and the sudden change to a smoother variation are all typical features of this condition, for other speakers as well as this particular subject. The Fx shown in (d) was derived from the Lx of a man with chronic laryngitis: the creaky voice quality at the beginning of the fall is not, in our experience, found normally (although it is normal in the trough of the fall) and the irregularity of the rise is typical of much of his Fx output. In (a), (b) and (c) the upper and lower marks to the left of each Fx graph correspond to 800 Hz and 200 Hz respectively. The marks by (d) correspond to 200 Hz and 50 Hz. All the Fx scales are logarithmic.

FUTURE WORK

The basic laryngograph is already established as a research tool, and we have provided it for use in laboratories abroad as well as in the United Kingdom; but we hope that the development of a suitable display will make it more widely useful, both for teaching (using Fx) and as an adjunct to normal diagnosis (using Lx). The digram structure of Lx periods may also prove to be a useful way of organizing large amounts of data in a meaningful way, since this too will be different for different conditions of larynx vibration, and work on its development is in progress. One other application of Lx, which is relevant here, relates to its use in delayed speech feed-back experiments; when speech itself is employed, this disturbs the monitoring control of speech articulation—delayed Lx feed-back appears to influence only speech rhythm and intonation.

ACKNOWLEDGEMENTS

I am glad to acknowledge the help I have received from Mr. John West and Miss Evelyn Abberton of University College, Mr. Rupert Donovan of St. Thomas' Hospital and Mr. Norman Punt of East Ham Memorial Hospital. The work has been partly supported by research grants from the Science Research Council and the Department of Education and Science.

REFERENCES

Asano, H. (1968) Application of the ultrasonic pulse method to the larynx, *Jap. J. Orol. (Tokyo),* **71,** 895–916.

Fabre, P. (1940) Sphygmographie par simple contact d'électrodes cutanées, introduisant dans l'artère de faibles courants de haute fréquence détecteurs de ses variations volumétriques, *C. R. Soc. Biol., (Paris),* **133,** 639–41.

Fabre, P. (1957) Un procédé électrique percutané d'inscription de l'accolement glottique au cours de la phonation, *Bull. nat. Méd.,* **141,** 66–69.

Fant, G., Ondráckovà, J., Lindqvist, J., and Sonesson, B. (1966) Electrical glottography. *STL-QPSR.,* **4,** 15–21.

Fourcin, A., and Abberton, E. (1971) First applications of a new laryngograph, *Med. biol. Ill.,* **21,** 172–82.

Fourcin, A., and West, J. (1972) Le laryngographe, *Rev. Ac.* [In press.]

Hollien, H. (1962) Vocal fold thickness and fundamental frequency of phonation, *J. Speech Res.,* **5,** 237–43.

Holmes, J. (1962) An investigation of the volume velocity waveform at the larynx during speech by means of an inverse filter, *Fourth Internat. Congr. Acoustics,* G13.

Lisker, L., Abramson, A., Cooper, F., and Schvey, M. (1969) Transillumination of the larynx in running speech, *J. acoust. Soc. Amer.,* **45,** 1544–6.

Michel, C., Pfister, K., and Luchsinger, R. (1970) Electroglottographie et cinématographie laryngée ultra-rapide, *Folia phoniat. (Basel),* **22,** 81–91.

Miller, R. (1959) Nature of the vocal cord wave, *J. acoust. Soc. Amer.,* **31,** 667–77.

Noll, M. (1967) Cepstrum pitch determination, *J. acoust. Soc. Amer.,* **41,** 293–309.

Sonesson, B. (1960) On the anatomy and vibratory pattern of the human vocal folds with special reference to a photo-electrical method for studying vibratory movements, *Acta oto-laryng. (Stockh.),* Suppl. **156,** 1–80.

DISCUSSION ON COMMUNICATION [19]

Klatt. I haven't seen this technique described in detail before, and I did not realize that you had to filter out the low-frequency components in the waveform. Now, as you said, one thing this does is to introduce a phase distortion—so that if you start with a waveform with a rather nice indication of a closed phase (with a simple flat top, for example) and you put it through a filter of the type you describe, that flat top will no longer look like a flat top.

Fourcin. It looks, in fact, like the waveforms I've been showing.

Klatt. Right. So why did you have to do the filtering at low frequencies? What sort of waveforms do you get if you simply use a low-pass filter and not a band-pass filter at the output stage of your device?

Fourcin. If you do not remove the low frequencies—in short, if you have the DC component being transmitted—then you have a wave form of the type I showed [see Fig. 19.4(e)], with very large excursions with the higher-frequency vocal fold vibrations superimposed on them. In order to examine the vocal fold component of that large variation in waveform, one has to have a system that has a very large dynamic range. In practice, it proves to be a prohibitively large dynamic range if one wishes to employ the apparatus for a range of subjects and not merely one. That is why it is necessary to filter.

Klatt. What was the frequency cut-off at the low frequency that you employed?

Fourcin. The frequency response of the laryngograph that we are using at present is about 3 decibels down at some 30 cycles.

Klatt. Could you have chosen 3 cycles, or something down there. The period of this low-frequency instability (the big swings) was rather long, and a cut-off below about one-tenth of the fundamental frequency would be expected to do little damage to the shape of the closed-phase component of the waveform.

Fourcin. If you want to examine the output from a particular subject for a particular sequence of utterances, there is nothing at all to stop you, in fact, having a DC output from the apparatus. I didn't mention this, simply because in most practical situations one does not want to set things up for particular people in particular circumstances. But as a research tool it is necessary, and it can be done.

Stevens. While we are on methodological questions, may I say that I am surprised that you find a 4 : 1 change in resistance on different occasions for the same subject. Maybe that is because of different degrees of electrode contact?

Fourcin. Yes. It is partly due to the contact resistance and the variations that this necessarily induces. It isn't a variability that can be substantially reduced by the employment of electrode jelly, for example, in our experience.

Wyke. Apropos of this point that Dr. Stevens has made, to what extent do you think these distortions that were referred to, and which you are trying to eliminate, are reflections of variations in skin resistance—or rather, of inter-electrode resistance?

Fourcin. Sorry, can I be quite sure what you mean? You have in mind the sort of distortion which could be associated with waveforms like this, this group here? (drawing on blackboard).

Wyke. No. I didn't make myself clear. I mean the instability of the output that you are trying to eliminate by incorporating your feed-back circuit. Because the question I wanted to ask is, do you find it necessary in using your apparatus to go in for a procedure of skin preparation other than merely rubbing electrode jelly on it? Do you, for example, abrade the epidermis with a high-speed drill, or use any of the other tricks that we neurologists sometimes use in other electrobiological recording situations.

Fourcin. Such as using electrodes with a multiplicity of sharp points, for example, so that one gets lower resistances? No, we do not do this. I am not convinced (although I really ought to speak from the point of view of quantitative evidence) that it would necessarily be a good thing in our case. It would certainly be a restriction on the use of the equipment if we were obliged to employ this type of procedure.

With regard to the comment that I think you are, perhaps, going to make.... 'If one were to employ these techniques, then the feed-back circuit certainly would not be necessary and one would get more reliable results from the whole equipment'. I think in some measure that comment, if you were to make it, would be a fair one. But it would not, in fact, make it unnecessary to use the feed-back technique, because the process of laryngeal adjustment that does take place all the time during phonation alters the structures surrounding the vocal folds and makes for an enormous difference in the tissue impedance which is between the electrodes—a difference that is not exclusively due to a difference in degree of surface contact. So it is still necessary to use the system.

Wyke. Dr. Fourcin pre-empted my line of thought exactly! . . . and I accept your very reasonable argument.

Anthony. I should say that I have been using one of Dr. Fourcin's laryngographs for about two years now. I feel privileged to have been able to use this instrument, and can certainly support the statement that it is a practical instrument to use—especially because when one has a lot of patients and one is looking at one patient after another, there is no need to make adjustments between patient and patient. But on the question of the odd waveform—where you get a sort of doubling of the waveform—I have only seen this once or twice when using the laryngograph. That is mainly because I have been using the laryngograph in a rather different way, and I haven't been looking for that particular waveform. I also use it as a means of controlling the stroboscope. However, I must say that I have seen this type of waveform once or twice; and I have also seen it in recordings from the cadaver larynx, where it seems to be two waves—or two motions occurring in the vocal folds that prevent them coming together again.

Fourcin. This is using the laryngograph on the cadaver larynx?

Anthony. This is looking at the cadaver larynx vibrating: when I have seen this, I couldn't see any obvious differences between the vocal folds.

Fourcin. I am not sure that a difference which would result in a change of natural frequency of one of the folds would necessarily be visually evident.

Brodnitz. I am interested, as always, in the practical aspect of such apparatus; and I was intrigued by that slide you showed of the patient with a polyp. I wonder whether you see any difference between patients with a solitary polyp and those with generalized polypoid thickening of one or both vocal folds. It would also be very interesting if one could see a similar investigation of such patients after operation, because it may help to explain to us laryngologists the disturbing problem that we never know in advance of operation what is going to happen. Sometimes you take off a small polyp and the patient develops a beautiful voice within a few days, after healing; but at other times he has considerable difficulty, and this is particularly true of generalized polypoid thickening. The worst cases are the ones with bilateral polypoid thickening of the vocal folds.

Fourcin. Thank you for your comments. I am unable to answer the first part of your statement, as I don't yet have enough experience. With regard to the last point, the presence of a hoarse voice can, I think, result not only from a change in the mass of the vibrating vocal folds—a differential change which causes irregular contact—but can also occur from inadequate or changed irrigation of the vocal folds such that the fluid which normally bathes them is much thicker, and so changed in surface tension. Dr. Otis made this point earlier on this morning [see p. 269]. If a patient does not have adequate mucosal irrigation, following an operation or whatever reason, I think this is necessarily going to lead to a hoarse voice. I have recently been running a small exploratory experiment on myself using a small intramuscular injection of atropine in order to dry up my vocal folds [see p. 330], and the effect of the drying is to produce hoarseness—and to produce, in fact, changes in the laryngograph waveform that are similar at first sight to those found when there is a unilateral paralysis; namely, asymmetrical vibrations of the folds.

Winckel. If you were to develop an electrical test programme to validate the electrical apparatus—that is, if you replace the natural glottic source by a pattern of pulses to simulate the different conditions of voice and speech—we would understand better the characteristics of the frequency and phase relations. One would then feel more sure of what can happen as a result of using the special circuits you have here.

Fourcin. I understand your question, Dr. Winckel. It seems to me that it has two parts. The first part we have dealt with right at the beginning. We have made, and we still employ, a dummy neck that is capable of having its impedance modulated at laryngeal frequency rates. In fact, it was this model which was employed to determine the frequency response of the whole equipment, in that it wasn't a signal so much that was injected but rather a change of impedance between the oscillator source and the laryngograph detecting circuits.

With regard to the second part of your question, for my part I would certainly not try to do this at all because I am far too ignorant of the processes that I think ought to be simulated by means of such an electronic model. I feel that we have an enormous amount to learn, before we are in a position to derive the right sorts of waveform. This sort of situation is one which one finds very often, for example, at a similar level in speech synthesis. It would, I think, be wrong to employ differing sorts of excitation waveform in an attempt to arrive at speech naturalness without any reference to the sorts of waveform one is intending to copy, and this is what we don't know with regard to this equipment.

Winckel. My wish is merely to understand the characteristics of the electrical circuits, and therefore to take a programme of pulses of different frequency order and put up its limits. One would feel better about the apparatus if one were to know beforehand its differing physical characteristics—because it is an important contribution, and it is a considerable responsibility to introduce this apparatus. For this reason, I would like to have such a test programme available.

Fourcin. I think, in fact, it would be an advantage—taking Dr. Klatt's points also into account—to employ the apparatus with the particular object (for example) of studying this sort of waveform, if one wanted to employ it in a non-DC mode of operation.

Ladefoged. I want to ask a little more about atropine, and inquire whether the effect wore off? Why I raise the question is that we have used injection of atropine for precisely the same purpose; namely, to dry the subject out. We used it, for instance, in the experiment I referred to earlier on (the one done by Chin Kim[1]) when we were measuring the degree to which the vocal folds were apart [see p. 314]. We were taking X-rays, and had to get the vocal folds in a state where we could photograph them nicely in a frontal view; and the only way we could do that was to dry them off and then spray them with a radio-opaque material so that they were nicely etched out. I have always worried about this experiment, as to whether in fact the subject was then talking in his normal way. Would you like to comment? Did the atropine effect wear off? Do you think one really needs to have wet vocal folds, or are they not wet enough by spraying them with our radio-opaque solution?

Fourcin. [*Editor's Note:* Dr. Fourcin played a tape recording of his own phonation at this point, to illustrate his answer.]

The first test recording was made an hour after I had had an atropine injection, and the second was an hour and a half after the injection. The one preceding it was normal, made just a few minutes after the injection. Because we could easily hear the difference between the first and the second episodes the effect in me was fairly evident, but not really marked because the vocal folds weren't completely dried up. This is just one little excerpt from a whole series of recordings which were made every half-hour during a period of two and a half hours—six sets of recordings in all. For at least one hour the speech quality was not normal, and the abnormality was due to drying of the vocal folds. If the vocal folds are not properly lubricated, their vibrations will be abnormal.

---

[1] Kim, C. (1970) A theory of aspiration, *Phonetica (Basel)*, **21**, 107–16.

Kirchner. I am not quite clear why Dr. Fourcin does not prefer to use the term 'glottograph' rather than 'laryngograph'.

Fourcin. I may be wrong in this, but my impression is that a glottograph should tell us something about the glottis. The glottis is, as I understand it, the space between the vocal folds. If an instrument does not tell us anything about variations in the space between the vocal folds, it is not really right to call it a glottograph: and this instrument doesn't do that.

Wyke. For the purposes of the record, may I ask the Professor of Anatomy present to define officially for us the term 'glottis'?

Ruth Bowden. The definition I would give would be the space between the free edges of the vocal folds.

Fourcin. If I may continue, I would like to ask Dr. Bowden a question too. Is 'laryngograph' then a better term?

Ruth Bowden. I would say 'laryngograph' would be the better term in the circumstances, because your apparatus does not give precise information about the area between the vocal folds—as you said.

Kirchner. May I add one minor point. We already have the term 'laryngogram'—which is what Dr. Ladefoged was getting when he coated the surfaces of the larynx with radio-opaque material and then took radiographic pictures of it. I wonder (and I don't mean to quibble about this) whether I now have to learn the difference between the two same words or should one speak of a laryngometric tracing rather than a laryngogram, in describing Dr. Fourcin's tracings?

Fourcin. I think the difficulty is very similar to that which exists with regard to the term 'oscillograph'. There are many 'oscillos' that can be graphed. The behaviour of the laryngeal apparatus can be described 'graphically' in a large number of different ways and so I think there is no possibility of escaping the need to employ a qualifying adjective.

Klatt. I want to take up your invitation to pursue mechanisms here, particularly in respect of the infant's cry. In this situation, any time that there is a significant rise in the laryngogram do you interpret that to mean that the vocal folds have at least in part touched one another in the midline?

Fourcin. Yes, I do. At the beginning of normal speech utterances (as with /ɑ/) you will recollect [see Fig. 19.3] that the laryngograph waveform showed nothing while speech was, in fact, beginning. This is, I think, because the model you described [see Communication 17] is in its mode of applicability, and the folds are vibrating *before* coming into contact. It is only when they are actually *in contact* that this apparatus gives information that is useful.

Klatt. To pursue the discussion about a possible mechanism for the multiple peaks during a period of infant vocalization, I think that Mr. Anthony [see page

328] did come close to what would be one possibility. However, the idea that the vocal folds are necessarily asymmetrical in their masses or in their controls seems to me to be rather unlikely. When you get a system like that vibrating, it is still coupled in such a way that everything is moving at the same frequency. Let me suggest an alternative hypothesis. When the vocal folds come together (particularly when seen with stroboscopic illumination) it is quite clear that when one vocal fold hits the other a 'squashing' of this vocal fold proceeds as a wave outward until it hits some harder structure, and then is reflected back. So one sees wave patterns on the top part of the vocal fold. If that wave comes back in phase—that is, at the same time as a fold is moving towards the midline—of course everything is going to work out fine and every pulse is going to be about the same; but if with alternating closing of the vocal folds the wave comes back in-phase and out-of-phase, one would get a type of vibration in which the speed with which the vocal folds came together was alternately increased or reduced. This would make a pretty small change in the actual vibration pattern, and yet might show up as a rather large change in the laryngogram.

Fourcin. If one looks at the speech pressure waveform that accompanies these differences in the laryngograph output, I would expect (if I follow your thoughts aright) to find that the speech pressure waveform carries on, so to speak, essentially unperturbed while the laryngogram waveform is grossly changed. But that is not the case.

Anthony. I haven't taken this particular waveform and put it into a speech synthesizer to see how it would affect the output—although I have done so with other cadaver larynx waveforms. I haven't seen this double action happen to the extent that the vocal folds did, in fact, just clash so that the cycle was really divided into two. It is a bit difficult to explain. It looks at first glance as if you have got double the frequency, but in fact you haven't—and when you listen to it the pitch is that of the total period, not the shorter period.

Fourcin. This, in a quite different mode of operation of the vocal folds, is similar to some of the creaky voice waveforms where one gets doublets of closure.

Ruth Bowden. Could I ask if you have a series of examinations in babies, because the character of the newborn cry is quite characteristic and can be recognized from that of a slightly older infant. Does the pattern change in early infancy?

Fourcin. I cannot give an answer at the moment; but in a month or so I hope we will have completed such a study.* Mrs. Anne Parker (of University College Hospital) is working with us on this particular point in the Maternity Unit. The slide I showed was just one of the very first waveforms we ever recorded in a newborn baby. Since then we have had a very large number of recordings of normal children, and we are studying their development. We hope then to examine abnormal children as well.

---

* *Note added in proof:* During the first 4 to 6 weeks of infancy the vibration pattern does tend to a more regular waveform—and the change would be readily perceptible to a listener.

Wyke. Being your amanuensis, can I elicit an answer to a question for the record? Dr. Klatt's original question to you contained the phrase 'towards the midline'. Would you agree that whilst the laryngogram may tell you *when* the vocal folds come into apposition, it doesn't tell you *where,* in spatial terms?

Fourcin. Yes, I would thoroughly agree. I think this is a very important point to bear in mind, when one considers the possibility of asymmetric vibrations. Then the vocal folds could well be making contact (and, indeed, I think they do) when they are well away from the midline position—as in the case of unilateral paralysis.

20 Some Ventilation-Phonation Relationships

A. B. OTIS

Department of Physiology, College of Medicine, University of Florida, Gainesville, Florida, U.S.A.

The primary function of breathing is to ventilate the lungs for respiratory purposes, but many animals have in the course of evolution developed ways of involving breathing in the production of sounds. This secondary use of breathing is especially well developed in man, who uses it in speaking, singing and in playing muscial instruments.

Although phonation and ventilation serve widely diverse purposes, these two functions may be carried out simultaneously with remarkably little conflict. Both ventilation and phonation can, to a considerable extent, be voluntarily controlled. However, the former usually, and the latter frequently, is controlled with a high degree of automaticity. In either case there is usually little obvious interference between these two functions, although some degree of compromise must exist.

A resting individual will, when speaking, tend to ventilate more than is optimal for respiratory purposes. Bunn and Mead (1971) have observed that sitting subjects, while reading aloud, increase their alveolar ventilation by an average of 27 per cent over the resting value, the magnitude of the increase being dependent in part on the nature of the text being read. All of the material was in English, and it would be interesting to make similar studies during the reading of other languages.

If the ventilation were stimulated to approximately double its resting value by inhalation of 3.5 per cent CO_2, Bunn and Mead (*loc. cit.*) found that reading reduced the ventilatory response. Similarly Bouhuys (personal communication) found that subjects walking on a treadmill ventilated less while reading aloud than while breathing without phonatory activity.

Thus it appears that if one is sitting at rest, continuous speech requires more ventilation than is optimal for respiratory purposes; whereas ventilation that is doubled in response to a respiratory drive (by CO_2) is greater than optimal for a reading task. Presumably there are intermediate conditions in which the ventilatory and phonatory demands are matched. Perhaps this is a reason (or at least a teleological explanation) why an individual who is speaking continuously, such as a lecturer, customarily does so while standing rather than sitting. This moderate amount of activity may be just sufficient to optimize ventilation for phonatory requirements.

The volume of air flow required for phonation depends on several factors. Thus, it is highly dependent on the particular material being phonated (Klatt, Stevens and Mead, 1968): for all vowel sounds have similar and relatively low air flow requirements, whereas consonants show a wide variation. The /*h*/ consonant sound requires four or five times the volume of the sounds associated with /*b*/, /*d*/, and /*g*/. In some recent studies of my own I find that, in counting aloud from one to ten, the volume of air as well as the peak flow rate may be considerably different from digit to digit. The requirement for a given digit also varies with pronunciation and degree of emphasis.

As Cavagna and Margaria (1968) have shown, there is no simple general relationship between air flow rate and loudness. Indeed they found that for a given tone, air flow rate might vary over a seven-fold range. However, for any specified tone a certain minimal air flow is required, and this minimal value is directly related to loudness. We (Otis and Clark, 1968) have observed that when ventilation is increased in response to muscular exercise, the loudness of the voice spontaneously rises. This probably is related to an elevation of subglottic pressure, as well as increased air flow.

Whispering requires an appreciably greater air flow than does normal speech. For example, counting from one to ten in a whisper requires a 40 to 80 per cent greater volume increment than counting at a similar rate in a normal voice.

Each individual has a natural spontaneous rate of phonation which seems to be set by some internal metronome. This rate varies considerably among individuals, but is fairly constant for a given individual under a given set of conditions. For example, one subject was instructed to count from one to ten at his own natural pace. During five consecutive trials the time required varied from 4.5 to 5.0 seconds, with a mean of 4.8 sec. This corresponds to a mean rate of 2.1 digits per second. Another subject required 7.0 to 8.0 seconds, with a mean of 7.6 sec or 1.3 digits per second. On a different day the latter subject took from 8 to 8.2 seconds for the same task, an average of 1.23 digits per second.

In a different experiment the same subject performed a series of ten trials in each of which he repeated a given digit six times at his natural rate. All digits from one to ten were used. The mean rate was 1.20 digits per second, with a range of 1.11 to 1.25. 'Threes' took the longest time, whilst 'fours,' 'sevens,' and 'tens' took the least.

The same subjects also performed trials of the one-to-ten counting task by whispering. The counting rates were not significantly different from those in which the natural voice was used, even though the air flow was 75 per cent greater during whispering.

Clark and I (1968) have previously demonstrated that the time required for simple speech tasks—such as counting, reciting the alphabet, or reciting a passage of memorized poetry—was essentially independent of the magnitude of the ventilation, even when the latter was increased as much as five times over its resting value.

The natural rate of phonation is, over a wide range of conditions, apparently regulated by some internal time-base rather than by the volume of air available. Of course, there must exist some minimal airflow for maintaining this rate, but increases above this minimum apparently have little influence on the natural rate.

Just as each individual has a natural rate of phonation for a given speech task, so does he have a natural frequency of breathing for a given respiratory requirement. This frequency appears to be set so as to require minimal power (Otis, 1954, 1964) or minimal mean force (Mead, 1960) from the respiratory muscles. That is, the body tends to optimize the frequency of breathing in terms of the economics of energy expenditure.

The function of phonation is to transmit information, and I would like to speculate that the rate of phonation is optimized so as to maximize information transfer. In the transfer of information, the rate-limiting step might occur in the generation or retrieval of information in the central nervous system, in the phonation process itself, or in the ability of the target individual to receive and assimilate.

If the generation of information is the limiting factor, the result is either awkward pauses in phonation or the interposition of nonsense syllables, such as '*er-er-er*', or in the current fashion, '*you know*'. If the ability of the receiving individual is limiting, the sender may have to repeat himself or else risk being misunderstood. If phonation rate is the limiting factor, some impatience or boredom on the part of the receiver may result.

I should like next to consider the maximal rates of phonation and of breathing. In 1938 Wallace Fenn pointed out that the maximal frequency of cyclic movements of the limbs and digits was related to the amplitude of motion, such that

$$t = t_o + bA$$

where t is the period or time for one cycle, A is the amplitude, b is a constant and t_o is the limiting period at zero amplitude. By extrapolation of experimental data, t_o appears always to have a finite value: b is dependent on load, whereas t_o is independent of load.

We (Otis and Guyatt, 1968) have confirmed a relationship of this type for the maximal frequency of breathing at various tidal volumes [see also Proctor and Hardy, 1949]. The constant, b, can be altered by gas density, but t_o shows relatively little change. It appears that b is determined by the mechanical and metabolic properties of the muscles and the load applied to them, whereas t_o is mainly a reflection of characteristics of the neuromuscular control system. The reciprocal of t_o may be thought of as the fastest frequency at which the system can operate. Our experimental estimates of t_o indicate that the upper limit for breathing frequency in man is of the order of five to seven per second. This is also the range of maximal frequency of a number of other to-and-fro movements that the body can perform, e.g. wagging a finger, swinging the forearm, etc. (Fenn, 1938).

It appears that the limiting frequency for phonation may also be of this order of magnitude. We have made some trials of counting aloud from one to ten as rapidly as possible. I was able to do this in 1.75 seconds and one of my students in 1.48 seconds, corresponding to frequencies of 5.7 and 6.8 per second respectively. Jere Mead (personal communication) tells me that he can make '*pah-pah-pah*' sounds up to about 7 per second but that he can double-tongue '*T-K-T-K*' at a rate of 12 per second. In this latter manoeuvre, however, the '*T*' and '*K*' respectively are generated in opposite phases of the same cycle of a tongue movement, so for a complete cycle the rate is 6 rather than 12 per second.

An interesting and pertinent study is that of Hudgins and Stetson (1937), who studied the maximal rates of each of several articulatory movements involved in speech. For nine subjects the average number of syllables per second was: tip-of-tongue, 8.2; jaw, 7.3; back of tongue, 7.1; velum, 6.7; lips, 6.7.

Thus phonation, like breathing, seems to be limited to an upper frequency of about 5 to 7 per second. This is not to imply that phonation rate is limited by the breathing mechanism, but rather to suggest that both the muscles of breathing and those of such phonatory mechanisms as the lips and tongue are regulated by control systems that set similar upper limits. Apparently the limiting frequency for muscular performance reflects some fundamental property of the central nervous system.

I wondered whether similar limitations existed on the sensory side of the verbal communication system, and Mr. Ronald Clark and I have done some experiments to determine the maximal rate at which subjects can count accurately the number of discrete sounds in a sequence. We prepared a tape recording, from which subjects were presented with a series of bursts of clicks. Each burst was 1 or 2 seconds in duration: the click rate within a given burst was constant, and varied among bursts from 2 to 20 per second. The interval between bursts was 2 seconds. During each interval, the subject recorded on paper the number of clicks counted in the previous burst. Our best performer counted accurately at rates up to 8 per second; the poorest made errors at frequencies as low as 4 per second. Our results are consistent with those of Garner (1951), who had already made a more elaborate study of a similar nature. He found that with less than five tones in a sequence the number could be counted at a repetition rate as high as 12 per second, but with more than five in a sequence counting errors occurred at rates of 6 or more per second. Another example was given by Helmholtz (1877), who observed that beats produced by two tones of differing frequency could be accurately counted only when they did not exceed 4 to 6 in a second.

The conclusion I draw from this is that the sending and receiving systems for verbal communication are rather well matched, and that we can hear about as fast as we can talk. This is not very fast. By the standards of modern technology this is an inordinately slow transfer of information. Two computers could be programmed to communicate with each other at a much higher rate. The conversation would be fast and accurate, but predictable and utterly devoid of imagination. Although the human brain, being slow and prone to error, is a lousy digital computer, it does have an imagination—which is one of its redeeming features.

ACKNOWLEDGEMENTS

This work was supported by Contract F44620–70–C–0109 with the U.S. Air Force Office of Scientific Research and by NASA Institutional Grant NSG-542. I would like to thank Mr. John Evans and Mr. Thom Tyler for their technical assistance.

REFERENCES

Bunn, J. C., and Mead, J. (1971) Control of ventilation during speech, *J. appl. Physiol.,* **31,** 870–2.

Cavagna, G. A., and Margaria, R. (1968) Airflow rates and efficiency changes during phonation, *Ann. N.Y. Acad. Sci.,* **155,** 152–64.

Fenn, W. O. (1938) The mechanics of muscular contraction in man, *J. appl. Physics,* **9,** 165–77.

Garner, W. R. (1951) The accuracy of counting repeated short tones, *J. exp. Psychol.,* **41,** 310–16.

Helmholtz, H. L. F. (1877) *On the Sensations of Tone as a Physiological Basis for the Theory of Music*, 2nd English Edition, reprinted by Dover Publications, Inc., 1954.

Hudgins, C. V., and Stetson, R. H. (1937) Relative speed of articulatory movements, *Arch. neerl. phonet. exp.*, **13,** 85–94.

Klatt, D. H., Stevens, K. N., and Mead, J. (1968) Studies of articulatory activity and airflow during speech, *Ann. N.Y. Acad. Sci.,* **155,** 42–55.

Mead, J. (1960) Control of respiratory frequency, *J. appl. Physiol.,* **15,** 325–36.

Otis, A. B. (1954) The work of breathing, *Physiol. Rev.,* **34,** 449–58.

Otis, A. B. (1964) The work of breathing, in *Handbook of Physiology, Section 3, Respiration,* Vol. 1, pp. 463–76, Ed. Fenn, W. O., and Rahn, H., Washington, D. C., American Physiological Society.

Otis, A. B., and Clark, R. G. (1968) Ventilatory implications of phonation and phonatory implications of ventilation, *Ann. N.Y. Acad. Sci.,* **155,** 122–8.

Otis, A. B., and Guyatt, A. R. (1968) The maximal frequency of breathing of man at various tidal volumes, *Resp. Physiol.,* **5,** 118–29.

Proctor, D. F., and Hardy, J. B. (1949) Studies of respiratory air flow, *Bull. Johns Hopk. Hosp.,* **85,** 253–80.

DISCUSSION ON COMMUNICATION [20]

Proctor. I agree with Dr. Otis that we all have our own internal metronome, and so have a natural rhythm of speaking. But I think that when it comes to training a singer or a speaker, just as the singer learns to vary the intensity of his sound in a song and makes it much more interesting this way, so the speaker learns to vary the rapidity with which he delivers one sentence (or one paragraph) or another and thus lends a great deal of interest and emphasis to his presentation. Do you agree that this can be learned: that is to say, can one learn to overcome the natural speaking frequency with which one was born?

Otis. I would agree that a great deal of it can be modified.

Stevens. Dr. Otis showed a graph indicating that with different amounts of ventilation the rate of speaking is about the same. So I wonder where does the air go when people speak? Presumably, independent of the amount of ventilation, they use about the same amount of air flow to say a vowel. Where does the rest of the air go at the end of a sentence?

Otis. I don't think the ventilation diminishes too much during speaking. People are somehow able to superimpose speech on top of the flowing air stream—which, I guess, is the 'shunt' that we were talking about earlier [see Communication 6]. There was, as I said, an increase in the loudness, which would mean there is some degree of higher pressure built up behind the glottis, perhaps. But I don't know how all the excess air is 'short-circuited'.

Stevens. My guess would be that even in normal conversation an appreciable fraction of the outflow of air occurs when one is not speaking at all.

Anthony. There is a difference in how you get rid of the air, and this seems to be where the difference lies between quiet breathing and spontaneous speaking. Speaking spontaneously, you take all sorts of opportunities to get rid of the air. When reading a text aloud, you immediately take in air when you have finished reading a phrase. In reading there is certainly a greater tendency to hypoventilation, because you can't get rid of the air. I would imagine, then, that when you are speaking spontaneously, you try as far as possible to keep this hypoventilation as small as possible—that is, you try to maintain your normal ventilation.

Otis. I don't think this has been studied adequately in spontaneous speech. Most of the reported studies have been made with prepared texts being read, instead of with spontaneous speech.

Anthony. I have done this[1] with six people speaking spontaneously (in a question-and-answer situation). A person reading a text may go well up in his vital capacity, whereas in spontaneous speech he will operate round about his tidal breathing level.

Otis. If you try reading when you are walking on a tread-mill moving at 50 metres a minute, you do find some difficulty in phrasing things right—you may

---

[1] Anthony, J. (1973) Respiration and the organization of airflow for speech [in preparation].

find yourself way down at the end of a breath before you are ready, as far as the reading is concerned. You have to *make* your reading follow the ventilatory pattern.

Anthony. Whereas in reading normally, it seems to be the opposite?

Otis. Normally it would be the other way round, at lower level ventilation.

Widdicombe. Can I go back to the question of ventilation, because you are one of the pioneers in the study of lung mechanics? Would you agree that even if we had the neurological apparatus that could make us breathe faster than five to seven times per second, the mechanical properties of the human respiratory apparatus are such that they wouldn't follow—or would only follow extremely uneconomically? If I remember correctly, the natural frequency of the lungs and chest wall is about six or seven cycles per second.

Otis. Yes indeed. I should have mentioned that—Arthur Du Bois[1] has measured it, and his estimate was about 5 to 7 cycles per second. It should be possible, however, to drive the system faster. The energy would seem to be available to do it, if one tried to drive it by electrical stimulation of inspiratory and expiratory muscles alternately.

Widdicombe. It would be very uneconomical?

Proctor. It might be, but I think it would be possible.

Brodnitz. I want to respond to Dr. Proctor's remark about the ability to 'teach' people breathing. There is a remarkable difference between the teaching of breath control and of control of the rate of speech. One can learn the economy of breathing so as to be able to draw out long phrases on one even breath—this can be taught. But when it comes to control of the rate of speech, every speech therapist always has trouble with that. It is, to a certain degree, possible to slow down a fast speaker—as in the case of a 'clutterer' (so-called tachylalia): if one makes the patient aware of his speech mechanism and induces him to concentrate on his articulatory movements while he speaks, he slows down—at first as a conscious process, and later as a conditioned habit. But it is almost impossible to speed up the extremely slow speakers—you cannot get them, by any training method, to give up their extremely slow rate of speech. This is where that inner metronome that Dr. Otis talked about clearly plays a big part.

Martin. Could I comment on the perceptual aspect, Dr. Otis? I was very interested in the experiments that you have done, because we have done some that are very similar. What Kemp and I did was to produce a series of audible pulses (pre-recorded on tape) that were electrically 10 msec in duration (so in fact they were a little bit longer than that). We started off by presenting these to our subjects at 5 per second, and found that it was very easy for them to count them, just as you said; but then we increased the rate and made the perceptual task very

---

[1] Du Bois, A. B., Brody, A. W., Lewis, D. H., and Burgess, B. F. (1956) Oscillation mechanics of lungs and chest in man, *J. appl. Physiol.*, **8**, 587–94.

much more difficult. We compared the number of actual pulses per second with the number estimated to have been heard by twelve individual subjects. Some of the subjects (about 4 or 5) were able to follow up to 10 per second, but beyond this their counts dropped off badly; whereas others could go further up, but developed a bigger scatter in their counts and began to be not very accurate. There were at least two individuals, however, who could follow accurately all the way up to the maximum frequency we have on the tape. We just didn't think people could do this; and this makes me wonder about the sensory 'matching' that you are postulating.

There is also another aspect of this 'matching'. You seemed to be saying that the semantic content of what has been said is identical with the individual phonemes or words or syllables—and I think a lot of people would rather question that.

Otis. If you remember, I said it was a very naïve conception. I am sure that there is more to it than at first appears. One cannot *count* a thousand cycles per second, but one can *distinguish* acoustically between a thousand and a thousand and two, or whatever. I am sure that one could learn to say that there were ten counts in a sequence, even though one couldn't actually count them—that is, one can learn somehow to make a recognition of over-all pattern. This is true of all our sensory activities, I guess. I suppose there is some upper limit for counting flashes of light: of course, you know that up to 25 or so you can say that there are distinct flashes, even though you can't count them, but beyond that they fuse—so there are several gradations of this sort of pattern recognition.

Ladefoged. I want to mention a couple of other points bearing on the notion raised by Mr. Martin. I don't think it is true that the rates of sending and receiving are well matched—because it is quite plain that you can receive far faster than you can put out. Fairbanks *et alii* (1957)[1] have done experiments where they play back speeded-up speech at rates that are impossible for a human being to produce, but nevertheless subjects can understand and receive the message—that is, they can receive it far faster than they can put it out. Equally, Dr. Stevens will probably remind us of the work of Bill Huggins[2], who showed that you could perceive very small differences in the duration of consonant sounds that are less than the differences that you can produce. So again, one's skill in perception is certainly above one's skill in production.

Otis. I introduced the idea mainly because I thought it was a nice simple-minded theory, and that maybe it would stimulate some discussion from people who know better—which it has!

Stevens. I would agree with Dr. Ladefoged. As Mr. Martin says, speech is highly redundant and therefore it is possible, when there is lots of context, to speed it up. In fact, this is done in devices for reading to the blind, because they like to get at it a little faster than just the normal rate of speaking.

---

[1] Fairbanks, S., Guttman, N., and Miron, M. (1957) Effects of time compression upon the comprehension of connected speech, *J. Speech Dis.*, **22,** 10–19.

[2] Huggins, A. W. F. (1968) How accurately must a speaker time his articulations?, *IEEE Trans. on Audio and Electroacoustics,* Vol. AU-16, No. 1., pp. 112–17.

Could I ask a question? This morning Dr. Ladefoged was saying [see p. 309] that the upper limit of the rate of motion of the structures involved in breathing is somewhat slower than the upper limit of moving the articulators, and that therefore changes in the respiratory system are not used segmentally in speech—but I think Dr. Otis is now implying that all these rates are of the same order of magnitude?

Otis. They seem to be.

Stevens. There seems to be a bit of a conflict here, then, and I wonder if it could be resolved. In fact, are the data correct (they seem to be) that the figure is 5 to 7 cycles per second for breathing and gets up to 8.2 for the tip of the tongue (which is not that much different)? I just wonder why, in fact, we don't use changes in the respiratory system to produce distinctive differences in speech segments.

Ladefoged. Let me put it another way, just to recalculate it. When I say [demonstrating] '*She sells sea shells on the sea shore*' and '*The sea shells that she sells are sea shells I'm sure*', that is at 20 segments per second. If you work it out, that is 50 segments in 2.5 seconds. Those of us who work with speech spectrograms know that one has a figure of 2.4 seconds as being exactly something that can go on a spectrograph drum, and I know I can get that on a spectrograph drum quite easily—and therefore it is at about 20 segments per second. I am not claiming, of course, that it is the same as a tap rate, because they are very diverse segments and one does it in the same kind of way as one taps two separate fingers on a piano. The indications are that movements of the tip of the tongue are much faster than the breathing mechanism. I was agreeing with Dr. Mead—he was saying 4 to 5 for the breathing mechanism, and I agree with that; and then I would go on to say 8 to 9 (almost double it, in fact) for the tip of the tongue. I don't think they are all of the same order of magnitude.

Otis. We found people who could make breathing movements at 7 per second at zero tidal volume.

Ladefoged. What is the task? You mean this? [Dr. Ladefoged demonstrates rapid panting—*Ed.*]. They can do that at 7 per second?

Otis. Yes. If you just connect a pressure gauge to the mouth and make breathing movements, the frequency can then be recorded.

Ladefoged. But I want to make quite sure that it is *breathing*. That is why I asked you particularly what the task was, to make sure that it wasn't something to do with valving with the larynx or valving with the tongue.

Otis. No. These are movements that would produce breathing. We also get a similar value for the maximal frequency of breathing from extrapolation of this curve [drawing on blackboard]. There is a rather nice linear relationship between tidal volume and cycle time over a range of tidal volumes going up to 70 per cent or so of the vital capacity; and if we extrapolate the line down to zero tidal volume, it appears to come out here at something of the order of 5 to 7 cycles per second, in different individuals.

Ladefoged. Can I press it a little bit more? Can you tell us exactly what is this task that people do 7 times a second?

Otis. They are told to breathe as fast as they can with very little breath—that is, with very tiny breaths [Dr. Otis demonstrates here—*Ed.*]. Whether you can count that or not, you can check it with a stopwatch. You can do the same thing with other tasks.

Ladefoged. I am sorry, but when I do something like this [demonstrating] I am, in fact, moving my larynx up and down. I am not, in fact, using the respiratory system in order to make these very small movements of air in and out of my mouth.

Fry. Can I just inject a comment here? Dr. Otis was certainly using movements of his abdominal wall. I could see them clearly from where I am sitting, and we can check the frequency because the sounds he made will be on the tape recording of this discussion.

Wyke. Mr. Chairman, I will see that the frequency is measured on the tape and will interpolate it into the printed record of the discussion. [*Editor's Note:* Measurements made on the slowed-up tape showed that Dr. Otis had made 22 respirations in 3.1 sec—that is, at a frequency of 7.1 per sec.]

Fourcin. I want to make one small comment. It is not just the maximum frequency of operation that is of importance here, but rather the precision of temporal adjustment. Dr. Ladefoged showed us electromyographic recordings of the activity of the intercostals during speech this morning [see Communication 18], from which it is apparent that it was precision of timing that was very efficient—and this really is what is involved in the control of speech sequences. I don't think that we should be misled by the near maximum frequency of the ability to operate—either with breathing or tongue-tipping, or what.

Otis. I am interested in it only because I think the limiting frequency (or minimal time) of operations points to something fundamental in the central nervous system.

Fourcin. It may well do. But this may not be something that is really crucial in speaking.

Otis. I don't expect it is, necessarily; but I think it represents the fastest rate at which the central nervous system is willing to act, so to speak. It may be the fastest rate at which co-ordinated movements are possible. In movements involving antagonists I think that you rarely have both sets of muscles firing at the same time—you fire one, and it cuts off before the other is fired. There is some regulatory mechanism that detests making antagonists work against each other. I don't know why. I am obviously not a neurophysiologist, otherwise I would not use such terminology.

Freedman. I can't remember the numbers on your plot of tidal volume against time, but it seemed to imply that very high ventilations were achieved that might not be available for more than a few seconds. In other words, you couldn't utilize them for speech in such a very short time.

Wyke. Apropos of the last comment from Dr. Otis, I think we should remind ourselves that the central nervous system has, in fact, a whole array of built-in rate-limiting devices at various stages in both input and output pathways. I suppose one of the most familiar at the output end is the Renshaw cell system; and if one distorts its operation with tetanus intoxication or strychnine poisoning, one can throw the whole rate-limiting business of motor control completely out of gear[1]. So from a neurophysiological point of view I would, perhaps, go along a little further with Dr. Otis than some of our respiratory colleagues might—because I think that the nervous system, if you like, is programmed genetically with built-in rate-limits and unfortunate things happen if the rate-limiting devices are thrown out of gear.

Widdicombe. Subject perhaps to Dr. von Euler's correction, I think that it is an oddity of the diaphragm that the phrenic motoneurones have not got a Renshaw cell system—so your rate-limiting process must be somewhere else in that particular instance.

von Euler. That is correct.

Wyke. I agree with that too—but Dr. Otis was mainly using his abdominal muscles in his performance (as our Chairman pointed out), and as far as I know no-one has suggested that abdominal motoneurones lack Renshaw loop systems.

Klatt. If this rate-limiting is a central process it must involve a feed-back path, because you showed earlier yourself, Dr. Otis, that to change the density of the gas slows down the rate of a given ventilation.

Cohen. With regard to a possible neurophysiological basis for the rate limit, I would like to suggest that the limiting frequency of respiration, although obviously related to the mechanical characteristics of the respiratory system, may also depend on the discharge characteristics of the neural oscillator. Even in the paralyzed animal (cat), using phrenic discharge as the indicator of the cycle, the highest frequency that I have observed during hypocapnic polypnoea was about 5 per sec (Cohen, 1964)[2]. It is interesting that there seems to be matching between mechanical and neural characteristics; and this was brought out long ago, in an old study by Gesell and Atkinson (1943)[3], who recorded diaphragmatic motor unit activity in several species ranging in size from horse to mouse. It was found that in the different species the number of spikes per inspiratory burst was about the same, since during the shorter inspiratory

---

[1] Wyke, B. D. (1969) *Principles of General Neurology. An Introduction to the Basic Principles of Medical and Surgical Neurology*, London.

[2] Cohen, M. I. (1964) Respiratory periodicity in the paralyzed, vagotomized cat: hypocapnic polypnea, *Amer. J. Physiol.*, **206,** 845–54.

[3] Gesell, R., and Atkinson, A. K. (1943) A comparison of motor integration in the mouse, rat, rabbit, dog and horse, *Amer. J. Physiol.*, **139,** 745–55.

durations there was a higher discharge frequency. Thus, in the smaller species with high respiratory frequencies (and thus shorter inspirations), the more rapid air flow was achieved by a higher discharge frequency of inspiratory motor units. Apparently, then, the natural frequency of the neural oscillator seems to be matched to the mechanical parameters of the system.

Anthony. I want to bring the discussion back to what happens in *normal* speech. The rates (as I have found[1] in six people either reading a text or doing continuous speech) were, in fact, much lower than those mentioned by Dr. Otis. They were surprisingly low, if you measure them in terms of syllables as a function of time. None of these people got much above an average rate of 5 syllables per second.

Otis. Were these people reading as fast as they could?

Anthony. They were just reading normally and just spoke normally, but none of them averaged much more than 5 syllables per second. The other side of that is also interesting, in that when you look at the average amount of air they use per syllable you find this is very constant, and seems not to depend on the nature of the spoken material. In spontaneous speech there seems to be a normal rate for the amount of energy used, in the long term.

Plum. Maybe Dr. Widdicombe could help us out here? It is my impression that in fast muscles during shivering one gets relatively high frequency rates of motor unit discharge. Furthermore, do you know anything about the comparative frequency of eye blinking? This is operated by small, fast, highly-innervated muscles at a maximal frequency close to that in the phonatory system—the muscles of which I assume to be somewhat similar, in terms of their structure and innervation ratios.

Otis. I have not tried measuring that.

Widdicombe. I think, although I forget the figures, that in general what you say is correct—but I am sure Dr. von Euler knows more about this than I do. The point that goes along with that idea is the interesting phenomenon that often when we shiver in a rigor we have a respiratory linkage with the shivering. We shiver in the inspiratory phase, and our respiratory muscles shiver as well. I have never measured this, but I would reckon that they are oscillating at something like 5 to 7 per second. That is the feeling of it: but possibly someone has measured it, and Dr. von Euler might know.

von Euler. Most shivering activity goes on at a rate of 10 to 12 per second, or in that order of magnitude.

Further to Dr. Wyke's comments on the Renshaw cell business, we should also keep in mind that the Renshaw cell is centrally controlled and is not a fixed circuit that limits things unalterably. It *is* a limiter of the frequency of individual motor unit activity, but one which can be centrally controlled.

---

[1] Anthony, J. (1973) Respiration and the organization of airflow for speech [in preparation].

Cohen. I would like to add something more about the relation between individual unit discharge frequency and frequency of the cycle. Of course, individual motoneurones (such as phrenic motoneurones) may discharge at a much higher frequency, up to 50 to 100 per second; but the frequency of the cycle is related only to whatever group of neurones is controlling it and thus as a parameter it may (hopefully) furnish information about the system. I would also like to repeat that apparently the neural frequency is matched to the natural mechanical frequency of the muscles—which one might think reasonable from an evolutionary or teleological point of view.

21 Interrelationships Between Voice and Laryngeal Mucosal Reflexes

WILBUR J. GOULD AND
HIROSHI OKAMURA*

*Department of Otolaryngology,
Lenox Hill Hospital, New York,
U.S.A.*

INTRODUCTION

Although initiated voluntarily, phonation is dependent mechanically upon precise subconscious integration and co-ordination of a large array of feed-back reflexes which constantly adjust the tone of the muscles involved in the production of sound. The existence in the larynx of what might be termed a 'servo-mechanism' has been postulated by a number of investigators, on the basis of both histological and neurophysiological studies. Wyke (1967, 1969), for example, has suggested that the so-called laryngeal servo-mechanism encompasses the following systems: the laryngeal myotatic reflexes, the laryngeal articular reflexes and the laryngeal mucosal reflexes.

Yet despite numerous investigations of the laryngeal servo-mechanism as such, there appear to be few experimental studies exploring the functional relationship between this reflex system or its component parts and phonation. For this reason, the authors have carried out preliminary studies designed to elucidate the precise contribution of the mucosal reflex system to the regulation of the human voice. Central to these studies were aerodynamic and electromyographic parameters measured during various types of phonation, both before and after anaesthetization of the laryngeal mucosa to inhibit or diminish the activity of mucosal receptors.

The results suggest that there exists a close relationship between the laryngeal mucosal reflex system and the regulation of voice. The purpose of the present Communication, therefore, is to outline these preliminary findings, particularly as they relate to the mechanisms for regulating the subglottal pressure during phonation and also as they relate, at least inferentially, to the laryngeal servo-mechanism.

---

* On leave of absence from the Department of Otolaryngology, Kyoto University, Japan.

METHODS

Before and after topical anaesthesia of the larynx, the subglottal pressure, air flow rate and volume of exhaled air were measured during various types of phonation on an adult male subject having a normal larynx, voice and lungs. In addition, electromyograms of the laryngeal and expiratory muscles and measurement of the movement of the chest wall and abdomen during each of these types of phonation were also done. The disposition of the instruments used in the investigation is shown schematically in Fig. 21.1.

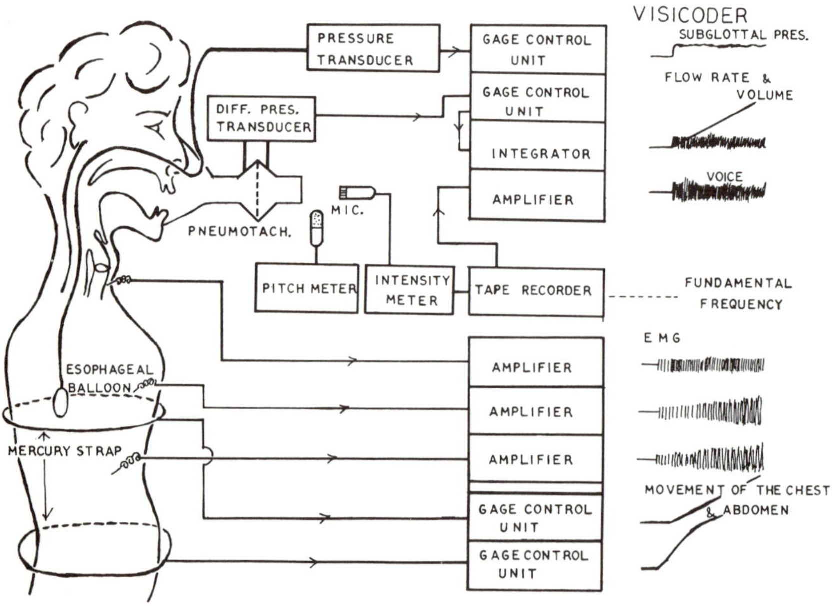

Fig. 21.1. Block diagram of experimental arrangement.

ANAESTHESIA

The topical anaesthetic agent used was 4 per cent *Xylocaine,* applied through the mouth. Care was taken to obtain deep anaesthetization of the subglottal area.

AERODYNAMIC PARAMETERS

The subglottal pressure during phonation was estimated with the oesophageal balloon technique. A balloon (5 cm in length) was inserted (via the nose) into the oesophagus, as far as the level of the bifurcation of the trachea. The balloon was connected to a pressure transducer via a vinyl catheter, and the signal transmitted from it was recorded on a *Visicorder*. Since the oesophageal pressure is the sum of the subglottal pressure and the elastic recoil pressure of the lungs, the elastic recoil of the lungs was also measured for the various lung volumes and this value was then subtracted from the oesophageal pressure to arrive at the subglottal pressure, in the manner outlined by Bouhuys *et alii* (1966).

A mesh-screen type pneumotachograph with face mask was used to measure the airflow rate. The air volume was measured by integrating the flow-rate electronically. Both of the signals used for these parameters were also recorded on a *Visicorder*. The mean flow rate was determined by dividing the exhaled volume of air by the phonation time.

GLOTTAL RESISTANCE

Mean glottal resistance was replaced by the ratio of the mean subglottal pressure to the mean flow rate, since it proved impossible to determine the phase-relationship between the latter two values with the instrumentation available.

ELECTROMYOGRAMS

Electrical potentials from the anterior cricothyroid, the internal intercostal (the sixth intercostal space at the mid-clavicular line) and from the rectus abdominis muscles were recorded. The motor unit potentials were picked up by 0.1 mm diameter bipolar concentric needles, amplified by means of solid state AC-DC amplifiers, and then recorded on a *Visicorder*. The writing speed was 100 mm/sec.

MOVEMENT OF THE CHEST WALL AND ABDOMEN

The movements of the chest wall and abdomen were determined from changes in the circumference of these cavities during phonation. Used to determine these changes as precisely as possible were small calibre vinyl tubes filled with mercury, one being placed around the subject at the level of the navel (for the abdomen measurements) and another at the midline of the chest in the plane between the nipple and the xiphoid process. Changes in the electrical resistance of the column of mercury in the tube, which varies inversely with its length, were recorded on a *Visicorder,* thus affording an accurate measure of changes in the circumference of the two cavities.

EXPERIMENTAL PROCEDURES

The subject was requested to phonate the vowel '*ah*', using the following types of vocalization: at normal (comfortable) pitch and normal intensity; at high pitch, at both loud and soft intensities; and at low pitch, at both loud and soft intensities. Each type of phonation was performed at least twice, both with and without mucosal anaesthetization, the subject being asked to reproduce the same level and quality of vocalization as nearly as he could in the separate trials.

Vocalization was monitored with both pitch and intensity meters, a Timcke stroboscope being employed for the former and a B & K 'Precise Sound Level Meter' for the latter. The microphone feeding the sound level meter was placed 18 cm from the outlet of the pneumotachograph. The voice signal was recorded on a dual-channel tape recorder, so that its exact pitch could be determined after the experiment: a *Visicorder* was also used to record the voice signal. The difference between the higher and the lower pitch used in phonation was one octave: the difference between the louder and the softer voice was 10 db. The mouth was opened to the same extent in each phase of the study, since the intensity of the voice is dependent upon the radiation area. As the aerodynamic determination required maximum phonation time, whereas the electromyographic measurements required phonation for only for a short time, these parameters were measured on separate occasions rather than simultaneously. Moreover, to afford a baseline for comparison purposes, all of these parameters were measured during a period of quiet respiration and during maximal expiration following maximal inspiration.

RESULTS

Aerodynamic analyses [Table 21.I and Fig. 21.2]. All parameters—namely, the mean subglottal pressure and the mean glottal resistance—had the same general configuration as that reported by other investigators (Van den Berg,

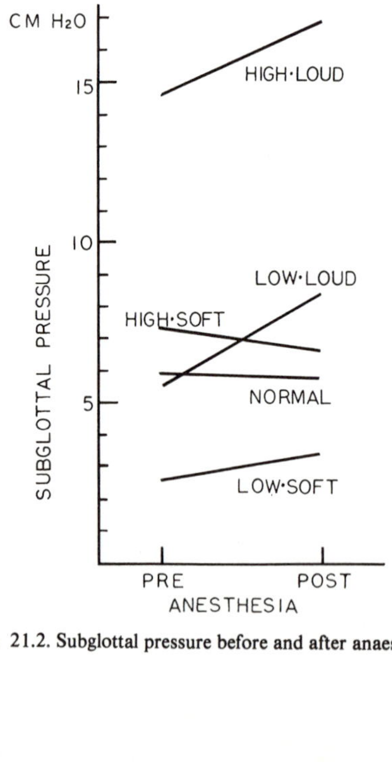

Fig. 21.2. Subglottal pressure before and after anaesthesia.

1956; Ladefoged and McKinney, 1963; Isshiki, 1964, 1965). Thus, to cite but one example, glottal resistance rose with increasing pitch and intensity of phonation.

During phonation at a comfortable pitch and intensity, it was found that none of these parameters was influenced by mucosal anaesthesia. With other types of phonation, however—except with softer voice at higher pitch—both glottal resistance and the subglottal pressure increased with anaesthesia.

Electromyographic analyses during brief phonation [Table 21.II and Fig. 21.3]. The anterior cricothyroid muscle showed a 210–740 msec period of activity preceding the onset of phonation.

At the higher pitch, regardless of intensity, the firing of the motor unit potentials was dominant. Anaesthesia did not affect the Δt* for the anterior cricothyroid muscle during any type of phonation.

TABLE 21.I

SUBGLOTTAL PRESSURE AND GLOTTAL RESISTANCE

TYPE OF VOICE	ANAESTHESIA	PITCH (Hz)	INTENSITY (db)	SUBGLOTTAL PRESSURE (cm H_2O)	GLOTTAL RESISTANCE (cm H_2O/l/sec)
Normal	pre	—	—	5.8	21.1
	post	—	—	5.7	21.1
High-soft	pre	215	80	7.3	35.8
	post	220	80	6.5	25.5
High-loud	pre	215	90	14.6	86.4
	post	215	90	16.8	107.7
Low-soft	pre	115	72	2.6	15.8
	post	110	70	3.4	20.2
Low-loud	pre	125	80	5.6	43.4
	post	125	82	8.4	49.5

* The time interval between the onset of a change in electrical activity in a particular muscle and the onset of phonation.

TABLE 21.II

TIME INTERVAL BETWEEN THE ONSET OF CHANGE IN ELECTRICAL ACTIVITY IN THE MUSCLES AND THE ONSET OF VOICE. (UNIT: MSEC)

TYPE OF VOICE	ANAESTHESIA	PITCH (Hz)	INTENSITY (db)	ANTERIOR CRICOTHYROID MUSCLE	INTERNAL INTERCOSTAL MUSCLE	RECTUS ABDOMINIS MUSCLE
Normal	pre	—	—	210	60	170
	post	—	—	220	70	170
High-soft	pre	240	80	460	110	370
	post	230	80	440	110	410
High-loud	pre	240	90	480	180	350
	post	240	88	470	180	420
Low-soft	pre	115	72	740	200	520
	post	110	71	760	220	670
Low-loud	pre	125	82	450	120	420
	post	125	80	500	130	490

The Δt of the internal intercostal muscle was 60 msec for normal phonation. For all other types of phonation, however, Δt was longer. The motor unit activity of this muscle became more pronounced as phonation progressed, and was most pronounced at the higher pitch and louder intensity. Anaesthesia did not appear to alter the value of Δt.

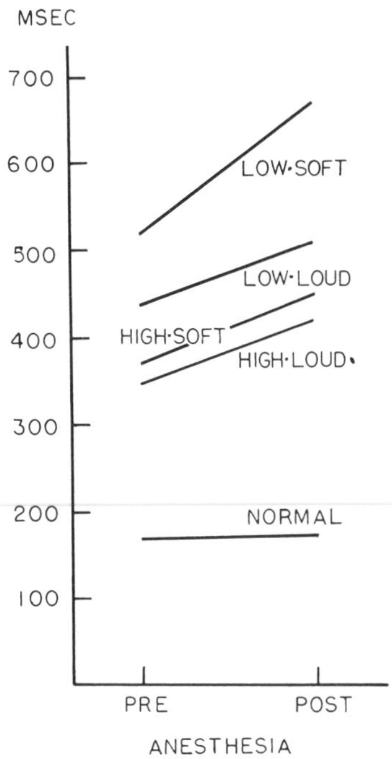

Fig. 21.3. The time interval between the onset of change in electrical activity in the rectus abdominis muscle and the onset of voice.

In general, the rectus abdominis muscle tended to show an onset of activity commencing before the onset of phonation, with the activity continuing and gradually increasing throughout phonation. Activity, moreover, was most pronounced during phonation in a loud voice at high pitch. The Δt of the abdominal muscles was lengthened by the application of mucosal anaesthesia (except during normal phonation), but not markedly so.

Movement of the chest wall and abdomen during brief phonation [Table 2.III and Fig. 21.4]. The chest wall showed a progressive diminution in circumference, commencing with the onset of phonation and continuing as long as it went on. Anaesthesia had no effect on the movement of the chest wall. The circumference of the abdomen, on the other hand, began to decrease before the onset of phonation, at first rapidly and then more slowly until the end of phonation. Anaesthesia

TABLE 21.III

TIME INTERVAL BETWEEN THE ONSET OF MOVEMENT OF THE ABDOMEN AND THE ONSET OF VOICE

TYPE OF VOICE	ANAESTHESIA	PITCH (Hz)	INTENSITY (db)	TIME INTERVAL (msec)
Normal	pre	—	—	140
	post	—	—	150
High-soft	pre	240	80	270
	post	230	80	340
High-loud	pre	240	90	300
	post	240	88	360
Low-soft	pre	115	72	400
	post	110	71	560
Low-loud	pre	125	82	290
	post	125	80	340

resulted in a very slight prolongation of the time interval between the beginning of deflation of the abdomen and the onset of phonation (except during normal phonation).

DISCUSSION

As noted earlier, there are three basic components of the servo-mechanism found in the larynx—the myotatic reflexes, the articular reflexes and the mucosal reflexes—and Wyke (1967, 1969) has suggested that these 'servo-reflexes' may be related to phonation. The purpose of the present study, consequently, was to determine to what extent the laryngeal mucosal reflex system—as a part of the total servo-mechanism—participates in the regulation of vocalization. Accordingly, we focused attention on the activity of the mucosal receptors in the larynx, since it has long been known that this organ is lined with mucosa containing slowly-adapting mechanoreceptors that respond to applied mechanical stress (Van Michel, 1963; Suzuki and Kirchner, 1968, 1969). It seems reasonable to postulate that such stimulation might bring about changes in the activity of the laryngeal musculature, as well as in the entire respiratory musculature, during speech—as Wyke (1967, 1969) has pointed out. In this connexion, Sampson and Eyzaguirre (1964) have shown that laryngeal mucosal

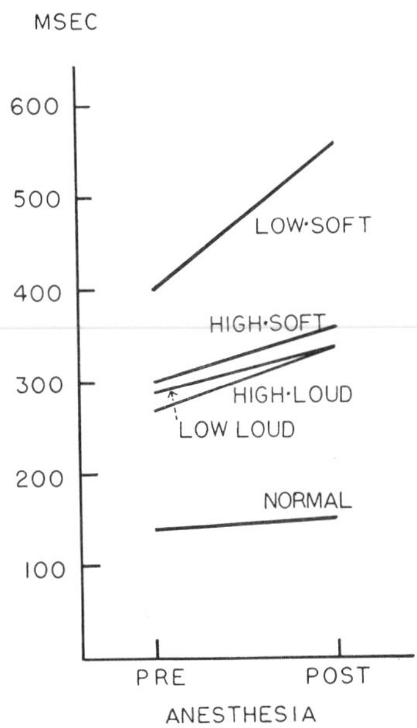

Fig. 21.4. The time interval between the onset of deflation of the abdomen and the onset of voice.

mechanoreceptors can be activated not only by direct touch but also by the oscillations of an air stream passing through the larynx. They concluded that the laryngeal mucosal mechanoreceptors would respond readily to contact with a foreign body, and that they might have a direct influence upon voice regulation. Tending to confirm this view is a study by Suzuki and Kirchner (1969), showing that the nerve impulses which a jet of air directed into the larynx of a cat evokes in afferent fibres in the laryngeal nerves can be inhibited or diminished by prior application of topical anaesthetic agents to the laryngeal mucosa.

In his study of the human voice, Husson (1951) reported that production of *falsetto* was impossible during local anaesthetization of the laryngeal mucosa. Iwamura (1967) found, however, that local anaesthesia had no effect upon the intensity of phonation. Using ultra-high speed cinematography, Zemlin (1969) found that the application of topical anaesthesia to the larynx did not affect the vibratory pattern of the vocal folds in women.

Aside from these few studies, however, there are apparently no other investigations touching upon the influence of the laryngeal mucosal mechanoreceptor reflexes on voice control.

As already noted, the results of our aerodynamic studies without anaesthesia were clearly consistent with those obtained by previous investigators. The application of local anaesthesia to the laryngeal mucosa, however, resulted in increased glottal resistance—except during the production of the normal voice, and of the softer voice at a higher than normal pitch. Accompanying this increase in glottal resistance, we also observed an increase in subglottal pressure. This would suggest that the laryngeal mucosal receptors are intimately involved in the regulation of the laryngeal activity required for voice production. As is known, the pitch and intensity levels of the voice are regulated by the expiratory force and the disposition of the vocal folds—that is, their mass, length and tension. Thus, it is apparent that subglottal pressure in particular plays a key rôle in regulating the pitch and intensity of the voice.

In our experiments, it seemed clear that the application of topical anaesthesia to the larynx raised the threshold for activation of the mucosal receptors of the larynx, and consequently also raised the pressure threshold for the firing of afferent nerve impulses therefrom. It can accordingly be concluded that a higher subglottal pressure is required for phonation at a given pitch and intensity during the anaesthetized state, as compared with the absence of anaesthesia. It seems fair to conclude further that increased glottal resistance is a prerequisite for the attainment of this increased subglottal pressure. Definite confirmation of these conclusions, however, must await further studies.

As noted, anaesthesia produced no apparent effects upon normal phonation. This nevertheless seems consonant with the foregoing hypotheses, since normal phonation is produced without any conscious modification of pitch and intensity levels.

The electromyographic studies, as noted, showed that the anterior cricothyroid muscle was activated shortly before the onset of phonation and further, that its motor unit activity during phonation was most pronounced with the higher pitch—findings in keeping with those of other investigators (Faaborg-Andersen, 1964; Hiroto, 1966). Local anaesthesia did not affect the activity of this muscle; and, in particular, it did not affect Δt.

Our findings with respect to the activity of the expiratory muscles are much like those of Draper *et alii* (1960), even though we were concerned only with short phonation times. The onset of brief phonation in our study started from around the end of a quiet inspiration in almost all instances. The Δt of the internal intercostal muscle was 50 msec in normal phonation, but was longer in the other types of phonation. The same trend was also observed in the anterior cricothyroid and the rectus abdominis muscles. Apparently vocalization at any pitch or intensity entails a certain time lag for the laryngeal apparatus to adjust for, and for the subglottal pressure to reach, the level required for the particular type of phonation. The motor unit activity of the internal intercostal muscle, we found, became more pronounced as phonation progressed, its intensity being most pronounced at the higher pitch and louder intensity. In addition, the point at which it became dominant occurred earlier with phonation at high pitch and loud voice than it did with other types of phonation. Local anaesthesia, however, did not affect the Δt of the muscle. But similar studies dealing with the myotatic reflexes and the articular reflexes may clarify the rôle of these muscles.

The activity of the rectus abdominis muscle also became more pronounced as phonation progressed. The most provocative finding with respect to the abdominal muscles, however, was that their activity set in before the onset of phonation—a finding that was corroborated by our measurements of changes in the circumference of the abdomen. Before the onset of phonation, the abdomen became deflated; whereas, by contrast, deflation of the chest wall began simultaneously with the onset of phonation. We also found that the onset of electromyographic muscular activity occurred somewhat before the beginning of deflation of the abdomen [Fig. 21.5].

It seems reasonable to assume that the early onset of activity seen in the rectus abdominis muscle and the early beginning of deflation of the abdomen represent a latency or preparatory period designed to raise the subglottal pressure to the threshold needed for proper vocalization. In general, anaesthesia of the laryngeal mucosa tended to lengthen the latency period or lag between the onset of abdominal muscle activity and the beginning of phonation, and to lengthen the duration of the pre-phonatory deflation of the abdomen, although in neither instance was the lengthening pronounced. Despite these observations, it should be noted that the increase in subglottal pressure seen with anaesthetization might be explicable in terms of the earlier onset of abdominal muscle activity: this latter, in other words, may compensate for the higher threshold of stimulation of the anaesthetized mucosal receptors.

Movement of the chest wall was not influenced by anaesthesia. Even so, the movement of the abdominal wall is of interest when we consider the mechanism for the production of subglottal pressure. The muscles of the abdominal wall, as we have seen, began to contract before phonation commenced. Electromyographic studies likewise showed a lag between the onset of abdominal muscle activation and the onset of phonation. The abdominal musculature continued to contract rapidly and forcefully once activation had begun, and continued to do so until attaining a certain peak, depending upon the type of phonation. It then remained at this level of activity for the entire period of vocalization. By contrast, chest wall deflation commenced simultaneously with the onset of vocalization—although the onset of activity in the intercostal muscle preceded the onset of vocalization, and continued at a steady rate so long as vocalization continued.

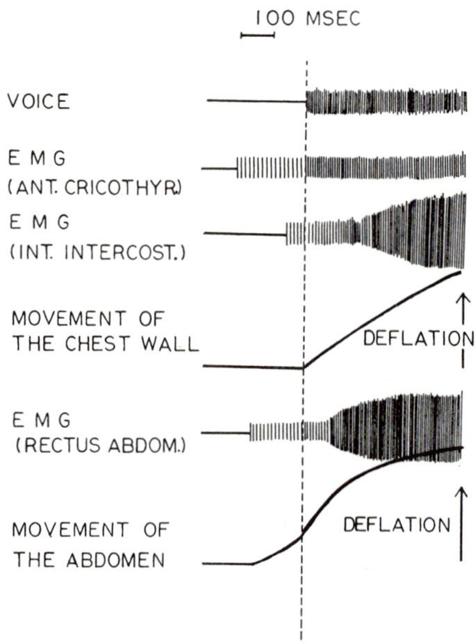

Fig. 21.5. Schema of several parameters at normal short phonation.

From these findings we conclude that the abdominal musculature plays a key rôle in producing the subglottal pressure necessary for phonation, as suggested by the lag between onset of the movement of the abdomen and the onset of phonation. The main function of the internal intercostal muscles during phonation may be maintenance of a steady airflow rate.

CONCLUSION

The principal purpose of this study was to investigate the contribution of the laryngeal mucosal reflex system to vocalization. After application of topical anaesthesia to the larynx, an increase in the subglottal pressure was observed, using the same pitch and intensity levels for phonation. Further, the latency period between the onset of activity in the abdominal muscles and the onset of phonation tended to lengthen with the application of anaesthesia, apparently as a result of the depressant effect of the anaesthetic agent upon the mechanoreceptors of the laryngeal mucosa and the consequent increase in pressure required to stimulate them. These results clearly imply a close reflex interconnexion between the laryngeal mechanoreceptors and the abdominal musculature. Our findings also clearly suggest that muscles involved in postural maintenance also influence vocalization. But, given the preliminary nature of our investigation, more extensive studies will be needed to determine the extent to which these findings are applicable to the population at large.

From the point of view of proprioceptive controls of voice production, further investigation of the initial phases of phonation would also appear to be indicated, since it is thought to be the crucial period in which the entire character of phonation may be determined.

REFERENCES

Bouhuys, A., Proctor, D. F., and Mead, J. (1966) Kinetic aspects of singing, *J. appl. Physiol.,* **21,** 483–96.

Draper, M. H., Ladefoged, P., and Whitteridge, D. (1960) Expiratory pressures and air flow during speech, *Brit. med. J.,* **1,** 1837–43.

Faaborg-Andersen, K. (1957) Electromyographic investigation of intrinsic laryngeal muscles in humans, *Acta physiol. scand.,* **41** (Suppl. 140), 1–148.

Hiroto, I. (1966) The mechanism of phonation: Its pathophysiological aspects, *Pract. Otol. (Kyoto),* **59** (Suppl. I), 229–91.

Husson, R. (1951) Physiologie: Etude stroboscopique des modifications réflexes de la vibration des cordes vocales declenchées par des stimulations expérimentals du nerf auditif et du nerf trijumeau, *C. R. Acad. Sci. (Paris),* **232,** 1247–9.

Isshiki, N. (1964) Regulatory mechanism of voice intensity variation, *J. Speech Res.,* **7,** 17–29.

Iwamura, S. (1967) An experimental study of control of vocal intensity, *J. Otolaryng. Jap.,* **70,** 728–44.

Ladefoged, P., and McKinney, N. P. (1963) Loudness, sound pressure, and subglottal pressure in speech, *J. acoust. Soc. Amer.,* **35,** 454–60.

Sampson, S., and Eyzaguirre, C. (1964) Some functional characteristics of mechanoreceptors in the larynx of the cat, *J. Neurophysiol.,* **27,** 464–80.

Suzukí, M., and Kirchner, J. A. (1968) Afferent fibres in the external branch of the superior laryngeal nerve in the cat, *Ann. Otol. (St. Louis),* **77,** 1059–70.

Suzuki, M., and Kirchner, J. A. (1969) Sensory fibres in the recurrent laryngeal nerve, *Ann. Otol. (St. Louis),* **78,** 21–31.

Van den Berg, J. (1956) Direct and indirect determination of the mean subglottic pressure, *Folia phoniat. (Basel),* **8,** 1–24.

Van Michel, C. (1963) Considerations morphologigues sur les appareils sensoriels de la muqueuse vocale humaine, *Acta anat. (Basel),* **52,** 188–92.

Wyke, B. D. (1967) Recent advances in the neurology of phonation: Phonatory reflex mechanisms in the larynx, *Brit. J. Disor. Commun.*, **2,** 2–14.

Wyke, B. D. (1969) Deus ex machina vocis. An analysis of the laryngeal reflex mechanisms of speech, *Brit. J. Disor. Commun.,* **4,** 3–25.

Zemlin, W. R. (1969) The effect of topical anaesthesia on internal laryngeal behaviour, *Acta oto-laryng. (Stockh.),* **68,** 169–76.

DISCUSSION ON COMMUNICATION [21]

Fry. I believe Dr. Brodnitz has some material which he thinks is particularly apposite to this paper so, with your permission, I will ask Dr. Brodnitz to speak first in the discussion.

Brodnitz. This material, as you will see, has only an indirect albeit an interesting connexion; but let me say, first, that this very interesting paper raises a number of important questions. I was particularly impressed by a fact that came up repeatedly in Dr. Gould's study; namely, the onset of abdominal muscle contraction (or activation) before the onset of phonation—because this is, in a way, a laboratory demonstration of what professional singers (with a very misleading term that they use a great deal) call the 'support of the voice'. When a singer gets set to sing he takes a good-sized breath and the abdominal wall moves outwards; but before the first tone comes out of the mouth, he perceives a certain amount of tension in the abdominal wall. If you could get a singer to perform for you in the laboratory you might see this even more dramatically: I have my doubts whether any singer would be willing to sit with a balloon in his oesophagus and needle electrodes stuck in all over his chest and abdomen—but you could try.

If you continue your studies as you plan, Dr. Gould, it would be very interesting if you could include an examination of the action of the strap muscles, and particularly the sternothyroid muscle, in relation to tensing of the vocal folds. In addition, maybe you could fulfil an old wish of mine that somebody would do a study of the back muscles in phonation, because singers always tell you that they feel certain tensions in their backs and all kinds of weird theories are bandied about by singing teachers in relation to this or that muscle. These longitudinal muscles probably play a part, too, in getting the singer into postural readiness for phonation, and somebody should investigate them in this respect.

The case that I want to introduce here has an indirect connexion with Dr. Gould's work, in so far as his study involved the abolition of mucosal reflexes by local anaesthesia. What I want to demonstrate to you now is something that seems to me unique—at least, I have never seen anything like it—namely, a patient who speaks with a method of voice production that puts everything upside down in relation to what we have been discussing at this Symposium by speaking fluently *on inspiration.*

This man had a severe automobile accident that threw him against the steering wheel, crushing his upper trachea and severing both recurrent laryngeal nerves. A tracheostomy cleared his airway, and an operation rejoined the stump of the trachea to the cricoid; but the surgeons couldn't find the ends of his recurrent nerves, so he was left with bilateral vocal fold paralysis—and that is what he still has, with a glottis that is 1 or 2 mm wide. Subsequent speech therapy did not enable him to produce any kind of voice; but then one day, out of the blue, he suddenly discovered (he is a very intelligent man) that he could make sounds on inspiration. He persisted with the development of this ability to the point where he is now a university lecturer, continuing to talk with an inspiratory voice.

What you will hear on this tape [*Editor's Note:* Dr. Brodnitz commented on a tape recording of his patient's voice at this point] are three sequences. First, in answer to questions, he describes his accident and how he discovered his ability to speak on inspiration: he does this in an astonishingly fluent voice that is interrupted by noisy exhalations after groups of words. Then he reads for a brief period, and you will notice that his voice is even better controlled. Finally he makes some attempts to speak on expiration but succeeds only in producing a weak and often 'cracking' voice.

Even after being informed about his accident and his success in developing an inspiratory voice I did not at first believe that what I heard was a pure inspiratory voice—until I asked him to take his jacket off, put my fingers on his chest and noted the expansion of his chest every time he spoke.

Kirchner. What do the cough-like sounds represent?

Brodnitz. They are noisy expirations. He speaks on inspiration and then quickly expels the inhaled air, producing a coughing noise. The situation has a lot of interesting implications about what it might do to his heart in the long run, and to the blood gas tensions and pH, because the whole physiological situation is upside down—and this is why I thought it might interest you to hear such a remarkable thing.

Fry. Of course, many of us have tried to do this kind of thing, and one's difficulty is in persuading oneself to get the air out of the lungs—those 'coughs' are your patient's device for actually doing that, so that he can start again from the bottom. When we talk, we want to breathe in and start. If you try to do what this man does—it is very difficult—the real difficulty is in getting the air out and *then* talking; and I think with that 'coughing' that is exactly what he is doing—he is getting rid of the air in order to start at the bottom of the cycle again.

Kirchner. I would like to ask Dr. Brodnitz to tell us what the patient's larynx looks like with a mirror.

Brodnitz. Bilateral paralysis, with both folds immobilized and just a small parallel chink between them about 1 or 2 mm wide.

Widdicombe. It did sound to me as though the expiratory efforts were not coughs. When one makes a forced expiration in coughing, you don't usually make that kind of noise—as when you blow straight into a microphone. Now we know you can get coughing from the larynx, but perhaps his larynx was changed in some way so that he was not really coughing. Are you quite sure it was not simply a hoarse expiration, rather than a cough?

Brodnitz. It is, as I said.

Fry. You notice that when he was reading he did not make that sound. Those 'coughs' were partly pausing phenomena, when he is thinking what to say next: where we might say '*um*' or '*er*', he partly fills in with that.

Kirchner. One of the key questions I want to ask Dr. Gould (and perhaps it is a little unfair because there may not be an answer to it) is how far below the mucosal surface do the effects of topical anaesthesia extend? This is one of the things that has concerned me in some of the experiments with mucosal receptors located below the vocal folds. With topical anaesthesia, is it not possible that enough absorption takes place from the surface to be affecting the underlying muscles and perhaps even deeper receptors in the larynx? I don't think this happens in fact, but do you have any information about it?

Gould. I have no further information on that point, other than the fact that the anaesthetic application was made to the mucosal surface. As I said before, one of the major things that we will have to do is to separate out possible effects on the various receptors, otherwise the observations are not adequately interpretable. The thing that was important to me was the substantiation in man of Dr. Wyke's animal experiments[1,2,3], and I would like to ask him if he would mention something about them so that the relation of our study to his will make sense to the audience.

Wyke. We have some information that may be relevant to the very pertinent question that Dr. Kirchner has asked—because this is a big problem with all topical anaesthesia experiments anywhere in the respiratory tact. It may be of some assistance if, as I am going to report tomorrow [see Communication 24], I say now that in the presence of topical anaesthesia of both the supra- and infraglottic mucosa one can still demonstrate myotatic and articular reflex activity in the nerves and intrinsic muscles of the larynx[4,5]. If that can be regarded as evidence of persisting activity in the related receptor systems with the methods of topical mucosal anaesthesia that we use (and, I believe, that Dr. Gould uses too), then this problem of deep diffusion of local anaesthetic agents and interference with deeply located nerve fibres and receptors might not be too significant.

But having said that, I would like to qualify it because I don't think the evidence I have just mentioned is as good as I would like—for the reason that the diameters of the afferent and efferent fibres that subserve the laryngeal myotatic stretch and arthrokinetic reflexes are larger than some of the small diameter fibres in the same area that Dr. Widdicombe mentioned earlier [see Communication 9] in connexion with other reflexes, the afferent fibres mediating which are present as well in the laryngeal nerves. So it might possibly be that whilst reflexes are obtainable in relatively large diameter fibre systems in the presence of topical anaesthesia, there might be some block of smaller diameter reflexogenic afferents—but this I just don't know. At least we have some experimental evidence that the whole laryngeal reflex system is not knocked out by topical mucosal anaesthesia.

---

[1] Wyke, B. D. (1967) Recent advances in the neurology of phonation: phonatory reflex mechanisms in the larynx, *Brit. J. Disor. Commun.* **2,** 2–14.

[2] Wyke, B. D. (1968) Effects of anaesthesia upon intrinsic laryngeal reflexes, *J. Laryng.*, **82,** 603–12.

[3] Wyke, B. D. (1969) Deus ex machina vocis: an analysis of the laryngeal reflex mechanisms of speech, *Brit. J. Disor. Commun.*, **4,** 3–25.

[4] Abo-El-Enein, M. A., and Wyke, B. D. (1966) Laryngeal myotatic reflexes, *Nature (Lond.),* **209,** 682–6.

[5] Kirchner, J. A., and Suzuki, M. (1968) Laryngeal reflexes and voice production, *Ann. N.Y. Acad. Sci.*, **155,** 98–109.

Anthony. I think there is a connexion between the case that Dr. Brodnitz introduced and the problem of subglottic pressure. I have made some subglottic pressure measurements on people speaking and reading on inspiration, and it looks as though the subglottic pressure variation is a mirror-image of what one has with expiratory speech.

I might just make the further point that women of between 50 and 60 years of age or so sometimes speak this way—I mean, they speak on inhaling, as a normal thing.

Gould. Where does that happen?

Anthony. In Edinburgh.

Murakami. Could I add a little information to your paper, and also to Dr. Kirchner's question? With Dr. Fukuda (who is now in Dr. Kirchner's Laboratory in Yale) I have tried to make recordings of afferent impulses in the internal branch of the superior laryngeal nerve (which is, of course, the most important afferent pathway from the larynx) during vocal fold mucosal vibration, produced by blowing air up from below the larynx after suturing the vocal folds to one another. The interesting thing is that the number of impulses then seen in the internal branch of the superior laryngeal nerve corresponds exactly to the cycle of the sound waves—that is, the number of impulses in the afferent nerve is related exactly to the mucosal vibrations. Another important thing is that this afferent discharge is very easily disturbed by and completely disappears after injection of *Xylocaine* into the mucosa or the submucosa. So the point is, I want to know whether this kind of afferent activity is really the afferent part of a mucosal reflex arc or not.

Freedman. As I understood it, Dr. Gould, you suggested that the abdominal muscles were starting to contract before phonation in order to build up subglottic pressure. The fact that there was no change in thoracic circumference when there was a change in abdominal circumference would imply, however, that either the diaphragm was contracting and you were taking up the slack, as it were; or that muscles other than the particular intercostal you were sampling were serving to hold the chest shape the same. But would you not expect, if the abdominal muscles were to contract and the diaphragm and the intercostals were to remain relaxed, that necessarily there would be some change in chest shape if you were going to transmit events up to the glottis?

Gould. It wasn't only the abdominal shape, but also the electrical activity of the abdominal muscles that changed. In other words, it was abdominal motor unit discharges that showed prephonatory activity.

Freedman. The abdominal muscles showed electrical activity prior to phonation?

Gould. Yes.

Freedman. And they showed some activity at a time when the intercostal muscle that you were sampling was silent?

Gould. Yes.

Freedman. In that case I would suggest that there is some other muscle —perhaps the diaphragm, perhaps an intercostal from further back—that was active at the time, in order to keep the chest shape from changing.

Klatt. I have another question on that point. Did your subject begin voicing with the so-called 'glottal stop' onset, so that he was able to build up pressure and then speak? That seems the most likely situation, since your pressure build-up preceded voicing by so much time.

Gould. Not as far as we could tell. There was probably some degree of glottal stop there, but we tried to keep it to a minimum.

Klatt. How far does the pressure build up before you begin to get an acoustic output?

Gould. I don't know off-hand. I would have to look up the actual figures, and show them to you later.

Klatt. My main question concerns the controls that one might apply to an experiment like this to determine whether the local anaesthesia was causing the events that you noted, or whether this was just a variation in the way people speak. In fact, we know that there are a lot of such variations. Can you give us some idea of how your results varied?

Gould. We asked the same question ourselves; and since our paper was written we have examined a group of about a dozen actors and singers (including one singer who has probably the most exquisite control of abdominal and thoracic breathing that I have ever seen), and the results thus far are very similar to those I reported. But I want to stress that this paper is to be regarded as a preliminary communication, as we clearly require many more data before we can determine the real significance of mucosal reflex effects in phonation.

Ladefoged. Obviously I find your observations very interesting indeed; but further to Dr. Klatt's question I would like to add something. In ordinary speaking activity (such as when we were asking about 12 people to say sentences like '*He wanted to be an elevator operator*' we noticed that there was a great deal of variation from individual to individual. Roughly speaking, 8 of the 12 had no rectus abdominis contractions in their ordinary conversation and 4 of the 12 did—so some people do one thing and some people do the other.

Gould. I want to stress that with normal conversation we did not notice any change in our experiments, and with soft voice at high pitch we did not notice anything either: it was the other variations of conversation I described that were affected. Essentially it is the voice at higher intensity and higher pitch that is affected in our procedure, and I don't want that to be mistaken.

Plum. Do you have recordings of this singer you mentioned who can differentiate between abdominal and chest muscles in voluntarily-controlled breathing, because that is a very interesting observation? Wade[1], many years ago, said he was able to observe this, and Stigol and Cuello[2] were able to record it in only one of eight people: but as far as I am aware, those are the only documented observations of individuals who can voluntarily change the programme of the pattern of breathing. Essentially, what you have done with your subject is to provide stronger evidence that descending pathways other than those subserving the automatic reticulospinal system programme can modify the act of breathing. This is why I would be so interested to hear a little more about the observation.

Gould. I have the recordings for which you ask. And could I add that the most interesting example I ever had in this respect was a young girl who came from Denver (Colorado), and therefore naturally had a very harsh, high-pitched voice. She came over here to the London School of Dramatic Art and was taught what she called 'proper breathing'; and after that it was like listening to two different people talking—she could switch from one to the other pattern of voicing at will.

Plum. The question then is, does this involve an alteration in the resistance against breathing or an alteration in general body posture; or is it, in fact, a change in the actual muscles used to carry out the breathing act without changing resistances and so forth? In other words, can she do this in quiet breathing, or just in speaking and singing?

Gould. Yes, she can do it in both sets of circumstances. As a matter of fact she is studying Yoga, and we had her locked in a room while we took recordings during complete relaxation—and she could do it in quiet breathing as well.

Plum. Then you should put this on record, because it is important.

Proctor. One suggestion that might help separate mucosal effects from possible effects on the laryngeal muscles of local anaesthesia would be to inject the superior laryngeal nerves, instead of using topical anaesthesia. Then, of course, you would probably not get your cricothyroid muscle operating—but the rest of the muscular activity should be all right.

Gould. That would be useful as a next step, so that we could compare the results with the data we already have.

Ruth Bowden. The suggestion that some speakers and singers might use their erectores spinae muscles beforehand (that is, prior to phonation) suggests that possibly the contraction of the rectus abdominis might be a secondary reflex to maintain posture, because it is one of the most important antagonists of the erectores spinae. Of course, investigating this means adding yet one more electrode to the unfortunate subject; and I would suggest an even more uncomfortable one, to make sure that intra-abdominal pressure is being raised, and that is to record from the levator ani muscle—because that should be contracting (that is, the pelvic floor should be contracting) as the pressure rises.

---

[1] Wade, O. L. (1954) Movements of the thoracic cage and diaphragm in respiration, *J. Physiol. (Lond.)*, **124**, 193–212.

[2] Stigol, L. C., and Cuello, A. C. (1966) Voluntary control of the diaphragm in one subject, *J. appl. Physiol.*, **21**, 1911–12.

Gould. That is very interesting, because as a matter of fact there is a neurologist in New York who has specialized in electromyographic studies of the levator ani, particularly in relation to breathing.

Wyke. In reply to the point made by Dr. Proctor about injection of the superior laryngeal nerves, I should mention for the record that we have, in fact, published[1,2] a comparable series of animal experiments employing unilateral and bilateral section of the internal laryngeal nerves and of the superior laryngeal nerves before and after testing the laryngeal myotatic reflex responses, comparing the effects with the reflex responses of the muscles to the same stretch stimuli before and after topical anaesthesia of the mucosa. I won't go into all the details here, as they are already in the literature [see also Abo-El-Enein, 1967[3]].

Beverly Bishop. I want to come back to something very arbitrary for a minute, and have Dr. Wyke describe for us what he means by the mucosal reflex. It seems to me that the context in which the term is used has become very complex, and I would like to be clued in on what the afferent and motor components are like.

Wyke. In answer to that, I don't think (and have never said) that there is any such thing as *the* mucosal reflex. I think there is a variety of glottic opening and closing reflexes to which mucosal receptors may contribute; but the one that I am talking about in relation to the experiments I have mentioned is that operated by low-threshold mechanoreceptors that are immediately submucosal. There are a lot of receptor systems at various depths within the mucosa—some of them corpuscular and some of them non-corpuscular—and I don't think we know much about what each individual morphological variety does. From a functional point of view, however, there is at least one clear population of very low threshold mechanoreceptors (distinct from the pain receptors)—which, for example, Drs. Kirchner and Suzuki[4] (as well as others[5]) have also demonstrated in operation. When we are talking about subglottic mucosal reflexes perhaps we should be more specific, and say that this particular reflex system—the low-threshold mechanoreceptor reflex system—is the one referred to.

Beverly Bishop. All right, that's fine. Then what do you see as the motor component of this reflex?

Wyke. What we find happens is that if you stimulate this reflex system by raising the subglottic pressure—either holding it raised, or pulsing the subglottic pressure—then you produce either a sustained increase in tension in the adductor muscles, or you produce a pulsed increase in adductor muscle activity. Other experiments with oscilloscopic recordings of the afferent discharges in the

---

1 Abo-El-Enein, M. A., and Wyke, B. D. (1966) Laryngeal myotatic reflexes, *Nature (Lond.)*, **209**, 682–6.

2 Wyke, B. D. (1973) Myotatic reflexogenic systems in the larynx, *Folia morphol. (Prague)*, **21**, 113–17.

3 Abo-El-Enein, M. A. (1967) *Functional Anatomy of the Larynx*, Ph.D. Thesis, University of London.

4 Kirchner, J. A., and Suzuki, W. (1968) Laryngeal reflexes and voice production, *Ann. N.Y. Acad. Sci.*, **155**, 98–109.

5 Sampson, S., and Eyzaguirre, C. (1964) Some functional characteristics of mechanoreceptors in the larynx of the cat, *J. Neurophysiol.*, **27**, 464–80.

laryngeal nerves lead us to believe that this is a reflex response operating from a low-threshold mechanoreceptor system in the subglottic mucosa up through some central relay mechanism in the brain stem (which we haven't yet traced out) to the adductor muscles of the vocal folds.

Beverly Bishop. I would now like to ask Dr. Gould if he sees the change in the time of appearance of abdominal rectus muscle activity in his experiment as an aspect of the motor output of this reflex system. I thought you were implying that.

Gould. I felt that this preceded mucosal stimulation, and therefore was possibly (as we postulated earlier this morning—see pp. 310–11) a preparatory act of the muscle for speech. The feeling I had was that if you had the afferent system depressed and so requiring more intense stimulation, then this would necessitate producing a higher pressure from the abdominal muscles.

Beverly Bishop. Well, this becomes anything but a simple problem!

Gould. No, it's not.

Beverly Bishop. Could I ask if Dr. Widdicombe would speak about the receptor response?

Widdicombe. (To Dr. Wyke). You left out some important information; namely, the histological identification of these receptors in relation to the particular stimuli that may excite them—we know that there are cough receptors and chemoreceptors, for example, in the laryngeal mucosa.

Wyke. I left it out because, as I said, we don't have enough combined histological and physiological data to be able to give you categoric information. I would like to reiterate my previous statement; namely, that there is an array of receptors that is morphologically identifiable within the mucosa, but at this stage I wouldn't like to attribute any particular physiological response to any one of the several varieties. That is something we may have further information about, I hope, by the end of this year.

Winckel. Another comment on the rectus abdominis muscles. In my own experience of singing training (and I had a very good teacher—he was also a doctor of medicine), it seems that with muscular tension of the rectus abdominis I have the right 'support' to use the air reserve—but not the whole reserve, as the mechanism is very elastic. This means that the control is so sensitive at the level of the glottis that one constantly adjusts between inspiration and expiration—or, in other words, the conditions of working are so sensitive that you have only small pressure changes going on. But you can only do this if you have an adequate air reserve; and if you are to do it with a loud voice and at a very high pitch, then you must have still more sensitive control—and this comes only after years of training. It is only possible if you have already trained the whole musculature of the body; and my teacher used to prepare us by saying that you must train until you feel the muscular 'strengthening' as far as the tips of your toes. But most of the control comes from the abdominal muscles; and therefore I say that you must have the correct posture, otherwise you fail to get the high tones—especially in opera singing.

As for the back muscles—and this includes the psoas—I learnt something else from my teacher. We went to the clinic and made radiographs of the back, and as a result he could decide 'This is a singer, and this is no singer', because the volume of the psoas shadow was more developed in singers than in normal persons.

Finally, I have a comment about the neck musculature. There was one paper on the first day relating to this [see Communication 1] and I wanted to comment, but there was not enough time. You will recall the neck posture of Caruso that you have seen in photographs; and my teacher used to say that you must have the posture of a boxer in order to get out sounds of sufficient intensity at high pitches.

Kirchner. There are a couple of points I want to get Dr. Gould's thoughts about. One is that with topical anaesthesia presumably all the receptors in the mucosa have been eliminated, and on his curves it was interesting to me that before anaesthesia and after anaesthesia the curve always went upward; and I gather from this that it always took more subglottic pressure to produce the tone that was eventually emitted nevertheless. I wonder if it might be that once the total receptor population in the mucosa is gone, what it then takes is stimulation of some other kind of receptors. Is it so that the subglottic pressure is always greater after local anaesthesia of the mucosa?

Gould. Yes, absolutely.

Kirchner. That deals with one question. The other is (I don't know whether these were trained singers whom you were testing) that it would be interesting (if they were) to see whether the quality of the refinements of voice production was changed after anaesthesia.

Gould. They were trained singers and actors, together with a very well-trained investigator from the University of Kyoto.

Kirchner. Was there any gross change in the quality of the vocal performance?

Gould. No, not as far as the audible tone was concerned. It required more subglottic pressure, but the tone came out with the same quality.

Wyke. I just want to respond briefly to Dr. Kirchner's question. My hypothesis in relation to the study that Dr. Gould has reported would be that the reflex effects from the various laryngeal receptors are centrally additive in terms of their control of motor unit activity in the muscles posturing the vocal folds during phonation[1,2,3,4], so that if you subtract the mucosal receptor mechanism

---

1 Wyke, B. D. (1967) Recent advances in the neurology of phonation: phonatory reflex mechanisms in the larynx, *Brit. J. Disor. Commun.*, **2**, 2–14.

2 Wyke, B. D. (1968) Effects of anaesthesia upon intrinsic laryngeal reflexes, *J. Laryng.*, **82**, 603–12.

3 Wyke, B. D. (1969) Deus ex machina vocis: an analysis of the laryngeal reflex mechanisms of speech, *Brit. J. Disor. Commun.*, **4**, 3–25.

4 Wyke, B. D. (1971) The neurology of stammering, *J. psychosom. Res.*, **15**, 423–32.

from the totality of the machinery then you have to build up a higher subglottic pressure in order to activate more forcefully the stretch receptors in the laryngeal musculature in order to maintain reflexly the vocal fold posture—that is, generate a more forceful myotatic reflex to maintain appropriate vocal fold adduction during phonation of a designated tone in the absence of the normally concurrent mucosal mechanoreceptor reflex effects. This is one possibility; and in this connexion you will remember reporting[1] an experiment that you did with Dr. Suzuki, in which you extracted in order various components from the total complex of reflex control systems in the larynx.

von Euler. Is the ability to maintain a constant pressure just the same in your subjects, Dr. Gould?

Gould. Yes, it is.

---

[1] Kirchner, J. A., and Suzuki, M. (1968) Laryngeal reflexes and voice production, *Ann. N.Y. Acad. Sci.*, **155**, 98–109.

22 Innervation of Intrinsic Laryngeal Muscles

RUTH E. M. BOWDEN

Department of Human Anatomy and Morphology,
Royal Free Hospital School of Medicine,
London, England

The control of the activity of the laryngeal muscles raises many interesting anatomical and physiological problems. Their actions are bilaterally symmetrical, and are precisely and delicately graded. Their phasic, tonic and/or volitional contractions form an integral part of complex patterns of activity in respiration, deglutition and phonation. The adductors of the vocal folds maintain and raise intrathoracic pressure during expulsive movements such as coughing, defaecation, micturition and parturition. They support and protect the vertebral column by raising intrathoracic pressure to as much as 100 cm H_2O during the initial and continuing phases of lifting and holding heavy weights (Davis, 1959a, b).

Such a high degree of integration suggests that central and peripheral mechanisms must be involved in the control of these small, but important muscles. Others at this Symposium are dealing with their central control mechanisms: in this paper attention is confined to the anatomy of the peripheral nerve trunks and the pattern of innervation of the muscles. These have been studied for over a hundred years by many workers using a variety of methods which now include histochemistry and neurophysiology.*

The topics dealt with here have long been controversial, and include the sites of the cells of origin of motor and afferent neurones; the sites and degrees of communication between ipsilateral and contralateral laryngeal nerves; the types of receptor within the muscles; the precise form of motor innervation of the muscles and the nature of the striated muscle fibres. The two latter points are of particular interest, in relation to the puzzling fact that the abductors of the vocal folds often appear to be most susceptible to paresis after partial injuries of the recurrent laryngeal nerve (Semon, 1881).

CELLS OF ORIGIN OF MOTOR AND AFFERENT NEURONES

The thyroid cartilage, epiglottis and vestibular folds are derived from the fourth branchial arch, and the cricoid and arytenoid cartilages and intrinsic muscles

*The bibliography therefore cannot be exhaustive, but the references cited give a wide coverage of the relevant literature.

arises from the sixth arch, the associated cranial nerves being the vagus and Cr XI respectively.

Lower motor neurones supplying the striated muscles innervated by IX, X and Cr XI nerves are located in the nucleus ambiguus. Chromatolytic studies have shown that multipolar cells in the lower part of the nucleus supply the larynx in man and chimpanzee (Obersteiner, cited by Strong and Elwyn, 1953; Bernard, cited by Hooper, 1887). In man and rhesus monkey some fibres from this nucleus cross the midline to emerge with nerves of the opposite side (Obersteiner, 1890; Mitchell and Warwick, 1955). In the cat, DuBois and Foley (1936) traced fibres in the superior laryngeal nerve to the most caudal vagal rootlets, and those of the recurrent to Cr XI rootlets.

Cell bodies of afferent neurones in the superior and recurrent laryngeal nerves of the rabbit are located mainly in the nodose ganglion, a few being in the jugular ganglion of the vagus (Mohlant, 1912; Evans and Murray, 1954). In the cat similar findings were made by Heinbecker and O'Leary (1933) and DuBois and Foley (1933, 1937).

Hakamada (1943; cited by Hosokawa, 1961) found groups of cells in vagal nerve rootlets in the rabbit and in the bulbar rootlets of the baboon's accessory nerve. He regarded the former as equivalent to cells in the jugular ganglion. The central processes of these afferent fibres end in various situations. Those subserving pain sensibility in the larynx probably end in relation to the spinal nucleus of the trigeminal nerve (Brodal, 1947); the few taste fibres from the epiglottis and vallecula end in the ipsilateral gustatory portion of the nucleus of the tractus solitarius, whilst others may terminate in the dorsal nucleus of the vagus (Mitchell and Warwick, 1955).

PERIPHERAL COMMUNICATIONS BETWEEN LARYNGEAL NERVES

Anatomically detectable communications occur at three levels—the highest, in or immediately below the jugular foramen; intermediate ones occur deep to the laryngeal mucosa; and the most distal occur within the mucosa and intrinsic muscles.

The Cr XI nerve communicates with the superior (jugular) ganglion and joins the main trunk of the vagus just distal to the inferior or nodose ganglion, its motor fibres being distributed through the pharyngeal and recurrent laryngeal branches of the vagus.

The submucosal plexus varies between orders and within a single species. In 26 species of Eutherian mammals (Withington, 1959; Bowden and Scheuer, 1961) there were four patterns of communication between intrinsic nerve trunks. In the first there was no gross link between the internal and recurrent nerves. In the second they were united by a ramus communicans. In the third, a median laryngeal nerve from the pharyngeal plexus of the vagus joined the superior laryngeal nerve which in turn was linked to the recurrent by a ramus communicans. In the fourth, a pararecurrent nerve ran parallel to the recurrent and joined the internal laryngeal nerve [see Fig. 22.1a, b, c and d] and Table 22.I.

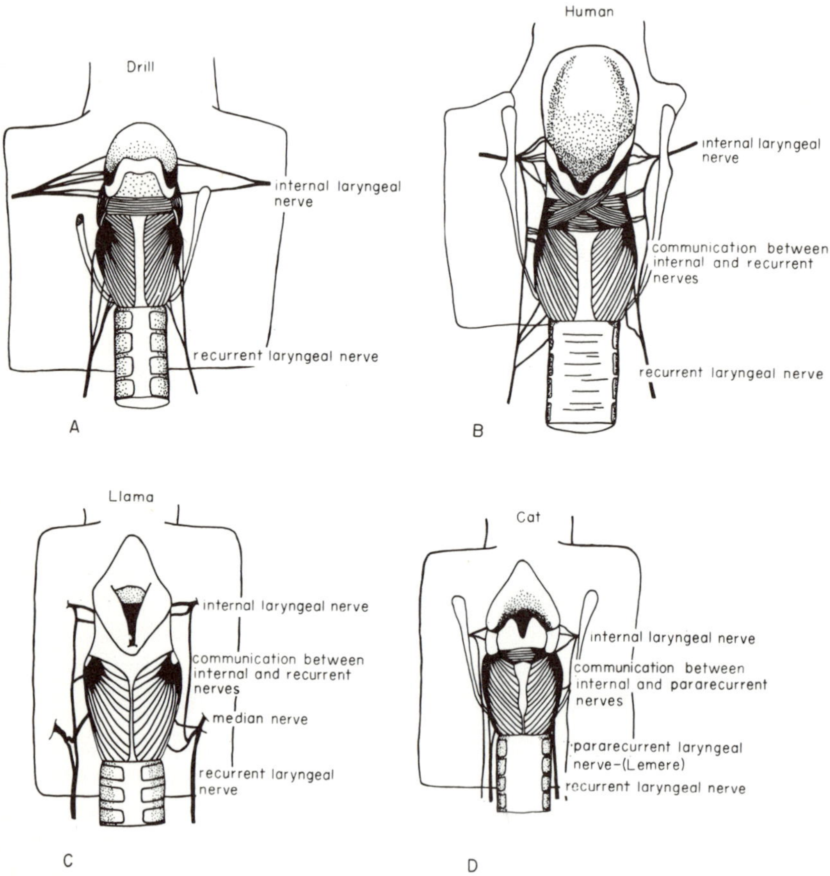

Fig. 22.1. Drawings of larynges to show communications between intrinsic nerve trunks (J. L. Scheuer). Type A, Drill; Type B, Man; Type C, Llama; Type D, Cat.

In this series two out of fourteen human specimens were in the first group (A) and twelve in the second (B): Nordland (1930) found a ramus communicans in only three out of nineteen human larynges, and Lemere (1932) found it in two out of twelve. In our series all six dogs were in the second group, and all five cats had a connexion between pararecurrent and internal laryngeal nerves, this being at

TABLE 22.I*

NERVE SUPPLY OF THE LARYNX IN EUTHERIAN MAMMALS

ORDER	NO. SPECIES	NO. SPECIMENS	TYPE OF COMMUNICATION (See Fig. 21.1)			
			A	B	C	D
Insectivora	2	3	3			
Chiroptera	2	2	2			
Lagomorpha	2	8	8			
Primates	9	25	10	14		1
Carnivora	5	15		7		8
Artiodactyla	6	10		3	7	
Totals	26	63	23	24	7	9

* From Bowden and Scheuer (1961)

variance with Lemere's (1932) observations. Seven rabbits and a single rhesus monkey were in the first group. Howell and Huber (1891) found a ramus communicans in the rabbit. There appears to be no clear-cut phylogenetic significance in these patterns, but the species and individual variations may account for some of the discrepancies in reports of the results of experimental stimulation or section of laryngeal nerves. Intramuscular plexuses have been found in rat, rabbit, cat and man (Feindel, Hinshaw and Weddell, 1952; Withington, 1959; Abo-El-Enein, 1967; Lucas Keene, 1961). There are also intramucosal and perivascular plexuses.

The distribution of the nerves of the two sides has been controversial. According to Longet (1842, cited by Mohlant, 1912) this was said to be strictly unilateral in dogs, the internal laryngeal nerve being purely afferent to the mucosa above the vocal folds, and the recurrent laryngeal supplying all the muscles except the cricothyroid, which was innervated by the ipsilateral external laryngeal nerve. Lemere (1932a, b) reported an additional supply for the cricothyroid from the middle laryngeal nerves in the dog. Exner (1884) stated that in the dog all nerves were distributed bilaterally, so that all muscles had a double supply from both right and left internal and recurrent laryngeal nerves.

Dilworth (1921) considered that in man the nerves of the two sides should be considered as a single vagal plexus.

Clinical, and most experimental, studies in man and animals support the statement that the motor supply is derived from the ipsilateral recurrent laryngeal nerve (the interarytenoid muscle receiving a supply from each nerve) and that the cricothyroid is supplied with motor fibres through the ipsilateral external laryngeal nerve, with an additional supply from another source. Feindel, Hinshaw and Weddell (1952) showed by degeneration studies in the rabbit that this second source of motor innervation came from the pharyngeal plexus.

The presence of intramucosal, perivascular and intramuscular plexuses neither proves nor excludes the possibility of bilateral communications between afferent nerves within the larynx. Although there is no definite anatomical proof of a contralateral afferent supply, Abo-El-Enein and Wyke (1966a, b) and Abo-El-Enein (1967) have provided interesting neurophysiological evidence for this in relation to certain proprioceptive impulses. They found that in the cat, the internal laryngeal nerves invariably contained proprioceptive afferent fibres from stretch-sensitive receptors located in the ipsi- and contralateral intrinsic laryngeal muscles. The recurrent laryngeal nerve contained proprioceptive afferents from mechanoreceptors in strictly ipsilateral muscles, there being no evidence of muscular proprioceptive afferent fibres in the external laryngeal nerves (although these nerves contain articular afferent fibres from receptors in the cricothyroid joints—see Suzuki and Kirchner, 1968).

THE INTERNAL TOPOGRAPHY OF LARYNGEAL NERVES

One traditional explanation of Semon's law (1881) is that the nerve supply of the abductors is in an exposed, and therefore more vulnerable, position within the recurrent nerve; others are that the nerve fibres are of smaller calibre, or that they enter the larynx in separate branches. There is no convincing evidence that the abductor nerve supply is in a more vulnerable position within the recurrent laryngeal nerve. Sunderland and Swaney (1952) have shown that the intraneural plexus makes it impossible to pick out a separate bundle to the abductors until the branch is given off.

Branching of the nerve may or may not occur prior to its entry through or beneath the lower border of the inferior constrictor of the pharynx [see Table 22.II] (Bowden, 1955). Whilst bundles of small fibres are present in the recurrent nerve in dog, cat and monkey (Howell and Huber, 1891; Lemere, 1932b; DuBois and Foley, 1936; Brocklehurst and Edgeworth, 1940; Murray, 1957; Abo-El-Enein, 1967), these are mostly distributed to the infraglottic, tracheal and oesophageal mucosa. Ogura and Lam (1953) found the large and small fibres intermingled in the superior laryngeal nerve and its two main branches.

TABLE 22.II*

EXTRALARYNGEAL BRANCHES OF THE RECURRENT LARYNGEAL NERVE

AUTHOR	NO. OF NERVES WITH EXTRALARYNGEAL BRANCHES TO LARYNX	NUMBER OF BRANCHES				
		2	3	4	5	6
Armstrong and Hinton (1951)	73	48	19	5	1	—
Dilworth (1921)	66	66	—	—	—	—
Weeks and Hinton (1942)	33	31	2	—	—	—
Rustad (1954)	86	67	9	7	2	1
Bowden (1955)	42	33	4	5	—	—
Totals	330	245	34	17	3	1

* From Bowden (1955)

THE FIBRE SIZE DISTRIBUTION IN LARYNGEAL NERVES

A bimodal fibre diameter spectrum and the presence of large diameter fibres in related nerves has been taken as strong evidence for the existence of mechanoreceptors in mammalian skeletal muscles, in which spindles and tendon organs are supplied with afferent fibres of Group I (12–22μ) and Group II (4–12μ) (Hunt, 1954). The presence of such receptors in intrinsic laryngeal muscles was denied because a unimodal distribution was found in the recurrent laryngeal nerves of rabbit and cat (Evans and Murray, 1954; Murray, 1957). In man, reports have been conflicting. A bimodal fibre diameter spectrum was found by Murtagh and Campbell (1951), with peaks at 1–2 μ and 4–6 μ; and Tomasch and Britton (1953) reported peaks at 1–2μ and 6–7μ. Faaborg-Andersen (1957) and Scheuer (1964) found a unimodal distribution, with the presence of fibres up to 18–20μ in diameter. When Scheuer combined the histograms of recurrent and internal laryngeal nerves the distribution was bimodal. Abo-El-Enein (1967) found (in two cats) that at 1 cm below the cricoid cartilage the recurrent nerve consisted of one large branch and one or more smaller branches. The diameter spectrum of individual branches could be uni- or bimodal; but when all branches of a nerve were combined the distribution was bimodal, with peaks at 2–4μ and 8–10μ or 10–12μ. The large branch contained fibres of 6μ or more in diameter, whilst those below 6μ were mostly in smaller branches. Three out of the four internal laryngeal nerves were bimodal. In cat I the diameter of myelinated fibres ranged from 2–16μ and in cat II from 2–14μ. The total fibre count for these animals is shown in Table 22.III, differences in numbers of fibres supplying the left and right sides of the larynx of the two animals being 384 and 265 respectively.

TABLE 22.III*

TOTAL NUMBERS OF MYELINATED FIBRES IN LARYNGEAL NERVES OF CAT

CAT I	LEFT	RIGHT
Recurrent laryngeal nerve	545	744
Internal laryngeal nerve	1234	751
Totals	1779	1495
CAT II	LEFT	RIGHT
Recurrent laryngeal nerve	650	962
Internal laryngeal nerve	1545	1398
Totals	2195	2360

* From Abo-El-Enein (1967)

The left and right external laryngeal nerves in cat II contained 69 and 205 fibres respectively, and both had a unimodal diameter spectrum. (It will be recalled that the cricothyroid muscle has an additional source of innervation.)

The fibre diameter spectra in cat and man are therefore compatible with the existence of classic mechanoreceptors in the intrinsic laryngeal muscles.

THE INTRAMUSCULAR PATTERN OF INNERVATION

There is a high density of innervation, and intramuscular and perivascular plexuses are invariably present in the intrinsic muscles in rat, cat and man.

RECEPTOR ENDINGS

Classical spindles were reported in some human intrinsic laryngeal muscles by Nakamura (1915), Sunder-Plassman (1933a, b), Göerttler (1950) and Paulsen

(1958a, b, c); and they have been found in all such muscles of the adult by Lucas Keene (1957, 1961) and in the foetus by Bowden (1963). Simpler types of claw and spiral endings have also been seen in all these muscles by several authors (Sunder-Plassman, 1933a, b; Lucas Keene, 1957, 1961; König and von Leden, 1961; Rudolph, 1961). Gracheva (1963) failed to find spindles in adult and foetal human material, but found spiral endings in the adult thyro-arytenoid muscle. Dichotomously dividing terminals were seen running between muscle fibres towards their tendons; and free endings were plentiful in the perichondrium, especially on the epiglottis and arytenoid cartilages.

Reports have conflicted in relation to subhuman forms. A few spindles were found in *Macaca mulatta* by Brocklehurst and Edgeworth (1940). In dog and cat Gracheva (1963) found none, but Withington (1959) found a few in the posterior crico-arytenoid of the cat. Abo-El-Enein (1967) found them in all the laryngeal muscles of the cat, although not always in each muscle of a single individual. He found spiral endings were more common, and that the related muscle fibre could be as much as 36μ in diameter. He was the first to find a simple type of tendon organ in laryngeal muscles, and he also confirmed the presence of the fine terminals described by Gracheva (1963).

MOTOR INNERVATION

A high density of motor innervation and scattered motor units was suggested by electromyography (Weddell, Feinstein and Pattle, 1944) and also by the rich intramuscular plexus found by many workers. Simple and compound motor endings have also been seen, and the presence of the latter has suggested that single muscle fibres might sometimes be innervated by separate parent neurones. Frequently the multiple plates are supplied by a single axone; but where this cannot be seen, pleurisegmental innervation of single muscle fibres cannot be assumed on histological evidence alone. However, Hunt and Kuffler (1954) demonstrated this neurophysiologically in the cricothyroid of the cat. Although Feindel, Hinshaw and Weddell (1952) found no histological evidence of double innervation of muscle fibres in the rabbit, they showed that this muscle was supplied from the external laryngeal nerve as well as the pharyngeal plexus.

Abo-El-Enein (1967) found a broad band of motor endings in the middle thirds of the lateral crico-arytenoid and thyro-arytenoid muscles. In the vocalis the band was very thin. Motor innervation zones were scattered in the posterior crico-arytenoid and cricothyroid muscles. He also found a correlation between the type of constituent muscle fibre and type of end-plate. Muscles composed mainly of fibres staining darkly with Sudan Black B carried compound motor endings and most fibres were multiply innervated. In muscles containing light, dark and intermediate fibres, endings were simple and multiple innervation of fibres was the exception rather than the rule.

The subneural apparatus was sharply defined in simple endings, but in compound endings it was less well defined and fragmented. These points are seen clearly in the lateral and vocalis portions of the thyro-arytenoid. Only simple endings were found in vocalis, cricothyroid and posterior crico-arytenoid muscles. Compound endings were found in crico-arytenoid lateralis and the lateral part of the thyro-arytenoid.

Histological investigations have not given evidence of a scatter of the muscle fibres constituting a single unit, but probably explain the differences in properties of contraction of the laryngeal muscles reported by Mårtensson and Skoglund (1964) and Hall-Craggs (1967).

The frequent, but not invariable, susceptibility of the abductors to paresis is not yet fully explained. The nerve supply does not appear to be in a more vulnerable position. If the fibres supplying these muscles are of larger calibre than those innervating the adductors (as suggested by Lemere (1934)), they would tend to be more susceptible to the effects of pressure (Seddon, 1942): if they are smaller, as suggested by Murtagh and Campbell (1951), they would be less vulnerable. This point requires further analysis, and there is need for additional study of the size of motor units and motor/afferent innervation ratios. The nature of the muscle fibres in these muscles is also relevant, and evidence is again conflicting (cf. Abo-El-Enein, 1967 and Lemere, 1942).

However, the relative bulk of the two groups of muscles may be significant, although it is unlikely to be the only factor. This has been studied by several workers, amongst them Faaborg-Andersen (1957), and a more extensive study was made in twelve human larynges by Bowden and Scheuer (1960). The total weight of the muscles on each side was comparable within the limits of experimental error. There was a highly significant difference between the bulk of adductors and abductors, for both wet and dry weights, the abductors being only 25 per cent of the weight of the adductors.

In conclusion, it m:y be said that anatomical studies have provided possible explanations for some discrepancies in the literature. They have provided conclusive proof of the presence of classical mechanoreceptors in the intrinsic laryngeal muscles, and this is supported by physiological evidence. The physiological characteristics of spindles, tendon organs and spiral endings in these muscles require further study. There is need for quantification and location of the afferent endings in relation to motor units within individual muscles. Further analysis of the differences between adductors and abductors is also required, and is in progress.

ACKNOWLEDGEMENTS

Tribute must be paid to the meticulous work of my colleagues Dr. Abo-El-Enein, Professor Lucas Keene and Dr. Scheuer (née Withington) who have contributed much to the resolution of long-standing problems of laryngeal function.

REFERENCES

Abo-El-Enein, M. A. (1967) *Functional Anatomy of the Larynx,* Ph.D. Thesis, University of London.

Abo-El-Enein, M. A., and Bowden, R. E. M. The pattern of innervation of laryngeal muscles. [In preparation.].

Abo-El-Enein, M. A., and Wyke, B. D. (1966a) Laryngeal myotatic reflexes, *Nature, (Lond.)*, **209,** 682–6.

Abo-El-Enein, M. A., and Wyke, B. D. (1966b) Myotatic reflex systems in the intrinsic muscles of the larynx, *J. Anat. (Lond.),* **100,** 926–7.

Armstrong, W. G., and Hinton, J. W. (1951) Multiple divisions of the recurrent laryngeal nerve *Arch. Surg.*, **62,** 532–9.

Bernard, C. (1887) Cited by Hooper, F. H. (1887).

Bowden, R. E. M. (1955) The surgical anatomy of the recurrent laryngeal nerve, *Brit. J. Surg.,* **43,** 153–63.

Bowden, R. E. M. (1963) Muscle spindles in the human foetus, *Acta biologica,* **9** (fasc. 1–4): 35–59.

Bowden, R. E. M., and Scheuer, J. L. (1960) Weights of abductor and adductor muscles of the human larynx, *J. Laryng.,* **74,** 971–80.

Bowden, R. E. M., and Scheuer, J. L. (1961) Comparative study of the nerve supply of the larynx in Eutherian mammals, *Proc. Zool. Soc. Lond.*, **136,** 325–30.

Brocklehurst, R. J., and Edgeworth, F. H. (1940) The fibre components of the laryngeal nerves of *Macaca mulatta, J. Anat. (Lond.),* **74,** 386–9.

Brodal, A. (1947) Central course of afferent fibres for pain in facial, glossopharyngeal and vagus nerves, *Arch. Neurol. Psychiat. (Chicago),* **57,** 292–306.

Buchthal, F. (1959) Electromyography of intrinsic laryngeal muscles, *Quart. J. exp. Physiol.,* **44,** 137–48.

Davis, P. R. (1959a) Posture of the trunk during the lifting of weights, *Brit. med. J.,* **1,** 87–9.

Davis, P. R. (1959b) The causation of hernia by weight lifting, *Lancet,* **2,** 155–7.

Dilworth, T. F. M. (1921) The nerves of the human larynx, *J. Anat. (Lond.),* **56,** 48–52.

Dubois, F. S., and Foley, J. O. (1936) Experimental studies on the vagus and spinal accessory nerves in the cat, *Anat. Rec.*, **64** (and supp.), 285–302.

Dubois, F. S., and Foley, J. O. (1937) Quantitative studies of the vagus nerve in the cat. II. The ratio of jugular to nodose fibres, *J. comp. Neurol.,* **67,** 69–87.

Evans, D. H. L., and Murray, J. G. (1954) Histological and functional studies on the fibre composition of the vagus nerve of the rabbit, *J. Anat. (Lond.),* **88,** 320–37.

Exner, B. (1884) Die Innervation des Kehlkopfes Sitzungsb. der Kais, *Akad. der Wissensch.,* **89,** 63–118.

Faaborg-Andersen, K. L. (1957) Electromyographic investigation of intrinsic laryngeal muscles in human. An investigation of subjects with normally movable vocal cords and patients with vocal cord paresis, *Acta. physiol. scand.*, **41** (supp. 140), 1–148.

Feindel, W., Hinshaw, J. R., and Weddell, G. (1952) The pattern of motor innervation in mammalian striated muscle, *J. Anat. (Lond.)*, **86,** 35–48.

Göerttler, K. (1950) Die Anordnung, Histologie und Histogenese der quergestreiften Muskulatur im menschlichen Stimmband, *Z. Anat. Entwickl.-Gesch.*, **115,** 352–401.

Gracheva, M. S. (1963) Sensory innervation of locomotor apparatus of the larynx. *Arkhiv Anat. Gristol. Embriol.* (Russian text). **44,** 77–86.

Hakamada, S. (1943) Cited by Hosokawa, H. (1961).

Hall-Craggs, E. C. B. (1967) Contraction times and metabolism in rabbit laryngeal muscles, *J. Anat. (Lond.)*, **101,** 181–2.

Heinbecker, P., and O'Leary, J. (1933) The mammalian vagus nerve. A functional and histological study, *Amer. J Physiol.*, **106,** 623–46.

Hooper, F. H. (1887) The anatomy and physiology of the recurrent laryngeal nerve, *N.Y. med. J.*, **46,** 29–36, 63–6, 99–103, 150–2, 188–90.

Hosokawa, H. (1961) Proprioceptive innervation of striated muscles in the territory of cranial nerves, *Tex. Rep. Biol. Med.*, **19,** 405–63.

Howell, W. H., and Huber, G. C. (1891) Physiology of the communicating branch of the superior and inferior laryngeal nerves, *J. Physiol. (Lond.)*, **12,** 5–11.

Hunt, C. C. (1954) Relation of function to diameter in afferent fibres of muscle nerves, *J. gen. Physiol.*, **38,** 117–31.

Hunt, C. C., and Kuffler, S. W. (1954) Motor innervation of skeletal muscle: multiple innervation of individual muscle fibres and motor unit function, *J. Physiol. (Lond.)*, **126,** 293–303.

Keene, M. F. Lucas, (1957) Nerve endings in human laryngeal muscles, *J. Anat. (Lond.)*, **91,** 590.

Keene, M. F. Lucas, (1961) Muscle spindles in human laryngeal muscles, *J. Anat. (Lond.)*, **95,** 25–29.

König, W. F., and von Leden, H. (1961) The peripheral nervous system of the human larynx. I. The mucous membrane, *Arch. Otolaryng.*, **74,** 1–14: II. The thyroarytenoid (vocalis) muscle, *Arch. Otolaryng.*, **74,** 153–63: III. The development, *Arch. Otolaryng.*, **74,** 494–500.

Lemere, F. (1932) Innervation of the larynx. I. Innervation of laryngeal muscles, *Amer. J. Anat.*, **51,** 417–37: II. Ramus anastomoticus and ganglion cells of the superior laryngeal nerve, *Anat. Rec.*, **54,** 389–407.

Longet (1842) Anatomie et physiologie du système nerveux. Cited by Mohlant (1912).

Mårtensson, A., and Skoglund, C. R. (1964) Contraction properties of intrinsic laryngeal muscles, *Acta physiol. scand.*, **60,** 318–36.

Mitchell, G. A. G., and Warwick, R. (1954) Nuclei of IX, X and XI complex of cranial nerves, *J. Anat. (Lond.)*, **88,** 569–70.

Mitchell, G. A. G., and Warwick, R. (1955) The dorsal vagal nucleus, *Acta anat. (Basel)*, **25,** 371–95.

Mohlant, M. (1912) Le nerf vague: Etude anatomique et expérimentale. II. Innervation motrice du larynx, *Le nervraxe*, **13,** 22–44.

Murray, J. G. (1957) Innervation of the intrinsic muscles of the cat's larynx by the recurrent laryngeal nerve: a unimodal nerve, *J. Physiol. (Lond.)*, **135,** 206–12.

Murtagh, J. A., and Campbell, C. J. (1951) The respiratory function of the larynx. III. The relation of fiber size to function in the recurrent laryngeal nerve, *Laryngoscope (St. Louis)*, **59,** 581–90.

Murtagh, J. A., and Campbell, C. J. (1952) Physiology of the recurrent laryngeal nerve: report on progress, *J. clin. Endocr.*, **12,** 1398–1401.

Nakamura, N. (1915) Cited by Hosokawa (1961).

Nordland, M. (1930) The larynx as related to the surgery of the thyroid, *Surg. Gynec. Obstet.*, **51,** 449–59.

Obersteiner, H. (1890) Cited by Strong and Elwyn (1953).

Ogura, J. H., and Lam, R. L. (1953) Anatomical and physiological correlations on stimulating the human superior laryngeal nerve, *Laryngoscope (St. Louis)*, **63,**947–59.

Paulsen, K. (1958a) Untersuchungen über das Vorkommen und Zahl von Muskelspindeln im M. vocalis des Menschen, *Z. Zellforsch.*, **47,** 363–6.

Paulsen, K. (1958b) Über Vorkommen und Zahl von Muskelspindeln in inneren Kehlkopfmuskeln des Meschen, *Z. Zellforsch.*, **48,** 349–55.

Paulsen, K. (1958c) Über die Bedeutung von Muskelnspindeln und Schleimhautreceptoren bei der Phonation, *Arch. Ohr.-, Nas.-, u. Kehlk.-Heilk.*, **173,** 500–3.

Rustad, W. B. (1954) Revised. anatomy of the recurrent laryngeal nerves. Surgical importance based on the dissection of one hundred cadavers, *J. clin. Endocr.*, **14,** 87–96.

Rudolph, G. (1961) Spiral nerve endings in human vocal muscles, *Nature (Lond.)*, **190,** 726–7.

Scheuer, J. L. (1964) Fibre size frequency distribution in normal human laryngeal nerves, *J. Anat. (Lond.)*, **98,** 99–104.

Seddon, H. J. (1942) A classification of nerve injuries, *Brit. med. J.*, **2,** 237–9.

Semon, F. (1881) Clinical remarks on the proclivity of the abductor fibres of the recurrent laryngeal nerves to become affected sooner than the adductor fibres, or even exclusively in cases of undoubted central or peripheral injury or disease of the roots, trunks of the pneumogastric, spinal accessory or recurrent nerves, *Arch. Laryng.*, **2,** 197–222.

Strong, O. S., and Elwyn, A. (1953) *Human Neuroanatomy*, 3rd ed., p. 212, Baltimore.

Sunder-Plassman, P. (1933a) Über den Nervenapparat des Musculus vocalis, *Z. Hals-, Nas.-u. Ohrenheilk*, **32,** 493–9.

Sunder-Plassman, P. (1933b) Über den Nervenapparat des menschlichen Glottisöffners, Musculus cricoarytenoideus posticus, *Z. Hals-, Nas., Ohrenheilk*, **32b,** 586–98.

Sunderland, S., and Swaney, W. E. (1952) The intraneural topography of recurrent laryngeal nerves in man, *Anat. Rec.*, **114,** 411–25.

Suzuki, M., and Kirchner, J. A. (1968) Afferent nerve fibres in the external branch of the superior laryngeal nerve in the cat, *Ann. Otol. (St. Louis)*, **77,** 1059–71.

Tomasch, J., and Britton, W. A. (1953) A fibre analysis of the laryngeal nerve supply in man, *Acta anat. (Basel),* **23,** 386–98.

Weddell, G., Feinstein, B., and Pattle, R. E. (1944) The electrical activity of voluntary muscle in man under normal and pathological conditions, *Brain,* **67,** 178–257.

Weeks, C., and Hinton, J. W. (1964) Extralaryngeal division of the recurrent laryngeal nerve, *Ann. Surg.,* **116,** 251–7.

Withington, J. L. (now Scheuer, J. L.) (1959) *Comparative Studies of the Nerve Supply of the Larynx with Special Reference to the Afferent Innervation of the Musculature,* Ph.D. Thesis; University of London.

DISCUSSION ON COMMUNICATION [22]

von Euler. That was a wonderful description of the innervation of the laryngeal tissues; but because of the controversy on the presence or absence of muscle spindles, I wish we had a special paper devoted to the question. Maybe it was due to my bad vision, but I wasn't absolutely convinced from the pictures you showed us that the receptors that were pointed out as muscle spindles really are muscle spindles—although one could certainly see something like nuclear bags in the slide you referred to as having been provided by Dr. Wyke [*Editor's Note:* Part of this slide is reproduced as Fig. 24.5A at p. 414]. Have you studied the motor innervation of the intrafusal muscle fibres in those preparations—so as to be able to convince us that these structures have all the characteristics of muscle spindles?

Ruth Bowden. Yes—but we need to do a good deal more because, as you will recall from the 1961 *Symposium on Muscle Receptors* in Hong Kong[1], there is just about every possible view on this—you name the pattern, and you can find it in a spindle somewhere. I *have* seen the *en plaque* type of ending, but I haven't seen a real 'trail' ending myself in these spindles. I agree it is always very difficult, especially in foetal material, to be convinced; but I can promise you that I have found nuclear bags. I went up and down the receptor with serial sections, and if a nuclear bag doesn't occur in every section then that is not a spindle for me—however much it might look like a nuclear bag in one section. Again, Withington[2] has made serial sections all the way through with similar results.

Widdicombe. I am afraid I am revealing my ignorance of the literature, but has anybody done the obvious experiment of making supra-nodose nerve sections and eliminating the motor fibres, so that you could divide the fibre spectrum up into afferent and efferent components?

Ruth Bowden. This was done by Evans and Murray (1954)[3] in the rabbit.

Widdicombe. I know their description of what happened to the pulmonary and cardiac innervation, but I don't recall their data on laryngeal innervation.

---

[1] Barker, D. [Ed.]. (1962) *Symposium on Muscle Receptors.* Hong Kong.

[2] Withington, J. Louise. (1959) *Comparative Studies of the Nerve Supply of the Larynx, With Special Reference to the Afferent Innervation of the Musculature,* Ph.D. Thesis, University of London.

[3] Evans, D. H. L., and Murray, J. G. (1954). Histological and functional studies on the fibre composition of the vagus nerve of the rabbit, *J. Anat. (Lond.),* **88,** 320–37.

Ruth Bowden. They did it for the larynx as well; but they didn't report that they studied every branch, and I think this is where, sometimes, it has been a bit misleading [see also Fernand and Young, 1951[1]]. As you saw, the fibre diameter spectrum is more likely to have a unimodal distribution if you take only one specimen. We have evidence in our material that one recurrent nerve was bimodal, but two were not. Join the two histograms of the recurrent and superior laryngeal nerves together, and you get the over-all picture.

Mårtensson. Conclusive evidence of the presence of muscle spindles in the intrinsic laryngeal muscles would be provided if in a cross-section you could find a conglomeration of nuclei indubitably belonging to one and the same intrafusal fibre. I am not an anatomist but I would imagine that you might find extrafusal fibres surrounded by connective tissue, thus mimicking intrafusal fibres. Therefore I would like to ask whether you have any slides demonstrating a conglomeration of nuclei all included in the same muscle fibre, thus indicating an intrafusal nuclear bag region. So far I think, all muscle spindles have been found to contain at least one intrafusal fibre that includes a nuclear bag.

Ruth Bowden. I showed one such slide in my Communication, and I have others.

Mårtensson. With high magnification, so we could see that all the nuclei are included in one and the same intrafusal fibre?

Ruth Bowden. I don't have the original histological section here, but I presented a slide showing the nuclear bag region in intrafusal fibres in a single spindle—you saw the cross-section of that.

Mårtensson. Isn't it possible to get a conglomeration of nuclei in a section that shows that they are in one and the same fibre?

Ruth Bowden. Certainly. But taking it in terms of one and the same fibre, your criterion would not be definitive for a spindle because one finds the same appearance at many motor end-plates. I agree that what one wants to see is a conglomeration of nuclei in an area where the muscle fibre cross-striations are faint: because as you know, with the electron microscope certainly (and even in good light microscope preparations) one sees faint cross-striation. In foetal muscles one certainly can trace intrafusal muscle fibres up and down in serial sections, because in foetal muscle spindles the intrafusal fibres are relatively large compared with the extrafusal. In the absence of a clearly-defined nuclear bag or chain seen throughout its full longitudinal extent, a spindle cannot be identified with certainty.

Kirchner. I would like to ask Dr. Bowden about the ramus communicans (or ramus anastomoticus in the old terminology). I wonder whether this is more common, or more commonly active, in the infant than it is in the adult. When we intubate infants for anaesthesia, or when we pass a bronchoscope and stretch the laryngeal lumen, we see all sorts of changes in the rate and rhythm of the heart. This is more common in the infant than in the adult; and one of the questions that this raises in my mind is whether the baroreceptor afferent fibres from some of

[1] Fernand, V. S. V., and Young, J. Z. (1951) The sizes of the nerve fibres of muscle nerves, *Proc. roy. Soc. Lond.*, **139B**, 38–58.

the intrathoracic structures that travel through the recurrent nerve may be present in the infant and perhaps not later on in life. Do you think this is possible?

Ruth Bowden. I have no evidence either way, but I would not think that the ramus communicans would disappear. One knows that dorsal root ganglion cells, for example, gradually degenerate throughout life, particularly in the elderly. This is probably, I think, secondary to ischaemic changes—but I don't think there is any evidence to support the idea that a whole nerve branch would disappear in that way.

Wyke. I would like to suggest that perhaps this argument that has been going on for so many years about whether or not there are muscle spindles in the intrinsic laryngeal muscles may, to some degree, be a red herring from a functional point of view. This is because whilst I am satisfied that there *are* a few muscle spindles in the intrinsic laryngeal muscles of some species (including the cat and man) I don't think, from a physiological point of view, that that is all that important—in view of the very much larger number of spiral nerve endings that are capable of functioning as stretch-sensitive mechanoreceptors[1,2,3]. In other words, my present argument would be that there are mechanoreceptors (stretch-sensitive receptors) in the intrinsic larygeal muscles, but that we are not yet in a position to attribute the afferent discharges from these receptors selectively to one or other of the two morphologically appropriate entities that we know to be present in these muscles. But I think Dr. von Euler and Dr. Mårtensson might be a little happier if they saw some neurophysiological evidence indicating that there are stretch-sensitive mechanoreceptors in the intrinsic laryngeal muscles—although, as I said, I can't tell you from which receptor type (spindles or spirals) the stretch responses are coming. Could I therefore have your permission, Mr. Chairman, to show two slides in physiological support of what Dr. Bowden has said?

This [Fig. D.14] is an oscilloscopic recording from the intact left recurrent laryngeal nerve of an anaesthetized cat. The upper tracing [A] shows the control background discharge, and the lower tracing [B] shows the effect of rapidly applying (at S) 5 gm of passive stretch to the detached anterior end of the ipsilateral vocalis muscle. This [Fig. D.15] is a recording (in the same animal) from the peripheral cut end of the same nerve (which has been partially teased out) when 5 gm of stretch has again been applied (at the signal) to the detached anterior end of the same vocalis muscle; and you can see that it evokes an afferent discharge at the moment the stretch is applied that decrements very slowly during maintained stretch. I should add (as I said before—see p. 362) that this discharge is unaltered by topical anaesthesia of the laryngeal mucosa, by disarticulation of the crico-arytenoid and cricothyroid joints, or by bilateral section of the superior laryngeal nerves; and therefore I suggest that part of the activity in Fig. D.14B and all of the activity in Fig. D.15 can reasonably be attributed to low-threshold, stretch-sensitive mechanoreceptors located in the vocalis muscle.

Ruth Bowden. Thank you very much. I would agree entirely with Dr. Wyke on this point of the stretch receptors not being spindles exclusively, because Sybil

[1] Abo-El-Enein, M. A., and Wyke, B. D. (1966) Myotatic reflex systems in the intrinsic muscles of the larynx, *J. Anat. (Lond.),* **100,** 926–7.

[2] Abo-El-Enein, M. A., and Wyke, B. D. (1966) Laryngeal myotatic reflexes, *Nature (Lond.),* **209,** 682–6.

[3] Wyke, B. D. (1973) Myotatic reflexogenic systems in the larynx, *Folia morphol. (Prague),* **21,** 113–17.

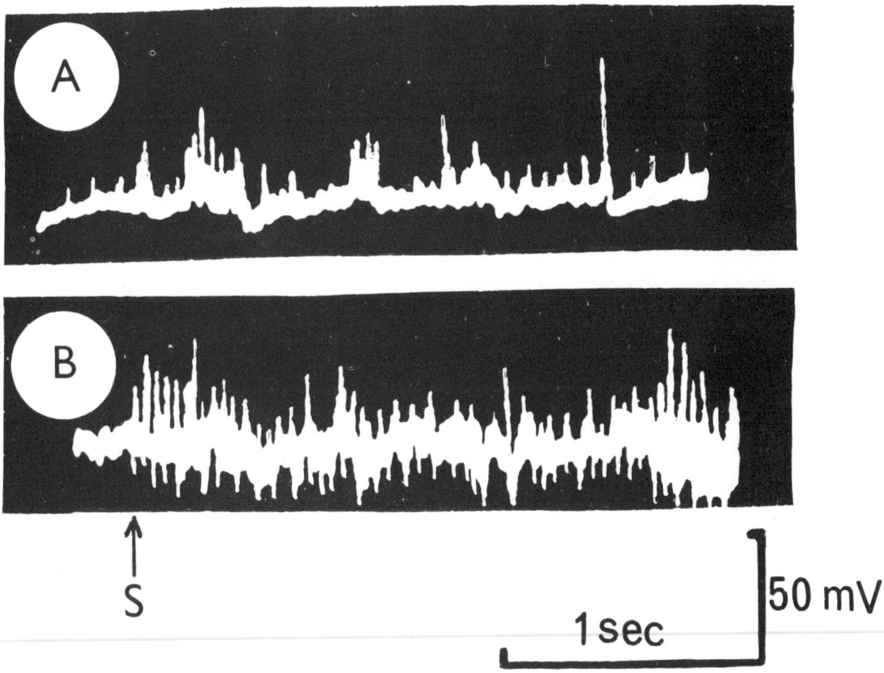

Fig. D.14. Activity of stretch-sensitive mechanoreceptors in the vocalis muscle of the cat. Description in text. [A]. Spontaneous activity in left recurrent laryngeal nerve (intact) prior to stretching of ipsilateral vocalis muscle. [B]. Activity provoked in the same nerve by application (at S) of 5 gm of sustained passive stretch to freed anterior end of the vocalis muscle.

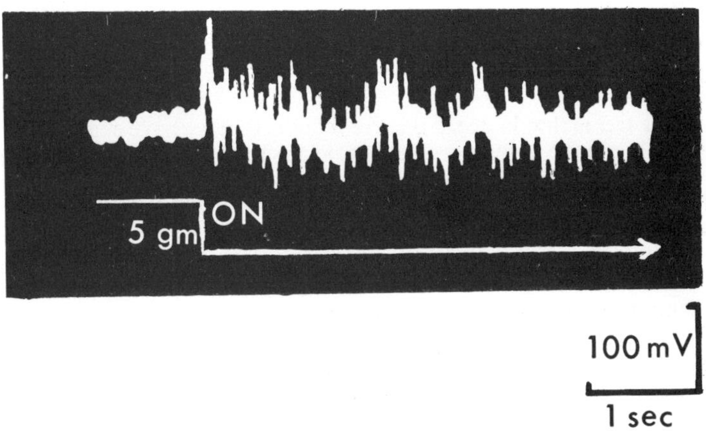

Fig. D.15. Afferent discharges from stretch-sensitive mechanoreceptors in the vocalis muscle of the cat. Description in text.

Cooper and her colleagues[1] demonstrated that in the external ocular muscles of the cat, from which stretch responses have been shown neurophysiologically to be elicitable, there are no spindles but only spiral nerve endings. Spiral endings identical with those found in the laryngeal muscles are present also in the facial muscles, and I have found them in the suboccipital muscles (which are crucial in the fine adjustment of the position of the head)—although in these muscles there are classical spindles as well as the spiral endings[2]. Spiral nerve endings have also been seen by Ann Alderson[3] in the intercostal muscles of the cat (and I have seen them in foetal intercostals), and they are present in the striated part of the urethral sphincter musculature.

[*Editor's Note:* Professor Bowden showed a series of slides illustrating spiral nerve endings in these various muscles at this point.]

Lindquist. I had a question for you about the physiological implications of muscle spindles in the intrinsic laryngeal muscles, but Dr. Wyke has already answered it.

Martin. Could I ask for the details of the distribution of the recurrent laryngeal nerve to the different muscles of the larynx. Your dissections apparently confirm the picture we usually have of two internal branches, designated posterior and anterior; and it has been stated, hasn't it, that the posterior (or medial) branch has on the whole a sphincteric closing action? Would your findings confirm or disprove that?

Ruth Bowden. I wouldn't be prepared to come down categorically on either side, because I think Dilworth[4] was right long ago in saying that unless one does degeneration experiments, it is impossible to follow the nerve branches precisely through to their terminations on account of the intramural and intramuscular plexus formation. I don't think you can say categorically that the medial branch goes to the adductors and the lateral branch to the abductors—but I wouldn't claim to have the last word on that.

Kirchner. There are several questions that are of interest to this group, it seems to me. First, Dr. Bowden, you said that in addition to the nerve supply from the external branch of the superior laryngeal nerve to the cricothyroid muscle in the rabbit, the pharyngeal plexus also contributes a branch—but has this ever been demonstrated in man?

Second, it seems that in every recurrent nerve that we have examined—either in the animal or in the human—where the vagus has been interrupted completely (we have had one in the human where the vagus had been cut inadvertently in a radical neck dissection, and another where it had been completely interrupted by a tumour at the base of the skull[5]) there are always some small diameter myelinated fibres surviving in the recurrent nerve. Am I correct in assuming this means that the cell bodies in the nodose ganglion still exist, and that this proves the existence of a communicating branch within the larynx between the recurrent

---

[1] Cooper, S., and Fillenz, M. (1955) Afferent discharges in response to stretch from the extra-ocular muscles of the cat and monkey, and the innervation of these muscles, *J. Physiol. (Lond.)*, **127**, 400–13.

[2] Bowden, R. E. M. (1963) Muscle spindles in the human foetus, *Acta biol. Szeged*, **9**, 35–59.

[3] Alderson, A. Personal communication.

[4] Dilworth, T. F. M. (1921) The nerves of the human larynx, *J. Anat. (Lond.)*, **56**, 48–52.

[5] Kirchner, J. A. (1966) Atrophy of laryngeal muscles in vagal paralysis, *Laryngoscope (St. Louis)*, **76**, 1753–65.

and the superior laryngeal nerves? Even though it may be impossible to demonstrate it by dissection, physiologically it must be there.

Ruth Bowden. I think that last suggestion is correct. I would also add—in relation to section of the nerve—that one does know that nerve fibres have the most incredible capacity for sprouting from strange places. Was any other nerve damaged? Was the section low down in the neck, so that some other nerve could have been damaged? Retrograde regrowth may occur if the peripheral migration of fibres is blocked by scar tissue.

Kirchner. It was in the mid-neck in the first patient I mentioned—it was certainly between the nodose ganglion and the take-off of the recurrent laryngeal branch from the vagal trunk in the neck.

Ruth Bowden. Then your interpretation is the one that would first come to mind; but I would want to be very certain that there hadn't been a chance of re-innervation of the distal stump by some neighbouring damaged nerve. There is evidence for this occurring, both in the experimental literature and clinically.

Murakami. As we know very well, the larynx has several kinds of physiological function—for example, respiration, phonation and reflex occlusion of the larynx (or reflex protection of the larynx). The point is whether these different physiological activities of the larynx are all regulated by the same type of nerve fibre, or by different ones. I am quite sure that as far as the posterior crico-arytenoid muscle is concerned (and this muscle is the only abductor muscle of the larynx) there are two different kinds of nerve fibres in the nerve twigs to this muscle. One is active during respiration, mainly in the inspiratory phase, and is inhibited by stimulation of the internal branch of the superior laryngeal nerve: another type of nerve fibre has no respiratory activity, and is activated by stimulation of the internal branch of the superior laryngeal nerve—which, I imagine, regulates the reflex contractions of the abductor muscle. So I think there has to be some kind of nerve fibre redistribution within the individual nerve twigs and not just in the nerve trunk.

Ruth Bowden. In every muscle in the larynx there is a nerve plexus. I am quite sure that your suggestion is one that it is important to follow up; and I think it would be necessary to do so not only on the afferent side but also on the efferent, because of the distribution of red, white and intermediate staining fibres in the abductor muscle. This slide [Fig. D.16] shows the results of some relevant, very recent work (actually finished 2 days ago by Dr. Malannino). You see the means of data obtained on five human post-mortem specimens, all taken from the same side and all pooled together, for the percentages of dark, intermediate and light fibre types in human intrinsic laryngeal muscles. The three populations are clearly evident, but there is a very high preponderance of the darkly-staining fibres in each muscle.

There is another feature of laryngeal activity which you did not mention that is very important in bipedal man, and that is protection of the vertebral column. As you know, when a heavy weight is picked up (and in the early stages of holding it up) a deep breath is taken and held: this produces pressures of about 100 cm H_2O, providing an air cushion above and a water cushion below the diaphragm to protect the vertebral column. There must therefore be different kinds of motor

Fig. D.16. Distribution of types of fibre in human intrinsic laryngeal muscles, as indicated by staining with Sudan Black B. Pooled data from counts on five adult larynges (post-mortem specimens). [■: dark staining fibres. ▤: intermediate staining fibres □: pale staining fibres.]
P.C.A.—posterior crico-arytenoid muscle.
L.C.A.—lateral crico-arytenoid muscle.
T.A.1—thyro-arytenoid muscle.
VOC.2—vocalis muscle.
T.A.3—transverse interarytenoid muscle.

unit activity, including both long-sustained and transient activity, in the laryngeal musculature. This has been confirmed both physiologically and histochemically—as we shall be hearing in much greater detail later this morning [see Communication 23].

Lindquist. What is the staining technique in your slide that you just showed?

Ruth Bowden. Sudan Black B. We also have (in the cat) succinic dehydrogenase preparations—but the ones I just showed are Sudan Black. This is post-mortem material—fresh, healthy tissue being unobtainable. We are in the process of doing biochemical controls on the differences between muscles treated immediately and after intervals equivalent to the periods between death and the staining of the specimens I showed.

von Euler. May I ask Dr. Wyke whether he also has evidence of some kind of central control of these laryngeal spindles? One special feature of muscle spindles is that they have a motor innervation and are therefore subject to centrifugal control.

Wyke. The answer to that is, as of now, no.

Godfrey. I think inside every one of us physiologists there is an anatomist trying to get out, and it appears that in the anatomists there is a physiologist trying to get out. I was interested in the discussion on spindles in the laryngeal muscles, and the innervation of these muscles; and I wonder what is the similarity, if any, between these muscles and, say, the diaphragm. The diaphram is pretty short on spindles—and from our work[1] on the curare experiments and Noble's work[2] with phrenic block these obviously are highly important structures in terms of respiration for generating respiratory sensation. Is there any similarity between the microstructure of laryngeal muscle innervation and that of the diaphragm?

Ruth Bowden. Really no. The work we have done on the diaphragm is somewhat limited, but there is a beautiful central band of subneural apparatuses. I agree that spindles are said to be sparse in the diaphragm, but this is something one has to take with a pinch of salt. There were no spindles in the extrinsic ocular muscles for 50 years, but now—at least in man—they are found to have a most dense population of spindles. The diaphragm, I believe (and I think there is somebody in the audience who will be much more capable of answering than I am) has a high proportion of pale muscle fibres, hasn't it?

Widdicombe. It is a mixture.

Ruth Bowden. Thank you. Now in regard to the innervation of the diaphragm, am I not right in saying that the lower six thoracic nerves come in on the periphery? Are we dealing with a double innervation again?

Godfrey. That may be true. But when you block the chest wall with local anaesthesia (as Eisele *et alii* (1969)[3] did), it seems to have no effect whatsoever on breath-holding or on the effects of added elastic or resistive loads on breathing. However, I know that Newsom Davis (1967)[4] did find that it affected respiratory sensation, so I wouldn't be at all displeased about that spinal innervation as a source of respiratory sensation.

Mårtensson. I wonder if it is possible that the recordings which Dr. Wyke showed might be derived from so-called 'pain' receptors. As you know, these are also sensitive to stretch and their afferent fibres belong to the Group II and Group III diameter ranges. The signal-to-noise ratio in your recordings is rather lower than one would expect with recordings from Group Ia fibres coming from muscle spindles, and is more like that seen in recordings from smaller diameter fibres.

Then there is one further question. I was working on this subject myself 6 or 7 years ago, and found it completely impossible to tell whether the receptors we were dealing with were located in the intramuscular portion of the larynx or not.

---

[1] Campbell, E. J. M., Godfrey, S., Clark, T. J. H., Freedman, S., and Norman, J. (1969) The effect of muscular paralysis induced by tubocurarine on the duration and sensation of breath-holding during hypercapnia, *Clin. Sci.*, **36,** 323–8.

[2] Noble, M. I. M., Eisele, J. H., Frankel, H. L., Else, W., and Guz, A. (1971) The role of the diaphragm in the sensation of holding the breath, *Clin. Sci.*, **41,** 275–83.

[3] Eisele, J. H., Trenchard, D., Burki, N., and Guz, A. (1969) The effect of chest wall block on respiratory sensation and control in man, *Clin. Sci.*, **35,** 23–33.

[4] Newsom Davis, J. (1967) Contribution of somatic receptors in the chest wall to detection of added inspiratory airway resistance, *Clin. Sci.*, **33,** 249–60.

Furthermore, I gave intravenous injections of succinylcholine or decamethonium—which, as you know, provoke contractions in intrafusal fibres—and in none of the nerve fibres from which I was recording did I get an increased discharge. This suggested to me that the afferent activity we were recording could not originate in muscle spindles.

Wyke. I would remind you that I carefully didn't attribute that afferent discharge specifically to muscle spindles. In fact, my own experience is that there is not a significant spindle afferent discharge for the reasons I have already mentioned—as well as those you give. I think the discharge comes (and this is pure speculation at the moment) more from the many spiral receptors than from the very few muscle spindles. As far as it emanating from 'pain' receptors is concerned, I don't think this likely in view of the fact that the applied stretch tension in the examples I showed was 5 gm—and this is well within the range of maximal twitch tensions found in the study you did in 1964 with Dr. Skoglund[1]. The fact that in the experiments that you describe that you did with succinylcholine you were unable to provoke discharges might be related to the fact that you were using dogs. . . . .

Mårtensson. I did it also on cats.

Wyke. You did it on cats? Good. Then in their muscles the number of spindles is so few that it might be rather difficult to detect any discharge provoked from them unless you actually had single fibre recordings from a spindle that had been exposed to the succinylcholine.

Could I repeat yet again that I don't think we are in a position to say whether the afferent discharges I showed you come from muscle spindles or from spiral endings, or from both. I would like to emphasize that what I (and I think Dr. Bowden) are trying to stress is that there *are* receptor endings in the intrinsic laryngeal muscles whose morphology and behaviour are appropriate for stretch-sensitive mechanoreceptors—but what their relative individual contributions in this regard may be is still undecided. This is an entirely different question from whether or not there are a few spindles in the intrinsic muscles of the larynx—because there are!

Widdicombe. In relation to what Dr. Godfrey and Dr. Mårtensson said, the diaphragm, it is true, has very few muscle spindles—according to Glebovskii (1963)[2] the cat diaphragm has about ten muscle spindles per hemi-diaphragm (although this estimate may not be very accurate). But there is a large number of Type 3 and Type 4 endings in the diaphragm, and I wonder whether you had spotted any of these in the laryngeal muscles and if so, whether Dr. Wyke's recordings might contain any contribution from them?

Ruth Bowden. What type of ending do you mean?

---

[1] Mårtensson, A., and Skoglund, C. R. (1964) Contraction properties of intrinsic laryngeal muscles, *Acta physiol. scand.*, **60,** 318–36.

[2] Glebovskii, V. D. (1963) Stretch receptors of the diaphragm. (Russian text). *Fiziol. Zh. Mosk.*, **48,** 545 (transl. in: *Fed. Proc.* 22/II, T405–T410).

Widdicombe. Either small diameter myelinated or non-myelinated fibres going to bare nerve endings.

Ruth Bowden. There are plenty of them.

Wyke. But could I repeat that I do not think an applied stretch tension of 5 gm would excite such endings significantly. Our own histological studies certainly agree with those of Dr. Bowden as to their presence, however.

23 Correlation Between Functional and Histochemical Properties of the Intrinsic Laryngeal Muscles in the Cat

L. EDSTRÖM, C. LINDQUIST AND A. MÅRTENSSON

Department of Physiology and Department of Neurology, Karolinska Institutet, Stockholm, Sweden

INTRODUCTION

Certain vocal fold adductors have a very brief contraction time which should make them well suited to produce rapid changes in pitch and also rapid closure of the glottis (Mårtensson and Skoglund, 1964). Since the intrinsic laryngeal muscles participate in respiration (Andrew, 1955; Green and Neil, 1955; Faaborg-Andersen, 1957; Mårtensson, 1963), they must also have a well developed capacity for prolonged work.

The contractile properties of the intrinsic laryngeal muscles prompted an investigation also of their histochemical characteristics, which so far do not seem to have been closely analysed. Some data are available from the rabbit (Hall-Craggs, 1968), but further studies were considered pertinent since recent experiments on the organization of single motor units in limb muscles have increased the possibility of correlating the histochemical and functional properties of muscle fibres (Edström and Kugelberg, 1968; Burke *et alii*, 1971; Kugelberg, 1971).

Using a large battery of staining methods, we found that some laryngeal muscle fibres in the cat have staining properties not previously described in striated muscles from mammals, i.e. fibres with reciprocal stainability for different oxidative enzymes. Moreover, we found a close parallelism between the percentage of fibres rich in myofibrillar ATPase and the contraction time in a specific muscle. After long-term stimulation of the motor nerve the decrease in muscle action potential and twitch tension is far less pronounced in the laryngeal muscles than in the anterior tibial muscle, and the depletion of glycogen is also less pronounced. The resistance to long-term stimulation may be related to a high content of oxidative enzymes and fat in the majority of intrinsic laryngeal muscle fibres.

METHODS

Experiments were performed on 19 cats weighing between 2 and 5 kg. Anaesthesia was obtained by intraperitoneal injections of pentobarbitone (40 mg/kg). The trachea was cannulated and the larynx partly freed from surrounding tissue. When necessary, a clamp was attached to the tracheal cannula to prevent excessive respiratory movements of the larynx. The temperature in the vicinity of the larynx was checked intermittently and maintained within a physiological range by means of a heating lamp.

A thin steel rod penetrating the thyroid cartilage gave firm fixation of the larynx. Isometric recordings were obtained from the thyro-arytenoid and lateral crico-arytenoid muscles by attaching a string to the arytenoid cartilage, and from the cricothyroid muscle by attaching a string to the cricoid cartilage. The initial muscle tensions were adjusted so as to give a maximal twitch contraction. When comparative experiments were made on the anterior tibial and soleus muscles, the tibia was rigidly fixed by steel rods drilled through the upper and lower ends of the bone. The tendons of the synergistic muscles in the lower leg were cut. The paraffin-bathed motor nerves to the respective muscles were cut and their peripheral ends stimulated supramaximally with 0.1 msec duration pulses through bipolar, chlorided silver-wire electrodes. By intermittently changing the stimulus intensity, the fact that the stimulus was supramaximal throughout the experiment was repeatedly checked. To enable recording from one single intrinsic laryngeal muscle with minimal interference, the nerve branches to the other muscles were cut. Recordings from the soleus were made after denervation of the gastrocnemius.

Muscle tension recordings were made with a Grass FT 03 force-displacement transducer in the larynx and with a Grass FT 10 in the lower leg. Muscle action potentials were recorded with two steel needles insulated except for the tip, or with DISA concentric needle electrodes. The bioelectric signals were fed into Grass P6 DC amplifiers and displayed on a dual-beam Tektronix 502 oscilloscope.

Stimuli were delivered at regular intervals and at frequencies varying between 5 and 30 per sec for 10–20 minutes. During the long-term stimulation periods, twitch contraction and Ap-amplitudes to 1/sec stimulation were checked for periods of 5–10 sec, 5–10 times in 10 minutes.

Histochemical procedure. The muscle biopsy material was immediately frozen in liquid nitrogen. Sections 15μ thick were cut in a cryostat.

$NADH_2$ tetrazolium reductase, succinic dehydrogenase, phosphorylase, myofibrillar A-band Adenosine triphosphatase (ATPase), lipids (Sudan Black B) and glycogen (PAS) were stained by the methods referred to by Nyström (1968).

By photographic registration and comparison between consecutive cross-sections stained with the different histochemical methods used, the staining properties of different muscle fibres could be studied.

RESULTS

HISTOCHEMICAL FIBRE TYPES IN DIFFERENT MUSCLES

Table 23.I summarizes the histochemical properties of all the muscle fibre types found in the intrinsic laryngeal muscles.

Figure 23.1 A shows a section from the lateral crico-arytenoid muscle stained for myofibrillar ATPase. Two arrows in the figure point to a fibre with a low staining level—a 'type I' fibre according to the nomenclature of Fenichel and Engel (1963) (cf. also Dubowitz and Pearse (1960)). A horizontal bar marks the position of two fibres rich in myofibrillar ATPase—a 'type II' fibre according to the same nomenclature.

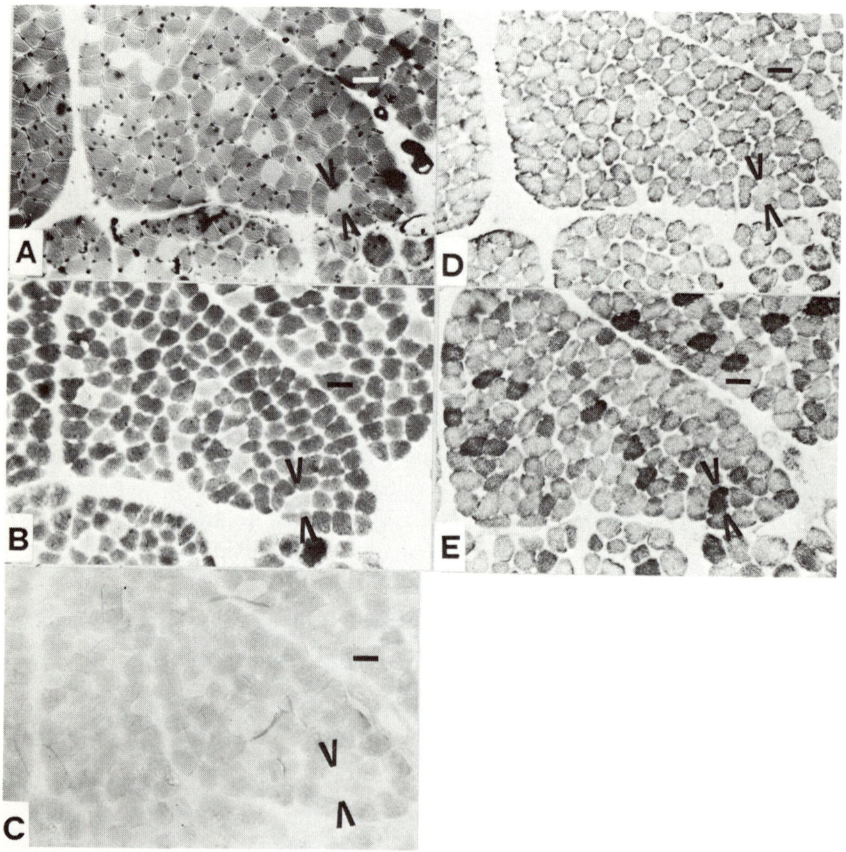

Fig. 23.1. Serial cross-sections from a lateral crico-arytenoid muscle. Sections A-E show the same field, stained for: [A] myofibrillar A-band ATPase; [B] phosphorylase; [C] glycogen; [D] succinic dehydrogenase; [E] $NADH_2$ tetrazolium reductase. A type I fibre is marked by two arrows in all sections, and two neighbouring type II fibres by a horizontal bar. Full description in text. (Magnification ×120.)

TABLE 23.I

INTRINSIC LARYNGEAL MUSCLE FIBRES*

REACTION	TYPE II			TYPE I	
ATPase	high	high	high	low	low
phosphorylase	high	high	high	low	low
glycogen	high	high	high	low	low
succ. dehyd.	high	low	interm. to high	low	interm. to high
$NADH_2$ reductase	low	low	interm. to high	high	interm. to high
fat	high	low	high	low	interm. to high
fibre type in skeletal muscle	—	IIA (FF)	IIB IIC (FR)	—	I (S)

* Fibre types in the intrinsic laryngeal muscles of the cat. Corresponding fibre types found in skeletal muscles are indicated according to the nomenclatures of Kugelberg (1971) and Burke (1971).

In Fig. 23.2, sections from the cricothyroid muscle (A) and the thyro-arytenoid muscle (D) stained for myofibrillar ATPase show the same pattern of lightly and darkly stained fibres as seen in Fig. 23.1A. In the thyro-arytenoid muscle [Fig. 23.2D] and the lateral crico-arytenoid [Fig. 23.1A] 10 per cent or less of the fibres are of type I. In the cricothyroid muscle [Fig. 23.2A] about 40 per cent are of type I. The same relative number of type I fibres as in the cricothyroid muscle is found in the posterior crico-arytenoid muscle. These findings may be related to differences in contraction time between the lateral crico-arytenoid and the thyro-arytenoid muscles on the one hand, and the cricothyroid and the posterior crico-arytenoid muscles on the other hand [see Discussion, p. 401].

The type I fibre in Fig. 23.1A (lateral crico-arytenoid), indicated by two arrows, shows a low staining level also in sections stained for phosphorylase

[Fig. 23.1B] and for glycogen [Fig. 23.1C]. These staining properties are characteristic of all type I fibres in the laryngeal muscles investigated, and also correspond to the general pattern of skeletal muscle type I fibres in the cat. The type II fibres marked by a horizontal bar in Fig. 23.1 have opposite staining characteristics; namely, high stainability for phosphorylase [Fig. 23.1B] and for glycogen [Fig. 23.1C]. These staining properties are characteristic of type II

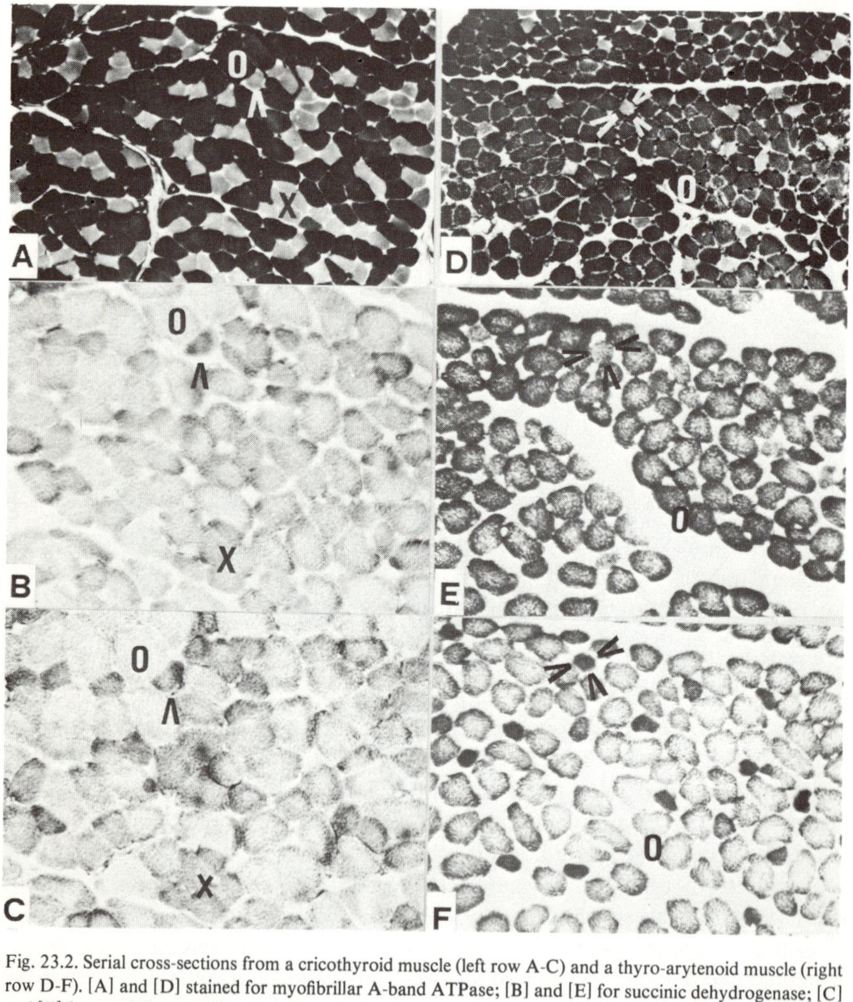

Fig. 23.2. Serial cross-sections from a cricothyroid muscle (left row A-C) and a thyro-arytenoid muscle (right row D-F). [A] and [D] stained for myofibrillar A-band ATPase; [B] and [E] for succinic dehydrogenase; [C] and [F] for $NADH_2$ tetrazolium reductase. (Magnification in A and D ×120; in B, C, E and F ×200). Compare A and D, and note the relatively large number of fibres with low ATPase content in the cricothyroid muscle. The 'O' in A-C marks a type II fibre. The arrow marks a single, and the cross a group of type I fibres in A-C. In D-F the three arrows indicate a type I fibre, and an 'O' marks two neighbouring type II fibres. Full description in text.

fibres in the internal laryngeal muscles, and also correspond to the staining properties of type II fibres in the skeletal muscles of the cat.

On the basis of differences in the content of succinic dehydrogenase and $NADH_2$ tetrazolium reductase, the type I fibre category can be further subdivided. In Fig. 23.1 (lateral crico-arytenoid), the type I fibre indicated by two arrows is characterized by a low content of succinic dehydrogenase (D) but a high content of $NADH_2$ tetrazolium reductase (E). Fig. 23.2D-F shows another fibre of this type in the thyro-arytenoid muscle (indicated by three arrows). Muscle fibres with these particular staining characteristics; namely, low stainability for phosphorylase, glycogen, myofibrillar ATPase and succinic dehydrogenase but high stainability for $NADH_2$ tetrazolium reductase have not previously been described in mammals [see Discussion, p. 401]. It is the most common type I fibre in all the intrinsic laryngeal muscles investigated, except in the cricothyroid muscle. However, in relation to the total fibre population, it is of quantitative importance only in the posterior crico-arytenoid muscle.

In Fig. 23.2 A-C (cricothyroid muscle), a group of type I fibres with an intermediate degree of activity of both succinic dehydrogenase (B) and of $NADH_2$ tetrazolium reductase (C) is indicated by a cross, and a type I fibre with high activity of these two enzymes is indicated by an arrow. These fibres, with parallel intermediate or high stainability for both succinic dehydrogenase and $NADH_2$ tetrazolium reductase, correspond to the features of the type I fibres of the skeletal muscles. They are found to a great extent in the cricothyroid muscle, but are comparatively few in number in the other laryngeal muscles.

The type II fibres in the laryngeal muscles also have different staining properties as regards their stainability for succinic dehydrogenase and $NADH_2$ tetrazolium reductase. In Fig. 23.2D-F (thyro-arytenoid muscle) a group of type II fibres is shown, indicated by an 'O', with high stainability for succinic dehydrogenase (E) and low stainability for $NADH_2$ tetrazolium reductase (F). In Fig. 23.1D and E, two type II fibres indicated by a horizontal bar show the same staining properties. Fibres with these staining characteristics are the most common type II fibres found in the thyro-arytenoid muscle, and also in the lateral crico-arytenoid muscle. Type II fibres with these staining properties have not been described in the skeletal muscles of mammals. In Fig. 23.2A-C (cricothyroid muscle), an 'O' indicates a group of type II fibres with low activity for both succinic dehydrogenase (B) and $NADH_2$ tetrazolium reductase (C). As shown in these Figures, there is also a large number of type II fibres with more intermediate or even high stainability for $NADH_2$ tetrazolium reductase and succinic dehydrogenase. Fibres with such parallel stainability for these two oxidative enzymes are the most common type II fibre found in the cricothyroid and the posterior crico-arytenoid muscles and, moreover, seem to correspond histochemically to the type II fibres of skeletal muscles. The percentage of fibres with intermediate or high stainability for both succinic dehydrogenase and

$NADH_2$ tetrazolium reductase is higher in the cricothyroid and the posterior crico-arytenoid muscles than in the anterior tibial muscle. It should thus be emphasized that in all the intrinsic laryngeal muscles of the cat the large majority of fibres have a high content of at least one of the two oxidative enzymes. Besides, as shown in Table 23.I, a high content of succinic dehydrogenase is paralleled by a high content of fat.

PERFORMANCE OF THE INTRINSIC LARYNGEAL MUSCLES DURING PROLONGED ACTIVITY

In a series of physiological experiments, the intrinsic laryngeal muscles and two extremity muscles with high and low fatigability respectively (the tibialis anterior and the soleus) were compared with respect to their ability to maintain contractile tension during prolonged activity. In Fig. 23.4, the twitch amplitudes of the cricothyroid (triangles) and the tibialis anterior (circles) have been plotted at different times after the onset of motor nerve stimulation at 5/sec. The cricothyroid maintains its contractile tension very well, even after 20 minutes of activity. In contrast, the twitch amplitude of the tibialis anterior is reduced by 50 per cent within 4 minutes, and when stimulation is interrupted after 10 minutes the muscle develops less than half of its initial force. A very small amount of twitch tension has been recovered 10 minutes after cessation of stimulation. In similar experiments, the thyro-arytenoid and the lateral crico-arytenoid muscles behaved in the same manner as the cricothyroid; and at the stimulus frequency used (5/sec), their performance was equal to that of the soleus.

Recordings of action potentials from the cricothyroid and the tibialis anterior muscles at different times after the onset of stimulation showed that the cricothyroid action potentials remained unchanged throughout stimulation, whereas the tibialis anterior action potentials declined only during the first minute and were then well maintained. Most of the fatigue observed in the tibialis anterior at 5/sec can thus be attributed to failure of the contractile mechanism.

Figure 23.5 shows typical experiments in which the same two muscles were stimulated at a frequency of 10/sec. At this rate, the contractile tension in the cricothyroid (triangles) slowly falls to a minimum, corresponding (after 8 min) to 75 per cent of the prestimulation value. The power of the tibialis anterior (circles) is reduced by 50 per cent during the first 3 minutes, and then declines further at a slower rate until, at 10 minutes, the muscle develops only 15–20 per cent of its initial twitch tension. In the poststimulation period, both muscles recover a very small amount of tension: the tibialis anterior has attained only 25 per cent of its prestimulation contractility after 7 minutes of rest. Whereas the tibial muscle is thus easily fatigued at 10/sec, the soleus proved to have about the same resistance to prolonged activity as the cricothyroid. In some experiments in which contractions were recorded from the thyro-arytenoid and the lateral crico-arytenoid muscles using a stimulus frequency of 10/sec, the performance of these muscles proved to be superior to that of the tibialis anterior and to be comparable to that of the fatigue-resistant soleus. There may be a differentiation in fatigue properties among the intrinsic laryngeal muscles, but further experiments are required to settle this question.

The action potentials of the cricothyroid remained unchanged throughout the 10 minute period of stimulation at 10/sec. The action potentials of the tibialis anterior, on the other hand, were considerably reduced, in particular during the first 4 minutes. This may partly account for the rapid initial fall in contractile tension observed in this muscle at stimulus frequencies of both 5 and 10 per sec. The cause of this action potential decrement has not been more closely investigated, but most likely it is due to failure of neuromuscular propagation. The action potential recordings from the thyro-arytenoid, the lateral crico-arytenoid and the posterior crico-arytenoid showed that the neuromuscular propagation to these muscles is unaffected by stimulation at a frequency of 10/sec. In similar experiments on the soleus, a small action potential decay was observed.

HISTOCHEMICAL CHANGES AFTER LONG-TERM MOTOR NERVE STIMULATION

The physiological experiments did not indicate that any contractile fatigue occurred in the intrinsic laryngeal muscles during 10–20 minutes of stimulation at 5/sec. Stimulation at 10/sec for 10 minutes did, however, induce a slight contractile fatigue in the intrinsic laryngeal muscles, as well as changes in the histochemical appearance of the muscle fibres. Figs. 23.3A and D show the same

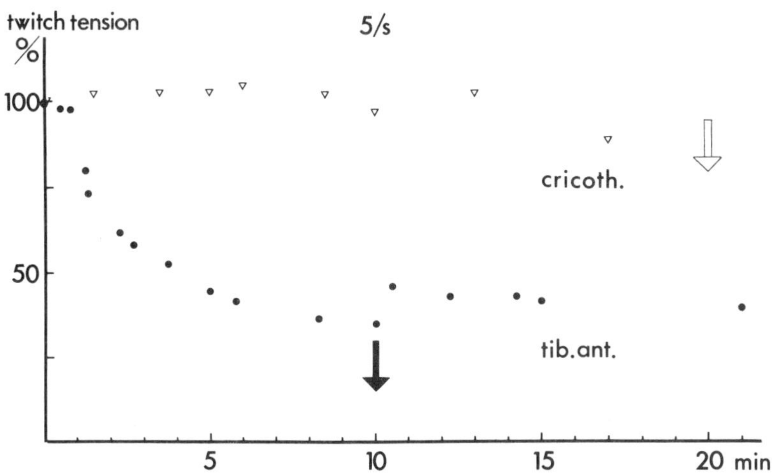

Fig. 23.3. Twitch contractions at l/sec elicited for brief periods during long-term motor nerve stimulation at 5/sec. Amplitude (*ordinate*) plotted against time (*abscissa*) after onset of stimulation. *Triangles:* cricothyroid; *circles:* tibialis anterior. Stimulus interrupted at arrows. Full description in text.

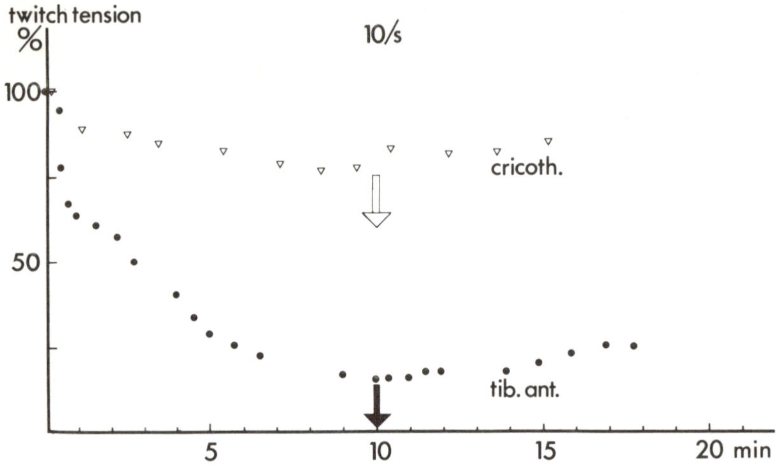

Fig. 23.4. Same as Fig. 23.3, except that the long-term stimulation frequency is 10/sec.

fields of sections from the cricothyroid muscles stained for $NADH_2$ tetrazolium reductase and glycogen respectively, after stimulation at 10/sec for 10 minutes. The activity of $NADH_2$ tetrazolium reductase in a muscle fibre was not changed by prolonged activity, and hence the stainability for this enzyme could be used for fibre classification. In A, two groups of fibres indicated by 'X' and 'O' show different stainability for $NADH_2$ tetrazolium reductase. In D it may be seen that both groups of fibres are depleted of stainable glycogen, which is typical. It is evident that the reduction in glycogen in the cricothyroid muscle is not restricted to a certain fibre type. These results apply to all four laryngeal muscles.

In comparative experiments on the anterior tibial muscle, stimulation at 5/sec resulted in a considerable reduction of twitch tension, and also in pronounced changes in the histochemical appearance of the muscle fibres. When the muscle was stimulated at 10/sec, conspicuous changes could be observed in the histochemical appearances, as shown in Fig. 23.5B-F. In the non-stimulated muscle, a fibre with a low activity of $NADH_2$ tetrazolium reductase is indicated by 'W' in B, and the same non-stimulated fibre stained for glycogen is shown in E: this fibre has a high glycogen content, and the same applies to the majority of fibres with low $NADH_2$ tetrazolium reductase activity. In C, 'W' indicates a fibre with low activity of $NADH_2$ tetrazolium reductase in the stimulated muscle. As shown in F, the glycogen content is low in this stimulated fibre; and in this respect the fibre is representative of most of the fibres with low $NADH_2$ tetrazolium reductase activity subjected to stimulation. In the anterior tibial muscle there is thus a marked reduction in the glycogen content after stimulation in most of the fibres that are poor in $NADH_2$ tetrazolium reductase.

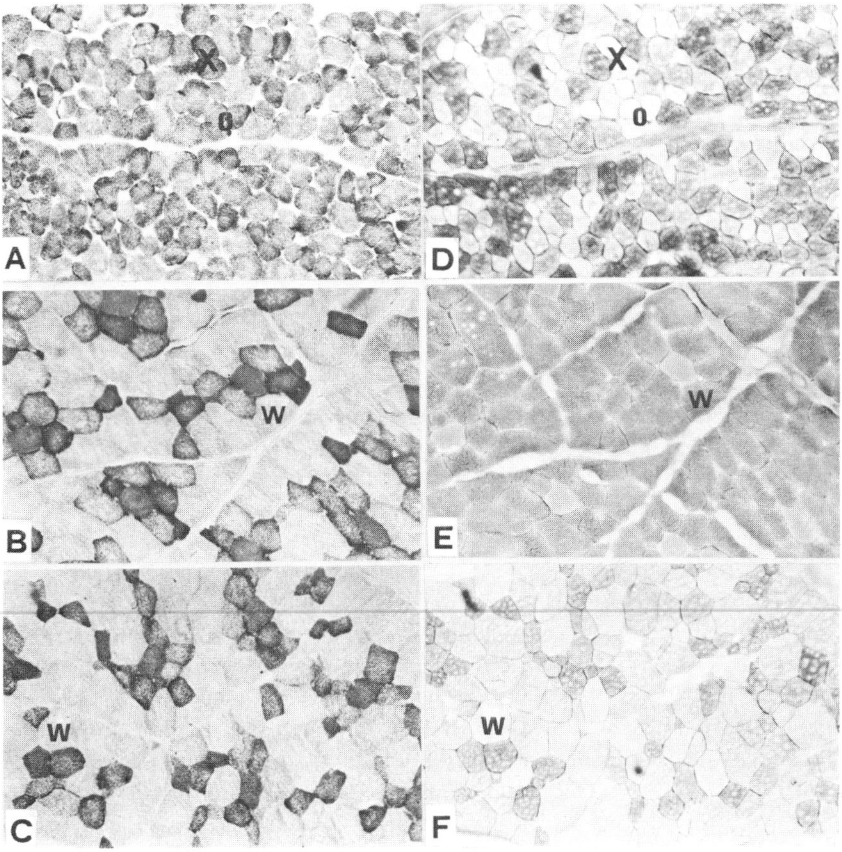

Fig. 23.5. [A] and [D] cross-sections from a cricothyroid muscle stimulated for 10 min at 10/sec, stained for $NADH_2$ tetrazolium reductase and glycogen, respectively. [B] and [E] cross-sections from non-stimulated, and [C] and [F] from stimulated (10 min, 10/sec) anterior tibial muscle, stained for $NADH_2$ tetrazolium reductase (B and C) and glycogen (E and F). Two groups of cricothyroid fibres depleted of glycogen following stimulation are indicated by 'O' and 'X' in A and D. In B and C 'W' marks a fibre poor in $NADH_2$ tetrazolium reductase: 'W' in E and F identical with fibres similarly marked in B and C respectively. Full description in text. (Magnification ×120.)

DISCUSSION

As shown in experiments on limb muscles in the rat (Kugelberg, 1971) and cat (Burke *et alii*, 1971), motor units composed of type I fibres (with low stainability for myofibrillar ATPase) have long contraction times, whereas type II units are much faster. In the present study about 40 per cent of the fibres in the cricothyroid and the posterior crico-arytenoid muscles were shown to be of type I;

but in the thyro-arytenoid and the lateral crico-arytenoid these fibres represent less than 10 per cent of the total. As shown in earlier experiments on cats, the cricothyroid and the posterior crico-arytenoid muscles have much longer contraction times than the thyro-arytenoid and the lateral crico-arytenoid muscles (Mårtensson and Skoglund, 1964). Thus, the correlation between stainability for myofibrillar ATPase and contraction time in limb muscles applies to the intrinsic laryngeal muscles as well.

In all the intrinsic laryngeal muscles investigated, some of the type I fibres were found to have a low enzyme and substrate content, and in this respect they resemble the slow tonic fibres of cold-blooded animals (cf. Lännergren and Smith, 1966; Lännergren, personal communication). However, the fibres of the laryngeal muscles differ from the latter type of fibre in that they have very intense stainability for $NADH_2$ tetrazolium reductase. Furthermore, intravenous injections of decamethonium did not result in a recordable tonic contraction of the thyro-arytenoid muscle. Thus this intrinsic laryngeal muscle in the cat does not seem to contain slow tonic fibres of the type found in cold-blooded animals and recently demonstrated also in mammalian extrinsic eye muscles (Hess and Pilar, 1963; Bach-y-Rita and Ito, 1966) and middle ear muscles (Fernand and Hess, 1969).

The predominant type II fibre in the anterior tibial muscle of the cat is a large fibre with low activity of oxidative enzymes and phosphorylase. In the experiments described above, this type of fibre was found to have a marked decrease in glycogen and phosphorylase stainability after 10 minutes of motor nerve stimulation at a frequency of 10/sec. Similar findings have previously been reported from experiments on the anterior tibial muscle of the rat (Kugelberg and Edström, 1968). Moreover, single-unit experiments from the same muscle in the rat have shown that fibres with these staining characteristics lose their glycogen content and their entire tension when stimulated at 10/sec (Edström and Kugelberg, 1968). Hence it seems reasonable to believe that the decrease in twitch tension of the anterior tibial muscle of the cat observed in the present investigation should at least in part be due to the exhaustion of the glycolytic mechanism observed in these fibres.

Fibres with similar staining properties are not found in the thyro-arytenoid and lateral crico-arytenoid muscles. They occur only in the cricothyroid and the posterior crico-arytenoid muscles; and in these muscles they are comparatively less numerous than in the anterior tibial muscle. Furthermore, stimulation (at 5/sec) of the motor nerve to the cricothyroid muscle does not result in glycogen depletion in these fibres—which indicates that they are functionally different from those found in the anterior tibial muscle, in spite of the histochemical similarities of these muscles.

Fibres with high or intermediate stainability for the oxidative enzymes examined in this investigation are more common in the laryngeal muscles than in the anterior tibial muscle. These staining characteristics may be related to a higher resistance to fatigue (Edström and Kugelberg, 1968).

Further comparisons between laryngeal and limb muscles with respect to their histochemistry and function are difficult, because many of the fibres in the former muscles are histochemically different from the fibres described in skeletal muscles and because the functional properties of the laryngeal muscles have not yet been studied at the single motor unit level. It must, however, be concluded that the majority of the motor units in the intrinsic laryngeal muscles must be fairly resistant to contractile fatigue at the frequencies used in this study.

In experiments on cats, Olson and Swett (1971) showed that the action potentials of motor units in the soleus and of 'slow' motor units in the gastrocnemius are less reduced in amplitude than 'fast' motor units in the gastrocnemius during motor nerve stimuation at 100/sec. This was taken as an indication of differences in neuromuscular transmission; and it was suggested that these differences 'reflect the morphological specialization of large nerve terminals known to exist at the neuromuscular junctions of the soleus'. The results from the experiments presented here show that action potentials recorded from the intrinsic laryngeal muscles retain their full amplitude throughout a stimulation period of 10 minutes at 10/sec, whereas action potentials recorded from the anterior tibial muscle are considerably reduced. This difference in fatigability is likely to be due to impairment of neuromuscular transmission in the anterior tibial muscle, that does not occur in the intrinsic laryngeal muscles. This fact may be of particular interest, since previous histological studies of the cricothyroid muscle in the rabbit (Feindel *et alii*, 1952) and of the thyro-arytenoid muscle in man (Rudolph, 1960) have demonstrated the existence of multiple end-plates.

REFERENCES

Andrew, B. L. (1955) The respiratory displacement of the larynx: A study of the innervation of accessory respiratory muscles, *J. Physiol. (Lond.)*, **130,** 474–87.

Bach-y-Rita, P., and Ito, F. (1966) *In vivo* studies on fast and slow muscle fibers in cat extraocular muscles, *J. gen. Physiol.*, **49,** 1177–98.

Burke, R. E., Levine, D. N., Zajac, F. E., Tsairis, P., and Engel, W. K. (1971) Direct correlation of physiological and histochemical characteristics in motor units of cat triceps surae, in *Abstracts of Communications and Invited Reports at the Fourth International Congress of Electromyography, Brussels,* pp. 21–22.

Dubowitz, V., and Pearse, A. G. E. (1960) A comparative histochemical study of oxidative enzymes and phosphorylase activity in skeletal muscles, *Histochemie,* **2,** 105–17.

Edström, L., and Kugelberg, E. (1968) Histochemical composition, distribution of fibres and fatiguability of single motor units. Anterior tibial muscle of the rat, *J. Neurol. Neurosurg. Psychiat.*, **31,** 424–33.

Faaborg-Andersen, K. (1957) Electromyographic investigation of intrinsic laryngeal muscles in humans, *Acta physiol. scand.*, **41** (Suppl. 140), 1–148.

Feindel, W., Hinshaw, J. R., and Weddell, G. (1952) The pattern of motor innervation in mammalian striated muscle, *J. Anat. (Lond.)*, **86**, 35–48.

Fenichel, G. M., and Engel, W. K. (1963) Histochemistry of muscle in infantile spinal muscular atrophy, *Neurology (Minneap.)*, **13**, 1059–66.

Fernand, V. S. V., and Hess, A. (1969) The occurrence, structure and innervation of slow and twitch muscle fibres in the tensor tympani and stapedius of the cat, *J. Physiol. (Lond.)*, **200**, 547–54.

Green, J. H., and Neil, E. (1955) The respiratory function of the laryngeal muscles, *J. Physiol. (Lond.)*, **129**, 134–41.

Hall-Craggs, E. C. B. (1968) The contraction times and enzyme activity of two rabbit laryngeal muscles, *J. Anat. (Lond.)*, **102**, 241–55.

Hess, A., and Pilar, G. (1963) Slow fibres in extraocular muscles of the cat, *J. Physiol. (Lond.)*, **169**, 780–98.

Kugelberg, E. (1971) Distribution, histochemical and physiological properties of normal and reinnervated rat motor units, in *Abstracts of Communications and Invited Reports at the Fourth International Congress of Electromyography, Brussels*, pp. 85–86.

Kugelberg, E., and Edström, L. (1968) Differential histochemical effects of muscle contractions on phosphorylase and glycogen in various types of fibre. Relation to fatigue, *J. Neurol. Neurosurg. Psychiat.*, **31**, 415–23.

Lännergren, J., and Smith, R. S. (1966) Types of muscle fibres in toad skeletal muscle, *Acta physiol. scand.*, **68**, 263–74.

Lännergren, J. Personal communication.

Mårtensson, A. (1963) Reflex responses and recurrent discharges evoked by stimulation of laryngeal nerves, *Acta physiol. scand.*, **57**, 248–69.

Mårtensson, A., and Skoglund, C. R. (1964) Contraction properties of intrinsic laryngeal muscles, *Acta physiol. scand.*, **60**, 318–36.

Nyström, B. (1968) Histochemistry of developing cat muscles, *Acta neurol. scand.*, **44**, 405–39.

Olson, C. B., and Swett, C. P. (1971) Effect of prior activity on properties of different types of motor units, *J. Neurophysiol.*, **34**, 1–16.

Rudolph, G. (1960) Multiple Innervation von Muskelfasern im M. vocalis des Menschen, *Experientia (Basel)*, **21**, 551–3.

DISCUSSION ON COMMUNICATION [23]

Murakami. I would like to make a comment from the physiological point of view. A further paper[1] on the respiratory activity of the posterior crico-arytenoid muscle is to be published by Dr. Kirchner and myself in the very near future [see also Suzuki and Kirchner, 1969[2] and Murakami and Kirchner, 1971[3]], and in

---

[1] Murakami, Y., and Kirchner, J. A. (1972) Respiratory movements of the vocal cords: an electromyographic study in the cat, *Laryngoscope (St. Louis)*, **82**, 454–67.

[2] Suzuki, M., and Kirchner, J. A. (1969) The posterior cricoarytenoid as an inspiratory muscle, *Ann. Otol. (St. Louis)*, **78**, 849–65.

[3] Murakami, Y., and Kirchner, J. A. (1971) Electrophysiological properties of laryngeal reflex closure, *Acta oto-laryng. (Stockh.)*, **71**, 416–25.

that study we found two kinds of respiratory activity in the nerve twig that innervates only the posterior crico-arytenoid muscle. One fibre system has very phasic activity and fibres only in the inspiratory phase of the respiratory cycle, whilst the other shows tonic activity that is independent of the phases of respiration; and it seems to me that this latter type of tonic activity closely resembles the nerve activity shown by Dr. Widdicombe a few days ago [see Communication 9].

May I add another comment, Mr. Chairman? We have to remember that the contractile force of any intrinsic laryngeal muscle falls off very rapidly, in the case of reflex contraction produced by stimulation of the internal branch of the superior laryngeal nerve. The contractile force of any intrinsic laryngeal muscle falls to 1/2 or 1/3 very quickly—that is, within 5 seconds or so; and these observations of course do not indicate the fatigability of the intrinsic laryngeal muscles, but a decrease in the excitability of the central nervous system.

Lindquist. I am not quite sure if, by your first comment, you were implying that there is tonic activity in nerves that may be related to the presence of tonic muscle fibres—is that correct?

Murakami. Yes.

Lindquist. Then you have to be very careful with the terminology, because there are twitch fibres which are tonically active—but in my Communication I was referring to the slow, tonic non-twitch fibres of the type that is found in amphibians, birds and in the extra-ocular muscles of cats. Of course, it is not possible to draw reliable conclusions as to the presence of such fibres merely from recordings of the motor nerve impulses. As you point out, there are many factors—including failure of synaptic transmission in the central nervous system—which, of course, have to be taken into account. We have also made some studies on reflex-elicited contractions and did not find any depletion of glycogen, which would imply that during reflex stimulation the intrinsic laryngeal muscles are not activated to the extent of depleting their glycogen stores. In such experiments, therefore, they should not be expected to show contractile fatigue.

Ruth Bowden. Might I ask Dr. Murakami whether these two populations of nerve fibres were studied in terms of their diameters? In avian muscle (which is so sensitive to temperature and readily goes into contracture) we find nerve fibres of smaller diameter in the muscles with a diffuse multiple innervation than in the ones with a zonal (or single) type of innervation. I should also add that others have found that you can alter the structural character of muscle fibres somewhat by cross-innervation, changing them from light to dark in appearance.

Murakami. I don't know about the relative sizes of the tonic activity type and phasic activity type of nerve fibre.

Kirchner. I cannot answer the question either, although I think it is a good one. But I can tell you that we have done the experiment with nerves going to individual muscles within the larynx in terms of conduction velocities. We record from a nerve twig to an individual muscle at the point where it enters the muscle while stimulating the peripheral stump of the vagus as close to the base of the skull as possible; and we have not found any significant differences in the conduction velocities in the nerves going to the various muscles in the larynx. So we have to assume that they are all in the same diameter range.

I would like to congratulate these Stockholm workers, because I think this is really a beautiful piece of work and it gives us some ideas for further experiments. I think it is particularly nice that you have correlated the ATPase activity with the contraction times measured by Drs. Mårtensson and Skoglund in 1964[1]. This provides us with yet another piece of evidence that there are different contraction times within these very fast muscles in the larynx.

Otis. Were these muscles all stimulated through their motor nerves, or directly?

Lindquist. All were stimulated through their motor nerves.

Otis. The blood supply to the muscles was normal?

Lindquist. We have also done experiments in preparations where we have deliberately impaired the circulation, because we are well aware of the fact that the circulation may change during such prolonged stimulations. When we occluded the carotid arteries, there was a faster drop—both of the contractile output and of the action potential—but we do not think that ischaemia alone accounts for the fatigue effects in the experiments on cats with unoccluded carotid arteries. One of the reasons we believe this is because the laryngeal muscles contract so rapidly, and we used low stimulus frequencies so as to avoid complete tetanus of the muscles: there is a sub-tetanic action all the time, which should be an efficient pump for the circulatory system within the muscles.

Wyke. I would like to add my congratulations. I think this is a splendid piece of work, especially as it answers a few questions that have long puzzled a lot of us who study the contraction characteristics of the laryngeal muscles. Even more, it may sort out a few problems about the motor unit potentials we record from these muscles—because I think I am not alone in saying I have always been intrigued by the variety of motor unit potential behaviour that one finds in the individual laryngeal muscles[1,2,3]. This has seemed to me to be somewhat wider than one finds in many limb muscles, and I think you may well have here a good explanation of that—particularly as you have unearthed two types of muscle fibre that don't exist in limb muscles. I would be very interested to know if you and Dr. Mårtensson and your colleagues are going on to see if these fibre types are distributed within specific motor unit populations: this is what one would really like to know.

---

[1] Mårtensson, A., and Skoglund, C. R. (1964) Contraction properties of intrinsic laryngeal muscles. *Acta physiol. scand.*, **60**, 318–36.

[2] Mårtensson, A. (1968) The functional organization of the intrinsic laryngeal muscles. *Ann. N.Y. Acad. Sci.*, **155**, 91–97.

[3] Hall-Craggs, E. C. B. (1968) The contraction times and enzyme activity of two rabbit laryngeal muscles. *J. Anat. (Lond.)*, **102**, 241–55.

Lindquist. We are planning to follow this up with studies at the single motor unit level.

Martensson. We[1] have done some studies of the distribution of the amplitude, shape and duration of individual motor unit potentials in the vocalis muscles in man. From these studies it appears that the motor unit potentials are of short duration, and that a fairly high percentage of the potentials have more than three phases. But the range of the distribution, as far as potential duration and amplitude go, is not greater in the laryngeal muscles than in limb muscles.

Murakami. I would like to make another comment. I quite agree that all of the intrinsic laryngeal muscles have different contraction times—but as far as vocal fold motion is concerned, the time relatioms are almost the same. For example, when you stimulate the thyro-arytenoid muscle alone the vocal fold adducts very well, and reaches the midline about 60 to 65 msec after stimulation; but when you stimulate the nerve twig to the abductor muscle alone, the vocal fold abducts very well and reaches its maximum displacement also after about 60 to 65 msec. I think the time relations depend rather on the inertia of the crico-arytenoid joint, or something like that.

Lindquist. May I ask you if summation of the muscle contractions is required to produce this movement of the vocal folds to the midline, or is it just twitch contractions that produce it?

Murakami. Just twitch contractions.

Fry. If there are no further comments, we can close this discussion. Let me just say as your Chairman that I hope it is perfectly clear to Dr. Lindquist and his co-authors that the reason why comments have dried up is that they have put such extraordinarily interesting and new material in front of us that people can't readily digest it. I am sure he understands that, but let me put it in for the record.

---

[1] Knutsson, E., Mårtensson, A., and Mårtensson, B. (1969) The normal electromyogram in human vocal muscles, *Acta oto-laryng. (Stockh.),* **68,** 526–36.

24 Respiratory Activity of Intrinsic Laryngeal Muscles: An Experimental Study

BARRY WYKE*

Neurological Laboratory, Royal College of Surgeons of England, London, England

INTRODUCTION

Although the activity of the extrinsic and intrinsic muscles of the larynx has long been known to vary in concert with that of the rest of the respiratory musculature (Negus, 1929, 1949; Pressman and Kelemen, 1970), there is considerable uncertainty regarding the neurological mechanisms responsible for controlling laryngeal muscular activity in spontaneous breathing, let alone in speaking and singing. For this reason, the nature of these mechanisms has been the subject of anatomical and physiological study in the Neurological Laboratory of the Royal College of Surgeons of England since 1963; and this Communication reports experimental observations that we have made that may throw some light on the mechanisms that regulate the activity of the intrinsic laryngeal musculature in spontaneous respiration, and (by implication) in phonation.

PROCEDURES

As the technical details of the various investigations on which this report is based are in all essentials similar to those that have been outlined in previous publications from my Laboratory (Kirchner and Wyke, 1964, 1965a, b; Abo-El-Enein and Wyke, 1966; Abo-El-Enein, 1967; Wyke, 1967a, b, 1969a), they need not be re-iterated here.

Suffice it to say that the activity of the individual intrinsic and extrinsic muscles of the larynx has been examined in synchronously recorded electromyograms, obtained in bipolar fashion from pairs of needle or fine wire electrodes inserted directly into the extremities of each muscle, in spontaneously-breathing cats maintained at various stages of continuously monitored bar-

* Assisted by grants from the British Postgraduate Medical Federation of the University of London, and the Camilla Samuel Fund.

biturate anaesthesia [see Wyke, 1955, 1957, 1969b]. The animals' breath-by-breath tidal volume, respiratory rate, minute volume and thoracic and abdominal respiratory movements were monitored at the same time, so that respiratory behaviour might be correlated with the activity of the laryngeal musculature; and comparisons were made between animals breathing *per vias naturales* (that is, through the larynx) and animals breathing through a low tracheostomy tube (that is, not through the larynx).

OBSERVATIONS AND DISCUSSION

TONIC POSTURAL ACTIVITY IN THE INTRINSIC LARYNGEAL MUSCLES

As a prelude to discussion of the relations of the activity of the intrinsic laryngeal muscles to the respiratory cycle, it is necessary to consider the question of background tonic (or 'postural') activity in these muscles—for whilst numbers of other workers over the years (Semon, 1890; Weddell *et alii*, 1944; Fink *et alii*, 1956; Faaborg-Andersen, 1957, 1964; Buchthal, 1959; Fabre *et alii*, 1962; Mårtensson, 1964; Kotby and Haugen, 1970a, b) have found such activity to be present, they have differed widely amongst themselves in respect of the particular muscles said to be thus involved. Furthermore, some of these workers have maintained that this activity must be of central origin from the respiratory nuclei in the brain stem, whilst others have assumed that it originates reflexly from some unspecified source within the lungs or in the laryngeal tissues themselves. I therefore wish to report here four sets of experimental observations that seem pertinent to this initial and fundamental problem.

In lightly anaesthetized animals breathing spontaneously through a tracheostomy tube, continuous motor unit activity is apparent to some degree in all the instrinsic laryngeal muscles without exception—and one example of such activity (in the cricothyroid muscle) is shown in the tracings in Fig. 24.1. However, this activity shows but little evidence of variation in amplitude or frequency in relation to the spontaneous respiratory cycle (contrary to an earlier statement by Nakamura *et alii*, 1958); and this is true of all the intrinsic laryngeal muscles in these circumstances. In animals breathing through the larynx, on the other hand, tonic postural activity is again present in all intrinsic laryngeal muscles; but, as Figs. 24.2 and 24.4 show, the motor unit discharges now show prominent phasic fluctuations in amplitude and frequency that are related to the respiratory cycle, which will be discussed in more detail later.

The upper electromyogram in Fig. 24.2 (and the lower electromyograms in Fig. 24.1A and 1B) also shows that if one end of an intrinsic laryngeal muscle be detached from its cartilaginous insertion, its tonic postural activity (as well as the

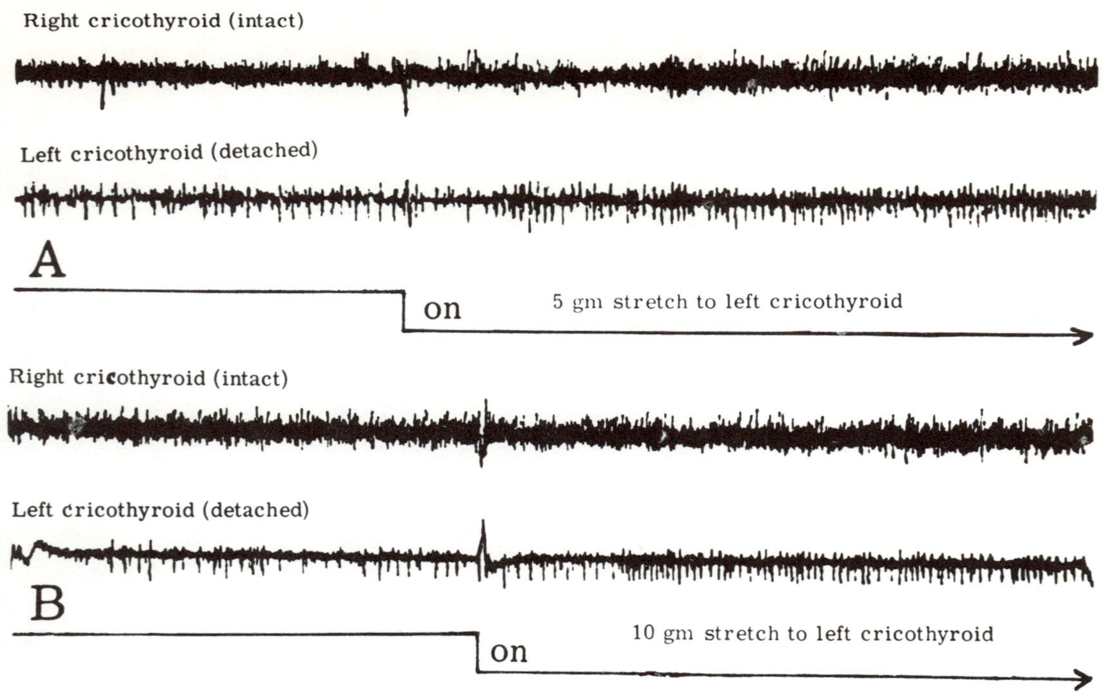

Fig. 24.1. Simultaneous electromyograms from the right and left cricothyroid muscles in an animal breathing through a tracheostomy tube (respiratory rate (f) 28/min; minute volume ($\dot{V}$) 1300 ml). The attachments of the right cricothyroid muscle are undisturbed, but the rostral end of the left cricothyroid muscle has been detached from the thyroid cartilage. In [A] and [B] tonic postural activity is much less marked in the detached than in the normally attached muscle, and there is little evidence of phasic respiratory fluctuations in motor unit activity in either pair of recordings. The signal lines indicate the application of 5 gm (in [A]) and 10 gm (in [B]) of longitudinal stretch to the rostral detached end of the left cricothyroid muscle, which results in an increase in tonic postural activity in each instance.

superimposed respiratory oscillations in motor unit discharge) is substantially reduced. Fig. 24.3 again shows (on the left of the lower tracings in [A] and in [B]) the residual, continuous low-frequency motor unit activity that is still present in a detached muscle (the left cricothyroid muscle, in this case) in comparison with the much greater activity in its still attached contralateral homologue [see also Fig. 24.1]; but it [and Fig. 24.1] also shows (on the right of the lower tracings in [A] and [B] in each illustration) that this activity may be progressively augmented during the application of graduated passive stretch to the detached end of the muscle. Fig. 24.4 shows elicitation of the same phenomenon in the vocalis musculature—for the upper tracing [A] depicts the continuous motor

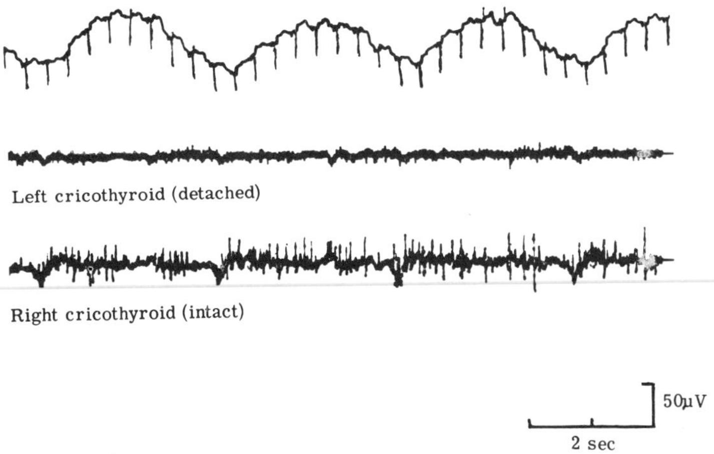

Fig. 24.2. Simultaneous electromyograms from the left and right cricothyroid muscles in an animal breathing through a pharyngeal airway, and through its larynx (respiratory rate (f) 22/min; minute volume ($\dot{V}$) 1280 ml). The attachments of the right cricothyroid muscle are undisturbed, but the rostral end of the left cricothyroid muscle has been detached from the thyroid cartilage. Respiratory recording (with superimposed electrocardiogram) in upper tracing (inspiration upwards). Oscillations of motor unit activity related to the respiratory cycle are now superimposed on the tonic postural activity—these are biphasic (see text) in the intact muscle, but are reduced and confined to the expiratory phase in the detached muscle.

unit activity that persists in the left thyro-arytenoid muscle after its separation from its anterior attachment to the thyroid cartilage, whilst the lower tracing [B] shows its augmentation (as well as that of the superimposed respiratory oscillations) during the application of passive stretch to its detached anterior end.

Many experiments of this type have convinced us that continuous motor unit activity is always present to some degree in all the intrinsic laryngeal muscles of lightly anaesthetized animals, irrespective of whether or not they breathe through the larynx. As Figs. 24.1, 24.2, 24.3 and 24.4 illustrate, however, the amplitude and frequency of this background activity is substantially reduced (although never abolished) whenever an intrinsic laryngeal muscle is freed at one end from the cartilaginous skeleton of the larynx; but it can be restored progressively, and eventually to its pristine pattern, by the application of adequate degrees of graduated stretch to the detached muscle.

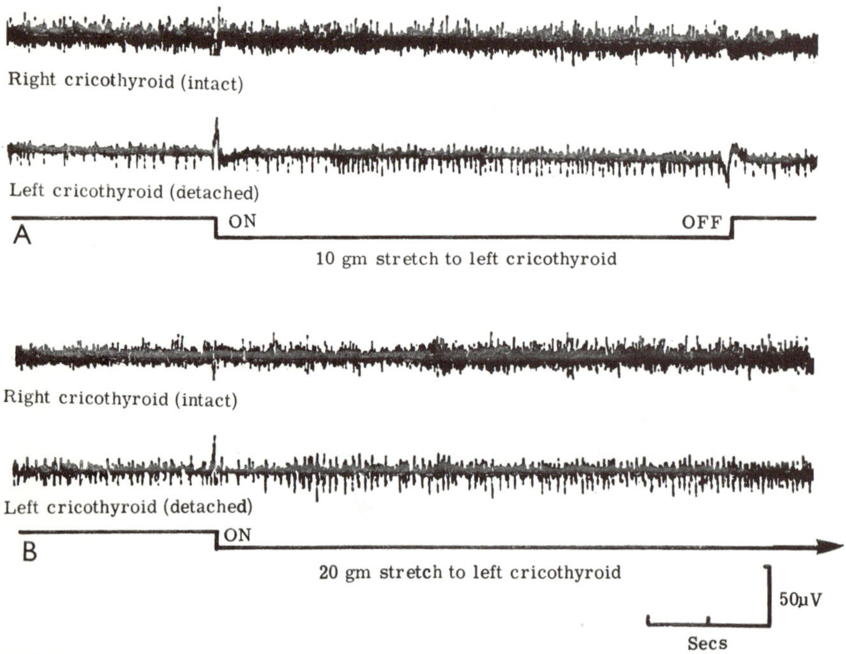

Fig. 24.3. Simultaneous electromyograms from the right and left cricothyroid muscle [c.f. Fig. 24.1] in an animal breathing through a tracheostomy tube (respiratory rate (f) 30/min; minute volume ($\dot{V}$) 1380 ml). The attachments of the right cricothyroid muscle are undisturbed, but the rostral end of the left cricothyroid muscle has been detached from the thyroid cartilage. The signal lines indicate the application of 10 gm (in [A]) and 20 gm (in [B]) of longitudinal stretch to the rostral detached end of the left cricothyroid muscle. Description in text.

Other experiments which we have reported previously (Wyke, 1968) have also shown that this tonic postural activity is reduced by increasing general anaesthesia—but to differing degrees and at different rates in the individual intrinsic muscles of the larynx. Thus we find that, in such circumstances, tonic postural activity begins to diminish first in the vocalis and lateral crico-arytenoid musculature. As anaesthesia is further increased, cricothyroid muscle activity is then affected; and finally, posterior crico-arytenoid activity begins to decrease—by which time, however, tonic postural activity has largely disappeared from most of the vocal fold adductor musculature [see also Murtagh and Campbell, 1956; Brewer and Dana, 1963; Suzuki and Kirchner, 1969].

It will be apparent that the findings I have described are not in accord with an early view (Semon, 1890) that the posterior crico-arytenoid muscle is the only intrinsic laryngeal muscle displaying tonic activity; nor do they support a later statement (Mårtensson, 1964) that tonic postural activity is absent from the lateral crico-arytenoid and thyro-arytenoid musculature, although present in the cricothyroid and posterior crico-arytenoid muscles. However, in the light of our observations on the effects of changing degrees of general anaesthesia (Wyke, 1968), it would seem that these discrepancies may be resolved by recognizing that such anaesthetic variations have differential effects on the neurological mechanisms controlling the individual laryngeal muscles (as is known to be the case with various groups of striated muscles elsewhere in the body (Wyke, 1969b)), such that the distribution of tonic postural activity throughout the laryngeal musculature is a function of the degree of anaesthesia prevailing at the time the observations are made.

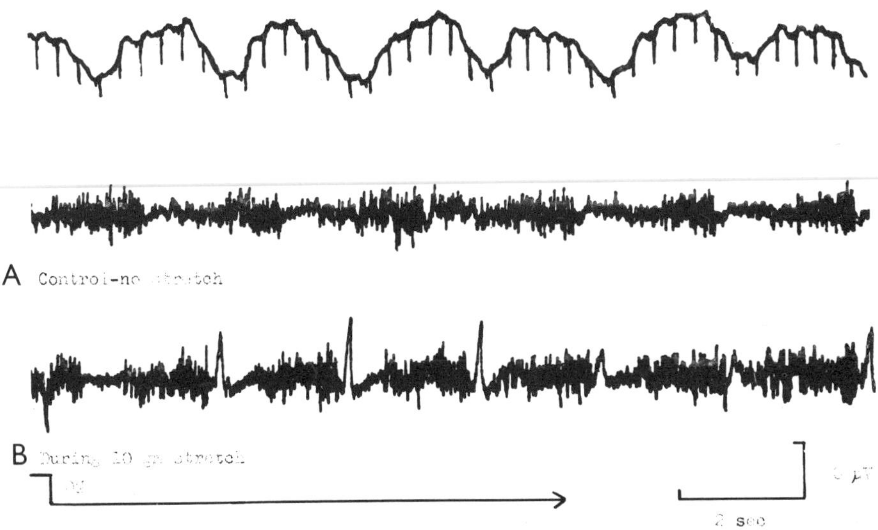

Fig. 24.4. Successive electromyograms from the left vocalis muscle (after separating it from its anterior attachment to the thyroid cartilage) before [A] and during [B] the application of 10 gm of longitudinal stretch to its detached anterior end, in an animal breathing through its larynx by way of an intrapharyngeal airway. Respiratory recording (with superimposed electrocardiogram) in upper tracing (inspiration upwards). Both superior laryngeal nerves were divided prior to the recordings. Description in text.

More germane to our present context, however, is the observation that tonic postural activity in the intrinsic laryngeal muscles is substantially reduced by freeing one end of such a muscle from its attachments, and that its motor unit activity can be restored progressively by the application of graduated stretch to the detached muscle. As Professor Bowden has already mentioned [see Com-

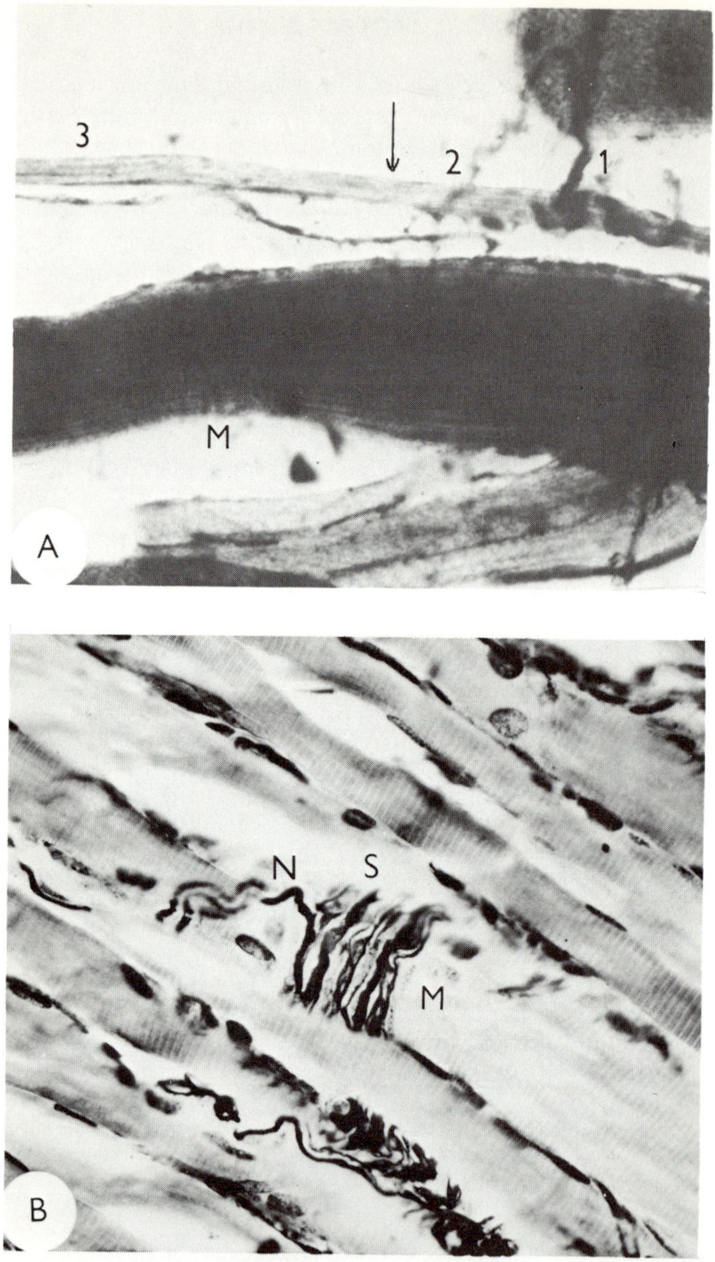

Fig. 24.5. Mechanoreceptor nerve endings in the intrinsic muscles of the larynx. [A]. Portion of a muscle spindle (indicated by arrow) lying between and in parallel with the fibres [M] of the lateral crico-arytenoid muscle. Large [1] and small [2] diameter myelinated fibres innervating the spindle can be seen, the former terminating in a primary ending in the equatorial region of the spindle near [1]. Three intrafusal muscle fibres can be identified within the thin capsule of the polar region of the spindle at [3]. Wyke gold chloride microdissection preparation. (Magnification ×1300). [B]. Spiral receptor nerve ending (indicated by [S]) coiled around one of the fibres [M] of the cricothyroid muscle: the parent myelinated afferent nerve fibre [N] can be seen running off to the left. Modified Glees' silver technique. (Magnification ×1200.)

munication 22] previous reports from her Laboratory (Lucas Keene, 1957, 1961; Withington, 1959; Bowden *et alii,* 1960; Bowden and Scheuer, 1961; Bowden, 1963) and from my own (Abo-El-Enein and Wyke, 1966a, b; Abo-El-Enein, 1967; Wyke, 1967, 1969a, 1971) have described receptor systems, comprising a few muscle spindles and many spiral nerve endings (examples of each of which are illustrated in Fig. 24.5), in the intrinsic laryngeal muscles of the cat (and man—see also Baken (1971)) that are morphologically appropriate as stretch-sensitive mechanoreceptors; and neurophysiological confirmation of their behaviour in this regard has also been provided in publications from my Laboratory (Abo-El-Enein and Wyke, 1966a, b; Abo-El-Enein, 1967; Wyke, 1967, 1969a, 1970, 1971), the details of which need not be recapitulated here [see also Suzuki and Kirchner, 1968].

In the light of these data, then, I should like to suggest that the basic, tonic postural activity of the intrinsic laryngeal muscles is determined primarily by reflexogenic afferent discharges generated from low-threshold mechanoreceptors located within each of the muscles, continuous excitation of which is normally provided by the slight stretch that obtains within each muscle between its attachments to the laryngeal cartilages. This myotatic reflex influence appears to be operative on laryngeal muscle motoneurones against a background of excitation thereof from neurones in the brain-stem reticular system (which others at this meeting have already described—see Cohen (1970) and Communication 12); and the oscillations of motor unit activity in the laryngeal muscles that are associated with the respiratory cycle (and which I wish to consider further in the remainder of this Communication) appear to be superimposed upon this background of tonic postural activity that is mainly of myotatic reflex origin. In this respect, then, the laryngeal muscles would seem to resemble the other striated respiratory muscles—and in particular, the intercostal muscles [see Corda *et alii,* 1965; von Euler, 1966; Campbell *et alii,* 1970].

RESPIRATORY OSCILLATIONS OF ACTIVITY IN THE INTRINSIC LARYNGEAL MUSCLES

I should now like to turn to more specific consideration of the rhythmical oscillations of motor unit activity that occur in the intrinsic muscles of the larynx during the spontaneous respiratory cycle, some examples of which are illustrated in Figs. 24.2, 24.4 and 24.11. The necessity for further experimental analysis of this matter is dictated by the fact that views as to the relation between the neurological mechanisms controlling the primary respiratory muscles and those controlling the activity of the laryngeal musculature are in a somewhat confused state—as exemplified by statements in the literature to the effect that variations in intrinsic laryngeal muscle activity are negligible in quiet breathing (Barwell, 1905; Fabre *et alii,* 1962), and only become apparent with augmented or forced ventilation (Semon, 1890; Negus, 1929; Fink *et alii,* 1956; Pressman and

Kelemen, 1970): or, that during expiration there is increased activity in the adductors of the vocal folds, which relax during inspiration (Grossmann, 1898; Negus, 1929; Murtagh, 1945; Green and Neil, 1955; Nakamura *et alii*, 1958): or conversely, that the vocal fold adductors relax during expiration, but contract during inspiration (Weddell *et alii*, 1944; Feinstein, 1946; Fink *et alii*, 1956; Faaborg-Andersen, 1957, 1964; Buchthal, 1959; Buchthal and Faaborg-Andersen, 1964; Hiroto *et alii*, 1964; Kotby and Haugen, 1970b).

PATTERNS OF RESPIRATORY OSCILLATIONS IN LARYNGEAL MOTOR UNIT ACTIVITY

The data already submitted in this Communication indicate that respiratory oscillations of motor unit activity in the laryngeal muscles are uniformly conspicuous in lightly anaesthetized animals breathing quietly through the larynx, but are relatively insignificant in tracheotomized animals maintained at the same stage of anaesthesia. Furthermore, the magnitude of such oscillations appears to be related to the degree of background tonic postural activity prevailing in the muscles—for if this latter be reduced by detaching one end of a muscle from its cartilaginous insertion [see Figs. 24.2, 24.4 and 24.11], or by increasing the degree of general anaesthesia (Wyke, 1968), the respiratory oscillations of motor unit activity are substantially reduced, or may even disappear.

In the case of animals breathing quietly through the larynx, the *cricothyroid muscle* displays low-grade biphasic changes of motor unit activity in relation to the spontaneous respiratory cycle, as illustrated in Fig. 24.2. These appear as grouped motor unit discharges, one burst occurring during each inspiration and another (sometimes shorter) burst at each expiration; and a recent report from Professor Kirchner's Laboratory in Yale has suggested that different motor unit populations within this muscle may be involved in the two instances (Suzuki *et alii*, 1970). However, when tracheostomy is performed, or when the degree of anaesthesia is increased slightly, the expiratory burst disappears—although a reduced inspiratory burst is still apparent in these circumstances: on the other hand, the inspiratory burst is selectively depressed by detaching one end of the muscle from the thyroid cartilage [Fig. 24.2]. The *thyro-arytenoid musculature* displays a more prolonged monophasic alteration in its activity, which is augmented at the onset of expiration and continues to be augmented as the expiratory phase develops until it terminates abruptly at, or shortly, after, the onset of inspiration. In very light anaesthesia, however (that is, when there is considerable background tonic activity in the muscle), the thyro-arytenoid muscle may also show a separate, very brief increment in its motor unit activity at the onset of inspiration [Figs. 24.4 and 24.11] that is synchronous with that occurring in the cricothyroid muscle. In contrast, and as illustrated in Fig. 24.6, the *posterior crico-arytenoid muscle* shows a more or less sinusoidal pattern of fluctuation in its activity, which begins to increase at the end of each expiration, continues to increment until just after inspiration reaches its peak, and then declines rapidly before the next expiration begins [see also Suzuki and Kirchner, 1969; Kotby and Haugen, 1970a, b).

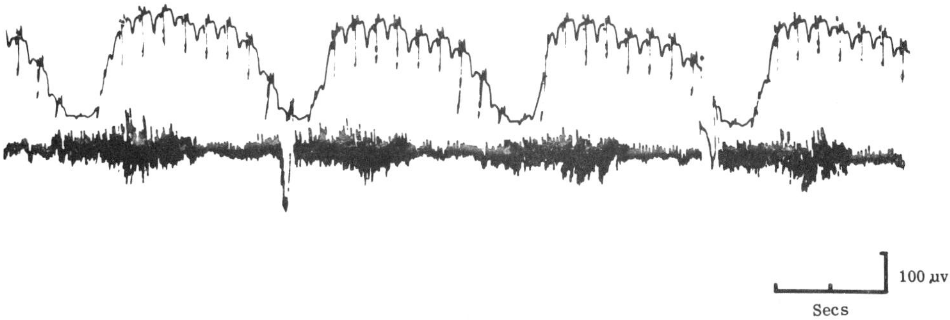

Fig. 24.6. An electromyogram of the left posterior crico-arytenoid muscle in an animal breathing through its larynx by way of an intrapharyngeal airway. Respiratory recording (with superimposed electrocardiogram) in upper tracing (inspiration upwards). Description in text.

SOURCES OF RESPIRATORY OSCILLATION IN LARYNGEAL MOTOR UNIT ACTIVITY

The question that now poses itself concerns the neurological mechanisms that are responsible for the co-ordinated patterns of change in the activity of the intrinsic laryngeal muscles that I have just described; and whilst it has to be admitted that we cannot provide any fully satisfactory explanation at this time, I should like to offer some observations that may contribute to this objective. Before doing so, however, may I remind you briefly of some of the views that have been held in this matter by other workers.

As might be expected, on *a priori* (if on no other) grounds, many early workers (Semon, 1890; Negus, 1929; Green and Neil, 1955; Andrew, 1955), as well as some later investigators (Rudomín, 1966b), have held that the activity of the laryngeal musculature involved in vocal fold adjustment during the respiratory cycle is controlled directly and exclusively from the respiratory nuclei in the brain stem, in synchrony with the rest of the striated respiratory musculature; and more recent studies (Bianconi and Raschi, 1959, 1964; Eyzaguirre *et alii,* 1966; Suzuki and Kirchner, 1969) have demonstrated the mutually synchronized discharges that occur in efferent fibres in the recurrent laryngeal and phrenic nerves [see also Guth *et alii,* 1960; Buchthal and Faaborg-Andersen, 1964). On the other hand, there are also numbers of reports in the literature implicating (to differing degrees) a variety of peripherally-operated reflexogenic mechanisms in such control—amongst which, pulmonary mechanoreceptors and laryngeal mucosal mechanoreceptors have been most often numbered.

Pulmonary Reflexogenic Influences

The relative significance of the first of these reflexogenic systems in relation to laryngeal muscular activity during respiration—that is, of the pulmonary mechanoreceptors—can be assessed from data already available in publications for other laboratories. Thus, workers in London (Green and Neil, 1955) and in Rome (Bianconi and Raschi, 1959; Molinari and Pivotti, 1962) found that alterations in vagal afferent discharge in response to passive inflation of the lungs may modify the patterns of reciprocally-related activity in vocal fold adductors and abductors, or the efferent discharges to these muscles in the recurrent laryngeal nerves [see also Joels and Samueloff, 1956; Eyzaguirre and Taylor, 1963); whilst Rudomín (1966a) in Mexico has shown that the same manoeuvre may depress the inspiratory and increase the expiratory activity of the cricothyroid muscle—as long as the vagus nerves are intact.

Nevertheless, we have found (as did Green and Neil (1955) previously) that spontaneous respiratory oscillatory activity of the vocal fold musculature persists in non-tracheotomized animals after bilateral vagal nerve section distal to the origin of the recurrent laryngeal nerves, even though passive inflation of the lungs then has little or no effect on intrinsic laryngeal muscular activity; and these observations, taken together, suggest that the rhythmical respiratory oscillations of activity that occur in the laryngeal musculature in quiet breathing are *not* produced primarily by activity in pulmonary afferent systems—although they may be modified by such activity, as Guz *et alii* (1970) and Widdicombe [see Communication 9] have shown.

Laryngeal Mucosal Reflexogenic Influences

Other observations that have been reported at intervals over the last 35 years (Aldaya, 1936; Campbell and Murtagh, 1956; Sampson and Eyzaguirre, 1964; Kirchner and Suzuki, 1968) have been interpreted as indicating that the respiratory activity of the intrinsic laryngeal muscles may depend upon afferent discharges from low-threshold mechanoreceptors located in the laryngeal mucosa (Matsumoto, 1950; Koizumi, 1953; Koizumi and Makami, 1953; König and von Leden, 1961; Van Michel, 1963), responding to the fluctuations in intralaryngeal (and especially subglottic) air pressure that occur during the phases of the respiratory cycle (van den Berg, 1956, 1964).

However, our own experiments and those of others have indicated that whilst such mucosal mechanoreceptor reflex mechanisms may modify the activity of the intrinsic laryngeal muscles in various circumstances, including phonation (Kirchner and Suzuki, 1968; Suzuki and Kirchner 1969; Wyke, 1967, 1968, 1969, 1971; Widdicombe, Communication 9), they are not primarily responsible for the respiratory oscillations of such activity that occur during quiet breathing—as evidenced [see Fig. 24.4, for example] by the persistence of such oscillations (albeit in reduced and/or modified patterns) following bilateral denervation of the laryngeal mucosa [see also Rudomín, 1966b: Suzuki and Kirchner, 1969), opening of the larynx to the atmosphere (Sampson and Eyzaguirre, 1964), or topical anaesthesia of the supraglottic and infraglottic mucosal lining.

Thus, although other publications from my Laboratory [see Wyke, 1971] have suggested that subglottic mucosal reflex mechanisms might play a significant rôle in the regulation of laryngeal muscular activity during phonation and singing (a suggestion, incidentally, that receives some support from the human experimental observations that Dr. Gould has described (Gould and Okamura, Communication 21), the studies to which I have just referred indicate that mucosal reflex mechanisms are *not* fundamentally involved in producing the lesser oscillations of motor unit activity that characterize the intrinsic laryngeal muscles during quiet breathing—perhaps, because of the lower range of subglottic pressure variation that then obtains, in comparison with the somewhat higher pressures (of 5–10 cm H_2O) that are usually associated with phonation (Ladefoged, 1960, 1968; Isshiki, 1964, 1965; Lieberman, 1968; Proctor, Communication 4).

Laryngeal Arthrokinetic Reflexogenic Influences

Following the anatomical and physiological identification of the laryngeal articular reflex system by Kirchner and Wyke (1964a, b, c, d, 1965a, b) in 1964 and 1965, my colleagues and I have naturally inquired [see Wyke, 1967, 1968, 1969a, 1970, 1971] whether afferent discharges from the corpuscular mechanoreceptors that are located in the fibrous capsules of each of the intercartilaginous joints in the larynx (a typical teased-out example of which is shown in Fig. 24.7) might contribute to the production of the changes in motor unit activity in the intrinsic laryngeal musculature that are associated with the respiratory cycle. As Fig. 24.8 shows, rapid passive movements of any individual isolated laryngeal joint (in this instance, the cricothyroid joint) in appropriate directions are certainly capable of producing arthrokinetic reflex changes in motor unit activity in the intrinsic laryngeal muscles; and the fact that such arthrokinetic reflex effects may be reciprocally co-ordinated as between vocal fold adductors and abductors is illustrated in Fig. 24.9, which shows that reflex thyro-arytenoid facilitation is accompanied by posterior crico-arytenoid inhibition when the cricothyroid joint is moved passively.

The fact that the laryngeal joints move during quiet breathing has long been familiar to laryngologists [see Negus, 1929, 1949; Fink *et alii*, 1956]; and their patterns of movement in this and other circumstances have been analysed in some detail in the radiological studies reported from Oxford by Ardran and Kemp (1966, 1967). However, our own studies (Kirchner and Wyke, 1964d, 1965a, b) have demonstrated that most of the mechanoreceptors in the laryngeal joints are rapidly-adapting in their behaviour, and that therefore their reflex effects on the laryngeal musculature are short-lived (as Fig. 24.10 indicates). For this reason, we suggest that whilst arthrokinetic reflex mechanisms may contribute significantly to the production of at least some of the changes in motor unit activity that are associated with respiration, in a manner similar to the influence that may be exerted by costovertebral articular mechanoreceptors on the respiratory activity of the intercostal muscles (Yamamoto *et alii,* 1956, 1960; Godwin-Austen, 1969), this contribution is transient and is restricted to the onset of certain of the oscillations of muscular activity described previously. Consis-

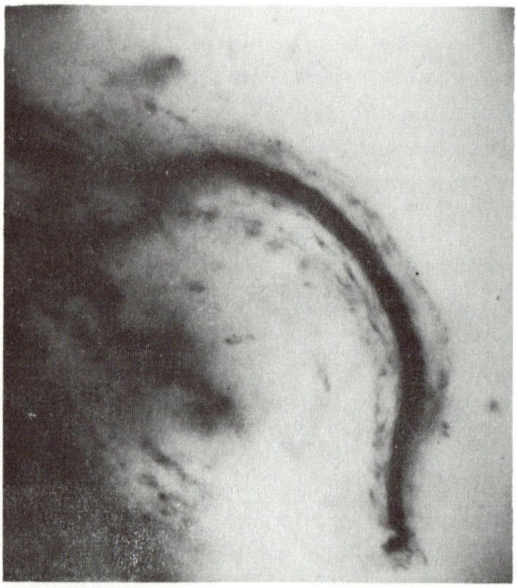

Fig. 24.7. A single mechanoreceptor corpuscle, teased out of the fibrous capsule of a cat's cricothyroid joint. Wyke gold chloride preparation (magnification ×1200). (From Kirchner and Wyke, 1965b.)

tent with this view is the observation that the respiratory oscillations of motor unit activity in the intrinsic laryngeal muscles are markedly reduced when ventilation occurs through a tracheostomy [Figs. 24.1 and 24.3]—in which circumstance, movements of the laryngeal joints are substantially lessened; and that bilateral local anaesthetic infiltration of the cricothyroid joints (or destruction of the receptors located therein by electrocoagulation of their capsules) significantly reduces (but does not abolish) the burst of activity that normally occurs in the cricothyroid muscles at the onset of inspiration, as tracheal traction rotates the cricothyroid joints [see also Mitchinson and Yoffey, 1947; Andrew, 1955; Murakami and Kirchner, Communication 25].

Laryngeal Myotatic Reflexogenic Influences

The final reflex system that requires consideration in the present context is that operated from the laryngeal myotatic mechanoreceptors, to which reference has already been made in connexion with the production of tonic postural activity in

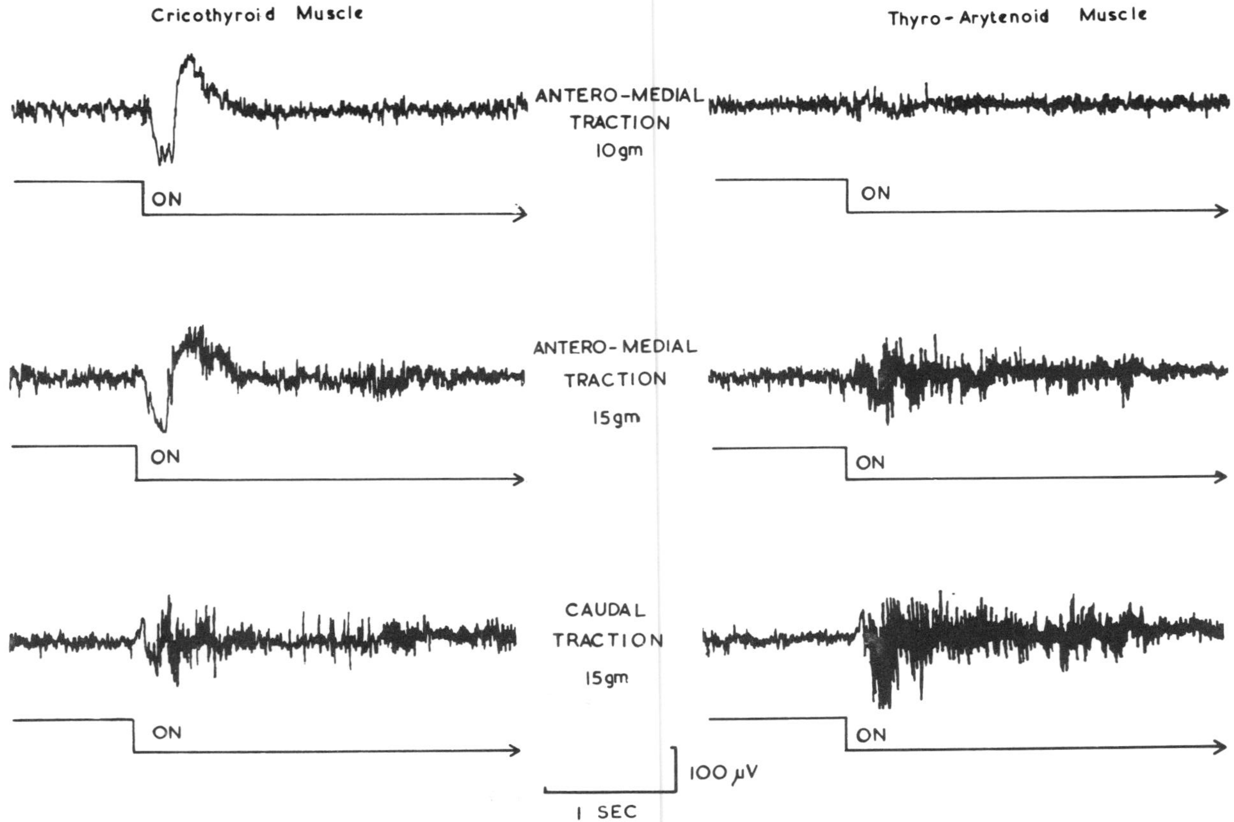

Fig. 24.8. Simultaneous electromyograms of the detached left cricothyroid (left-hand column) and thyro-arytenoid (right-hand column) muscles, showing co-ordinated arthrokinetic reflex responses to differing degrees and directions of displacement (at signals) of the ipsilateral (isolated) cricothyroid joint. (From Kirchner and Wyke, 1965b.)

the intrinsic muscles of the larynx. It was then indicated (*loc. cit.*) that detaching one end of an individual intrinsic laryngeal muscle from the laryngeal cartilages results in profound attenuation of the oscillations in its motor unit activity that are normally present during quiet breathing *per vias naturales*, in association with the reduction of its tonic postural activity that then occurs. Furthermore, and again as previously pointed out and as is additionally illustrated in Fig. 24.11, these respiratory oscillations may be progressively restored, along with the background tonic postural activity, by the application of graduated passive stretch to the detached muscle.

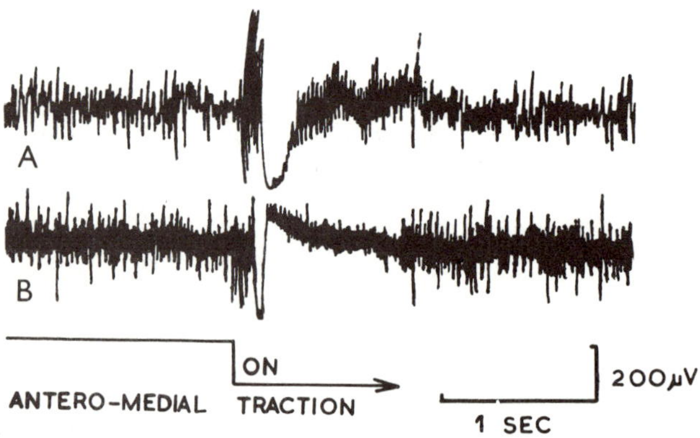

Fig. 24.9. Simultaneous electromyograms of the detached left thyro-arytenoid [A] and cricothyroid [B] muscles, showing reciprocally co-ordinated arthrokinetic reflex responses to sudden displacement of the ipsilateral (isolated) cricothyroid joint. (From Wyke, 1967b.)

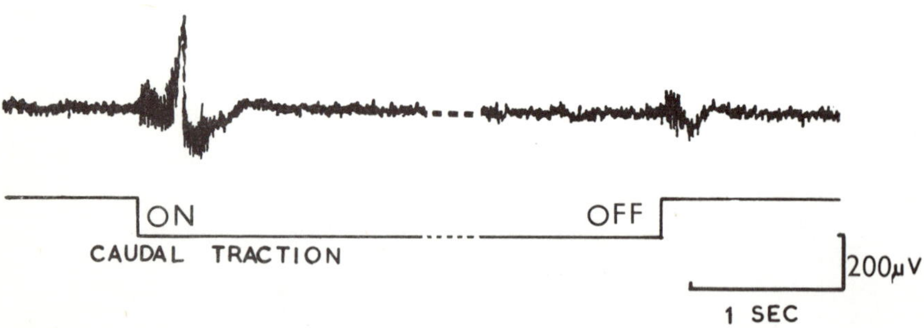

Fig. 24.10. Electromyogram of the detached left cricothyroid muscle, showing the phasic nature of the arthrokinetic reflex responses to sudden displacements (at signal) of the ipsilateral (isolated) cricothyroid joint: the interrupted line indicates a 10 sec period of continuously applied displacement. (From Kirchner and Wyke, 1965b.)

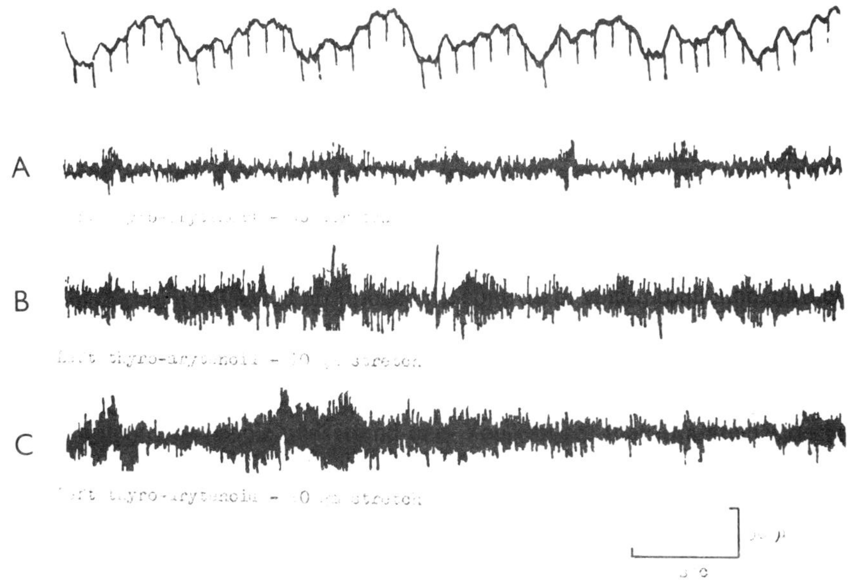

Fig. 24.11. Successive electromyograms from the left vocalis muscle (after separating it from its anterior attachment to the thyroid cartilage) before [A] and during the application of 10 gm [B] and then 20 gm [C] of longitudinal stretch to its detached anterior end, in an animal breathing through its larynx by way of an intrapharyngeal airway [see also Fig. 24.4]. Respiratory recording (with superimposed electrocardiogram) in upper tracing (inspiration upwards). Description in text.

CONCLUSION

If I now attempt to integrate these various observations into a coherent theoretical schema, then it seems that the following are the essential data for this purpose:

1. Whether air flows through the larynx or not in quietly breathing animals, tonic postural activity is present throughout the intrinsic laryngeal musculature.
2. Only when the respired air ebbs and flows through the larynx are prominent rhythmical oscillations of motor unit activity superimposed upon the tonic postural activity in the intrinsic muscles.
3. Removal of the anatomical stretch to which the intrinsic muscles are normally subjected between their cartilaginous attachments markedly reduces (although it does not abolish) this oscillatory activity, *pari passu* with reduction in the background tonic postural activity.
4. Artificial restoration of this stretch to individual intrinsic laryngeal muscles

restores their oscillatory respiratory activity, at the same time and to the same extent as it restores their tonic postural activity.

5. Afferent discharges from articular mechanoreceptors, provoked by the movements of the laryngeal joints that occur during quiet breathing, may reflexly re-inforce some of the initial components in the episodes of oscillatory motor unit activity that are asociated with the ebb and flow of air through the larynx.

6. Afferent discharges from pulmonary mechanoreceptors, and from laryngeal mucosal mechanoreceptors, do not appear to exert dominating influences on intrinsic laryngeal muscular activity in quiet breathing—although they may markedly modify or distort such activity in other ventilatory situations, including phonation.

7. Waxing and waning discharges of nerve impulses in the axones of laryngeal motoneurones located in the nucleus ambiguus are synchronized, during quiet spontaneous breathing, with discharges from phrenic motoneurones.

In the light of these data, I should like to make the following proposal regarding the nervous control of the oscillations of motor unit activity that occur in the intrinsic muscles of the larynx during the respiratory cycle in quiet breathing—with the proviso that it represents an operant hypothesis that constitutes merely a theoretical 'best fit', in the present state of knowledge. What I suggest as a basis for discussion and further research, then, is that the *primary drive* to the laryngeal motoneurones is provided from the respiratory nuclei in the brain stem, whose oscillating activity [see Cohen, 1970; Communication 12] generates slowly-alternating depolarization and hyperpolarization in the cells of the nucleus ambiguus, such as occurs (Sears, 1964, 1966; Campbell *et alii*, 1970) in the motoneurone pools innervating the diaphragm and intercostal muscles; but that the moment-to-moment effectiveness of this influence in evoking the discharge of impulses to the laryngeal muscles is critically dependent upon the level of excitability prevailing in the motoneurone pools of these muscles in response to the degree of on-going activity in the myotatic reflex afferent systems operated from the mechanoreceptors located within the intrinsic laryngeal muscles themselves. This central excitatory state in the motoneurone pools of the laryngeal muscles [see also Murakami and Kirchner, 1971) is also phasically influenced by transient afferent inputs emanating from the mechanoreceptors in the laryngeal joint capsules; but it is only slightly affected by afferent discharges from pulmonary mechanoreceptors or laryngeal mucosal mechanoreceptors in quiet breathing—although such reflexogenic systems may modify laryngeal motoneurone activity significantly in circumstances of altered ventilation, and in speaking and singing. In addition, in these latter two circumstances still further modifications of laryngeal motor unit activity may be imposed voluntarily, by way of corticobulbar projections to the motoneurones of the nucleus ambiguus [see Rudomín, 1964; Dunker, 1968; Brodal, 1969)—as suggested in previous publications from my Laboratory (Wyke, 1967, 1969, 1970, 1971).

ACKNOWLEDGEMENTS

Thanks are due to Professor John Kirchner (Yale University, New Haven, Conn., U.S.A.) and Dr. M. A. Abo-El-Enein (Al-Azhar University, Cairo, U.A.R.) for their skilled co-operation in some of the earlier investigations in this study, and to Mr. J. Kitau, Mr. J. Ramcharan, A.I.S.Y. and Mrs. S. des Tombe, B.Sc. for their technical assistance.

REFERENCES

Abo-El-Enein, M. A. (1967) *Functional Anatomy of the Larynx,* Ph.D. Thesis, University of London.

Abo-El-Enein, M. A., and Wyke, B. D. (1966a) Laryngeal myotatic reflexes, *Nature (Lond.),* **209,** 682–6.

Abo-El-Enein, M. A., and Wyke, B. D. (1966b). Myotatic reflex systems in the intrinsic muscles of the larynx, *J. Anat. (Lond.),* **100,** 926–7.

Aldaya, F. (1936) Le contrôle réflex de la respiration par la sensibilité du larynx, *C. R. Soc. Biol. (Paris),* **123,** 1001–2.

Andrew, B. L. (1955) The respiratory displacement of the larynx: a study of the innervation of accessory respiratory muscles, *J. Physiol. (Lond.),* **130,** 474–87.

Ardran, G. M., and Kemp, F. H. (1966) The mechanism of the larynx. I. The movements of the arytenoid and cricoid cartilages, *Brit. J. Radiol.,* **39,** 641–54.

Ardran, G. M., and Kemp, F. H. (1967) The mechanism of the larynx. II. The epiglottis and closure of the larynx, *Brit. J. Radiol.,* **40,** 372–89.

Baken, R. J. (1971) Neuromuscular spindles in the intrinsic muscles of a human larynx, *Folia phoniat. (Basel),* **23,** 204–10.

Barwell, H. B. (1905) Laryngeal paralyses and their diagnostic value, *Lancet,* **1,** 1481–4.

Bianconi, R., and Raschi, F. (1959) Osservazioni sull' attivita elettrica dei nervi ricorrenti nel gatto, *Bull. Soc. ital. Biol. sper.,* **35,** 1917–18.

Bianconi, R., and Raschi, F. (1964) Respiratory control of motoneurones of the recurrent laryngeal nerve and hypocapnic apnoea, *Arch. ital. Biol.,* **102,** 56–73.

Bowden, R. E. M. (1963) Muscle spindles in the human foetus, *Acta biol. (Szeged),* **9,** 35–39.

Bowden, R. E. M., Lucas Keene, M. F., Gooding, M., Mahran, Z. Y., and Withington, J. L. (1960) The afferent innervation of facial and laryngeal muscles, *Anat. Rec.,* **136,** 168.

Bowden, R. E. M., and Scheuer, J. L. (1961) Comparative studies of the nerve supply of the larynx in Eutherian mammals, *Proc. zool. Soc. Lond.,* **136,** 325–30.

Brewer, D. W., and Dana, S. T. (1963) Investigations in laryngeal physiology: the canine larynx. II, *Ann. Otol. (St. Louis),* **72,** 1060–75.

Brodal, A. (1969) *Neurological Anatomy in Relation to Clinical Medicine,* 2nd ed., New York.

Buchthal, F. (1959) Electromyography of intrinsic laryngeal muscles, *Quart. J. exp. Physiol.,* **44,** 137–48.

Buchthal, F., and Faaborg-Andersen, K. (1964) Electromyography of laryngeal and respiratory muscles: correlation with phonation and respiration, *Ann. Otol. (St. Louis),* **73,** 118–23.

Campbell, C. J., and Murtagh, J. A. (1956) Electrical manifestations of recurrent nerve function, *Ann. Otol. (St. Louis),* **65,** 747–65.

Campbell, E. J. M., Agostoni, E., and Newsom Davis, J. (1970) *The Respiratory Muscles: Mechanics and Neural Control,* 2nd ed., London.

Cohen, M. I. (1970) How respiratory rhythm originates: evidence from discharge patterns of brainstem respiratory neurones, in *Breathing: Hering-Breuer Centenary Symposium*, pp. 125–50, ed. Porter, R., London.

Corda, M., von Euler, C., and Lennerstrand, G. (1965) Proprioceptive innervation of the diaphragm, *J. Physiol. (Lond.),* **184,** 898–923.

Dunker, E. (1968) The central control of laryngeal function, *Ann. N.Y. Acad. Sci.,* **155,** 112–21.

von Euler, C. (1966) The control of respiratory movements, in *Breathlessness,* pp. 19–32, ed. Howell, J. B. L., and Campbell, E. J. M., Oxford.

Eyzaguirre, C., Sampson, S., and Taylor, J. R. (1966) The motor control of intrinsic laryngeal muscles in the cat, in *Nobel Symposium. I. Muscular Afferents and Motor Control*, pp. 209–25, ed. Granit, R., Stockholm.

Eyzaguirre, C., and Taylor, J. R. (1963) Respiratory discharges of some vagal motoneurones, *J. Neurophysiol.,* **26,** 61–78.

Faaborg-Andersen, K. L. (1957) Electromyographic investigation of intrinsic laryngeal muscles in humans: an investigation of subjects with normally movable vocal cords and patients with vocal cord paresis, *Acta physiol. scand.*, **41** (Suppl. 140), 1–148.

Faaborg-Andersen, K. L. (1964) Electromyography of the laryngeal muscles in man, in *Research Potentials in Voice Physiology,* pp. 105–29., ed. Brewer, D. W., New York.

Fabre, P., Husson, R., and Roëlens, R. (1962) Physiologie phonatoire laryngée: separation expérimentale des fonctions cloniques et toniques de la musculature laryngée par la glottographie électrique dite 'respiratoire', *C.R. Acad. Sci. (Paris),* **255,** 1526–8.

Feinstein, B. (1946) The application of electromyography to affections of the facial and intrinsic laryngeal muscles, *Proc. roy. Soc. Med.,* **39,** 817–19.

Fink, B. R., Basek, M., and Epanchin, V. (1956) The mechanism of opening of the human larynx, *Laryngoscope (St. Louis),* **66,** 410–25.

Godwin-Austen, R. B. (1969) The mechanoreceptors of the costovertebral joints, *J. Physiol. (Lond.),* **202,** 737–53.

Green, J. G., and Neil, E. (1955) The respiratory function of the laryngeal muscles, *J. Physiol. (Lond.),* **129,** 134–41.

Grossmann, M. (1898) Experimentelle Untersuchungen über die funktionelle Ausschaltung einzelner Musklen, *Archiv. ges. Physiol.,* **73,** 184–218.

Guth, L., Soutter, L., Frank, K., Campbell, J. B., and Lloyd, J. B. (1960) Diaphragmatic function following anastomosis of recurrent laryngeal and phrenic nerves, *Exp. Neurol.*, **2,** 251–60.

Guz, A., Noble, M. I. M., Eisele, J. H., and Trenchard, D. (1970) The role of vagal inflation reflexes in man and other animals, in *Breathing: Hering-Breuer Centenary Symposium.* pp. 17–40, ed. Porter, R., London.

Hiroto, I., Hirano, M., Toyozumi, Y., and Shin, T. (1964) Function of laryngeal muscles, *Oto-rhino-laryngol. Clinics, Kyoto,* **57,** 1–9.

Isshiki, N. (1965) Vocal intensity and airflow rate, *Folia phoniat. (Basel),* **17,** *Res.,* **7,** 17–29.

Isshiki, N. (1965) Vocal intensity and airflow rate, *Folia phoniat. (Basel),* **17,** 92–104.

Joels, N., and Samueloff, M. (1956) The activity of the medullary centres in diffusion respiration, *J. Physiol. (Lond.),* **133,** 360–72.

Kirchner, J. A., and Suzuki, M. (1968) Laryngeal reflexes and voice production, *Ann. N.Y. Acad. Sci.,* **155,** 98–109.

Kirchner, J. A., and Wyke, B. D. (1964a) Innervation of laryngeal joints and laryngeal reflexes, *Nature (Lond.),* **201,** 506.

Kirchner, J. A., and Wyke, B. D. (1964b) The innervation of the laryngeal joints in the cat, *J. Anat. (Lond.),* **98,** 684.

Kirchner, J. A., and Wyke, B. D. (1964c) Laryngeal articular reflexes, *Nature (Lond.),* **202,** 600.

Kirchner, J. A., and Wyke, B. D. (1964d) Electromyographic analysis of laryngeal articular reflexes, *Nature (Lond.),* **203,** 1243–5.

Kirchner, J. A., and Wyke, B. D. (1965a) Afferent discharges from laryngeal articular mechanoreceptors, *Nature (Lond.),* **205,** 86–87.

Kirchner, J. A., and Wyke, B. D. (1965b) Articular reflex mechanisms in the larynx, *Ann. Otol. (St. Louis),* **74,** 749–68.

Koizumi, H. (1953) On sensory innervation of larynx in the dog, *Tohoku J. exp. Med.,* **58,** 199–210.

Koizumi,H., and Mikami, S. (1953) On innervation of mucous membrane of larynx in canine fetus, *Tohoku J. exp. Med.,* **58,** 217–21.

König, W. F., and von Leden, H. (1961) The peripheral nervous system of the human larynx. I. The mucous membrane, *Arch. Otolaryng.,* **73,** 1–14.

Kotby, M. N., and Haugen, L. K. (1970a) The mechanics of laryngeal function, *Acta oto-laryng. (Stockh.),* **70,** 203–11.

Kotby, M. N., and Haugen, L. K. (1970b) Critical evaluation of the action of the posterior crico-arytenoid muscle, utilizing direct EMG-study, *Acta oto-laryng. (Stockh.),* **70,** 260–8.

Ladefoged, P. (1960) The regulation of sub-glottal pressure, *Folia phoniat. (Basel),* **12,** 169–75.

Ladefoged, P. (1968) Linguistic aspects of respiratory phenomena, *Ann. N.Y. Acad. Sci.,* **155,** 141–51.

Lieberman, P. (1968) Vocal cord motion in man, *Ann. N.Y. Acad. Sci.,* **155,** 28–38.

Lucas Keene, M. F. (1957) Nerve endings in human laryngeal muscles, *J. Anat. (Lond.),* **91,** 590.

Lucas Keene, M. F. (1961) Muscle spindles in human laryngeal muscles, *J. Anat. (Lond.),* **95,** 25–29.

Mårtensson, A. (1964) Proprioceptive impulse patterns during contraction of intrinsic laryngeal muscles, *Acta physiol. scand.,* **62,** 176–94.

Matsumoto, T. (1950) Innervation, especially sensory innervation, of laryngeal mucosa except the epiglottis, *Tohuku med. J.,* **45,** 11–18.

Van Michel, C. (1963) Considerations morphologiques sur les appareils sensoriels de la muqueuse vocal humaine, *Acta anat. (Basel),* **52,** 188–92.

Mitchinson, A. G., and Yoffey, J. M. (1947) Respiratory displacement of the larynx, hyoid bone and tongue, *J. Anat. (Lond.),* **81,** 118–20.

Molinari, G. A., and Pivotti, G. S. (1962) Sul reperto di fibre tensocettrici polmonari nel nervo laringeo ricorrenta, *Valsalva,* **38,** 207–14.

Murakami, Y., and Kirchner, J. A. (1971) Electrophysiological properties of laryngeal reflex closure, *Acta oto-laryng. (Stockh.),* **71,** 416–25.

Murtagh, J. A., (1945) The respiratory function of the larynx: some observations on laryngeal innervation, *Ann. Otol. (St. Louis),* **54,** 307–22.

Murtagh, J. A., and Campbell, C. J. (1956) The effect of the barbiturates on laryngeal function, *Ann. Otol. (St. Louis),* **65,** 545–58.

Nakamura, F., Uyeda, Y., and Sonoda, Y. (1958) Electromyographic study on respiratory movements of the intrinsic laryngeal muscles, *Laryngoscope (St. Louis),* **68,** 109–19.

Negus, V. E. (1929) *The Mechanism of the Larynx,* London.

Negus, V. E. (1949) *The Comparative Anatomy and Physiology of the Larynx,* London.

Pressman, J. J., and Kelemen, G. (1970) *Physiology of the Larynx,* 2nd ed., rev. Kirchner, J. A., Rochester, Minnesota, American Academy of Ophthalmology and Otolaryngology.

Rudomín, P. (1964) Some aspects of the reflex activation of the recurrent laryngeal motoneurones of the cat, *Bol. Inst. Estud. méd. biol. (Mex.),* **22,** 341–52.

Rudomín, P. (1966a) The electrical activity of the cricothyroid muscles of the cat, *Arch. int. Physiol.,* **74,** 135–53.

Rudomín, P. (1966b) Some aspects of the control of cricothyroid muscle activity, *Arch. int. Physiol.,* **74,** 154–68.

Sampson, S., and Eyzaguirre, C. (1964) Some functional characteristics of mechanoreceptors in the larynx of the cat, *J. Neurophysiol.,* **27,** 464–80.

Sears, T. A. (1964) The slow potentials of thoracic respiratory motoneurones and their relation to breathing, *J. Physiol. (Lond.),* **175,** 404–24.

Sears, T. A. (1966) The respiratory motoneurone: integration at spinal segmental level, in *Breathlessness,* pp. 33–47, ed. Howell, J. B. L., and Campbell, E. J. M., Oxford.

Semon, F. (1890) On the position of the vocal cords in quiet respiration of man and the reflex tonus of their abductor muscles, *Proc. roy. Soc., Lond.,* **48B,** 156–9; 403–5.

Suzuki, M., and Kirchner, J. A. (1968) Afferent nerve fibers in the external branch of the superior laryngeal nerve in the cat, *Ann. Otol. (St. Louis)*, **77,** 1059–70.

Suzuki, M., and Kirchner, J. A. (1969) The posterior crico-arytenoid as an inspiratory muscle, *Ann. Otol. (St. Louis)*, **78,** 849–64.

Suzuki, M., Kirchner, J. A., and Murakami, Y. (1970) The cricothyroid as a respiratory muscle, *Ann. Otol. (St. Louis)*, **79,** 976–84.

Van den Berg, J. W. (1956) Direct and indirect determination of the mean subglottic pressure, *Folia phoniat. (Basel)*, **8,** 1–24.

Van den Berg, J. W. (1964) Some physical aspects of voice production, in *Research Potentials in Voice Physiology*, pp. 63–101, ed. Brewer, D. W., New York.

Weddell, G., Feinstein, B., and Pattle, R. E. (1944) The electrical activity of voluntary muscles in man under normal and pathological conditions, *Brain*, **67,** 178–257.

Withington, J. L. (1959) *Comparative Studies of the Nerve Supply of the Larynx with Special Reference to the Afferent Innervation of the Musculature*, Ph.D. Thesis, University of London.

Wyke, B. D. (1955) Electrographic monitoring of barbiturate narcosis: studies of cortical function, *Anaesthesia*, **10,** 247–59.

Wyke, B. D. (1957) Electrographic monitoring of anaesthesia: clinical and experimental studies of cerebral function during barbiturate narcosis, *Anaesthesia*, **12,** 157–73; 259–75.

Wyke, B. D. (1967a) The neurology of joints (Arris and Gale Lecture), *Ann. roy. Coll. Surg. Engl.*, **41,** 25–50.

Wyke, B. D. (1967b) Recent advances in the neurology of phonation: phonatory reflex mechanisms in the larynx, *Brit. J. Disor. Commun.*, **2,** 2–14.

Wyke, B. D. (1968) Effects of anaesthesia upon intrinsic laryngeal reflexes, *J. Laryng.*, **82,** 603–12.

Wyke, B. D. (1969a) Deus ex machina vocis: an analysis of the laryngeal reflex mechanisms of speech (Jansson Memorial Lecture). *Brit. J. Disor. Commun.*, **4,** 3–25.

Wyke, B. D. (1969b) *Principles of General Neurology. An Introduction to the Basic Principles of Medical and Surgical Neurology*, London.

Wyke, B. D. (1970) Neurological mechanisms in stammering—an hypothesis, *Brit. J. Disor. Commun.*, **5,** 6–15.

Wyke, B. D. (1971) The neurology of stammering, *J. psychosom. Res.*, **15,** 423–32.

Yamamoto, S., Miyajima, M., and Urabe, M. (1960) Respiratory neuronal activities in spinal afferents of cat, *Jap. J. Physiol.*, **10,** 509–17.

Yamamoto, S., Sugihara, S., and Kuru, M. (1956) Microelectrode studies on sensory afferents in the posterior funiculus of cat, *Jap. J. Physiol.*, **6,** 68–85.

25 Respiratory Activity of the External Laryngeal Muscles: An Electromyographic Study in the Cat

Y. MURAKAMI AND J. A. KIRCHNER*

Department of Otolaryngology,
Yale University School Of Medicine,
New Haven, Connecticut, U.S.A.

INTRODUCTION

In addition to their supporting rôle in phonation, the external laryngeal muscles are credited with an accessory respiratory function. Mitchinson and Yoffey (1947) studied the respiratory movements of the larynx by means of X-ray photography, and concluded that the thyrohyoid and suprahyoid muscles counteract the tension produced by tracheal pull during inspiration. Andrew (1955), studying the respiratory displacement of the larynx in the rat, observed electrical activity during inspiration in the sternothyroid, genioglossus, styloglossus and stylopharyngeus muscles, but no muscular activity during expiration.

The expiratory ascent of the hyoid bone and thyroid cartilage is generally attributed to release from tracheal tension or oesophageal pull. Muscle activity in the expiratory phase has not been conclusively demonstrated. Fink (1962) reported increased thyrohyoid muscle activity during expiration, which might effect closure of the larynx by bringing the thyroid cartilage nearer to the hyoid bone and thereby cause an inward displacement of the pre-epiglottic soft tissues. Such activity, as shown in our experiments, may be more 'phonatory' than 'expiratory', since it appeared only under very light anaesthesia accompanied by muscular straining.

Using X-ray photography, Fink (1956) also demonstrated an increased distance between the hyoid bone and thyroid cartilage during inspiration in the human. With motion pictures, he showed that the glottic opening during quiet respiration is effected by a lateral sliding movement of the arytenoid cartilages at the crico-arytenoid joint, usually with little or no outward rotation of the vocal processes. This lateral sliding movement was synchronous with the separation of the hyoid bone from the thyroid cartilage, whether it occurred in the human subject during ordinary respiration or in the cadaver when the trachea and the hyoid were pulled in opposite directions.

---

* Supported by U.S.P.H.S. Grant No. 2. RO 1-N5 05465.

Many (but not all*) have maintained that the intrinsic laryngeal muscles are devoid of muscle spindles and that, therefore, no afferent information originates in the muscles themselves [see Andrew, 1954; Mårtensson, 1963, 1964; Sampson and Eyzaguirre, 1964]. Andrew (1955), studying reflex mechanisms in the external laryngeal muscles that might control the position of the larynx during respiration, concluded that a feed-back mechanism did not play a leading part. On the other hand, receptors in the joints and mucous membrane of the larynx have been described, and these might influence the reflex excitability of some laryngeal motoneurones (Andrew, 1954; Mårtensson, 1964; Sampson and Eyzaguirre, 1964; Kirchner and Wyke, 1964, 1965; Wyke, 1967, 1969; Suzuki and Kirchner, 1969).

The main purpose of this study was to observe the respiratory function of each external laryngeal muscle and to investigate the possible existence of a feed-back mechanism which, activated by movement of the hyoid bone or the larynx, might regulate or augment the respiratory discharge pattern of these muscles. Such activity could presumably control the displacement of the laryngeal components.

METHODS

Fifty-one adult cats were studied in these experiments. The innervation of each extralaryngeal muscle was observed by dissection under the operating microscope and by electrical stimulation of each nerve [Table 25.I]. Our findings agree with those of Lemere (1932), Hwang (1948), Lanz (1955) and Miller *et alii* (1964).

Each cat was anaesthetized by intraperitoneal injection of 25–30 mg/kg of sodium pentobarbitone and maintained in relatively light anaesthesia (Stage II or III, plane 1) with repeated intravenous doses of the drug as required. The cat was placed in the supine position. Electromyographic (EMG) activity of the diaphragm was recorded as a monitor of respiration, using a concentric unipolar needle electrode inserted into the abdominal cavity through a vertical, midline skin incision. An 8 cm vertical midline skin incision was made in the neck. The platysma was separated in the midline. Almost all of the external laryngeal muscles were easily exposed without sacrifice of their motor nerves, which come mainly from the cervical plexus. It is essential in this kind of experiment to avoid trauma of any kind to the muscles.

To expose the deeper geniohyoid and hyoglossus musles, however, it was necessary to cut partway through the more superficial mylohyoid muscle. Abnormal fibrillation of the muscles usually stopped after they were immersed in warm mineral oil. Tracheostomy was performed low in the neck, and a vinyl

---

* See Communications 22 and 24 in this volume [Ed.].

tracheal cannula was inserted. Hyperpnoea was produced by a rebreathing bag. Inspiratory or expiratory dyspnoea was produced by a one-way valve connected to the tracheal cannula.

TABLE 25.I

MUSCLE	INNERVATION
Mylohyoid	Mylohyoid twig of mandibular branch of trigeminal nerve
Stylohyoid	Twig from facial nerve
Digastric (anterior belly)	Twig from mandibular branch of trigeminal nerve
Digastric (posterior belly)	Twig from facial nerve
Geniohyoid	Twig from hypoglossal nerve, containing C.1 and C.2 components of cervical plexus
Hyoglossus	Same as Geniohyoid
Sternohyoid	Cervical plexus, C.2, C.3 components (descending cervical root)
Sternothyroid	Same as Sternohyoid, but mainly C.3
Thyrohyoid	Cervical plexus, C.1, C.2 components (join hypoglossal nerve)
Hyopharyngeus (middle pharyngeal constrictor)	Pharyngeal plexus, containing mainly a vagal component and a few fibres from superior cervical ganglion
Thyropharyngeus	Same as Hyopharyngeus
Cricopharyngeus	Same as Hyopharyngeus, but not completely identified

1. *Respiratory movements of the hyoid bone and thyroid cartilage.*

One end of a 3–0 silk thread was sutured to the midpoint of the hyoid bone or to the superior edge of the thyroid cartilage, the other end being tied to a force-displacement transducer (Grass FT 10C) set over the sternum. Gain in the transducer was observed on an oscilloscope (Tektronix 564) through a carrier amplifier (Hewlett-Packard 8805A), monitored by another oscilloscope, then recorded by kymograph camera (Grass).

2. *Electromyography of the external laryngeal muscles.*

The muscles studied in this report were:

a. Suprahyoid group (mylohoid, stylohyoid, digastric, geniohyoid and hyoglossus).
b. Infrahyoid group (sternohyoid, sternothyroid and thyrohyoid)
c. Posterior laryngeal group (middle pharyngeal constrictor, thyropharyngeus and cricopharyngeus).

The omohyoid muscle is not developed in the cat. The genioglossus muscle is poorly developed and very thin, and could not be studied. The stylopharyngeus and palatopharyngeus muscles are regarded as elevators of the pharynx and larynx, but could not be exposed without severe damage to the other muscles.

The pterygopharyngeus muscle in the cat is probably analogous to the superior pharyngeal constrictor muscle in the human, but it cannot be examined without damage to other muscles or nerves.

The external laryngeal muscles are numerous, and in the restricted suprahyoid region cannot be individually studied without concentric bipolar electrodes. The vigorous respiratory movements of these muscles, in addition, adds to the difficulty of studying them with conventional needle electrodes.

Specially made unipolar or bipolar copper wire electrodes (0.2 mm apart) were placed on the surface of the muscle to be examined. The wires were enamel-insulated except at the cut ends, and were wound as spirals to prevent shifting and consequent artifacts. When a unipolar wire electrode was used, the indifferent electrode was placed on the neck incision.

Single or multi-unit discharges were displayed on the oscilloscope after pre-amplification (with Tektronix 122), monitored by an audiosystem (Grass AM5) and a second oscilloscope, then recorded continuously by kymograph camera. Results obtained were always checked with those obtained by concentric bipolar needle electrodes.

The cricopharyngeus muscle in the cat is too thin to be examined by needle electrode. Furthermore, it is very close to the posterior crico-arytenoid muscle, which is very active during inspiration. For this reason, discharges from the posterior crico-arytenoid muscle were recorded simultaneously with a needle electrode as a control.

To examine the effects on muscle activity of artificial movement of the hyoid bone or the thyroid cartilage, a 3–0 silk thread was sutured to the midpoint of the hyoid bone or the thyroid cartilage just beside the thread tied to the force-displacement transducer. Pulling the thread cranially or caudally along the longitudinal axis of the larynx effectively stretched or shortened the infra- or suprahyoid muscles.

3. *Relationship between the external laryngeal muscles and the respiratory centre.*

The following experiments were designed to determine whether the external laryngeal muscles are driven primarily by the respiratory centre:

a. EMG activity in these muscles was recorded during electrical stimulation of the central segment of the internal branch of the superior laryngeal nerve (SLN) (Bipolar platinum electrode and Grass S-4 stimulator).
b. Inflation of the lung stimulates pulmonary stretch receptors and induces reflex apnoea (Widdicombe, 1954). The inspiratory phase of respiration is inhibited and all inspiratory muscles, including the diaphragm, become inactive. Deflation of the lung activates inspiratory muscle activity by afferent stimuli carried by the vagus nerve (Adrian, 1933; Paintal, 1957). Effects of inflation of the lung on external laryngeal muscle activity were studied by blowing 150–200 ml of room air into the lung through the tracheal cannula. Effects of deflation were studied by connecting a suction tube to the tracheal cannula. Observations were made before and after bilateral section of the recurrent laryngeal (RLN) and vagus nerves to eliminate afferent pathways from the thorax. After these nerves were sectioned, inflation with room air was repeated to increase PO_2, and the rebreathing bag was connected to the tracheal cannula to check the effect of PCO_2.

To study the effect of artificial movements of the hyoid bone on muscular activity, the respiratory centre was rendered inactive by high frequency stimulation of the SLN or by excessive increases in PO_2.

RESULTS

RESPIRATORY MOVEMENTS OF THE HYOID BONE AND THE THYROID CARTILAGE

At light or moderate levels of anaesthesia, all cats showed various degrees of displacement of the hyoid bone and the thyroid cartilage; namely, descent during inspiration and ascent during expiration. These movements were most marked during hyperpnoea and during inspiratory or expiratory dyspnoea.

The deeper the anaesthesia, the less was the degree of movement. In the late plane of Stage III there was very little movement, even during hyperpnoea or complete obstruction of the tracheal cannula.

Under very light anaesthesia, on the other hand, the animal whined and strained with nearly every expiration, but the basic pattern of ascent and descent continued. Since the movements indicated strain, coughing and swallowing, they were recorded simultaneously as controls with the EMG in all subsequent experiments.

ELECTROMYOGRAPHY OF THE EXTERNAL LARYNGEAL MUSCLES

Suprahyoid Group

1. Mylohyoid muscle:

In five of seven cats under light anaesthesia, tonic-type discharges occurred throughout respiration. In the remaining two, there was no activity whatever. During hyperpnoea, the action potentials gradually disappeared. During inspiratory dyspnoea a slight decrease in frequency developed in the inspiratory phase.

2. Stylohyoid muscle:

Two kinds of multi-unit discharges were recorded in this muscle in all seven animals examined. The first was a tonic-type discharge apparent throughout respiration, and the second a larger phasic one that fired only during expiration. The continuous tonic-type discharge was very unstable during hyperpnoea, and completely disappeared in dyspnoeic respiration. The phasic activity was not affected by artificial movement of the hyoid bone.

3. Digastric muscle:

In the cat this muscle has no direct connexion with the hyoid bone. The posterior belly arises from the jugular process of the occiput: the anterior belly inserts into the ventral portion of the mandible. These two bellies, separately innervated, showed completely different respiratory discharge patterns. In all seven cats examined, the anterior belly was silent, even under very light anaesthesia. The posterior belly, on the contrary, was usually active (five cases in seven) throughout respiration, with maximum activity near the end of the inspiratory phase. This discharge pattern was unchanged by inspiratory or expiratory dyspnoea. Hyperpnoea resulted in recruitment of new units during the expiratory phase.

4. Geniohyoid and hyoglossus muscles:

These two suprahyoid muscles showed almost the same discharge pattern, except that the hyoglossus muscle was a little more active than the geniohyoid. Two cats, under deep anaesthesia, showed no activity. In the remaining five, during quiet respiration, several units discharged constantly with maximal activity near the end of the inspiratory phase. During hyperpnoea or inspiratory dyspnoea discharges disappeared, except at the very end of inspiration when a remarkable increase in frequency was observed. During expiratory dyspnoea a slight decrease in frequency occurred at the beginning of inspiration. Pulling the hyoid bone cranially so as to shorten the muscle resulted in complete inhibition that was independent of the respiratory cycle. Pulling it caudally to stretch the muscle resulted in a considerable increase in muscle activity, even during inspiration.

Infrahyoid Group.

1. Sternohyoid muscle:

This muscle, unexpectedly, was inactive throughout respiration in moderate or deep anaesthesia. Under very light anaesthesia, irregular discharges appeared continuously throughout respiration without any relationship to the respiratory cycle [Fig. 25.1]. During hyperpnoea, and inspiratory or expiratory dyspnoea,

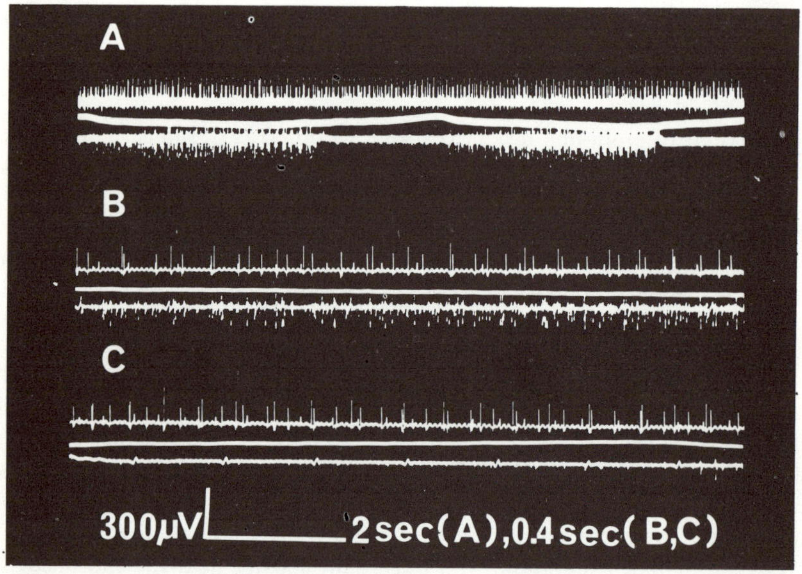

Fig. 25.1. EMG activity of the sternohyoid muscle during quiet respiration, under very light anaesthesia.
Upper trace: Sternohyoid activity.
Middle trace: Respiratory movements of the hyoid bone (descent of the line indicates descent of the bone).
Lower trace: EMG of the diaphragm as an indicator of the respiratory cycle.
[A]. Motor unit activity with no relation to the respiratory cycle.
[B]. Same, at higher recording speed during inspiration.
[C]. Same, at higher recording speed during expiration.
Note inactivity of the diaphragm during continued sternohyoid activity.

there was usually no change in this discharge pattern. Pulling down the hyoid bone to shorten the muscle inhibited its activity. Pulling the hyoid upward resulted in an increase in frequency of spontaneously discharging potentials. The activity of this muscle seems to be of a 'postural tonic' type, and any relationship to the activity of the respiratory centre is uncertain.

2. Sternothyroid muscle:

In lightly anaesthetized preparations, single or multi-unit discharges appeared during quiet respiration, mainly in inspiration [Fig. 25.2]. In all cases, the discharge pattern continued during inspiratory and expiratory dyspnoea. During hyperpnoea the frequency of these action potentials gradually decreased and then completely disappeared.

Discharges were markedly inhibited by passive movement of the thyroid cartilage. Muscle activity could not be induced even when the muscle was stretched. In moderate to deeper anaesthesia, the muscle showed no activity under any condition.

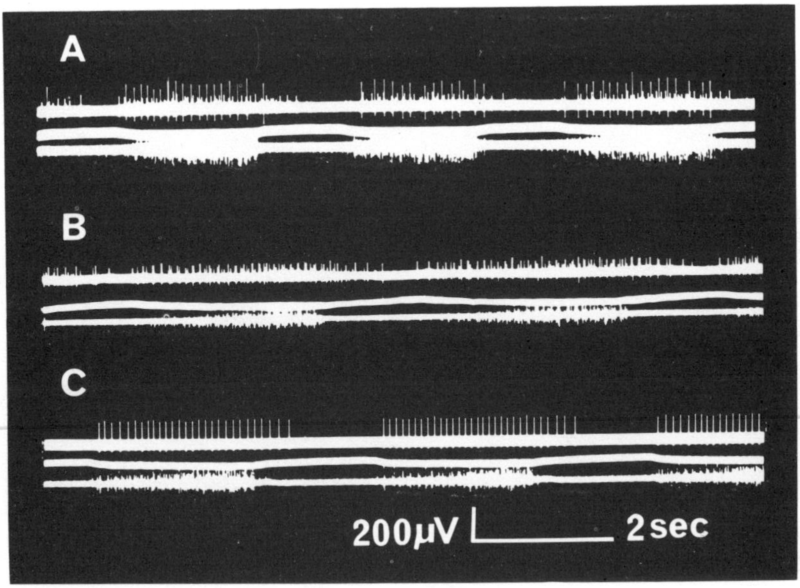

Fig. 25.2. EMG of the sternothyroid muscle in three different cats during quiet respiration.
Upper trace: Sternothyroid.
Middle trace: Movement of hyoid bone.
Lower trace: Diaphragm.
[A, B]. Multi-unit discharges, mainly in inspiratory phase.
[C]. Single unit discharges in a third animal.
This pattern continued unchanged after the muscle was detached from the thyroid cartilage.

3. Thyrohyoid muscle:

The discharge pattern of this muscle was generally similar to that of the mylohyoid muscle. The frequency of the discharge decreased in the mid-inspiratory and late expiratory phases. During hyperpnoea, discharges gradually disappeared, beginning with inspiration. During inspiratory or expiratory dyspnoea, a marked decrease in frequency occurred in the inspiratory phase.

These discharges were markedly inhibited by artificial movement of the hyoid bone either upward or downward [Fig. 25.3]. In deeper anaesthesia (beyond Stage III, plane 1), we observed no discharges even during vigorous dyspnoea.

Decrease in activity of this muscle in the inspiratory phase appears to be regulated by local feed-back mechanisms. This would passively assist the downward displacement of the thyroid cartilage and, consequently, widen the distance between the thyroid and the hyoid.

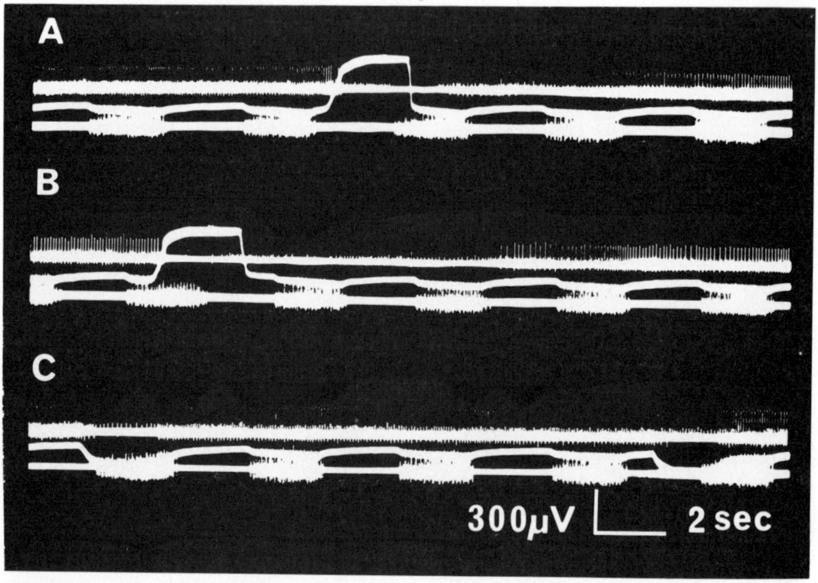

Fig. 25.3. Effect of manual displacement of the hyoid bone or EMG activity of the thyrohyoid muscle during quiet respiration.
Upper trace: Thyrohyoid muscle.
Middle trace: Movement of the hyoid bone.
Lower trace: Diaphragm.
[A, B]. Cranial displacement of the hyoid produces marked inhibition, especially during inspiration.
[C]. Caudal displacement also inhibits activity in this muscle.

Posterior Laryngeal Group

1. Middle pharyngeal constrictor (hyopharyngeus muscle):

In the cat this muscle is separated into two parts—the chondropharyngeus and keratopharyngeus muscles. The condropharyngeus muscle (the larger part), arising from the lateral surface of the thyrohyoid bone, was easily exposed, whilst the keratopharyngeus muscle, arising from the keratohyoid bone, could not be examined without sectioning the stylohyoid muscle.

In all eleven cats examined these two parts showed the same activity pattern, many of the motor units firing as expiration proceeded and ceasing abruptly at the beginning of inspiration. This phasic rhythm was well maintained during hyperpnoea, inspiratory or expiratory dyspnoea [Fig. 25.4], and even during deep anaesthesia (Stage III, plane 1). Increased frequency and recruitment were observed under these conditions.

Artificial movement of the hyoid bone did not affect the discharge pattern in this muscle.

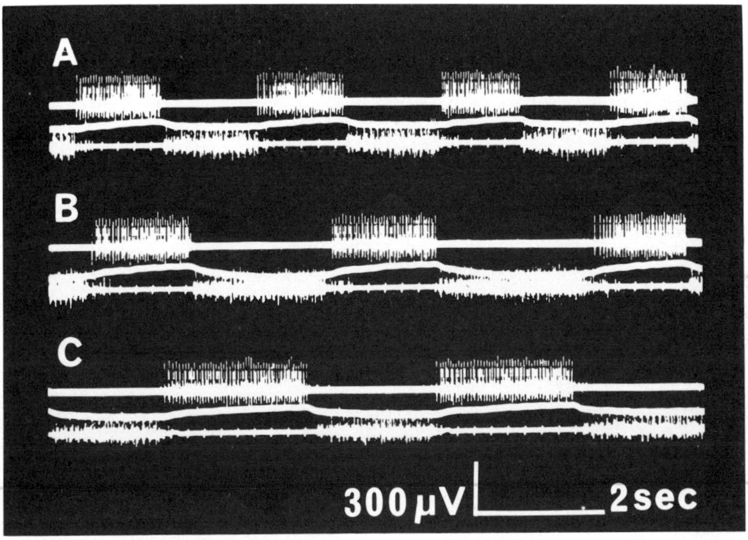

Fig. 25.4. EMG activity of middle constrictor muscle (hyopharyngeus).
Upper trace: Middle constrictor muscle.
Middle trace: Hyoid bone movements.
Lower trace: Diaphragm.
[A]. During hyperpnoea.
[B]. During inspiratory dyspnoea.
[C]. During expiratory dyspnoea.
This pattern of expiratory activity continues under deep anaesthesia, and is not affected by artificial movements of the hyoid bone.

2. Thyropharyngeus muscle:

The discharge pattern of this muscle was very similar to that of the stylohyoid muscle: (a) a tonic type, active throughout respiration; and (b) a larger phasic discharge, active during expiration. Anatomic distribution of these two functional types of fibres differed, the expiratory type being most numerous in the cranial part of the muscle.

Tonic-type discharges were unstable and disappeared completely during hyperpnoea, dyspnoea or deep anaesthesia. The expiratory discharges showed no response to artificial movement of the thyroid cartilage.

3. Cricopharyngeus muscle:

Single or multi-unit action potentials were recorded in lightly anaesthetized preparations. The discharge pattern showed no relationship to the respiratory cycle in any of eleven cats examined, firing tonically in both inspiratory and expiratory phases during quiet respiration, and even during hyperpnoea or dyspnoea [Fig. 25.5].

After both recurrent laryngeal nerves were sectioned, the discharge pattern showed no change. This indicates that the origin of these potentials is actually in the cricopharyngeus muscle itself, and not in the posterior crico-arytenoid muscle.

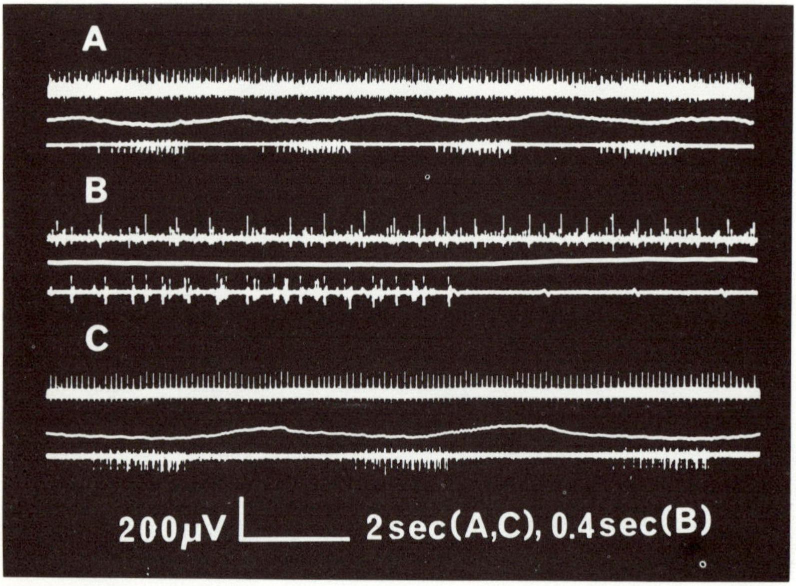

Fig. 25.5. EMG of the cricopharyngeus muscle during quiet respiration under light (Stage II) anaesthesia.
Upper trace: Cricopharyngeus.
Middle trace: Hyoid bone movements.
Lower trace: EMG of posterior crico-arytenoid muscle.
[A]. Multiple units discharging tonically with no relationship to respiratory cycle.
[B]. Same, recorded at higher speed. Discharges continue after the end of the inspiratory phase (indicated by posterior crico-arytenoid activity).
[C]. Single motor unit activity in another animal, showing activity during both inspiratory and expiratory phases.

The thyro-arytenoid muscle, on the other hand, was completely silent—as shown by simultaneous recording.

In deeper anaesthesia (Stage III, plane 1), EMG activity in the cricopharyngeus usually ceased. Occasionally, multi-unit discharges appeared in inspiration, but these were of very low voltage (60 microvolts or less).

It is possible that the cricopharyngeus muscle may contain both inspiratory and expiratory fibres, so that tonic activity of this muscle would result under light anaesthesia. It is interesting that the cricopharyngeus muscle showed a completely different discharge pattern from that of the thyropharyngeus muscle, although both constitute the inferior pharyngeal constrictor [Tables 25.II and 25.III].

RELATIONSHIP BETWEEN THE EXTERNAL LARYNGEAL MUSCLES AND RESPIRATORY CENTRE ACTIVITY

Experiments had to be done under deeper anaesthesia (Stage III, plane 1) in order to reduce gag and cough reflexes. Most muscles examined showed no spontaneous activity at this deeper stage of anaesthesia—except for the middle pharyngeal constrictor (hyopharyngeus muscle), which fired during expiration.

Expiratory Group

The hyopharyngeus (middle constrictor) exhibited activity only during expiration. Low frequency (l/sec) electrical stimulation to the internal branch of the superior laryngeal nerve caused temporary cessation of activity in both the hyopharyngeus and the diaphragm. Stimulation at higher frequency (10–20/sec) produced complete inactivity in both muscles, indicating inhibition of the respiratory centre. Artificial movement of the hyoid bone during this silent period failed to produce any activity in the hyopharyngeus muscle.

Considerable activation of hyopharyngeus activity and instant cessation of diaphragmatic activity were induced by inflation of the lung. Deflation of the lung, on the contrary, was followed by immediate activation of diaphragmatic activity and instant cessation of hyopharyngeal discharges. After both the recurrent laryngeal and vagus nerves were sectioned, the respiratory rate became slower but the basic pattern of respiratory discharges from both muscles remained unchanged. However, neither the hyopharyngeus nor the diaphragm was now influenced by inflation or deflation of the lung. Only when inflation was repeated often enough to increase PO_2 did diaphragmatic activity gradually decline and disappear. Nevertheless, the hyopharyngeus activity continued, even after cessation of inflation.

A very high concentration of O_2 produced complete inactivity in both muscles, indicating inhibition of discharges from the respiratory centre. Artificial movement of the hyoid bone during this silent period again failed to induce any activity in the hypopharyngeus muscle. Increased activity in both muscles and recruitment of new units followed an increase in PCO_2 produced by a rebreathing bag. This occurred in spite of bilateral vagal section, indicating that the mechanism is independent of stimuli originating in intrathoracic structures.

Under light anaesthesia, the stylohyoid, posterior belly of the digastric, the thyropharyngeus, the geniohyoid and the hyoglossus muscles showed almost the same pattern as the hyopharyngeus. Considerable activity was induced by inflation of the lung before sectioning the recurrent laryngeal and vagus nerves.

TABLE 25.II

EMG DISCHARGE ACTIVITY—LIGHT ANAESTHESIA

MUSCLE	QUIET RESPIRATION	HYPERPNOEA	INSPIRATORY DYSPNOEA	EXPIRATORY DYSPNOEA	ARTIFICIAL MOVEMENT OF THE HYOID BONE	NOTES
Mylohyoid	tonic-type throughout respiration, slightly decreased in mid-expiration	disappeared	slightly decreased in mid-inspiration	slightly decreased in mid-expiration	completely inhibited	silent in deeper anaesthesia
Stylohyoid	tonic-type throughout respiration and larger phasic only in expiration	phasic in expiration	only in expiration	only in expiration	tonic-type inhibited, phasic not responsive	silent in deeper anaesthesia
Digastric (posterior belly)	throughout respiration but mainly in expiration	recruitment in expiration	mainly in expiration	mainly in expiration	not responsive	anterior belly of the digastric was inactive even in light anaesthesia
Geniohyoid and Hyoglossus	mainly in expiration	disappeared in most part of inspiration, increased mainly in expiration	disappeared in most part of inspiration, increased mainly in expiration	slightly decreased, mainly in inspiration	inhibited by shortening, activated by stretching	silent in deeper anaesthesia
Thyrohyoid	tonic-type throughout respiration	disappeared	decreased in inspiration	slightly decreased in inspiration	inhibited	silent in deeper anaesthesia
Hyopharyngeus (middle pharyngeal constrictor)	only in expiration	recruitment in expiration	only in expiration	only in expiration	not responsive	active even in deeper anaesthesia
Sternohyoid	irregularly tonic throughout respiration	irregularly tonic	irregularly tonic	increased in expiratory phase in a few cases	inhibited by shortening, activated by stretch, but not reciprocal innervation	active only in very light anaesthesia
Sternothyroid	only in inspiration	disappeared	only in inspiration	only in inspiration	inhibited	silent in deeper anaesthesia
Thyropharyngeus	tonic-type throughout respiration, and larger phasic only in expiration	only in expiration	only in expiration	only in expiration	tonic-type inhibited, phasic not responsive	tonic-type inactive in deeper anaesthesia
Cricopharyngeus	tonic throughout respiration	tonic throughout respiration	tonic throughout respiration	tonic throughout respiration	not responsive	silent in deeper anaesthesia

TABLE 25.III

EMG ACTIVITY—DEEP ANAESTHESIA

MUSCLE	INFLATION OF LUNG	DEFLATION OF LUNG	INCREASED PO_2	INCREASED PCO_2
Mylohyoid	silent	silent	silent	silent
Stylohyoid	induced only in expiration	silent	silent	silent
Digastric (posterior belly)	induced only in expiration	silent	silent	silent
Geniohyoid and Hyoglossus	induced only in expiration	silent	silent	silent
Thyrohyoid	silent	silent	silent	silent
Hyopharyngeus (middle pharyngeal constrictor)	activated	disappeared	continued then disappeared	recruitment in expiratory phase
Sternohyoid	silent	silent	silent	silent
Sternothyroid	silent	induced only in inspiration	silent	induced only in inspiration
Thyropharyngeus	induced only in expiration	silent	silent	silent
Cricopharyngeus	induced in expiration	induced in inspiration	silent	induced in inspiration

Deflation of the lung, on the other hand, did not induce any activity in these muscles, suggesting that they were originally expiratory muscles—though they were not activated by increases in PO_2 or PCO_2.

Inspiratory Group

Under light anaesthesia, the sternothyroid showed maximum activity during inspiration. With deeper anaesthesia the muscle was silent, and there was no response to electrical stimulation of the superior laryngeal nerve or inflation of the lung. Deflation of the lung, on the contrary, induced considerable activity in this muscle during the inspiratory phase.

After both recurrent laryngeal and vagus nerves had been cut bilaterally to eliminate afferent pathways from the thorax, no activation could be induced in the muscle even with extreme negative pressures in the lung [Fig. 25.6].

Increased PO_2 produced inactivity in the diaphragm, but induced no sternothyroid activity. Artificial movement of the hyoid bone during this silent

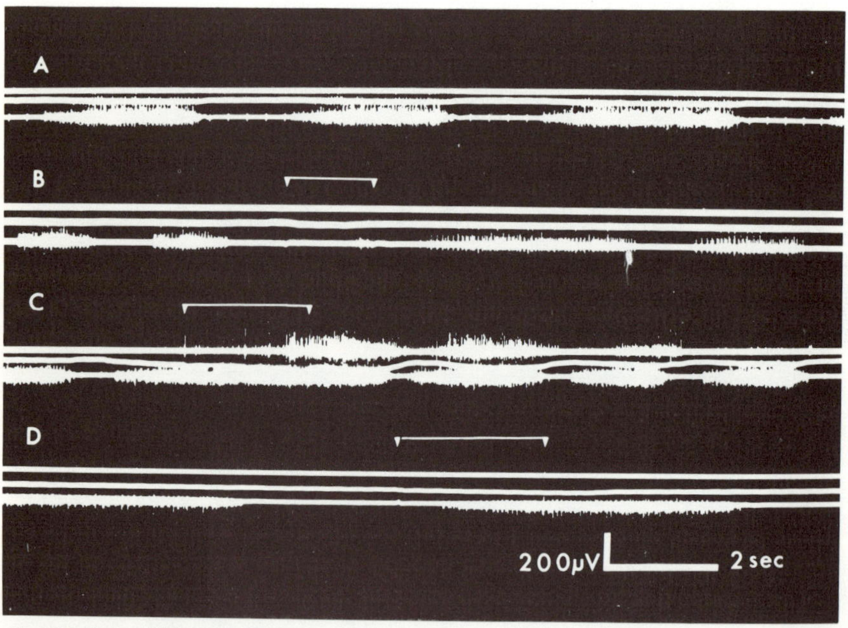

Fig. 25.6. Effect of inflation and deflation of lung on sternothyroid muscle activity.
Upper trace: Sternothyroid.
Middle trace: Movement of hyoid bone.
Lower trace: Diaphragm.
[A]. Deep anaesthesia, with no spontaneous activity in the sternothyroid muscle.
[B]. Inflation of the lung produces no effect on sternothyroid activity, but inhibits that of the diaphragm.
[C]. Deflation of the lung activates both muscles in the inspiratory phase.
[D]. No activation of either muscle during deflation of the lung after the vagus and recurrent laryngeal nerves have been sectioned bilaterally to eliminate afferent pathways from the thorax.

period produced no sternothyroid activity. With increased PCO_2, on the other hand, both the diaphragm and sternothyroid became active during inspiration, indicating control by the respiratory centre.

Tonic Group

The discharge pattern of the cricopharyngeus muscle under light anaesthesia showed no relationship to the respiratory cycle, firing tonically in both inspiratory and expiratory phases. The experiments were done after sectioning the recurrent laryngeal nerves bilaterally to eliminate possible artifact (current spread from the internal laryngeal muscles), and also sectioning the internal branch of both superior laryngeal nerves—because the cricopharyngeus contracted readily even with minimal stimulation of this latter nerve.

Inflation of the lung induced considerable activity of the muscle in expiration, with immediate inactivation of the diaphragm. Deflation of the lung, on the other hand, showed immediate activation of both muscles, the cricopharyngeus firing mainly during inspiration.

The results indicate that the cricopharyngeus contains both inspiratory and expiratory fibres and that the effect of their combined activity is tonic contraction, at least under lighter anaesthesia. After the vagus nerves were sectioned bilaterally, there was no response to inflation or deflation of the lung.

The mylohyoid and thyrohyoid muscles, which showed a tonic pattern throughout respiration, were not influenced by inflation of the lung. This type of experiment would require detaching the muscles from the hyoid bone because these muscles are easily inhibited by excessive movement of the larynx. It was, however, technically impossible to cut the muscle wihout producing abnormal fibrillation.

The sternohyoid muscle was not affected by inflation or deflation of the lung, nor by increases in PO_2 or PCO_2. The relation of its activity to that of the respiratory centre is uncertain. Activation of the muscle during expiration may be the result of a local feed-back mechanism.

DISCUSSION

The lower end of the trachea moves caudally during inspiration and cranially during expiration. The caudal movement is probably due: (1) to tension in the elastic fibres of the bronchi during expansion; and (2) to downward movement of the diaphragm transmitted through the heart, aorta and oesophagus. Cranial movement results from release of these forces.

Concerning respiratory movements of the larynx and hyoid, reports differ, particularly as to how much of the movement is due to muscular activity.

In our studies, all of the cats examined showed a distinct movement of the hyoid bone and the thyroid cartilage, with descent during inspiration and ascent during expiration. Inspiratory descent appeared to be due chiefly to tracheal and oesophageal tension.

During spontaneous respiration (as contrasted with phonation) the strenohyoid and sternothyroid muscles appear to have little if any effect in depressing the hyoid bone or thyroid cartilage. The sternohyoid muscle did not display its maximal activity in inspiration, nor was the displacement pattern of the hyoid bone changed by detachment of the sternohyoid muscle.

Similarly, the sternothyroid muscle is of doubtful importance in inspiratory descent of the larynx, because such movement continued unchanged after this muscle was detached from the thyroid cartilage.

Widening of the glottis can be produced by the unfolding of the vestibular and vocal folds that results when the thyroid cartilage is pulled away from the hyoid bone. This is true whether the separation is produced by passive traction on these structures, or by the action of the external laryngeal muscles (Fink, 1956). Discharges appearing near the end of inspiration in several of the suprahyoid muscles would appear to counteract the inspiratory descent of the hyoid bone. In this regard, stretching or shortening the suprahyoid muscles by passive displacement of the hyoid bone results in augmentation or inhibition of this discharge pattern. This suggests that the downward inspiratory movement of the larynx is limited somewhat by contraction of the suprahyoid muscles through an afferent feed-back mechanism.

In summary, inspiratory contraction was observed only in those muscles which raise the hyoid bone or which lower the thyroid cartilage.

The thyrohyoid muscle, by means of its tonic activity, maintains a constant distance between the hyoid and thyroid cartilages during normal quiet respiration. During hyperpnoea or dyspnoea, however, this tonic activity is inhibited, so that the thyroid cartilage can descend, the thyrohyoid distance increase, and the glottic aperture widen.

Similarly, the thyropharyngeus muscle, whose inspiratory contraction helps stabilize and support the thyroid cartilage, showed complete cessation of inspiratory activity during hyperpnoea or dyspnoea. Thus, the thyroid cartilage is allowed to descend and move anterior to the hyoid bone so that the glottic aperture is increased.

The stylohyoid, thyropharyngeus and middle constrictor muscles showed maximal EMG activity during expiration. The activity was constant, and was not affected even by excessive manual displacements of the hyoid or thyroid cartilage: it probably serves to elevate the hyoid and thyroid on expiration.

The geniohyoid and hyoglossus muscles (hyoid elevators) showed a marked increase in EMG activity when they were stretched by downward traction applied to the hyoid bone. Similarly, sternohyoid activity was increased by cranial traction on the hyoid bone. However, when both the supra- and infrahyoid groups of muscles were stretched simultaneously by passive elevation of the mandible, sternohyoid activity was inhibited completely. This observation suggests a complicated feed-back mechanism, such as reciprocal innervation with the suprahyoid muscles.

When the larynx was isolated by cutting through the trachea and oesophagus just above the level of the tracheostome, the hyoid bone and thyroid cartilage continued to descend during inspiration—although to a lesser degree. Moreover, the sternohyoid and sternothyroid muscles—which had been cut across at the same level–continued to exhibit the same discharge pattern as before.

Similarly, those external laryngeal muscles which exhibited EMG activity during expiration were activated by inflation of the lung or by increased PO_2. Those external laryngeal muscles which exhibited inspiratory activity were activated by deflation of the lung and by increased PCO_2.

The activity of the external laryngeal muscles resembles skilled movements in other striated muscle systems of the body, and is probably controlled by a pool of information in the brain, including contributions by the *gamma* system.

SUMMARY AND CONCLUSIONS

Respiratory activity of the external laryngeal muscles in the cat was studied by electromyography, particularly as it relates to displacement of the hyoid bone and thyroid cartilage.

The hyoid bone and thyroid cartilage descended in inspiration and ascended in expiration, with maximal displacement during hyperpnoea or dyspnoea.

Descent was produced mainly by tracheal tension and oesophageal pull. Ascent was produced mainly by muscular activity in the suprahyoid and posterior laryngeal groups.

The discharge pattern is characteristic of each muscle, in terms of whether maximal activity occurs during inspiration, expiration or both. The sternothyroid is the only typical inspiratory muscle. The hyopharyngeus (middle constrictor) is the most typical of the expiratory group. The cricopharyngeus muscle, which fired tonically throughout respiration, showed inspiratory activity during deflation and expiratory activity during inflation of the lung.

Inspiratory activity was observed only in those muscles which raise the hyoid or which lower the thyroid cartilage, thus assisting in the opening mechanism of the glottis.

The phasic activity of each external laryngeal muscle is governed by the respiratory centre and is modulated by mechanical and chemical information originating to a large degree in the thorax. The delicate interrelationship of these various muscular activities also controls the relationship of the hyoid bone to the thyroid cartilage, and in this way assists the opening and closing mechanism of the larynx.

The discharge patterns of the various muscles are also accentuated during hyperpnoea or dyspnoea by local feed-back mechanisms originating in these muscles as they are lengthened or shortened by movements of the larynx. This is particularly true of those muscles innervated by the cervical plexus.

REFERENCES

Adrian, E. D. (1933) Afferent impulses in the vagus and their effect on respiration, *J. Physiol. (Lond.),* **79,** 332–58.

Andrew, B. L. (1954) Proprioception at the joint of the epiglottis of the rat, *J. Physiol. (Lond.),* **126,** 507–23.

Andrew, B. L. (1955) The respiratory displacement of the larynx: A study of the innervation of accessory respiratory muscles, *J. Physiol. (Lond.),* **130,** 474–87.

Fink, B. R. (1962) Tensor mechanism of the vocal folds, *Ann. Otol. (St. Louis),* **71,** 591–9.

Fink, B. R., Basek, M., and Epanchin, V. (1956) The mechanism of opening of the human larynx, *Laryngoscope (St. Louis),* **66,** 410–25.

Hwang, K., Grossman, M. I., and Ivy, A. C. (1948) Nervous control of the cervical portion of the esophagus, *Amer. J. Physiol.,* **154,** 343–57.

Kirchner, J. A., and Wyke, B. D. (1964) Innervation of laryngeal joints and laryngeal reflexes, *Nature (Lond.),* **201,** 506.

Kirchner, J. A., and Wyke, B. D. (1965) Articular reflex mechanisms in the larynx, *Ann. Otol. (St. Louis),* **74,** 749–68.

Lemere, F. (1932) Innervation of the larynx, *Amer. J. Anat.,* **51,** 417–32.

Mårtensson, A. (1963) Reflex responses and recurrent discharges evoked by stimulation of laryngeal nerves, *Acta physiol. scand.,* **52,** 248–69.

Mårtensson, A. (1964) Proprioceptive impulse patterns during contraction of intrinsic laryngeal muscles, *Acta physiol. scand.,* **62,** 176–94.

Miller, M. E., Christensen, G. C., and Evans, H. E. (1964) *Anatomy of the Dog,* Philadelphia.

Mitchinson, A. G., and Yoffey, J. M. (1947) Respiratory displacement of the larynx, hyoid bone and tongue, *J. Anat. (Lond.),* **81,** 118–20.

Paintal, A. S. (1957) The location and excitation of pulmonary deflation receptors by chemical substances, *Quart. J. exp. Physiol.,* **42,** 56–71.

Sampson, S., and Eyzaguirre, G. (1964) Some functional characteristics of mechanoreceptors in the larynx of the cat, *J. Neurophysiol.,* **27,** 464–80.

Suzuki, M., and Kirchner, J. A. (1969) Sensory fibres in the recurrent laryngeal nerve, *Ann. Otol. (St. Louis),* **78,** 21–31.

von Lanz, T. (1955) *Praktische Anatomie,* Berlin.

Widdicombe, J. G. (1954) Respiratory reflexes excited by inflation of the lung, *J. Physiol. (Lond.),* **123,** 105–15.

Wyke, B. D. (1967) Recent advances in the neurology of phonation: phonatory reflex mechanisms in the larynx, *Brit. J. Disor. Commun.,* **2,** 2–14.

Wyke, B. D. (1969) Deus ex machina vocis: an analysis of the laryngeal reflex mechanisms of speech, *Brit. J. Disor. Commun.,* **4,** 3–25.

DISCUSSION ON COMMUNICATIONS [24] AND [25]

Widdicombe. May I ask Dr. Wyke about a problem that he touched on—which is a discrepancy between his results (which relate to one particular aspect of the pattern of discharge in the intrinsic laryngeal muscles) and what I think is a consensus of opinion among people who record from single nerve fibres going to the larynx. If I am correct, Dr. Wyke showed that in tracheotomized animals there was virtually no respiratory-modulated discharge in these muscles, and that only when the larynx was put into the circuit did you begin to get respiratory modulation—presumably by feed-back from some afferent input from the larynx. Now, I think nearly everybody who has recorded from single fibres in the recurrent laryngeal nerve has found that the vast majority have respiratory modulation. The research workers with me, Drs. Stransky and Szereda-Przestaszewska, have in the last few months recorded from well over 120 single laryngeal motoneurones, and probably less than 5 per cent have continuous, non-respiratory-modulated activity whilst about 30 per cent are silent fibres that are brought in by respiratory acts like coughing; but the majority, and the silent ones (when they are brought in), have shown respiratory modulation. I don't understand this difference, and I don't know if you can explain it —perhaps I should say that all our experimental preparations have tracheostomies.

Wyke. As you say, I am aware of this difference between our results recording from motor unit populations in the laryngeal muscles and some of the earlier results obtained with recordings from single fibres in the recurrent laryngeal nerve[1,2,3,4]. But from the observational point of view, there is no doubt that our results (as you have seen) with tracheostomy and with breathing through the larynx are very different from one another; and I must make the point absolutely clear that in the tracheotomized animals (particularly when one puts the activity on an oscilloscope rather than on a pen-writer) the respiratory oscillations of motor unit activity are *not entirely* absent. What I said was that they were substantially reduced by tracheostomy.

Nevertheless, the point you made about the high proportion of neurones in the recurrent laryngeal nerve that were shown by your colleagues to have respiratory-linked potential oscillations is a matter that puzzles me—except for the fact that, as Dr. Bowden has shown, there are over 1,000 fibres in the recurrent laryngeal nerve and you said your people had sampled some 120-odd of these fibres. It would be tendentious to say that they may have sampled those fibres that have this activity, or that some of the fibres may have been afferents, or efferents going to the oesophagus or trachea rather than to the laryngeal muscles. Certainly I cannot explain the discrepancy at the present time (although I have some ideas about it that will have to await testing in further experiments).

---

[1] Eyzaguirre, C., and Taylor, J. R. (1963) Respiratory discharge of some vagal motoneurones, *J. Neurophysiol.*, **26,** 61–78.

[2] Bianconi, R., and Raschi, F. (1964) Respiratory control of motoneurones of the recurrent laryngeal nerve and hypocapnic apnoea, *Archiv. ital. Biol.*, **102,** 56–73.

[3] Eyzaguirre, C., Sampson, S., and Taylor, J. R. (1966) The motor control of intrinsic laryngeal muscles in the cat, *Nobel Symposium on Muscular Afferents and Motor Control*, ed. Granit, R., pp. 209–25. Stockholm.

[4] Suzuki, M., and Kirchner, J. A. (1969) The posterior cricoarytenoid as an inspiratory muscle, *Ann. Otol. (St. Louis),* **78,** 849–65.

Widdicombe. On this point, I must just support my colleagues and say that they certainly did not select nerve fibres on the basis of respiratory rhythm or no respiratory rhythm. Of course, they might have selected on the basis of large fibres compared with small ones (this one can't always avoid doing), but anything that has been sending clear impulses along single fibres, they have isolated and studied.

Wyke. For the record, Mr. Chairman, may I say that there was no methodological criticism implied in my remarks.

Ruth Bowden. Were you recording, Dr. Widdicombe, from random samples of the nerve fibre population? I am not quite sure about this, because certainly as you get near the larynx there is a whole selection of bundles of small fibres that are derived from the laryngeal, tracheal and oesophageal mucosa.

Widdicombe. We recorded from near the larynx—and by that, in the cat, I mean 0.5 to 1 cm from the larynx. We were fully aware of the patterns of activity of oesophageal and tracheal fibres, because these have been studied separately. I am not saying that some of these were not included in the sample I mentioned—but the majority of the fibres we studied were, I think, destined for the larynx.

Mårtensson. I think the experiments of Sampson and Eyzaguirre (1964)[1] suggest that the tidal air passing through the larynx may be sufficient to make receptors in the laryngeal mucosa discharge; and when they discharge, there is reason to believe that there is cyclic reflex activation of laryngeal motoneurones in the brain stem. If you let the tidal volume pass through the larynx then you have this effect, whereas if you let the tidal volume pass through a tracheostomy you won't have it. I think that would explain the difference that you have been discussing.

Wyke. I wish it did, but I don't think it is the whole story. You will recall that I said that topical anaesthesia of the whole laryngeal mucosa does reduce the respiratory oscillations of laryngeal muscle activity in a non-tracheotomized animal—but it does not do so to the same extent as performance of tracheostomy in the same animal; and you will also recall the fact that it is variations in the degree of stretch applied to a muscle that exerts the major influence on its oscillatory (as well as its tonic) activity—even with a tracheostomy, or with the larynx open to the atmosphere. So I don't think the difference is solely dependent on the presence or absence of stimulation of laryngeal mucosal receptors by the tidal air flow.

Murakami. In my opinion, the best way to solve this problem would be to make a recording from an individual nerve twig—that is, not from the recurrent nerve trunk but from a nerve twig that independently innervates either the abductor or an adductor muscle. We made our nerve recordings in Yale[2] with such twigs.

---

[1] Sampson, S., and Eyzaguirre, C. (1964) Some functional characteristics of mechanoreceptors in the larynx of the cat, *J. Neurophysiol.*, **27**, 464–80.

[2] Murakami, Y., and Kirchner, J. A. (1972) Respiratory movements of the vocal cords: an electromyographic study in the cat, *Laryngoscope (St. Louis)*, **82**, 454–67.

von Euler. May I ask Dr. Widdicombe a question? How many of the population of nerve fibres that you talked about could be afferents—especially pulmonary afferents?

Widdicombe. We haven't eliminated these by cutting the vagus below the origins of the recurrent laryngeal nerve—and we should test this. All I can say is that on the basis (a) of the pattern of activity, (b) of the way that the discharge is changed by events that either change the pattern of breathing or produce coughing, and (c) of the fibre responses to inflation and deflation of the lungs, we are fairly sure that these are efferents, and not afferents from the lungs.

Kirchner. In application, perhaps, to what Dr. Murakami has said I think it is important not only to record from twigs going to individual muscles, but also to clean off the recurrent nerve from along the trachea and the oesophagus to interrupt these afferents. It is obvious that one cannot go down into the chest and pick up and cut the afferents coming in, say, from mechanoreceptors in the thorax; but at least one can eliminate a great many of these afferents in the recurrent trunk by carefully cleaning it off the trachea before preparing it for recording.

Widdicombe. I quite agree with you. But in practice, if you are recording from these fibres there is no problem about distinguishing efferents from afferents. They are quite different in their patterns of discharge: by inflating the lungs or changing blood pressure, and so changing afferent activity, you can spot the difference straight away.

Otis. I would like to ask how the air stream through the larynx serves as a stimulus—what are the receptors? Is this a mechanical thing, or is it changes in chemical composition in the inflowing and outflowing gas—or is it none of these?

Wyke. To say none of them, or one rather than the other, I think would be premature at this stage—especially in view of Dr. Widdicombe's observations [see Communication 9]; but if you are inviting me to speculate about it I will.

The working hypothesis, at the moment, is that it is the expiratory pressure of air up through the larynx that, in fact, alters the stretch on the intrinsic laryngeal muscles (in addition to affecting low-threshold mucosal mechanoreceptors and articular receptors). You may remember that Ardran and Kemp[1] showed in their radiographic studies of human subjects (whether this is true in the cat or not, I don't know) that during the inspiratory phase of quiet breathing the vocal folds are not only abducted but are also rotated upwards, and that during the expiratory phase of quiet breathing they rotate downwards and are then adducted. Our present suggestion would be that a central respiratory discharge of the sort that Dr. Widdicombe was mentioning might be responsible for the initial downward displacement of the vocal folds, and then as the air stream comes up against them during developing expiration it applies stretch to the intrinsic myotatic receptors which then further activates reflexly the adductor

---

[1] Ardran, G. M., and Kemp, F. H. (1966) Mechanism of the larynx. I. The movements of the arytenoid and cricoid cartilages, *Brit. J. Radiol.*, **39**, 641–54.

mechanism. We have a little indirect evidence (and it is indirect) of this effect; namely, that if you anaesthetize the whole of the laryngeal mucosa above and below the larynx (or divide both superior laryngeal nerves) then you still get the same alterations in respiratory response when you detach the vocalis muscle at one end. That is to say, if you have an animal breathing through its larynx and its whole laryngeal mucosa has been anaesthetized topically and you then detach a muscle, its residual respiratory oscillations diminish considerably—although they don't entirely disappear; but if you then apply stretch to the muscle, the respiratory oscillations come back in proportion to the degree of applied stretch in the presence of continued mucosal anaesthesia. But I want to emphasize for the record that what I have suggested is pure speculation, at present.

Ladefoged. Can I ask Dr. Wyke to continue speculating in regard to that? Namely, is there any evidence of greater laryngeal muscular activity with greater force of expiration—because it would seem to follow from what you have just said that with forced expiration you should get greater activity. Now putting in another little bit, and referring it to speech: colleagues in my Laboratory (Drs. Hirano and Ohala[1,2]) have, in fact, been recording the laryngeal muscle activity which goes with highly stressed syllables, and their results seem to indicate that when you get the peaks of pressure with a stressed syllable (which we were talking about yesterday—see Communication 18) then perhaps as a deliberate act or perhaps not (we don't know) you get increased laryngeal muscular activity as well. So now I wonder if this is due to a reflex increase with forced expiration.

Wyke. I am delighted with that little piece of information, because it fits in with some of our observations which I didn't refer to in my Communication but which I mentioned in an earlier discussion [see p. 366]; namely, that if one exposes (in experimental animals) the laryngeal muscles to increasing subglottic pressures then one can show that there is a relationship between the increased (integrated) motor unit activity in the adductors of the vocal folds and the degree of change of pressure in the subglottic space, even in the presence of topical anaesthesia of the laryngeal mucosa. There is a fairly close relationship between them, if one plots it out; and I would suspect (again, without any direct evidence at the moment) that this is what might happen (supplemented by the effects of mucosal receptor stimulation) in at least certain circumstances of phonation, where you get an abrupt increase in muscular activity with phonemes whose utterance is associated with sudden elevations of subglottic pressure. I would like to know whether this might be so?

Ladefoged. That might well be so. But now the next question to go on with is what is the response time of these systems? Remember speech happens pretty quickly and one event happens pretty soon after the next (again as we were talking about yesterday—see Communication 18): so it seems likely, because speech happens so quickly, that the muscular activity has to be pre-programmed, rather than merely being a response to an increase in pressure. In the data that you have just been reporting, did you note what the response time was between the increase in pressure and the increase in laryngeal muscle activity?

---

[1] Hirano, M., and Ohala, J. (1969) Use of hooked-wire electrodes for electromyography of the intrinsic laryngeal muscles, *J. Speech Res.*, **12,** 362–73.

[2] Hirano, M., Ohala, J., and Vennard, W. (1969) The function of the laryngeal muscles in regulating fundamental frequency and intensity of phonation, *J. Speech Res.*, **12,** 616–28.

Wyke. It ought to be easy to answer that question; but, in fact, it isn't—because we have other evidence[1,2,3] that these myotatic reflexes from the laryngeal muscles (as well as the mucosal and articular reflexes) are probably polysynaptic. For this reason, in our experimental animals the time between the application of stretch to a muscle and the motor unit response to that stretch varies considerably with the degree of anaesthesia[2]: therefore, what we have in our data is a distribution of response times that are individually related to the degree of anaesthesia rather than simply to the degree of applied stretch. I don't think our figures would really be meaningful in the context in which you are interested, because they were obtained in anaesthetized animals and they don't relate in any temporal way to what happens in the unanaesthetized state. But this isn't to say one couldn't get those data on unanaesthetized human beings—if one had subjects prepared to volunteer for the investigation.

Fry. Whom are you looking at?

Wyke. The other thing I might add is that I am not sure whether the phonatory pre-programming that you mentioned yesterday [see Communication 18], and about which we then had a little discussion [see pp. 310–11], isn't in fact one of the functions of the corticobulbar projection system (to which I made a passing reference right at the end of my Communication). I have suggested elsewhere[4,5] that this initially sets the phonatory programme for the muscles; and then, if you like, the detailed execution of the programme during phonation is continuously modulated by the activity of the local circuit reflex systems I have described[4,5,6]. Anyway, this is the kind of operational model, as applied to phonation, that we are trying to test.

Stevens. With regard to the mechanism whereby the receptors might be stimulated, there are two possibilities. One is that you can have a sort of DC pressure applied to the muscles and the mucosa—that is, an increased subglottal pressure (which is the one you were talking about); but also when one has air flow over a surface or past an obstruction there is turbulence in the air flow and there are therefore fluctuating forces on the tissues—if you like, there is a turbulence fluctuating force which is dependent on the velocity of flow and not on any pressure drop. This has been studied by physicists examining the aerodynamics of air flow over the wings of aeroplanes, and so on. In this case it is the velocity of the flow that is contributing to the fluctuating forces, and not simply a pressure drop across the glottis.

Wyke. Might I respond to that, because I think it is a very important point. I have to confess to you that we have not been able to find any myotatic receptors in the muscles that are rapidly adapting. All the stretch-receptor behaviour that

---

[1] Abo-El-Enein, M. A., and Wyke, B. D. (1966) Laryngeal myotatic reflexes, *Nature (Lond.)*, **209**, 682–86.

[2] Wyke, B. D. (1968) Effects of anaesthesia upon intrinsic laryngeal reflexes, *J. Laryng.*, **82**, 603–12.

[3] Dunker, E. (1968) The central control of laryngeal function, *Ann. N.Y. Acad. Sci.*, **155**, 112–21.

[4] Wyke, B. D. (1967) Recent advances in the neurology of phonation: phonatory reflex mechanisms in the larynx, *Brit. J. Disor. Commun.*, **2**, 2–14.

[5] Wyke, B. D. (1969) Deus ex machina vocis: an analysis of the laryngeal reflex mechanisms of speech, *Brit. J. Disor. Commun.*, **4**, 3–25.

[6] Wyke, B. D. (1972) Laryngeal myotatic reflexes and phonation, *Folia phoniat. (Basel)* [In press].

we have been able to identify in the intrinsic laryngeal muscles has been very slowly adapting[1,2]; and this is another reason why we think that maybe the spiral endings—of which there are large numbers—might be mainly responsible for this effect, and that the very small proportion of muscle spindles (which ought to be capable of displaying both phasic and tonic stretch responses[3]) might, in fact, provide a slight background of phasic activity that we have not been able to detect in our recordings. But I would prefer to think that the rapidly-fluctuating turbulence effects to which you refer might be picked up, as it were, by the rapidly-adapting receptors which we know[4,5,6] to be present in the laryngeal joint capsules—in other words, the turbulence might cause rapidly oscillating discharges from the mechanoreceptors in the capsules of the readily-mobile intercartilaginous joints. I just don't know at the moment, but I would like to think about it a little and see if we could get some recordings.

Would you allow me, Mr. Chairman, to put a question to Dr. Stevens to assist me in this regard? Is there any way, from a physical point of view, in which we could deliberately create an experimental situation that would set up controlled turbulence in a space like that below and in the glottis—in other words, control the vortical velocity?

Stevens. I think it might be possible. You might insert some sort of an obstruction to direct the air flow against the surface. You could possibly make the air flow more or less turbulent by putting a variety of obstructions in the trachea underneath the larynx.

Plum. It struck me that the observations that were made here about man could be tested experimentally in animals, in an effort to find out whether corticobulbar pathways are involved in the modifications you were postulating. That would mean putting your electrodes into a curarized animal and delivering the appropriate stimulus, and then seeing if the thing was repeatable in the decerebrate animal. In both instances one would be free of anaesthesia, which I agree makes it very difficult to interpret some of the records which have been provided; and one could clearly separate what is forebrain activity from what arises from more caudal automatic structures. In so behavioural an instrument as the larynx, it might provide a better approach to what is otherwise an elusive experimental problem.

Beverly Bishop. I would like to come back to the point that Dr. Stevens brought up. I think what I want to say is pertinent, although completely remote (geographically speaking) from his approach. I got my own start in physiology studying the reflex control of the external anal sphincter in Professor Garry's

---

[1] Abo-El-Enein, M. A., and Wyke, B. D. (1966) Laryngeal myotatic reflexes, *Nature (Lond.)*, **209,** 682–6.

[2] Wyke, B. D. (1973) Myotatic reflexogenic systems in the larynx, *Folia morphol. (Prague)*, **21,** 113–17.

[3] Matthews, P. B. C. (1972) *Mammalian Muscle Receptors and Their Central Actions,* London.

[4] Kirchner, J. A., and Wyke, B. D. (1964) Electromyographic analysis of laryngeal articular reflexes, *Nature (Lond.)*, **203,** 1243–5.

[5] Kirchner, J. A., and Wyke, B. D. (1965) Afferent discharges from laryngeal articular mechanoreceptors, *Nature (Lond.)*, **205,** 86–87.

[6] Kirchner, J. A., and Wyke, B. D. (1965) Articular reflex mechanisms in the larynx, *Ann. Otol. (St. Louis)*, **74,** 749–69.

Laboratory in Glasgow many years ago. One of the segmental controls of that particular sphincter (which is continuously active, until inhibited by reflex responses) was found to be exerted by pressure developed in the lower gastro-intestinal tract (which would relax it) or by movement of solid materials or fluid through the lower gastro-intestinal tract (which also inhibited it). We are able to sort out different receptor systems involved in this effect. Inflating the colon with a balloon relaxed the sphincter; but after anaesthetizing the mucosa of the colon with local application of *Xylocaine,* the so-called 'movement receptors' were still functional in the absence of the response to inflation of the balloon[1].

Now, going to the other end of the gastro-intestinal tract, as it were, I wonder if it is possible that there are similar populations of receptors that have not yet been identified histologically but might be defined neurophysiologically in an analogous way. Many of the experiments reported this morning have been done on animals with a tracheal cannula in place: is there a marked difference in response between the tracheotomized animal and one in which air flows through the larynx? If so, that could give us a clue: otherwise could we design experiments that have already been suggested here?

Widdicombe. I just want to tell Dr. Bishop that her early physiological experience might not be quite irrelevant to phonation. There was an article in the *Guardian* newspaper a couple of years ago—it was not a very scientific article—that described a French music hall comedian who, if he had an appropriate intake of air into his colon, could then phonate with his anus (in French, of course)! This is surely a subject that is worthy of more study!

Wyke. In my Communication I reported that we had in fact made a comparison between two parallel sets of experiments (and sometimes between two experiments in the same animal) in which, on the one occasion we had airflow proceeding spontaneously through the larynx (these were not artificially ventilated animals, but anaesthetized animals breathing spontaneously) and on other occasions there was no air flow through the larynx because the animals were breathing through a tracheostomy tube. In some of the animals we had a valved tracheostomy tube, so that in one and the same animal we could on some occasions have air flow through the larynx and on other occasions no flow through the larynx and could switch from one to the other situation. This, of course, brings us back to the point that Dr. Widdicombe made earlier; namely, that whenever we switch the air flow, even in the same animal, from out of the larynx into the tracheostomy tube there is a marked reduction in the motor unit activity in the laryngeal muscles that is oscillating in synchrony with the phases of the respiratory cycle. I hope I have established that observation to everybody's satisfaction; but I also emphasized that these differences in response could *not* be attributed *solely* to variations in subglottic mucosal mechanoreceptor stimulation.

Mårtensson. As the different laryngeal structures are so closely related mechanically it is very difficult to determine the location of the receptors from which activity recorded in the laryngeal nerves arises. I would like to ask Dr.

[1] Bishop, B., Garry, R. C., Roberts, T. D. M., and Todd, J. K. (1956) Control of the external sphincter of the anus in the cat, *J. Physiol. (Lond.),* **134,** 229–40.

Wyke about his experiments with stretching of the laryngeal muscles, and inquire if he can conclude that the receptors are localized in the muscles themselves. Even small changes of tension in the muscles (for example, of the order of 5 to 10 dynes) are capable of changing the discharge frequency of receptors located outside the muscles. This is a dilemma.

Wyke. The point that Dr. Mårtensson is making is crucial to workers in this field, particularly in relation to some of the experiments that he himself reported[1] some years ago that involved probing around various parts of the tissues of the larynx. Therefore I think it is very important for me to justify my statement that I believe these stretch responses to come from mechanoreceptors located in the laryngeal muscles; and although we have published[2,3,4] most of the relevant information, I think I ought to summarize it here.

First, the stretch reflex response in an individual laryngeal muscle (measured in terms of its motor unit discharge) has a threshold of about 3 gm of linear applied stretch, and rises with increasing stretch in the same sort of relationship as one gets with a static stretch response in a limb muscle—that is, the integrated EMG curve has a similar (although not identical) slope. Second, if one applies topical anaesthesia to the whole of the laryngeal mucosa, then this does not modify in any way the stretch reflex response. Third, if one infiltrates the capsules of the laryngeal joints bilaterally with a local anaesthetic agent, or destroys the receptors located therein by electrocoagulation, this again does not abolish the tonic stretch reflex responses—although it abolishes most of the phasic components in the muscle response to stretch (which some people, including myself, believe to come from the laryngeal articular receptors). In addition, one can disarticulate the crico-arytenoid and cricothyroid joints, and this again makes no difference to the tonic stretch responses of the individual laryngeal muscles.

So by a process of exclusion, if you like, we have arrived at a position such that I cannot think of where the stretch reflex responses may come from, if it is not from some kind of mechanoreceptor system in the muscles—and I would be grateful for a suggestion as to where the relevant receptors might be, if they are not there. We have repeated Dr. Mårtensson's (*loc. cit.*) probing experiments many times, and I agree with him that one can probe all round the place in the larynx and get responses that are rapidly-adapting or slowly-adapting, or a combination of the two, depending on where you poke the tissue; and I would remind you that Dr. Kirchner has also published[5] some experiments from his Laboratory in Yale in which he has shown that if you take the global afferent discharge in a laryngeal nerve you can subtract some of these various components one by one from the totality of the afferent discharge by carrying out *seriatim* some of the various procedures that I have just mentioned.

---

[1] Mårtensson, A. (1963) Reflex responses and recurrent discharges evoked by stimulation of laryngeal nerves, *Acta physiol. scand.*, **57,** 248–69.

[2] Abo-El-Enein, M. A., and Wyke, B. D. (1966) Myotatic reflex systems in the intrinsic muscles of the larynx, *J. Anat. (Lond.),* **100,** 926–7.

[3] Abo-El-Enein, M. A., and Wyke, B. D. (1966) Laryngeal myotatic reflexes, *Nature (Lond.)*, **209,** 682–6.

[4] Wyke, B. D. (1973) Myotatic reflexogenic systems in the larynx, *Folia morphol. (Prague)*, **21,** 113–17.

[5] Kirchner, J. A., and Suzuki, M. (1968) Laryngeal reflexes and voice production, *Ann. N.Y. Acad. Sci.*, **155,** 98–109

Lindquist. Before we end this discussion, I should like to mention a paper by Testerman (1970)[1] describing some experiments on artificially-induced vocalization in the cat. If I remember correctly, he was unable to alter the electromyographic activity obtaining in the cricothyroid muscle during vocalization by stretching it, whereas inflation of the lungs markedly affected the activity of the muscle—which might interest Dr. Widdicombe.

Plum. Could we now turn to Dr. Kirchner's Communication. I was intrigued by and admired his tracings, and I want to ask him if he believed, as his discussion would imply, that these muscles either act or don't act. That is to say, single unit analysis has great virtues—but can it be taken to record the behaviour of the total muscle in all circumstances? Do the extralaryngeal respiratory muscles act as global units, or are they fractionated into different functional components within the individual muscles?

Kirchner. I think the answer is yes, they are fractionated. This can be observed occasionally in the patient—it can be demonstrated by X-ray studies; and the single unit tracings were merely obtained incidentally. More often we had multi-unit activity from these muscles. Some of these muscles are very consistent in their activity in relation to the respiratory cycle: the middle constrictor, for example, is active even under very deep general anaesthesia and its activity is not influenced by movements of the hyoid bone. So I think, certainly in the cat, that these tracings do in fact represent the gross activity of the muscle. We get this same type of activity from different parts of the muscle—if you sample the cranial end of the muscle versus the lower (caudal) end of the muscle a pattern seems to emerge that is very consistent.

I didn't get in to discuss one of the questions that came up earlier, but Dr. Otis brought it up; namely, what actually excites the generation of afferent impulses with the passage of air through the larynx? Dr. Suzuki[2], in our Laboratory in Yale, has in several cats done the experiment of blowing very gentle jets of air across the subglottic mucous membrane without distorting it (at least, so as not to distort it grossly, as observed with the operating microscope) while recording from the peripheral cut end of the recurrent laryngeal nerve after taking it apart into filaments. He picked up some single units that showed activity when air passed over the mucous membrane which was abolished after topical anaesthesia was applied to the mucous membrane. Now whether stronger air jets would produce an increase in the frequency of these discharges I am not prepared to say, but it can be observed with air and with tiny wisps of cotton just touched to the surface. This may bear on some of the questions that have been raised.

Ruth Bowden. Have you tried passing air at different temperatures across the mucous membrane? I seem to remember that somebody has done this and found activity in the muscles. Also, have you tried introducing irritant material into the inspired air?

---

[1] Testerman, R. L. (1970) Modulation of laryngeal activity by pulmonary changes during vocalization in cats, *Exp. Neurol.*, **29,** 281–97.

[2] Kirchner, J. A., and Suzuki, M. (1968) Laryngeal reflexes and voice production, *Ann. N.Y. Acad. Sci.*, **155,** 98–109.

Kirchner. No. We have not done either of those things.

Widdicombe. I can give a direct answer to that, if you like. We have tried, not by recording from laryngeal afferents but by seeing if you can get a reflex response to cold air in the larynx of cats, and found you can't. Of course, chemical and mechanical irritants elicit powerful reflex responses.

Ruth Bowden. It seems to me that there are some extremely important features in this work with the supra- and infrahyoid muscles. These muscles suspend the larynx both from the base of the skull and from the jaw (posteriorly from the styloid process and the mastoid process) and anchor it below (anteriorly to the sternum and posteriorly to the scapula), so forming a sling system that is involved in positioning the larynx.

These same muscles also play an important part in the laryngeal protective mechanism during swallowing, by pulling the larynx forward and tipping it so that the airway and the food passage are not quite so much in parallel. Could I ask Dr. Kirchner if he has worked on the swallowing reflex in very lightly anaesthetized animals?

On another point, the muscles involved have an innervation which extends from the midbrain (in terms of the proprioceptive nuclei for those muscles that are innervated by the mandibular division of the trigeminal nerve—namely, the mylohyoid and the anterior belly of the digastric muscle) right down to the upper four segments of the spinal cord. There must be an enormous field of input and of output extending throughout the length of the brain stem.

Kirchner. Of course, but that is a big question! We have not studied the function of deglutition in relation to the external laryngeal muscles. It is interesting, in connexion with the last part of Dr. Bowden's remarks, that the various parts of the muscles—for example, the anterior and posterior bellies of the digastric—have entirely different patterns of innervation. This is true, even in respect of the relationship of the activity of this muscle to the respiratory cycle.

General Discussion on Communications

CHAIRMAN: *Professor D. B. Fry*
London, England

Fry. We will now go on to the General Discussion of the Communications—which, of course, embraces not only all the papers that have been given but also any other relevant topic that anyone wishes to raise. But to begin with, I have the names of three people who wish to make supplementary contributions; and I call Dr. Viljanen as the first of them, to make his supplementary statement.

SUPPLEMENTARY STATEMENT BY A. A. VILJANEN

The regulation of breathing can be considered in terms of the block diagram in Fig. D.17 (cf. Campbell and Howell, 1962[1]).

To begin with, I will try to clarify the relationship between the electrical and mechanical activities of the respiratory muscles. As mentioned earlier [see Communication 8], several workers have shown that a close relationship exists between the electrical and mechanical activities of the respiratory muscles, both in man and in animals. In our earlier studies (Viljanen, 1967[2]) we established that in man a linear relationship exists between:

1. EMG impulse frequency recorded from the external intercostal muscles at the end of inspiration [f_d] and the muscular inspiratory work [W] [Fig. D.18]. This can be expressed by the following formula—

$$f_d = b_d W + c_d$$

where b_d and c_d are constants.

2. The total number of EMG impulses recorded during inspiration [n] and the first time integral of the muscular inspiratory work [H_t] [Fig. D.19]. Thus—

$$n = b_n H_t + c_n$$

3. Alveolar relaxation pressure [$P_{st(rs)}$] and EMG impulse frequency determined four seconds after the end of inspiration, with the subject holding his breath with an open glottis [f_s] [Fig. D.20].

$$f_s = b_s P_{st(rs)} + c_s$$

4. The number of EMG impulses [n] and the pressure impulse produced by the inspiratory musles [I] during one and the same inspiration [Fig. D.21].

$$n = b_i I + c_i = b_i \int P_{mus}\, dt + c_i$$

This means that during deepening inspiration the discharge frequency in the EMG is directly proportional to the pressure produced by the inspiratory muscles.

---

[1] Campbell, E. J. M., and Howell, J. B. L. (1962) Proprioceptive control of breathing, in *Ciba Foundation Symposium on Pulmonary Structure and Function*, ed. de Rueck, A. V. S., and O'Connor, M., pp. 29–45, London.

[2] Viljanen, A. A. (1967) The relation between the electrical and mechanical activity of human intercostal muscles during voluntary inspiration, *Acta physiol. scand.*, Suppl. **296,** 1–61.

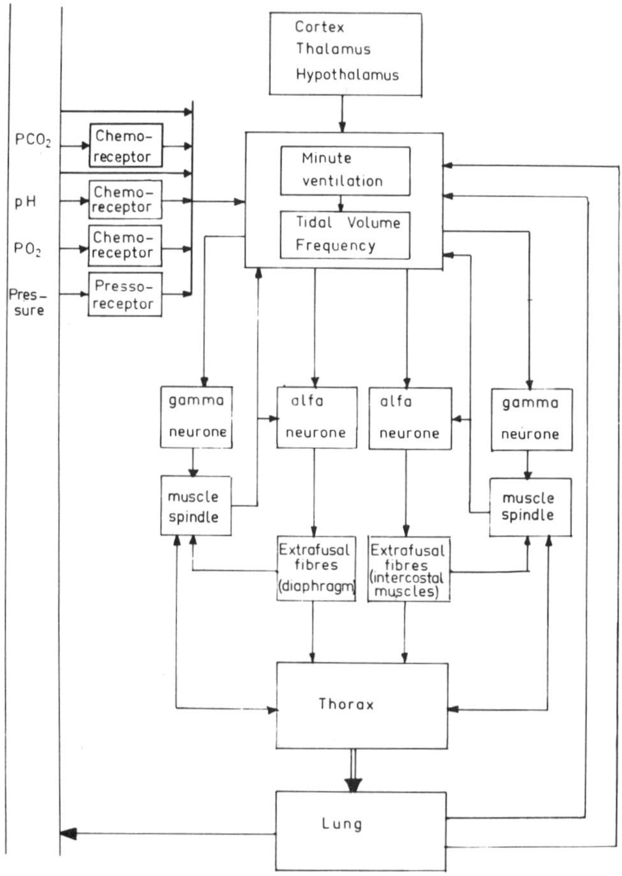

Fig. D.17. Block diagram of the regulation of breathing.

It is difficult to say how constant the relationship between the electrical and mechanical activities is in different circumstances. I should like to call attention to the phenomenon observed during carbon dioxide stimulation, during which the respiratory muscles seem to produce the same physical action with a smaller number of impulses (as recorded from the external intercostal muscles) than before and after such stimulation [see Fig. D.22].

The above mentioned findings suggest that the respiratory centres determine in advance the nature of the next inspiration. We have also studied the relationship between the electrical activities of the external intercostal muscles and of the diaphragm during one and the same inspiration, and have found a linear relationship between these activities both in man (Viljanen and Halttunen,

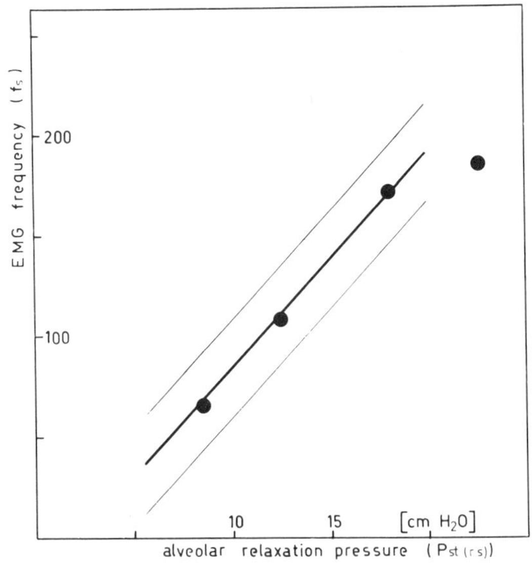

Fig. D.18. EMG impulse frequency [f_d] at the end of inspiration as a function of inspiratory muscular work [W]. The points represent the means of at least 6 values of the variables in columns with a class interval of 10 cm H_2O x l. There are 42 values of the variables in the range 0–50 cm H_2O x l, the regression coefficient ($b_{.i}$) is 4.0 cm H_2O^{-1} x sec^{-1} x l^{-1}, the constant (c_d) 38.6 cps, the coefficient of the linear correlation (r) 0.90 and the modified standard error of estimate ($s_{y \cdot x}$) ± 27.9 cps.

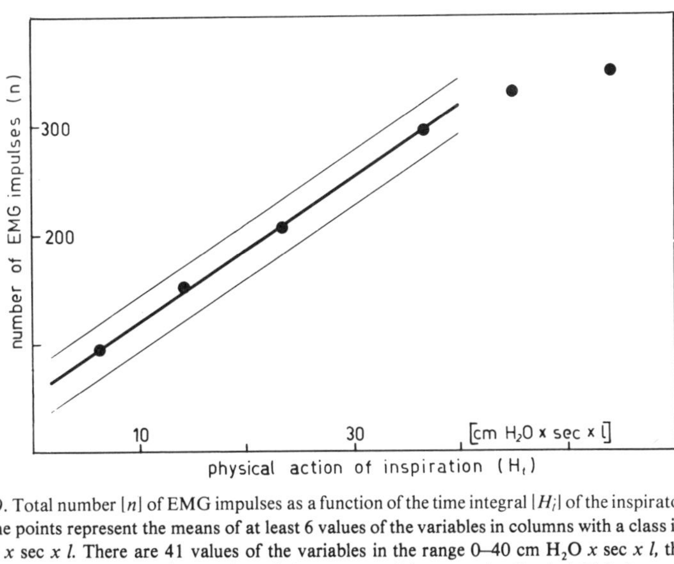

Fig. D.19. Total number [n] of EMG impulses as a function of the time integral [H_i] of the inspiratory muscular work. The points represent the means of at least 6 values of the variables in columns with a class interval of 10 cm H_2O x sec x l. There are 41 values of the variables in the range 0–40 cm H_2O x sec x l, the regression coefficient (b_n) is 6.7 cm H_2O^{-1} x sec^{-1} x l^{-1}, the constant of the regression line (c_n) 51.1, the coefficient of the linear correlation (r) 0.94, and the modified standard error of estimate ($s_{y \cdot x}$) ± 25.4.

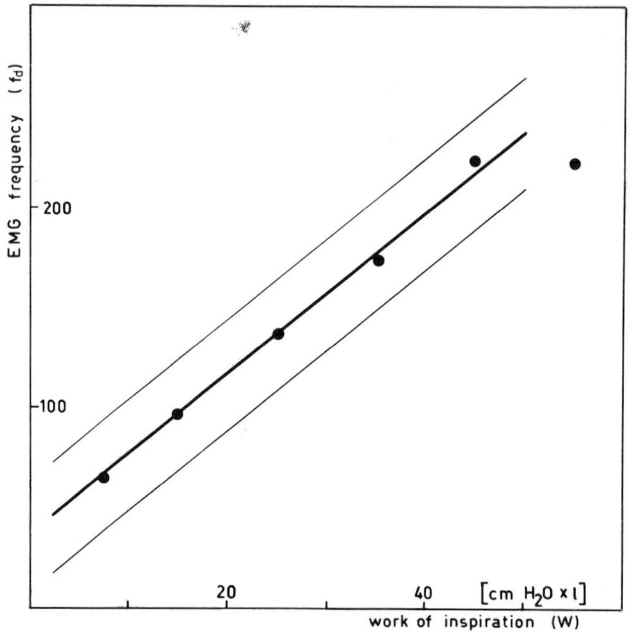

Fig. D.20. EMG impulse frequency four seconds after the end of inspiration [f_s] (the subject holding his breath) as a function of the alveolar relaxation pressure [$P_{st(rs)}$]. The points represent the means of at least 6 values of the variables in columns with a class interval of 5 cm H_2O. There are 39 values of the variables within the range 0–20 cm H_2O, the regression coefficient (b_s) is 10.8 cm H_2O^{-1} x sec^{-1}, the constant of the regression line (c_s) −24.5 cps, the coefficient of the linear correlation (r) 0.78, and the modified standard error of estimate ($s_{y;x}$) ±22.0 cps.

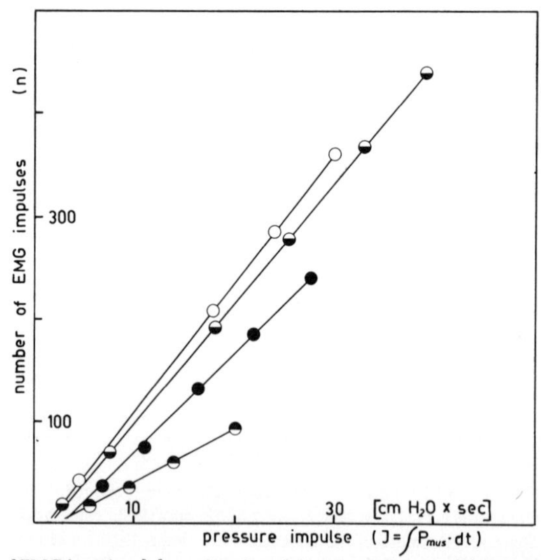

Fig. D.21. Number of EMG impulses [n] as a function of the pressure impulse [I] produced by the inspiratory muscles during one and the same inspiration. The figure relates to four different inspirations.

1967[1]) and in the cat (Viljanen *et alii*, 1968[2]). In man this relationship can be voluntarily regulated to a certain degree, so that the linearity remains relatively unaffected (c.f. Viljanen and Poppius, 1968[3]) [Fig. D.23]. In the cat, the proportion of external intercostal electrical activity decreases with increasing depth of aneasthesia [Fig. D.24].

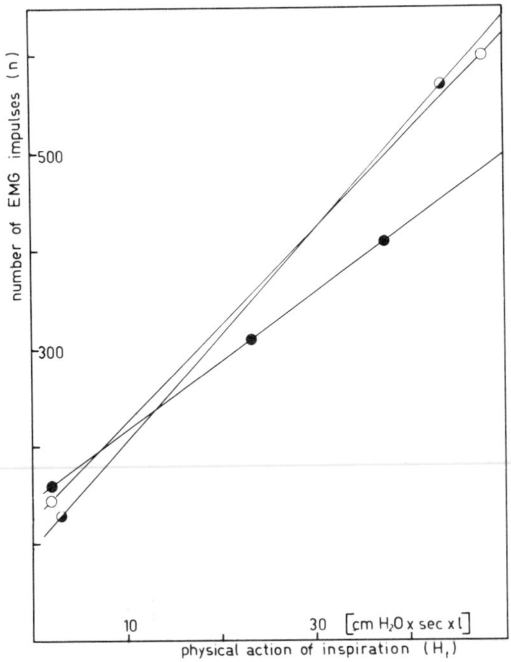

Fig. D.22. Total number of electromyogram impulses [*n*] as a function of the first time integral of the inspiratory muscular work [H_I] before (o–o–o), during (●–●–●) and after (◐–◐–◐) 5.0 per cent CO_2 stimulation.

At this point I should like to refer to an early report of Anderson and Lindsley (1935)[4], who demonstrated (with a single motor unit recording technique) that the external intercostal muscles on both sides of the thorax are equally activated in normal and in unilaterally pneumonectomized cats, both with regard to the number of units involved and to the average frequency of the responses. We have also recorded in man the activities of both sides of the thorax using surface electrodes, and found these activities to be closely similar (unpublished data). The

[1] Viljanen, A. A., and Halttunen, P. K. (1967) Comparison of electromyogram activities of the diaphragm and the intercostal muscles in man during voluntary inspiration, *Acta neurol. scand.*, **43** (Suppl. 31), 191.

[2] Viljanen, A. A., Salorinne, Y., Laitinen, L. A., and Paloheimo, M. (1968) The effect of ether anaesthesia on the electrical activity of the diaphragm and intercostal muscles during inspiration in the cat, *Scand. J. clin. lab. Invest.*, **21**, (Suppl. 101), 64.

[3]. Viljanen, A. A., and Poppius, H. (1968) The voluntary regulation of contraction of diaphragm and intercostal muscles in man during inspiration, *Scand. J. clin. lab. Invest.*, **21** (Suppl. 101), 63–64.

[4] Anderson, F. M., and Lindsley, D. B. (1935) Action potentials from intercostal muscles before and after unilateral pneumectomy, *J. lab. clin. Med.*, **20**, 623–8.

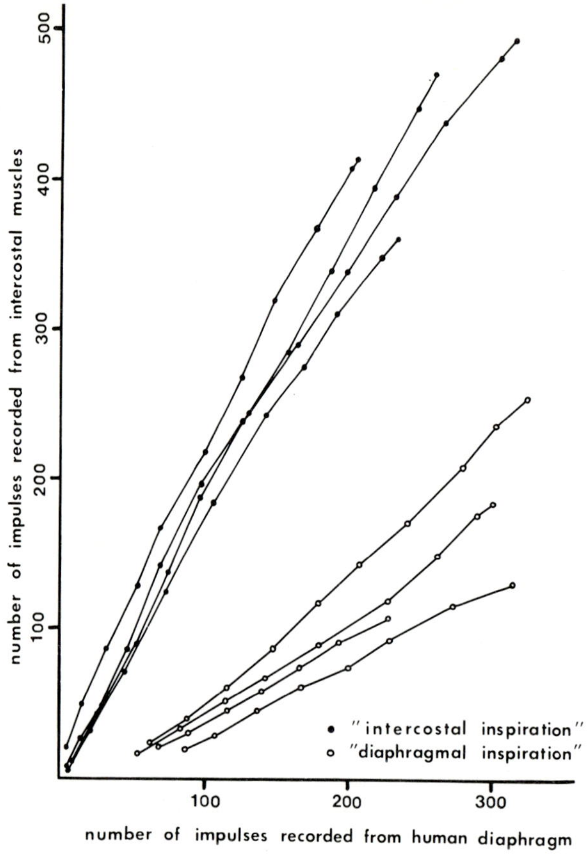

Fig. D.23. Number of impulses recorded from intercostal muscles as a function of the number of impulses recorded from the diaphragm. The figure relates to four different 'intercostal inspirations' (●—●—●) and to four 'diaphragmatic inspirations' (o—o—o) recorded from man during voluntary breathing.

possibility of voluntarily separate regulation of different parts of the thorax has not been studied.

I would also like to refer to the work of Boyd and Basmajian (1963)[1], who implanted multiple clip electrodes in the diaphragm of rabbits at open surgery and found no obvious general pattern of difference in the electrical activity of different muscular slips in one and the same rabbit's diaphragm.

Finally, I would like to express the opinion that integration of the electrical activity of the respiratory muscles is likely to cause part of the information to be lost, compared with the information available from counting separate impulses.

Fry. Thank you, Dr. Viljanen. May I now call on Dr. Widdicombe to give his supplementary contribution.

---

[1] Boyd, W. H., and Basmajian, J. V. (1963) Electromyography of the diaphragm in rabbits, *Amer. J. Physiol.*, **204,** 943–48.

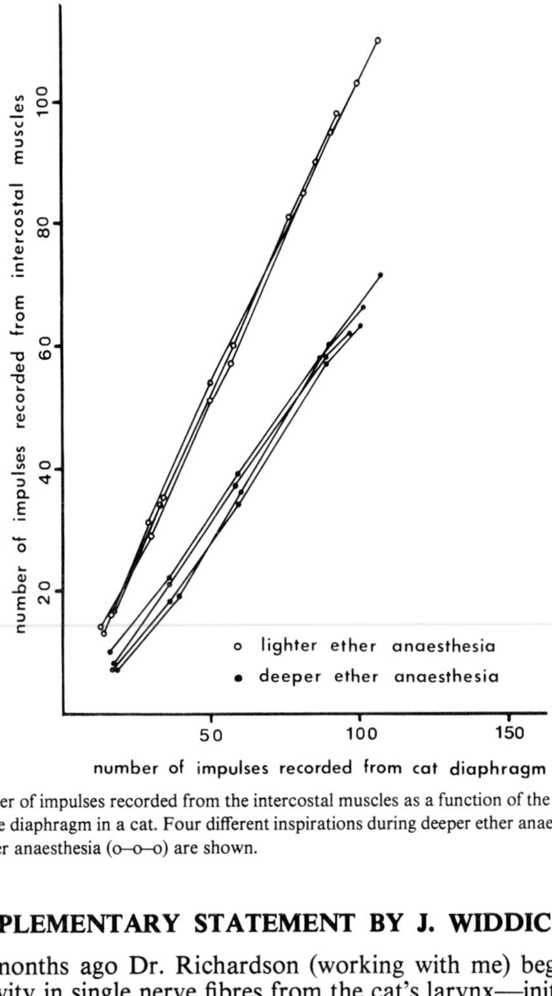

Fig. D.24. Number of impulses recorded from the intercostal muscles as a function of the number of impulses recorded from the diaphragm in a cat. Four different inspirations during deeper ether anaesthesia (●–●–●) and four during lighter anaesthesia (o–o–o) are shown.

SUPPLEMENTARY STATEMENT BY J. WIDDICOMBE.

Some 6 months ago Dr. Richardson (working with me) began a study of afferent activity in single nerve fibres from the cat's larynx—initially repeating what other people had done, and what we have heard about today to some extent—in particular relation to the action of irritant gases inside the larynx. I am not going to show you slides dealing with this, but the findings are relevant in a couple of ways. He recorded from quite a large number of fibres, and divided them up into those coming from receptors which seemed to be in the mucous membrane (as judged by stimulation with cotton wool or by light jets of air, as Dr. Kirchner has mentioned) and from receptors which seemed to be deeper and only responded when there was clear movement of a laryngeal joint or deformation of the laryngeal muscles—he had to push fairly hard to get such stimulation.

He found two types of receptor in the mucous membrane. One, the commoner type, had myelinated afferent fibres and was rapidly-adapting, and their behaviour was very similar to that of the epithelial receptors people have recorded from in the nasopharynx, and in the trachea and the bronchi: they seemed to be epithelial endings (and thus presumably irritant receptors, as in other parts of the respiratory tract) but were not at all like the mucosal mechanoreceptors that Dr. Kirchner and Dr. Wyke have described. And then there was a second group of responses (which I don't think anybody has studied before) that came from endings innervated by non-myelinated fibres. Of course, these were more difficult to study and we have only got a few records of them, but they seemed to be very similar to the nociceptive endings that you get in most tissues, and their behaviour was very similar to that of the J receptors in the lungs.

Having studied the responses of these receptors to irritants, he and Dr. Boushey then decided they really did not know enough about laryngeal reflexes and went on to study[1] the effect of irritant gases on reflexes from the laryngeal mucous membrane—and I want to show you three slides on that. They applied to the laryngeal mucous membrane weak ammonia gas down to one part in ten million (which is adequate for stimulation), cigarette smoke, various other chemical irritants and, more recently, riot control (CS) gas—which latter has fascinating implications that I do not have time to tell you about.

First slide please [Fig. D.25]. With irritant gases, you get two types of response. Obviously sometimes you get coughing, and this is an example.

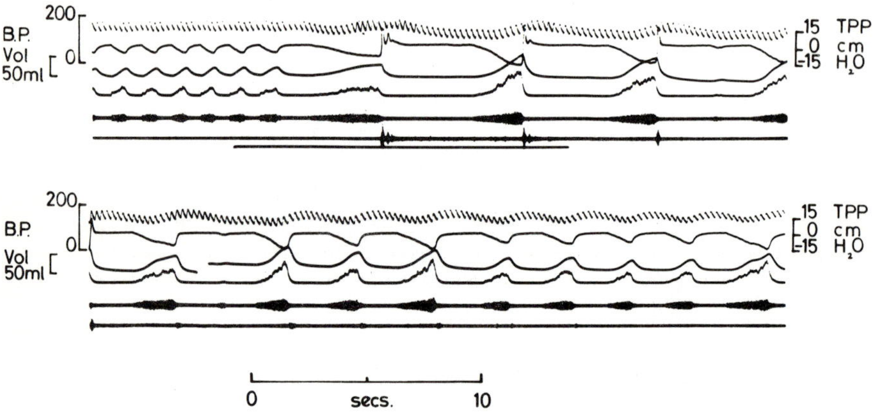

Fig. D.25. Respiratory response to cigarette smoke blown into the larynx. Traces from above down: blood pressure, transpulmonary pressure, tidal volume, integrated phrenic nerve activity, phrenic electroneurogram, lumbar electroneurogram and signal. The two sets of recordings are continuous. Cigarette smoke in the larynx caused coughing (inspiratory and expiratory efforts), followed by slower and deeper breathing.

During the signal, cigarette smoke was blown into the larynx and the first change—the change in the pattern of inspiratory activity—is slower deeper breathing, and a slower build-up of phrenic activity compared with the controls. This is rather interesting from the point of view of the Clark-von Euler analysis, but we won't go into that. This is followed by vigorous expiratory efforts—so

[1] Boushey, H. A., Richardson, P. S., and Widdicombe, J. G. (1972) Reflex effects of laryngeal irritation on the pattern of breathing and total lung resistance, *J. Physiol. (Lond.)*, **224**, 501–13.

this is coughing, which is what you would expect. Notice that long after the coughing has stopped, you still get slower and deeper breathing—although the stimulus has been switched off. The breathing change can go on for a minute or so.

Next slide please [Fig. D.26]. Given a weak enough concentration of the irritant, either cigarette smoke or ammonia gas, you can get the slow, deep breathing without any expiratory efforts. This is highly characteristic and, in fact, is far more common than coughing in our experimental conditions. You will notice here that there are now no expiratory efforts, either on the pressure trace or on the volume trace or on the silent lumbar nerve trace (although we know this was a living preparation, because when there was a vigorous cough stimulus it produced abdominal muscle activity from these nerves).

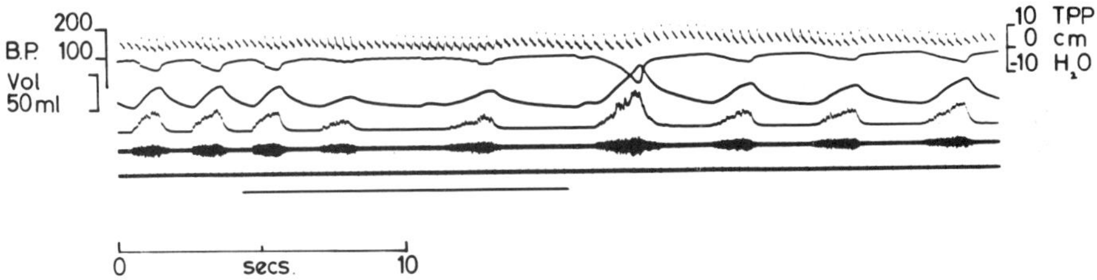

Fig. D.26. Tracings as in Fig. D.25, but a weaker stimulus with cigarette smoke causes only slower breathing, without expiratory efforts.

Next slide please [Fig. D.27]. Then, to get slightly more physiological, it occurred to us to repeat the experiment that Kratschmer (1870)[1] did a long time ago. He blew 30 per cent carbon dioxide into the larynx and got respiratory tract reflexes; and everybody, I think, has assumed that these are like the cough reflex and are only brought in in non-physiological conditions. However, Drs. Richardson and Boushey[2] have been trying far lower concentrations of carbon dioxide, down to 5 to 6 per cent, to see what this does to reflexes from the larynx. This is a record using 20 per cent carbon dioxide, to show a clear response. Quite clearly there is very considerable slowing of breathing, and the breathing is deeper; and again, there is no coughing with carbon dioxide. They got exactly the same response with 5 to 6 per cent carbon dioxide, but of course the record is less dramatic and so is more difficult to take in at first sight.

So it looks as though even physiological concentrations of carbon dioxide can act on the larynx to produce reflex changes in the pattern of breathing. What relation this may have to the normal control of breathing is rather obscure: whether or not the swings in laryngeal CO_2 concentration from 0 to 6 per cent that happen all the time in breathing have an important reflex input into respiration, we don't know. All we can say is that, if in the cat you change the laryngeal

---

[1] Kratschmer, F. (1870) Über Reflexe von der Nasenschleimhaut auf Atmung und Kreislauf, *Sber. Akad. Wiss., Wien*, **62**, 147–70.

[2] Boushey, H. A., and Richardson, P. S. (1973) The reflex effects of intra-laryngeal carbon dioxide on the pattern of breathing, *J. Physiol. (Lond.)*, **228**, [In press].

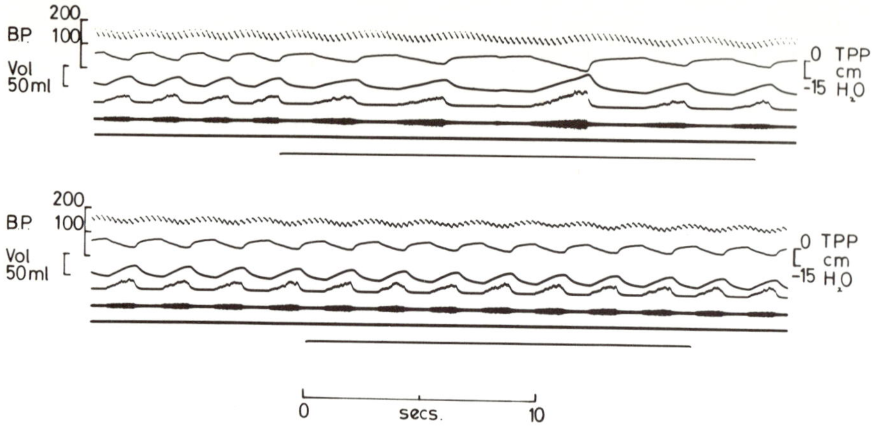

Fig. D.27. Tracings as in Fig. D.25, but 20 per cent CO_2 blown into the larynx causes slower deeper breathing without expiratory efforts when the larynx is innervated (upper recording). The lower recording shows the absence of such responses after both superior laryngeal nerves had been cut.

luminal environment from 0 per cent CO_2 to 5–6 per cent CO_2, then this is a powerful enough reflex stimulus to produce measurable changes in the pattern of breathing.

Fry. Thank you, Dr. Widdicombe. May we now have the third supplementary contribution from Dr. Cohen.

SUPPLEMENTARY STATEMENT BY M. I. COHEN

I wish to describe additionally some of my observations on laryngeal efferent activity that are relevant to the theme of this Symposium. I undertook this study as an adjunct to my studies of central respiratory activity (described in Communication 12), in order to ascertain the over-all pattern of recurrent laryngeal discharge for comparison with the discharge patterns of medullary respiratory neurones that might be involved in the control of the laryngeal system. The experiments were conducted on spontaneously-breathing, tracheotomized, decerebrate cats.

The tracings in Fig. D. 28 show efferent activity in the phrenic and recurrent laryngeal nerves (whole-nerve recordings). The original recordings (on the left) show that the phrenic inspiratory burst starts sharply after the period of phrenic silence in the expiratory phase. In contrast, the laryngeal inspiratory burst seems to start less sharply, and is superimposed on tonic activity that persists during the expiratory phase. A more precise specification of discharge pattern is obtained by summing the action potentials (after half-wave rectification) with an average-response computer, the computer sweeps being synchronized to the start of the inspiratory phase. The resulting tracings of summed activity (on the right) show that the phrenic discharge starts with a sharp burst of activity, and thereafter the discharge is augmenting in pattern, a maximum being reached at the end of the inspiratory phase: whilst the laryngeal nerve discharge pattern is trapezoidal in form, reaching a maximum at the start of the inspiratory phase and thereafter tending to decrement as inspiration progresses. The time relation of the two activities at the start of the inspiratory phase is shown in tracings made on an expanded time scale (bottom right), and it can be seen that laryngeal nerve activity starts to increase a considerable time (*circa* 60 msec) before the sharp onset of the phrenic burst. The delay between these two activities ranged between 20 and 100 msec in different cats.

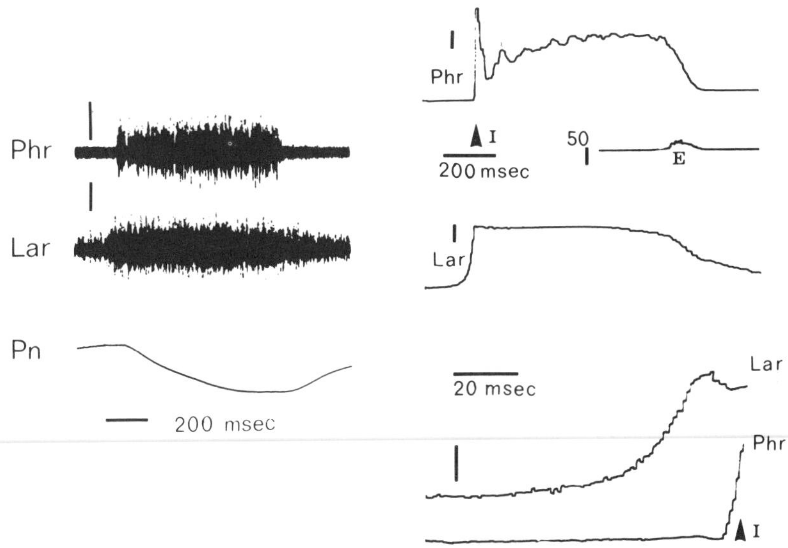

Fig. D.28. Comparison of phrenic and recurrent laryngeal inspiratory discharge patterns.
Left: oscilloscope tracings of activity during one inspiratory phase. *Phr:* efferent phrenic nerve potentials. *Lar:* efferent recurrent laryngeal nerve potentials. *Pn:* pressure recording from chest pneumograph. Voltage calibrations (vertical bars): 50 μV.
Right: average discharge patterns: summed activity in 400 respiratory cycles (analogue readouts from average response computer). *Top:* slow time base (5 msec bins); tracings start 198 msec before I pulse (arrow). *E:* distribution of E pulses (marker, 50 pulses). *Bottom:* fast time base (1 msec bins); traces start 98 msec before I pulse (arrow). Voltage calibrations (vertical bars): *Phr*, 1.5 μV msec; *Lar*, 1.0 μV msec.

The differences between phrenic and laryngeal nerve discharge patterns may be related teleologically to their differing motor functions. The augmenting character of diaphragmatic discharge serves to overcome the elasticity of the chest and lungs. The laryngeal inspiratory discharge results in dilatation of the glottis; and measurement of glottic dilatation with a balloon (Bianconi *et alii*, 1963[1]) has shown a trapezoidal pattern similar to the nerve discharge pattern—i.e., maximal dilatation is reached at the beginning of inspiration and is

---

[1] Bianconi, R., Corazza, R., and Raschi, F. (1963) Attività elettrica dei muscoli intrinseci della laringe e posizione delle corde vocali nella iperventilazione con aria e nell'apnea ipocapnia, *Arch. Sci. biol. (Bologna)*, **47**, 376–89.

maintained throughout the phase. This pattern of action, in which maximal dilatation is reached near the time of onset of air flow, is thus useful as a way of minimizing airway resistance during inspiration.

The occurrence of a relatively long delay between the onset of phrenic and of laryngeal inspiratory discharge means that the phase-switching mechanism is acting differentially on the two motor systems. This delay (20–100 msec) is far too long to be due solely to differences in axonal conduction time (thus, the latency of a phrenic response to medullary stimulation is only 3–6 msec), and therefore must be attributed to synaptic mechanisms of temporal summation.

Another major difference between phrenic and laryngeal inspiratory discharges is in their Hering-Breuer reflex responses. The changes in both activities as a result of airway occlusion at minimal lung volume (at the start of the inspiratory phase) are shown in Fig. D.29. Such occlusion results in prolongation

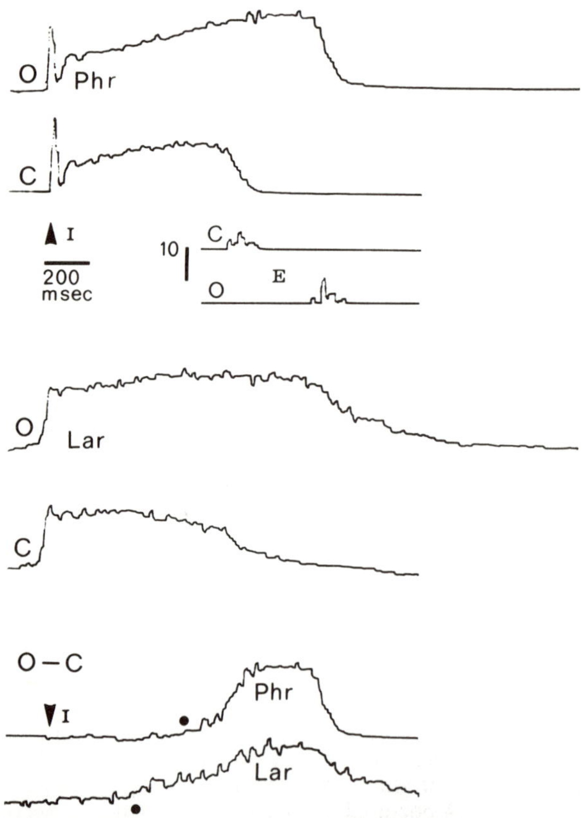

Fig. D.29. Effects of airway occlusion at minimal lung volume (started by I pulse, ended by E pulse) cn phrenic (*Phr*) and laryngeal (*Lar*) discharge patterns. Each tracing, starting 198 msec before the I pulse (arrow) and ending 198 msec before the termination of the shortest cycle in the sample, is the computer analogue readout of summed activity (N= 20, 15 msec bins). [O]: occlusion. [C]: control. [E]: distribution of E pulses in C and O cycles (marker, 10 pulses). [O–C]: activity in 20 'O' cycles minus activity in 20 'C' cycles. Dots: inflection points, when mean activity during occlusion starts to increase above the level of control activity.

of both types of inspiratory activity, due to the 'disinhibition' caused by prevention of the normally-occurring vagal afferent input during lung expansion. The details of timing of the effects are seen in the [O-C] tracings (bottom), which show the changes in activity caused by airway occlusion (activity during occlusion minus control activity). It can be seen from the inflection points on the tracings (indicated by dots) that laryngeal activity starts to increase about 200 msec before phrenic activity—in other words, disinhibition starts earlier. This delay is similar to that observed in other cats. Thus in spontaneous breathing, vagal afferent input starts to exert an inhibitory effect on laryngeal discharge a considerable time before its effect on phrenic discharge becomes apparent. This observation further indicates the great differences in synaptic organization between the two motor systems controlling the laryngeal muscles and the diaphragm.

Fry. Thank you, Dr. Cohen. Now, ladies and gentlemen, we will continue the general discussion. I will simply leave it to you to introduce whatever topics you want to bring up—providing they relate to the theme of our Symposium.

Klatt. I would like to talk, for just a minute or two, about the relationship between this Conference and the previous one that took place in New York in 1966[1] and ask the question: what have we learned as a shared experience about the relation between speech and respiration in the intervening period? It seems to me that there are a couple of things.

We can look at the subglottic system and the lungs and say, perhaps, that from my viewpoint—the speech production aspect—we now have some understanding of what muscles are involved, and what happens in the subglottic system during speech to maintain pressure. I think I understand the laryngeal mechanisms far less—although one of the things that has happened, perhaps stimulated by the previous New York Conference, is that many more people have become very interested in the larynx and in trying to determine which muscles do what during speech production, as well as in breathing.

Some of this interest is, I think, reflected in the work of people present here; but there is, in fact, a large number of investigators now using techniques recording from muscles during speech production—workers like Hirano and, of course, those in the Haskins Laboratory, as well as several investigators in Japan who are not represented here. I think most of us here from the speech production side are a little bit too remote from this research fully to appreciate the interactions that we should so much like to establish between the two systems of breathing and speaking.

Wyke. I think I owe it to Dr. Klatt and his speech production colleagues to explain why some of the gentlemen mentioned are not, in fact, present. It does not mean that they were not invited; and, indeed, this applies also to a number of distinguished people in other related fields. It simply turned out that of those who were invited there were (inevitably) a number who had other commitments, or who were not able to be present for various personal reasons. I entirely agree with you that this aspect of the work is crucial as a background to quite a lot of our discussions, and I personally regret very much that more workers in the

---

[1] Bouhuys, A. [Ed.] (1968) *Sound Production in Man*, New York.

speech production field were not able to be with us. But I should apologize—as I was most specifically instructed to do by those invited—on their behalf for the fact that they could not be present, and say that they would very much have liked to be here with us.

Kirchner. In regard to Dr. Klatt's question and comment, there is one thing that I think ought to be brought out in relation to some of the studies that are currently being done on the activity of the laryngeal muscles during speech. This is that in order to introduce electromyographic electrodes into the various individual muscles of the larynx, some of the studies that I have had an opportunity of looking at have involved taking the subject and premedicating him with a sedative and atropine (which dries all the secretions on the laryngeal surfaces) and introducing hook electrodes into the muscles after topical anaesthesia of the mucosal surface (which, of course, eliminates all the surface receptors, and introduces the factor of trauma to the muscles). I therefore think that the information that can be gained from some of these experiments is only of the grossest kind; and I really am appalled at some of the objectives that are mentioned in connexion with some of these human experiments. I think that the fine adjustment mechanisms of the vocal folds are not likely to be elucidated by this type of study. It really worries me to see people expect to get anything but the grossest kind of information from this type of study in the human.

Ladefoged. I feel I really must comment on that, since after all it seems to relate to the studies that Dr. Hirano and Dr. Ohala were conducting in my Laboratory (and are now conducting elsewhere). I don't know whether they are the kind of studies you refer to—but we don't in fact, inject atropine. We occasionally use topical anaesthesia; but as the experiments go on for a couple of hours or so (and I personally have been a subject for them), any topical anaesthetic effect is well gone by the end of that time. I feel that the subjects are behaving in a fairly normal way, and are giving a very valuable record of things like the reciprocal innervation behaviour of different laryngeal muscles—how they come in and out in different speech acts. It is only through studying speech, which is the fastest controlled movement that you can get, that you can obtain this kind of information.

Looking at cats and things is fine, but cats can't talk; and you must be able to get a voluntary action going on so that you can say what happens when you suddenly change the pitch—and a human being can voluntarily change his pitch suddenly. We can also study what happens when you change from a voiced to a voiceless sound, and all these different kinds of things. Surely all these are new kinds of information that have recently come up.

Finally, I would like to put in a plea here. Our own activity is now extending and seems to be working upwards, in that having dealt with the respiratory system and the larynx we are now working on the tongue; and let's hope that at any future conference we might include the actions of the tongue as one of the subjects under discussion.

Mårtensson. There is a saying that the cat does not talk because he has nothing to say! But I think that as far as the contraction properties of the muscles and the arrangement of the receptors in the larynx and so forth are concerned, there need not be many dissimilarities between the species.

Wyke. Apropos of this somewhat philosophical (but nonetheless relevant) point: I hesitate to debate linguistic matters with Dr. Ladefoged, but I would, with respect, say that while a cat can't talk and may have nothing to say even if it tried, it can certainly communicate vocally—particularly with other cats of the opposite sex. I would therefore suggest, from an experimental point of view, that to investigate (for example) cortical influences on laryngeal motoneurones in a cat by the introduction of electrodes into the corticobulbar tract system and the nucleus ambiguus with the usual neurophysiological techniques might not be so irrelevant to our needs as it might at first seem. In principle, I would go along with Dr. Mårtensson (and Dr. Kirchner) here.

Ladefoged. May I make a quick comment? Obviously I didn't mean to decry these procedures, from which one can get an enormous amount of information. But one thing that we find looking at human beings relates to a remark I was going to make earlier when we were talking about the hyoid bone, and which it is now appropriate to make. We have been studying the position of the hyoid—after all, when we went from the larynx to the tongue we had to pass the hyoid region—and one of the things we have found is that if you take six human beings performing given voluntary actions like saying a given sentence, they behave in very different ways from one another—so much so, that very often you can sort them out and say these three human beings choose this mode of voluntary action to get the tongue in the right position whereas the other three behave differently. For example, in a set of vowel utterances, three human beings had the tongue height correlated with the hyoid height, and three human beings had it inversely correlated (speaking from memory). It is only by examining voluntary actions in a number of subjects that one can see these kinds of things happening: therefore while perhaps the cat has its place in the scheme of things, still let's look at voluntary actions in human subjects.

Murakami. I want to hear some opinions from Dr. Bowden, Dr. Wyke and Dr. Mårtensson, and from Dr. Kirchner too, concerning my idea that the cricothyroid muscle should be regarded as an extrinsic laryngeal muscle rather than an intrinsic laryngeal muscle—at least from the functional standpoint. I know very well that my opinion is apparently against the general viewpoint of laryngology—but can you give me your view from a neurological standpoint, Dr. Wyke?

Wyke. It is inappropriate for me to respond to that type of question regarding anatomical nomenclature. This is really a matter for the International Committee on Anatomical Nomenclature, and so I think Dr. Bowden should issue any directives on this. But looking at it from the empirical point of view of a neurological clinician and experimentalist, I would say that it is neither intrinsic nor extrinsic. but both.

Because your own experiments, Dr. Murakami[1]. have clearly shown that there are at least two distinct motor unit populations in the cricothyroid muscle that operate in entirely different circumstances [see also Rudomin, 1966[2]; Sutton *et*

[1] Suzuki, M., Kirchner, J. A., and Murakami, Y. (1970) The cricothyroid as a respiratory muscle, *Ann. Otol. (St. Louis),* **79,** 976–84.

[2] Rudomin, P. (1966) Some aspects of the control of cricothyroid muscle activity, *Archiv. int. Physiol.,* **74,** 154–68.

alii, 1972[1]], I would suggest that we might consider that this argument is in fact a non-argument, and that we are dealing (perhaps for evolutionary or embryological reasons) with a muscle that has developed to serve at least two (and probably more) completely different functions so that it just happens, if you like, that the motor units subserving these disparate functions are mixed up together in the same anatomical muscular entity. For these reasons, I will not let myself be involved in an argument as to whether the cricothyroid muscle is intrinsic or extrinsic—for me it is a laryngeal muscle that does a whole lot of things in a wide variety of circumstances.

Kirchner. I think that is a very good answer.

I would now like to bring up one question that I haven't heard discussed during this week. Dr. Mårtensson has just remarked that the cat doesn't talk because he has nothing to say; but I heard a discussion recently by some people who understand these things (Lieberman[2], for example) to the effect that the chimpanzee would never be able to learn to speak because of the shape of its supraglottic air tract. That is to say, the nodal positions and the damping and so on that result from the shape of the supraglottic air column in the chimpanzee would make it impossible for him to make certain of the sounds that would be necessary in any kind of speech. Now, is this logical? It would be interesting to hear some comments from the speech experts on this.

Klatt. I was involved with some of that work, and the anatomical differences between chimpanzee and humans are striking. However, I would hesitate to conclude that verbal communication in some ways similar to human communication is logically impossible merely because of these differences.

Plum. I was hoping that the same expert would tell us if he is that different from a parrot!

Mårtensson. If the functional organization of their cerebral cortex were more similar to ours there would, of course, be a better chance for chimpanzees to learn to talk.

Ruth Bowden. I am no expert, but I am told by one of the leading workers in this country that although it has no speech such as we understand it, the chimpanzee is an extremely eloquent animal; and in a Zoological Society Symposium in 1962[3] there were some interesting records presented of noises made by various primates. It might be interesting to relate these to the individual structures of their larynges. There is, of course, the question of the air sacs in certain of the large primates which complicates the issue of the influence of the shapes of the air chambers—but there are some data of an objective kind in the acoustic analyses of the noises produced.

---

1 Sutton, D., Larson, C. R., and Farrell, D. M. (1972) Cricothyroid motor units, *Acta oto-laryng. (Stockh.)*, **74,** 145–51.

2 Lieberman, P. (1968) Primate vocalizations and human linguistic ability, *J. acoust. Soc. Amer.*, **44,** 1574–84.

3 Napier, J., and Barnicot, N. A. [Eds.]. (1963) *The Primates*, Symposia of the Zoological Society of London, No. 10: pp. 1–285.

Stevens. I think we should not leave Dr. Kirchner's question entirely unanswered. The way the reasoning goes is that in sub-human primates the larynx is located much higher in the vocal tract, so that there is not the bend at the pharyngeal meeting with the oral cavity that exists in man. Looking at the way the tongue works, it can be shown that the range of vowel sounds that the animal would be capable of generating is much less than we, as humans, are able to generate. The animals do not have the same pharyngeal cavity, and therefore do not get the same variety or range of cavity resonances; and in addition, they would not be able to produce the variety of consonants that we have. But I don't know whether that proves anything.

Mårtensson. You can always make a communication that is sophisticated to a similar degree by writing. The hands of primates are probably well suited to writing, but the reason that they don't write may be that their cerebral cortex is less well developed than that in man.

Fourcin. I think that the current position is that the limitation is thought to be due to the animal's inability to articulate and vocalize in the way that we do. Aren't experiments at present being carried out with a view to ascertaining the extent to which language development in some primates can be based on a system of gestural organization? I think that the experiments are already in their second or third year, and quite a sophisticated degree of communication is beginning to emerge.

Ladefoged. I think what might be better, for my money, than the gesture experiments are the ones related to so-called 'writing' by chimpanzees performed by Premack (1971[1]). The 'writing' consists of taking plastic blocks and putting them on different parts of a magnetic blackboard. His chimpanzee (*Sarah*) has a vocabulary of the order of 200 words. It can also answer questions of the typical psychologist's type like 'Is the red square on top of the green triangle?' The answer is 'yes' or 'no'; but the chimpanzee apparently has a preference for answering 'yes', in that she will write 'yes' and then move the red square to a position above the green triangle!

Wyke. Could I address a question to Dr. Kirchner? You showed this very interesting result that you and Dr. Murakami had obtained whereby if you stretch the subhyoid muscles one activates a myotatic reflex response in them, and if you stretch the suprahyoid muscles you get a similar thing; but if you stretch the two together (while attached to their various origins and insertions) you in fact inhibit what should otherwise be a myotatic reflex response. Now this fascinates me, because I would like to remind Dr. Kirchner of something he may have forgotten; namely, that when he and I were engaged together in our original experiments on laryngeal articular reflexes in my Laboratory here in 1963 and 1964[2,3,4] we did some experiments on passive manipulation of the cricothyroid

---

[1] Premack, D. (1971) Language in Chimpanzee?, *Science,* **172,** 808–22.

[2] Kirchner, J. A., and Wyke, B. D. (1964) Innervation of laryngeal joints, and laryngeal reflexes, *Nature (Lond.)*, **201**, 506.

[3] Kirchner, J. A., and Wyke, B. D. (1964) Laryngeal articular reflexes, *Nature (Lond.)*, **202,** 600.

[4] Kirchner, J. A., and Wyke, B. D. (1964) Electromyographic analysis of laryngeal articular reflexes, *Nature (Lond.)*, **203,** 1243–5.

joint in which we found, to our astonishment, that movement of this joint in certain directions inhibited the activity of the geniohyoid and other submandibular muscles while facilitating that of the sternohyoid muscle. You may recall, Dr. Kirchner, that we didn't publish these observations in detail amongst our other data[1], because we couldn't understand them; and I wonder whether you could not now be tempted to speculate as to why you think this inhibitory response was generated in the suprahyoid muscles—because it is very curious.

Kirchner. It is. We also have this in our recordings in Yale. It was obviously impossible to show all the tracings in our Communication, but this is something that Dr. Murakami has observed and repeated in a number of animals. I don't know, Dr. Murakami, whether you would like to comment on possible reasons for the inhibition of the sternohyoid muscle when both the sternohyoid and the suprahyoid muscles are stretched together?

Murakami. I think it might be better answered by Dr. Widdicombe or Dr. Mårtensson, because I am not an expert on the muscle spindles or tendon organs in the suprahyoid and infrahyoid muscles.

Widdicombe. I can't think of a solution to this. It is a fascinating observation that could be studied, if one could think of a way of testing the mechanism; but if such experiments haven't been done, I would rather not speculate. I would prefer to see the experiments done.

Murakami. In the limbs, there is the familiar reciprocal inhibition between the flexor muscles and the extensor muscles. But as far as the muscles external to the larynx are concerned, we can't say which is 'flexor' and which is 'extensor' in this sense—although it may be that the suprahyoid and infrahyoid muscles have a reflex relationship just like the flexor and extensor muscles of the limbs. But I don't know.

Wyke. Can I be a nuisance and ask another question of the speech experts about the relationship between speech and changes in subglottic pressure? A number of references have been made to this by Dr. Proctor and Dr. Ladefoged, and I now find myself completely confused about the matter because Dr. Proctor, if I understood him correctly, said that in phonation (and I wasn't sure whether he meant continuous speech or phonation of a single sound) the subglottic pressure ran at somewhere about 5 cm H_2O or so. But if I understood Dr. Ladefoged, he said a little later that only in special phonatory cirumstances was there much of a significant elevation in subglottic pressure. I wonder if my confusion could be resolved?

Proctor. I gave a much wider range than that. You can phonate with pressures as low as 2 or 3 cm H_2O, and in speech I think you rarely go above 12 or 15 cm H_2O—although perhaps Dr. Ladefoged will come up with a different answer. In singing, on the other hand, it isn't too uncommon to go up to 20 cm H_2O or more; and, as I said, on rare occasions with a very loud *crescendo* a value as high as 70 cm H_2O is the highest we have ever recorded.

---

[1] Kirchner, J. A., and Wyke, B. D. (1965) Articular reflex mechanisms in the larynx, *Ann. Otol. (St. Louis)*, **74**, 749–69.

Ladefoged. I would entirely agree that is very rare to get up to 70 cm H_2O. Where I was talking about variations, I was referring to changes superimposed upon a comparatively steady base line of approximately 7 cm H_2O—and I am probably talking now at something like that. On top of that, when I exclaim '*NOW*' then I am going up maybe to 10 cm H_2O; but there is a steady base line always of the order of 7 cm H_2O at an ordinary level of conversation such as I am now using.

Stevens. Could I make a comment while we are talking about pressures in speech. I am sure there is no simple answer to it—although there might be at some future conference. As Dr. Ladefoged has said, at one level speech can be considered as a sequence of sounds—*/e/*, */a/*, */t/* and so on—but some of these sounds are, in fact, themselves sequences of events, yet for some reason we regard the sequence as one sound. We call them a unit in the act of language—the sound */t/* is one example and the sound '*ch*' is another example of such a sequence of events. It is interesting to note that in these kinds of sounds the first act—the first part of the gesture—involves creating a sudden change in pressure or in air flow in the mouth (usually) or in some part of the vocal tract, and only then is the second part of the gesture carried on.

I wonder if some of the ideas that have been raised here about reflex actions might be involved in such sequences of events. Even though on the surface it looks like a sequence of gestures, the whole process seems to be triggered off by the first part of the gesture which creates some change in air flow or pressure; and perhaps that, in turn, could trigger off the next part of the sequence. There are, of course, certain units in speech that are just steady sounds and don't have any sequences—but there are these other units that behave as one sound but, in fact, are sequences of events.

Klatt. If no one else has anything to say on the matter, we are left at a loose end with respect to sub-human primate vocalization. As I understand it, such animals have a fairly large inventory of sounds which they use to communicate with one another; but these (by and large, if not exclusively) involve changes at the larynx or below, and do not involve adjustments of the vocal tract during speech—although primates are certainly capable of making such adjustments. There is nothing neurologically deficient that would prevent them making something like a '*BAA*' sound, because they use similar gestures when they chew; but during speech production it has been observed in these animals that the vocal tract above the larynx remains open and more or less stationary.

Moran. May I ask a small question of Dr. Widdicombe, and follow up what he had to say about CO_2 receptors? Would you care to speculate on what effect these receptors might have on the pattern of ventilation in the human subject with some sort of dead space volume containing carbon dioxide.

Widdicombe. These are the laryngeal ones I mentioned earlier?

Moran. Yes. In other words, in a situation where when you start to breathe in, instead of 'seeing' fresh air you would 'see' carbon dioxide for a while.

Widdicombe. You have put your finger on the reason why we are interested in these receptors, because respiratory physiologists keep giving people CO_2 to breathe and they draw conclusions about the control of breathing and assume everything comes from chemosensitive cells in the medulla—but maybe there are other mechanisms involved as well. There have been long searches for lung chemoreceptors; but I think most people now believe they don't exist—except in birds, which have a curious 'negative' chemoreceptor that is switched off by carbon dioxide in the respiratory tract. These avian receptors inhibit breathing, so 'double inhibition' means that CO_2 stimulates breathing from the lung chemoreceptors of birds. We wondered whether or not there might be respiratory tract chemoreceptors in mammals, and that is why we have been looking for them.

All I can say, to judge from the reflex effects we have got from the larynx, is that if you give somebody a CO_2 mixture to breathe (or increase their dead space, so that they have a relatively longer time with a high CO_2 in their respiratory tract or a relatively higher mean concentration of CO_2 in their respiratory tract), then—if they behave like our cats—the result would be a tendency to slower and deeper breathing. But, of course, this would interact with various feed-backs you get from the CO_2 itself once you began to absorb it: this is just the primary response, and the end-result would depend on these other feed-backs coming in as well.

Freedman. I would like to follow that up briefly in regard to the experiments you mentioned of blowing CO_2 on to the isolated larynx rather than through to the lungs—because, as you probably know, in Purves's laboratory it has been suggested that CO_2 may, in fact, alter the threshold (or the sensitivity) of the pulmonary stretch receptors. But I think your observation is very interesting in the light of the work you probably know about by Sterling (1969[1]), who showed that if you do inhale CO_2 it affects total airway resistance in man.

Widdicombe. Can I just point out— I haven't said this before, but since you introduced the point I am saying it now—that we know that these particular reflex responses which I mentioned as causing slower and deeper breathing with mild concentrations of irritants may also cause laryngeal constriction. So when you stimulate the larynx in this way, you not only get this slower, deeper pattern of breathing but the larynx is also constricted; and this may well, as you say, explain some of Sterling's (*loc. cit.*) observations on total respiratory tract resistance during CO_2 breathing.

Godfrey. That is fine—but as you have great cause to know yourself, when you block the vagus nerves you flatten the response to CO_2. It has always mystified me as to how this can possibly occur. If the receptors in the lung are actually making the breathing increase, that would be very nice; but what about the receptors in the larynx? How do you explain the flattening of the CO_2 response curve in vagal block?

Widdicombe. Do you want to answer that for me, Dr. von Euler?

von Euler. In animal experiments you don't see this effect in records of phrenic nerve activity. Does that answer your question?

---

[1] Sterling, G. M. (1969) The mechanism of decreased specific airway conductance in man during hypercapnia caused by inhalation of 7 per cent CO_2, *Clin. Sci.*, **37**, 539–48.

Godfrey. You see it in the live respiratory physiologist sitting down in the laboratory!

von Euler. Yes, because you have something interposed between the phrenic nerve activity and the mechanical events.

Godfrey. Are you suggesting that translation of electrical energy into ventilation at the level of the muscles is affected by vagal block? I don't see how that could occur.

von Euler. Yes, I am. After vagotomy or vagal block the demand for tidal volume, as evidenced from phrenic activity, goes up in response to CO_2 since frequency doesn't change very much; but tidal volume and thereby ventilation does not follow suit because of increasing mechanical limitations.

Widdicombe. I think that blocking the vagus nerves is, as we all know, an extremely complicated type of intervention; and Dr. von Euler's analysis of the patterns of breathing in vagal-intact and vagal-blocked animals explains a great deal about why the pattern of breathing CO_2 is different after a vagal block in man. To get back to the larynx, what I would like to see done (and what may have been done, somebody can tell us) is to find out what happens if you anaesthetize the mucous membrane of the larynx and then give CO_2 and do a CO_2 response curve. If these laryngeal receptors have any importance in this particular type of response, then this should be the way of testing them. I don't know if Dr. Wyke has done this, or Dr. Kirchner?

Wyke. We haven't done that—but it is a good idea.

Kirchner. We haven't done it either.

Fourcin. I wonder if it would be possible to eliminate one of the variables by using a tracheotomized animal, employing an ordinary nitrogen-O_2 mixture, and altering the CO_2 concentration passing the glottis separately. You could then observe the effect of any laryngeal response quite independently of the effect of the increased CO_2 in the lungs.

Widdicombe. This is what we intend to do in 10 days time!

Freedman. As we are coming to the end, may I just say that workers in the ventilatory field are always interested in systems that may be capable of altering the rates of air flow and the pressures within the respiratory system; and therefore I think it has been very profitable for us to hear about the intrinsic mechanisms of the larynx, so that we may take away a lot of new ideas and speculate teleologically about the importance of the larynx in controlling breathing.

Fourcin. I want to ask about the possibility of a slight further elaboration of the experiment that Dr. Widdicombe proposes to do in 10 days time. That is, to investigate also the effects of subglottic pressure changes and the effect of changing the velocity of the air stream impinging on the vocal folds, with regard to the possible separate influences of pressure change and air velocity. Of course, the air can or cannot contain irritant materials; and in this way one would have a complex of experiments, in which the variables would be not less controllable than is the case at present.

Widdicombe. I have learnt enough in the last few days to keep me busy for years!

Winckel. I would like to put again my question of yesterday. What could be the function of the neck muscles in the combination of respiration and phonation? Perhaps we have now progressed a bit since Dr. Kirchner spoke about the extrinsic laryngeal muscles; but can anyone contribute anything further on this matter?

Gould. This is not an answer, but rather a further question in relation to your question. At a time when one has a 'stiff' neck there is a definite impairment of the vocal quality—but I do not know why. I assume that there is some connexion with the altered state of the neck muscles—but are there any data on this.

Wyke. In this connexion, it might also be relevant to record that it is the frequent experience of patients who have to wear cervical support collars for cervical spondylitis that if they are public speakers (and I have had a couple of patients who are in this position) they take the collar off when they have to speak, because they say that in some way (which I am not able to specify myself, because I am not a speech expert) their voice is not the same, and they can't talk as well. This is not merely a matter of simple mechanical interference with the movements of their jaws and larynx and so on. It is an alteration in the quality of the voice and the way in which they can manipulate their voices that they are aware of, but which I find very hard to analyse by ear when I listen to them.

Fry. May I interject here that you may be changing the character of the side-tone that they are getting. What they hear of their own voice is going through the skull, and this acoustic pathway may well be modified by the collar.

Ruth Bowden. Could I ask Dr. Wyke whether his patients were wearing a collar that came down over the manubrium or not? I have been a victim of this situation myself, and I agree that it certainly does affect one's capacity to speak in public.

Wyke. All the patients that I am associated with have their cervical support collars down over the manubrium sterni and clavicle, because otherwise the cervical support collar is merely a cosmetic (or rather, an anti-cosmetic) device and has no mechanical influence on the cervical joints.

Fry. Well, ladies and gentlemen, it is now time to terminate our discussions, and I shall finally call upon Dr. Gould to give the closing address.

Closing Remarks

W. J. Gould (New York City, U.S.A.)

Mr. Chairman, Ladies and Gentlemen: On behalf of Barry Wyke and myself as co-conveners of this Symposium, I would like to thank each of the Section Chairmen, and the participants and guests at the various sessions, for the excellent presentations and the stimulating discussions throughout each day.

At the initial preparation of this Symposium, Professor Wallace Fenn had agreed to be Chairman of the entire proceedings, and I would like to express our mutual great sorrow with respect to his recent death. We miss him very much.

The holding of an International Symposium devoted to the interrelationship of these two systems—ventilatory and phonatory—one basic to man's physiologic or 'animal' being, the other the foundation of all that is specifically human (that is, speech and communication) testifies, it would seem to me, to the growing awareness and appreciation of the fact that an understanding of the one system is essential to an understanding of the other. At the same time, the mere fact that a Symposium such as this can be convened, and the fact that it has called forth some of the best minds working in the relevant fields from around the world, attests to the progress that we have made and are making in our understanding of these relationships. And this latter, the convening of leaders in their fields from around the globe, surely attests the promise of still more progress and understanding to come.

When we consider the tremendous scope of the papers that have been delivered, as well as the wide-ranging material presented in the discussions, it is difficult to conceive how one is to summarize our proceedings in a few moments.

I think perhaps that a typical case of mine comes to mind which, at least for me, illustrates the mutual involvement that we have of our various disciplines. This patient was a college professor, who was unable either to project or sustain voice. She could lecture only with the aid of a public address system, and then she had to pause occasionally for rest. Phoniatric evaluation showed her voice quality to be excessively breathy, and her vocal function was severely disordered in all acoustic and aerodynamic analyses. Pulmonary function tests showed a markedly depressed breathing capacity, because of bronchial asthma and secondary physiological emphysema. Her reserve volume was markedly elevated (181 per cent of the predicted value) and, concomitantly, the RV/TLC ratio was extremely high (194 per cent of the predicted value). With proper pulmonary therapy for two months, the RV fell from 181 to 128 per cent, whilst there was marked subjective improvement in voice, as well as measurably in acoustic and aerodynamic analyses. Why? How did this happen? Voice therapy alone would have been useless. Only what we have been devoting our time to at this Symposium on Ventilatory and Phonatory Control Systems could in any way explain this result.

We started our proceedings with a discussion of the basic principles of ventilatory and phonatory behaviour, and the relationship of the abdominal and intercostal musculature during voice production, as well as consideration of the behaviour of the entire ventilatory mechanism. It was stressed that posturally erect man must be considered simultaneously in relation to his breathing mechanism and voice control. At the conclusion of this first session, consideration was given to the so called 'shunt' or 'air-escape' which can so alter the efficiency of this system in illness, whether of physical or hysterical nature. This emphasized the requirement for efficiency of this system.

At the next session we continued with the study of mechanisms influencing ventilatory behaviour—how the depth and rate of respiratory movements relate to one another; how the expiratory duration is dependent on inspiratory duration; what relative importance the Hering-Breuer threshold curve may have in relationship to the entire pulmonary reflex mechanism in ventilatory regulation.

Relating back to the case that I mentioned, the irritant receptor reflexes in the asthmatic patient were shown to cause laryngeal constriction and abdominal muscle relaxation. How does this relate to voice improvement upon correction of these factors? What rôle does this relationship play in control of these mechanisms?

In speech, there is a change of resistance or load at the glottis. When studied, it was demonstrated that the reflex system does not take responsibility for the character of the first loaded breath. How does this relate to our total control mechanism, if we consider that in such control the entire sensation of breathing must be taken into account? Afferent discharges from the pulmonary stretch receptors, or from receptors in muscles, tendons or joints, must surely relate to sensory awareness—although, as carefully pointed out in discussion, afferent impulses are not necessarily sensory.

Any consideration of mechanisms influencing ventilatory and phonatory behaviour must consider central control, as was the case in one study. As the meeting developed, it seemed more and more that the integration processes at central—or at least brain stem—level would be a primary target for future studies. The central integration was analysed on the basis of three types of experimental input—carbon dioxide tension, lung inflation and brain stem electrical stimulation, especially of the rostral pontine pneumotaxic system. On the basis of such data, hypotheses were formulated about the control functions of various classes of neurone. We then considered how this may vary in injury and illness.

Our next consideration was the discussion of the physical aspects of ventilatory and phonatory behaviour. Can we predict and evaluate ventilatory problems relative to chemical control mechanisms by computer models? This would add further understanding of control mechanisms in general. Can we analyse and predict, by electronic models of the acoustic processing of the voice? This would add still more to our understanding of control mechanisms—ventilatory as well as phonatory. Can we make a physical model? How would it help? This was not resolved.

We concluded our sessions with a detailed study of the mechanisms influencing phonatory behaviour. In considering all the available information relative to current models of sound sources for speech, we concluded that there are two generator sources for sound—either at the level of the vocal fold with its vibratory motion, or the turbulent noise generators at the supraglottal level. What variables must be considered for phonatory production? For example, there is a limit to the vibratory frequency in the respiratory system; but also there are variables to be considered in the nature of the subglottic pressures, particularly in voicing stressed syllables. We were shown that we can create an electrical pattern of vocal fold movements with the laryngograph; but we must in the future define its place in our studies.

Breathing serves both to ventilate the lungs and to produce sound. What is the efficiency of the ventilatory function in relation to this production of sound? Obviously, referring back to the college professor that I first mentioned, without proper ventilation there is improper sound production. If irritant receptor stimulation in asthma reduces the abdominal muscle efficiency, perhaps the suggested pre-phonatory 'set' of the abdominal musculature becomes impossible.

The function of phonation is to convey information and, as was stated, the rate of phonation and the rate of breathing are individual functions. Apparently the frequency of breathing is regulated to minimize energy expenditure, whereas the phonation rate is adjusted to maximize the transfer of information. This succinctly summarizes our basic thoughts relative to the efficiencies of the control mechanisms.

In addition, when do each of the muscle groups start their action? Is there a pre-setting of either abdominal or laryngeal musculature prior to the onset of speech? Obviously, in the case of our now familiar college professor, the induced inability to use the abdominal musculature properly would lead to further inefficiency in phonatory control.

Without understanding of the laryngeal musculature, both intrinsic and extrinsic, the understanding of control in relation either to breathing or voice production would be impossible. The fact that there is an efficient proprioceptive control of this musculature is an essential part of our understanding of this matter. The fact that the abductor muscles are one quarter of the weight of the adductors would re-emphasize the expiratory-directed force of the larynx. The fact that there are varied types of slowly acting fibres in the intrinsic laryngeal musculature would help to explain the ability to maintain prolonged contraction for both respiratory and speech requirements. In both the intrinsic and extrinsic laryngeal muscular activity related to respiration, it would seem that there is a central area of co-ordination, possibly in the brain stem reticular system. Phonation could then be conceived of as altered ventilation.

During inspiration the larynx descends, and ascends in expiration. The combined effect of muscular action in this ascent is not only a respiratory, but also a phonatory function: the skilled movement implied would require contributions from control systems such as the fusimotor neurone system.

As we have seen, this Symposium has been devoted to a study of all the systems relating to ventilatory and phonatory control. The two functions mentioned, whether in health or in sickness, cannot be separated one from the other. Therefore those of us that pay attention to isolated areas (as in my case, the larynx alone) find that we cannot function to the greatest benefit, either of understanding in the theoretical sense or of providing adequate clinical care for patients such as the one I described, without at least some knowledge of what other disciplines can and do contribute.

ENVOI

Barry Wyke (London, England)

Ladies and gentlemen, that brings us to the end of our scientific proceedings this week; and I would like to repeat now what I said to you at the very beginning—that I am extremely grateful to every one of you for the co-operation that you offered in the preparation of this Symposium. I would especially thank you most sincerely for the immense amount of work that you have all put in to preparing your Communications; for as a result, I think this Symposium may have achieved a certain significance for later years, both in documentation of available data and (even more important) in laying the groundwork for further studies on ventilatory and phonatory control systems. I think that this week we have defined for ourselves a number of areas where we just do not have adequate or reliable data—and I hope we will now go home and get them! With that exhortation, and with my warmest thanks for your contributions, may I now, on behalf of the Royal College of Surgeons of England, bid you an official farewell.

Author Index

Note: Bold type figures indicate complete chapters. Italics indicate discussion on communications.

Subject Index

This Index includes matters dealt with in the Discussions, as well as in the formal Communications.